Dependable IoT for Human and Industry

Modeling, Architecting, Implementation

RIVER PUBLISHERS SERIES IN INFORMATION SCIENCE AND TECHNOLOGY

Indexing: all books published in this series are submitted to the Web of Science Book Citation Index (BkCI), to CrossRef and to Google Scholar

The "River Publishers Series in Information Science and Technology" covers research which ushers the 21st Century into an Internet and multimedia era. Multimedia means the theory and application of filtering, coding, estimating, analyzing, detecting and recognizing, synthesizing, classifying, recording, and reproducing signals by digital and/or analog devices or techniques, while the scope of "signal" includes audio, video, speech, image, musical, multimedia, data/content, geophysical, sonar/radar, bio/medical, sensation, etc. Networking suggests transportation of such multimedia contents among nodes in communication and/or computer networks, to facilitate the ultimate Internet.

Theory, technologies, protocols and standards, applications/services, practice and implementation of wired/wireless networking are all within the scope of this series. Based on network and communication science, we further extend the scope for 21st Century life through the knowledge in robotics, machine learning, embedded systems, cognitive science, pattern recognition, quantum/biological/molecular computation and information processing, biology, ecology, social science and economics, user behaviors and interface, and applications to health and society advance.

Books published in the series include research monographs, edited volumes, handbooks and textbooks. The books provide professionals, researchers, educators, and advanced students in the field with an invaluable insight into the latest research and developments.

Topics covered in the series include, but are by no means restricted to the following:

- Communication/Computer Networking Technologies and Applications
- Queuing Theory
- Optimization
- Operation Research
- Stochastic Processes
- Information Theory
- Multimedia/Speech/Video Processing
- Computation and Information Processing
- Machine Intelligence
- Cognitive Science and Brian Science
- Embedded Systems
- Computer Architectures
- Reconfigurable Computing
- Cyber Security

For a list of other books in this series, visit www.riverpublishers.com

Dependable IoT for Human and Industry

Modeling, Architecting, Implementation

Editors

Vyacheslav Kharchenko

National Aerospace University KhAI
Ukraine

Ah Lian Kor

Leeds Beckett University
UK

Andrzej Rucinski

University of New Hampshire
USA

LONDON AND NEW YORK

Published 2019 by River Publishers
River Publishers
Alsbjergvej 10, 9260 Gistrup, Denmark
www.riverpublishers.com

Distributed exclusively by Routledge
4 Park Square, Milton Park, Abingdon, Oxon OX14 4RN
605 Third Avenue, New York, NY 10158

First published in paperback 2024

Dependable IoT for Human and Industry Modeling, Architecting, Implementation / by Vyacheslav Kharchenko, Ah Lian Kor, Andrzej Rucinski.

Routledge is an imprint of the Taylor & Francis Group, an informa business

Publisher's Note
The publisher has gone to great lengths to ensure the quality of this reprint but points out that some imperfections in the original copies may be apparent.

ISBN: 978-87-7022-014-9 (hbk)
ISBN: 978-87-7004-380-9 (pbk)
ISBN: 978-1-003-33784-3 (ebk)

DOI: 10.1201/9781003337843

Contents

Preface xxi

Acknowledgments xxiii

List of Contributors xxv

List of Figures xxxiii

List of Tables xlv

List of Abbreviations xlvii

Dependable IoT for Human and Industry: Introduction and Book Scope 1

1. Internet of Important Things 1
2. Internet of Things and Collaboratory 7
3. Main Topics and Scope 12

PART I: Internet of Vital and Trust Things

1 Disruptive Innovation in Vital Embedded Systems and the Internet of Vital Things 17

Ted Kochanski

1.1 Introduction and Brief History 17
1.1.1 Embedded Systems 17
1.1.2 Critical Embedded Systems 20
1.1.3 The Internet of Things in Context 21

1.1.4 Some Observations of the Status Quo and the Near Term 24
1.2 Internet of Vital Things (IoVT) 25
1.2.1 Conception of IoVT 25
1.2.2 Historic Example: SAGE 26
1.3 The SAGE Air Defense System 27
1.4 Evolution of Disruptive Innovation in the Design of Microelectronic Systems and the IoVT 27
1.4.1 Creation of CIDLab 28
1.4.2 Concepts of GAIN and Global Systems Engineering Education 29
1.4.3 I-GEMS and the Virtual Design Universe 30
1.4.4 Design for Globalization 31
1.4.5 Vital Electronics 33
1.5 Vital-iSolve and the Internet of Vital Things (IOVT) 35
1.5.1 Vital-iSolve Fundamentals 36
1.5.2 Vital-iSolve Ingredients 38
1.5.3 Example From an E-health Ambulatory Sensor Application: Heart Sensor 39
1.6 Conclusion 39
1.6.1 Summing up the Recent Disruptive Innovation in Microelectronics Systems Education 39
1.6.2 The Big Question Which Needs to be Addressed 40
References 41

2 How to Support Creativity in the Complex IoT with Ethics and Trust for Users 47

Raymond J. Garbos

2.1 Introduction 48
2.2 Architecting the Future 50
2.2.1 Conceptual Architects 50
2.2.2 System-of-Systems 52
2.2.3 Trusted Interfaces 54
2.3 Conclusion 57
References 58

PART II: Modelling and Assessment

3 Design and Simulation of an Energy-efficient Sensor Network Routing Protocol for Large-scale Distributed Environmental Monitoring Systems **63**

Awais Ahmad, Muhammad Adeel Pasha, Shahid Masud and Axel Sikora

3.1 Introduction 64
3.1.1 Context and Motivation 64
3.1.2 Contributions 67
3.2 Related Work 67
3.3 Proposed Protocol EESNR 70
3.3.1 Network Topology Model for EESNR 70
3.3.2 Path Loss/Fading 73
3.3.3 Radio and Data Transmission Model 73
3.4 Simulation Setup and Results 74
3.4.1 Simulation Setup 74
3.4.2 Simulation Results 76
3.5 Conclusions 79
References 80

4 Modeling and Assessment of Resource-sharing Efficiency in Social Internet of Things **83**

Kashif Zia, Arshad Muhammad and Dinesh Kumar Saini

4.1 Introduction 83
4.2 Related Work 86
4.3 Motivation 88
4.4 The Proposed Model 89
4.4.1 P2P Resource Sharing Specifications 89
4.4.2 Agent-Based Model of Peers in Competitive Mode . 90
4.4.3 Agent-Based Model of Peers in Cooperative Mode . 92
4.5 Simulation and Results 93
4.5.1 Simulation Setup 93
4.5.2 Simulation Results 95
4.6 Conclusions 100
References 100

5 Modeling and Availability Assessment of Mobile Healthcare IoT Using Tree Analysis and Queueing Theory **105**
Anastasiia Strielkina, Dmytro Uzun, Vyacheslav Kharchenko and Artem Tetsky
5.1 Introduction 106
5.1.1 Motivation 106
5.1.2 State of the Art 107
5.1.3 Aim and Objectives 107
5.2 Healthcare IoT Infrastructure 108
5.3 Applicable Approaches and Methods for Modeling and Simulation of Healthcare IoT 110
5.3.1 Fault Tree Analysis for Failure Occurrence Nature of Healthcare IoT 110
5.3.2 Justification of Applicability of the Queueing Theory 114
5.4 Case Study: Modeling of Healthcare IoT Using Queueing Theory 115
5.4.1 Initial Model "Birth–Death" 115
5.4.2 The Model Considering Attacks on Vulnerabilities . 116
5.4.3 The Model Considering Elimination of Vulnerabilities 117
5.4.4 Discussion of the Simulation Results 122
5.5 Conclusions 122
References 124

6 PSMECA Analysis of IoT-based Physical Security Systems **127**
Al-Khafaji Ahmed Waleed, Vyacheslav Kharchenko, Dmytro Uzun, Oleg Illiashenko and Oleksandr Solovyov
6.1 Introduction 127
6.1.1 Motivation 127
6.1.2 The Objectives, Approach and Structure 128
6.2 IoT-based Physical Security System 131
6.3 Establishment of the Models of PSS 134
6.3.1 Models of Functions and Components of PSS . . . 135
6.3.2 Fault Models of Physical Security System 137
6.3.3 Investigation and Analysis of the Occurrence of Failures in PSS 138
6.4 Conducting of PSMECA 139

6.4.1 An Example of PSMECA Tables for the Case of CCTV Subsystem Functioning in Normal Operation Mode 139
6.4.2 Discussion of the PSMECA 142
6.5 Conclusions and Future Steps 142
References 143

7 IoT Security Event Correlation Based on the Analysis of Event Types 147
Andrey Fedorchenko and Igor Kotenko
7.1 Introduction 148
7.2 State of the Art 150
7.3 Approach to Security Event Correlation 152
7.3.1 Security Correlation and Sources of Information . . 152
7.3.2 Events, Event Types, and Properties 153
7.3.3 Correlation Method Based on Analysis of Event Types 154
7.3.4 Input Data Requirements 156
7.4 Implementation and Experiments 157
7.5 Conclusion 165
References 166

8 Investigation of the Smart Business Center for IoT Systems Availability Considering Attacks on the Router 169
Maryna Kolisnyk, Vyacheslav Kharchenko and Iryna Piskachova
8.1 Introduction 170
8.2 Security Challenges for IoT Technologies 172
8.2.1 Technologies and Features to Create IoT Systems . 172
8.2.2 Vulnerabilities and Types of Attacks in Wireless IoT Systems 174
8.2.3 Security Issues of Some Wireless Technologies of IoT 175
8.2.3.1 ZigBee technology 175
8.2.3.2 Z-wave technology 176
8.2.3.3 Long-Term Evolution/Long-Term Evolution Advanced (LTE/LTE-A) technologies . . 177
8.2.3.4 Low-power Wide-area Network (LoRAWAN) technology 178

8.2.3.5 Radio Frequency IDentification (RFID) technology . . . 179
8.2.3.6 Bluetooth Low Energy technology (BLE) 179
8.2.4 Spyware in IoT . . . 180
8.3 The Markov Model of the SBC Router States . . . 182
8.3.1 Assumptions and Initial Data for Modeling . . . 182
8.3.2 Description of the SBC Router States' Graph . . . 185
8.3.3 Simulation Results . . . 187
8.4 Conclusion . . . 190
References . . . 191

9 An Internet of Drone-based Multi-version Post-severe Accident Monitoring System: Structures and Reliability 197

Herman Fesenko, Vyacheslav Kharchenko, Anatoliy Sachenko, Robert Hiromoto and Volodymyr Kochan

9.1 Introduction . . . 198
9.1.1 Motivation . . . 198
9.1.2 State of the Art . . . 198
9.1.3 The Goals and Structure . . . 201
9.2 Principles of Creating an Internet-of-drones-based Multi-version Post-severe Accident Monitoring System . . . 201
9.2.1 Structure . . . 201
9.2.2 Principles . . . 202
9.3 Reliability Models for the Internet-of-drones-based Multi-version Post-severe Accident Monitoring System . . . 204
9.3.1 Simplified Structure . . . 204
9.3.2 Subsystems' Reliability Models . . . 206
9.3.3 System Models . . . 210
9.4 Simulation . . . 212
9.5 Conclusion . . . 214
References . . . 215

PART III: Architecting and Development

10 Virtualization of Embedded Nodes for Network System Characterization in IoT Applications 221

Manuel Schappacher, Artem Yushev, Mahbuba Moni and Axel Sikora

10.1 Introduction 222
10.2 Related Work 222
10.2.1 System Level Simulation 223
10.2.2 Network Level Simulation 223
10.2.3 Network Level Emulation 224
10.3 Requirements 225
10.4 Background 226
10.4.1 The emb::6 Networking Stack 226
10.4.2 TTCN-3 227
10.5 VTENN Basics 227
10.5.1 General Architecture 227
10.5.2 Node Virtualization 228
10.5.3 Virtual Radio and Channel 229
10.5.4 Virtual Topologies 230
10.5.5 Monitoring and Control 231
10.6 Design and Implementation 232
10.6.1 Test Executor 232
10.6.2 Network Manager 234
10.6.3 Virtual Nodes and Virtual Channels 234
10.6.4 Sample Test Cases 235
10.7 VTENN in IoT Applications 237
10.8 Conclusion and Future Work 238
References 239

11 IoT Meets Opportunities and Challenges: Edge Computing in Deep Urban Environment 241

Marta Chinnici and Saverio De Vito

11.1 Introduction 242
11.2 The Role of Big Data in IoT Era 244
11.2.1 Big Data Generation 244
11.2.2 IoT Data and Big Data Analytics 245
11.2.3 IoT System Architecture 246
11.3 Deep Urban Environment 248
11.3.1 Urban Paradigm 248
11.3.2 Urban IoT Applications 250
11.4 The Emergence of Edge Computing in Urban Context . . . 253
11.4.1 Edge Vision 255
11.4.2 Application in Urban Environment: Pollution Monitoring 256

11.4.3 Network Load Improvements 259
11.4.4 Network Local Estimation of Concentration for Immediate Exposure Feedback 260
11.4.5 Dependability: Reliability, Security, and Maintenance 263
11.5 Challenges 264
11.6 Conclusion 268
References 269

12 Hybrid Control System of Mobile Objects for IoT 273
Anzhelika Parkhomenko, Dmytro Kravchenko, Oleksii Kravchenko and Olga Gladkova
12.1 Introduction 273
12.2 Related Work 275
12.3 Methodology 276
12.4 Implementation and Evaluation of the Hybrid Control System 278
12.4.1 Subsystem of Remote Control 278
12.4.2 Subsystem of Autonomous Control 283
12.5 Results and Further Work 288
12.6 Conclusion 289
References 290

13 Software Architecture for Smart Cities and Technical Solutions with Emerging Technologies' Internet of Things 293
Dinesh Kumar Saini, Kashif Zia and Arshad Muhammad
13.1 Introduction 293
13.1.1 Challenges in a Smart City 294
13.1.2 Software Architecture for a Smart City 294
13.1.3 Smart City Governance: Example of Oman 295
13.1.4 Examples of Services Like Intelligent Transport System or Smart Transportation 295
13.1.5 Smart Urban Modeling 297
13.2 Security in a Smart City 297
13.2.1 Attack Analysis 299
13.2.2 Cyber-Physical Systems in Smart Cities 300
13.3 IoT Solutions for a Smart City 302
13.4 Conclusion 303
References 304

14 Approaches and Techniques to Improve IoT Dependability **307**

Nikolaos G. Bardis, Nikolaos Doukas, Vyacheslav Kharchenko, Vladimir Sklyar and Svitlana Yaremchuk

14.1 Introduction 308

14.1.1 Motivation 308

14.1.2 Objectives and Structure 310

14.2 Secure Implementation of Modular Arithmetic Operations for IoT and Cloud Applications 311

14.2.1 Modular Arithmetic Operation for IoT and Cloud Security 311

14.2.2 Shortfalls of Methods for Secure Remote Implementation of Modular Exponentiation 313

14.2.3 Secure Parallel Modular Exponentiation 313

14.2.4 Secure Modular Exponentiation in Cloud Infrastructure 314

14.3 Security and Safety Case Driven Design for IoT Systems . . 316

14.3.1 Concept of Assurance Case Driven Design 316

14.3.2 Approach to Implement ACDD 318

14.4 Software Requirements Correctness Improvement for IoT Reliability 320

14.4.1 Challenges for Software Systems' Reliability 320

14.4.2 Methods of SWS Requirements' Correctness Improvement 321

14.4.3 Proposed Metric for Requirement Complexity Evaluation 323

14.5 Conclusion 324

References 325

PART IV: Implementation and Industry Cases

15 Holistic Systems Engineering Methodology for Intelligent Energy Systems – with a Case Study from "ruhrvalley" **331**

Carsten Wolff, Torben Lippmann and Uwe Jahn

15.1 Introduction 332

15.2 Systems Engineering for Intelligent Energy Systems – Literature Review and State of the Art 333

15.3 Case Study: ORC Turbine 338

15.4 Software Architecture for an IoT System based on OCM . . 340

15.5 Smartification Process for Intelligent Technical Systems . . 344
15.6 Conclusion 347
References 348

16 Smart Waste Management System: A Fusion of IoT and Blockchain Technology 351

Manish Lamichhane, Oleg Sadov and Arkady Zaslavsky

16.1 Introduction 352
16.2 IoT, Blockchain and Dependability 353
16.3 Background and Motivation 354
16.4 Related Work 355
16.4.1 Sensing Waste Levels 356
16.4.2 Sorting Waste at Source 356
16.4.3 WSN-based Architectures 356
16.5 Architecture Overview 357
16.5.1 Blockchain 358
16.5.1.1 The bank 358
16.5.1.2 Community DAO 359
16.5.1.3 SGBFactory 360
16.5.2 SWM Server 360
16.5.2.1 MongoDB 360
16.5.2.2 Telegram Bot 360
16.5.3 SGB Simulation 361
16.5.4 User Domain 361
16.5.4.1 Telegram 361
16.5.4.2 Web application 361
16.6 MQTT Message Exchange Architecture 361
16.6.1 MQTT Topic Hierarchy 362
16.6.2 Authentication 362
16.6.3 Waste 362
16.7 Limitation and Future Work 363
16.8 Conclusion 363
References 364

17 Automation of Control Processes in Specialized Pyrolysis Complexes Based on Industrial Internet of Things 367

Yuriy Kondratenko, Oleksiy Kozlov, Andriy Topalov, Oleksiy Korobko and Oleksandr Gerasin

17.1 Introduction 368

17.2 Industrial Internet of Things Approach and Its Implementation . . . 369
17.3 Generalized IIoT-Based Pyrolysis Complex Control System . . . 372
17.4 Implementation of the IIoT System for the SPC MCP-5 . . . 376
17.4.1 Functional Structure of the IIoT System for the SPC MCP-5 . . . 376
17.4.2 Description of the MCP-5 IIoT System Hardware . . . 378
17.4.3 Description of the MCP-5 IIoT System Software . . . 379
17.4.4 HMI of the MCP-5 IIoT System . . . 381
17.5 Conclusion . . . 383
References . . . 384

18 Cloud-based IT Infrastructure for "Smart City" Projects 389

Oleksii Duda, Nataliia Kunanets, Oleksandr Matsiuk and Volodymyr Pasichnyk

18.1 Introduction . . . 390
18.2 Socio-Communicative Component of the "Smart Cities" Projects . . . 391
18.3 Information-Technological Resources for the "Smart Cities" Projects Implementation . . . 392
18.4 The "Smart City" Project Tasks which can be Solved on the Basis of Cloud Computing . . . 394
18.5 Generalized Architecture of Information-Technological Support of the "Smart City" . . . 395
18.6 Infrastructure Platform for Cloud-Based "Smart City" Projects . . . 396
18.7 Architecture of the Center for the "Smart City" Analytical Data Processing . . . 399
18.8 Computing Architecture for Providing Information and Technology Services in the "Smart Cities" . . . 402
18.9 Conclusions and Further Investigation . . . 404
References . . . 405

19 A Framework for Real-Time Public Transport Information Acquisition and Arrival Time Prediction Based on GPS Data 411

Inna Skarga-Bandurova, Marina Derkach and Artem Velykzhanin

19.1 Introduction . . . 411

19.1.1 Real-Time Public Transport Information Service Infrastructure 412
19.1.2 Objective and Challenges 413
19.2 Arrival Time Prediction Models 415
19.2.1 Prediction Methodology 415
19.2.2 Kalman Filtering 417
19.2.3 Accuracy Metrics 419
19.3 Case Study 419
19.3.1 General Strategy of Public Transport Information Service Delivering 420
19.3.2 The Remote Access Configuration 420
19.3.3 GPS Data Acquisition 421
19.3.4 Assigning Route Number to Each Trolleybus 421
19.3.5 Calculate the Predicted Arrival Time 422
19.3.6 Data Acquisition and Information Processing Algorithm 424
19.3.7 Experimental Results and Model Predictions for Trolleybus Arrival Time 425
19.4 Conclusion and Future Work 429
References 429

20 Scalable Smart Transducer Networks Using Power-over-Ethernet and Neural Networks 433

Ivan Lobachev

20.1 Introduction 434
20.2 Research Objectives and Related Work 434
20.3 Advantages and Improvements 436
20.4 System Architecture 437
20.4.1 System Requirements 438
20.4.2 Sensor Hub Classes 439
20.4.3 Configuration 441
20.4.4 Requirements and Modes of Operation 442
20.4.5 Parameters, Organization, and Data Processing . . 443
20.4.6 Data Processing and Presentation 444
20.4.7 The Hierarchy 445
20.4.8 Incorporation of Neural Networks 446
20.5 Testing 447
20.6 Conclusion 449

20.7 Future Work 451
References 451

21 IoT Systems of the AAL Sector: Application, Business Model, and Data Privacy **455**

Jelena Bleja, Uwe Grossmann, Bettina Horster, Andree Roß, Enrico Löhrke, Christof Röhrig, Jan Oelker, Aylin Celik and Reiner Hormann

21.1 Introduction 455
21.2 Smart Service Power (SSP) – Ambient Assisted Living for Elderly People 458
21.2.1 The Ambient Assisted Living (AAL) of SSP 458
21.2.2 Smart Service Power Top-Level System Architecture 459
21.3 Solion – A Radio-based Assistance System 462
21.4 Covibo – Vital Data Acquisition 464
21.4.1 System for Vital Data Acquisition 465
21.4.2 Communication Structure 465
21.5 Business Models for AAL Applications 465
21.6 Data Privacy and Data Usage Control 470
21.7 Conclusion 472
References 473

PART V: Education and Training

22 Internet/Web of Things: A Survey of Technologies and Educational Programs **479**

Volodymyr Tkachenko and Eugene Brezhniev

22.1 Introduction 479
22.1.1 Motivation 481
22.1.2 State of Art 481
22.1.3 Goals and Structure 483
22.2 Survey of IoT/WoT Technologies 484
22.2.1 IoT Global Network Architecture 485
22.2.2 Web of Things 488
22.3 Structure of the Training Program “Technologies and Tools for Developing WoT Applications” 490
22.4 Conclusion 494
References 495

23 Prospects for Constructing Remote Laboratories to Study Cognitive IoT Systems 503
Mykhailo Poliakov, Karsten Henke and Heinz-Dietrich Wuttke
23.1 Introduction 504
23.2 State of the Art 505
23.3 Cognitive Control System Model 505
23.4 Prospects for Constructing Remote Laboratories 508
23.5 Conclusion 511
References 512

24 Project-Oriented Teaching Approach for IoT Education 515
Peter Arras, Dirk Van Merode and Galyna Tabunshchyk
24.1 Introduction 516
24.2 Remotely Controlled Experiments 517
24.2.1 Informational Systems on Reliability Tasks-lab (ISRT) 518
24.2.2 Computer-Aided Learning Module (CALM) 521
24.2.2.1 Aims and usage of the CALM 521
24.2.2.2 Project-oriented approach 523
24.3 IoT Projects for Education 524
24.3.1 Smart-campus Project 524
24.3.2 Re-engineering of Existing Engineering Software for a New Platform (COPTURN Project) 526
24.4 The Embedded Factory as a Tool for Implementation 527
24.5 Conclusions 530
References 531

25 Internet of Things for Industry and Human Applications: ALIOT-Based Vertically Integrated Education 535
Artem Boyarchuk, Oleg Illiashenko, Vyacheslav Kharchenko, Dmytro Maevsky, Chris Phillips, Anatoliy Plakhteev and Lolita Vystorobska
25.1 Introduction 536
25.1.1 Motivation 536
25.1.2 State of the Art and Publication Statistics 536
25.1.3 Objectives and Approach 542
25.2 The Aliot Project for Vertically Integrated Education 543
25.2.1 Challenges and Rationale 543
25.2.2 Innovative Character 545

25.2.3 Project Activities and Methodology 546
25.2.4 Expected Impact of the Project 547
25.2.5 ALIOT Curriculum 549
25.3 Overview of the IoT Courses in Europe and the United States 549
25.3.1 Overview of IoT Courses in ALIOT Project Partners 549
25.3.2 Metrics-Based Approach of IoT Courses Analysis . 550
25.4 ALIOT Project Case Studies 553
25.4.1 Control Unit for Mini Plotter 553
25.4.2 Control Unit for the LED Ribbon with Pixel Addressing 556
25.5 Conclusions 558
References 559

Index **561**

About the Editors **565**

Preface

Book “Dependable IoT for Human and Industry” covers the main aspects of Internet of Things and IoT based systems such as global issues of applications, modeling, development and implementation of dependable IoT for different human and industry domains. The book contains 25 Chapters divided into 5 parts (Introduction in IoT, Modelling and assessment, Architecting and development, Implementation and business, Education and training).

Technical topics discussed in the book include:

1. Introduction in Internet of vital and trust Things (Chapters 1, 2)
2. Modelling and assessment techniques and tools for dependable and secure IoT systems (Chapters 3–9)
3. Architecting and development of IoT systems (Chapters 10–14)
4. Implementation of IoT for smart cities and drone fleets; business and blockchain, transport and industry (Chapters 15–20)
5. Training courses and education experience on Internet and Web of Things (21–25).

The book chapters have their roots in the International Conference IDAACS 2017, Workshop on Cyber Physical Systems and IoT Dependability CyberIoT-DESSERT 2017 and Workshops (technical meetings, trainings, schools) on EU funded project ERASMUS+ ALIOT “Internet of Things: Emerging Curriculum for Industry and Human Applications” http://aliot.eu.org/

Acknowledgments

The editors of this book would like to express their deepest appreciation and gratitude to Prof. Anatoly Sachenko, a great friend, collaborator, the founder and chair of the IDAACS conferences, in particular IDAACS2017 http://idaacs.net/2017 where an idea and plan of the preparation of this book was discussed. Without his engagement and vision, this publication would have been impossible.

We would like to thank all organisers and participants of 2nd International Workshop on Cyber Physical Systems and Internet of Things Dependability (WS CyberIoT-DESSERT) http://www.idaacs.net/2017/WS-CyberIoT who supported preparation of this book.

We thank colleagues who develop the project ERASMUS+ ALIOT "Internet of Things: Emerging Curriculum for Industry and Human Applications" http://aliot.eu.org/ and participate in discussions of topics related to IoT during a few meetings and schools in Sweden (Stockholm, December 2016), Ukraine (February 2017, 2018, Chernivtsi; May 2017, Mykolaiv; May 2018, Kyiv), Portugal (Coimbra, October 2017), United Kingdom (Newcastle-Leeds, July 2018).

List of Contributors

Al-Khafaji Ahmed Waleed, *Department of Computer Systems, Networks and Cybersecurity, National Aerospace University "KhAI", Kharkiv, Ukraine; E-mail: eng_ahmed.waleed@yahoo.com*

Anastasiia Strielkina, *Department of Computer Systems, Networks and Cybersecurity, National Aerospace University "KhAI", Kharkiv, Ukraine; E-mail: a.strielkina@csn.khai.edu*

Anatoliy Plakhteev, *Department of Computer Systems, Networks and Cybersecurity, National Aerospace University "KhAI", Kharkiv, Ukraine; E-mail: a.plahteev@csn.khai.edu*

Anatoliy Sachenko, *1. Ternopil National Economic University, Ternopil, Ukraine;*
2. Kazimierz Pulaski University of Technology and Humanities in Radom, Radom, Poland; E-mail: as@tneu.edu.ua

Andree Roß, *VIVAI Software AG, Betenstrasse 13-15, 44137 Dortmund, Germany; E-mail: Andree.Ross@vivai.de*

Andrey Fedorchenko, *1. St. Petersburg Institute for Informatics and Automation of Russian Academy of Sciences, Saint Petersburg, Russia;*
2. St. Petersburg National Research University of Information Technologies, Mechanics and Optics, Saint Petersburg, Russia;
E-mail: fedorchenko@comsec.spb.ru

Andriy Topalov, *Admiral Makarov National University of Shipbuilding, Mykolaiv, Ukraine; E-mail: topalov_ua@ukr.net*

Anzhelika Parkhomenko, *Software Tools Department, Zaporizhzhia National Technical University, Zaporizhzhia, Ukraine;*
E-mail: parhom@zntu.edu.ua

Arkady Zaslavsky, *Senior Principal Research Scientist, Data61, CSIRO, Australia; E-mail: Arkady.Zaslavsky@data61.csiro.au*

Arshad Muhammad, *Faculty of Computing and Information Technology, Sohar University, Sohar, Oman; E-mail: amuhammad@su.edu.om*

Artem Boyarchuk, *Department of Computer Systems, Networks and Cybersecurity, National Aerospace University "KhAI", Kharkiv, Ukraine; E-mail: a.boyarchuk@csn.khai.edu*

Artem Tetskyi, *Department of Computer Systems, Networks and Cybersecurity, National Aerospace University "KhAI", Kharkiv, Ukraine; E-mail: a.tetskiy@csn.khai.edu*

Artem Velykzhanin, *Volodymyr Dahl East Ukrainian National University, Severodonetsk, Ukraine; E-mail: velykzhanin@snu.edu.ua*

Artem Yushev, *Institute of Reliable Embedded Systems and Communication Electronics (ivESK), Offenburg University of Applied Sciences, Offenburg, Germany; E-mail: artem.yushev@hs-offenburg.de*

Awais Ahmad, *Department of Electrical Engineering, Syed Babar Ali School of Science and Engineering, Lahore University of Management Sciences, Lahore, Pakistan; E-mail: 15060032@lums.edu.pk*

Axel Sikora, *Institute of Reliable Embedded Systems and Communication Electronics (ivESK), Offenburg University of Applied Sciences, Offenburg, Germany; E-mail: axel.sikora@hs-offenburg.de*

Aylin Celik, *University of Applied Sciences and Arts Dortmund, Emil-Figge-Strasse 42, 44227 Dortmund, Germany; E-mail: Aylin.Celik@fh-dortmund.de*

Bettina Horster, *VIVAI Software AG, Betenstrasse 13-15, 44137 Dortmund, Germany; E-mail: bettina.horster@vivai.de*

Carsten Wolff, *Fachhochschule Dortmund, Dortmund, Germany; E-mail: carsten.wolff@fh-dortmund.de*

Chris Phillips, *University of Newcastle upon Tyne, United Kingdom; E-mail: chris.phillips@newcastle.ac.uk*

Christof Röhrig, *University of Applied Sciences and Arts Dortmund, Emil-Figge-Strasse 42, 44227 Dortmund, Germany; E-mail: cristof.roehrig@fh-dortmund.de*

Dmytro Kravchenko, *Software Tools Department, Zaporizhzhia National Technical University, Zaporizhzhia, Ukraine; E-mail: dmitrykravchenko95@gmail.com*

Dinesh Kumar Saini, *Faculty of Computing and Information Technology, Sohar University, Sohar, Oman; E-mail: dinesh@soharuni.edu.om*

Dirk Van Merode, *Department of Electronics–ICT, Artesis-Plantijn University College, Belgium; E-mail: dirk.vanmerode@ap.be*

Dmytro Maevsky, *Odessa National Polytechnic University, Ukraine; E-mail: dmitry.a.maevsky@opu.ua*

Dmytro Uzun, *Department of Computer Systems, Networks and Cyber-security, National Aerospace University "KhAI", Kharkiv, Ukraine; E-mail: d.uzun@csn.khai.edu*

Enrico Löhrke, *inHaus GmbH, Grabenstrasse 70, 47057 Duisburg, Germany; E-mail: Loehrke@inhaus.de*

Eugene Brezhniev, *National Aerospace University "KhAI," Kharkiv, Ukraine; E-mail: e.brezhnev@csn.khai.edu*

Galyna Tabunshchyk, *Software Tools Department, Zaporizhzhia National Technical University, Ukraine; E-mail: galina.tabunshchik@gmail.com*

Heinz-Dietrich Wuttke, *Ilmenau University of Technology, TU Ilmenau, Ilmenau, Germany; E-mail: dieter.wuttke@tu-ilmenau.de*

Herman Fesenko, *1. Department of Computer Systems, Networks and Cybersecurity, National Aerospace University "KhAI", Kharkiv, Ukraine;*
2. O. M. Beketov National University of Urban Economy in Kharkiv, Kharkiv, Ukraine; E-mail: h.fesenko@khai.edu

Igor Kotenko, *1. St. Petersburg Institute for Informatics and Automation of Russian Academy of Sciences, Saint Petersburg, Russia;*
2. St. Petersburg National Research University of Information Technologies, Mechanics and Optics, Saint Petersburg, Russia;
E-mail: ivkote@comsec.spb.ru

Inna Skarga-Bandurova, *Volodymyr Dahl East Ukrainian National University, Severodonetsk, Ukraine; E-mail: skarga_bandurova@ukr.net*

Iryna Piskachova, *Department of Computer Science and Control System, Ukrainian State University of the Railway Transport, Kharkiv, Ukraine; E-mail: ipiskacheva@gmail.com*

Ivan Lobachev, *1. Intel Corporation, USA;*
2. Odessa National Polytechnic University, Odesa, Ukraine;
E-mail: lobachev@ieee.org

Jan Oelker, *University of Applied Sciences and Arts Dortmund, Emil-Figge-Strasse 42, 44227 Dortmund, Germany; E-mail: Jan.Oelker@fh-dortmund.de*

Jelena Bleja, *University of Applied Sciences and Arts Dortmund, Emil-Figge-Strasse 44, 44227 Dortmund, Germany; E-mail: jelena.bleja@fh-dortmund.de*

Karsten Henke, *Ilmenau University of Technology, TU Ilmenau, Ilmenau, Germany; E-mail: karsten.henke@tu-ilmenau.de*

Kashif Zia, *Faculty of Computing and Information Technology, Sohar University, Sohar, Oman; E-mail: kzia@soharuni.edu.om*

Lolita Vystorobska, *Odessa National Polytechnic University, Ukraine; E-mail: lolitav1998@gmail.com*

Mahbuba Moni, *Institute of Reliable Embedded Systems and Communication Electronics (ivESK), Offenburg University of Applied Sciences, Offenburg, Germany; E-mail: mahbuba.moni@hs-offenburg.de*

Marta Chinnici, *ENEA, R.C. Casaccia, Energy Technologies Department, ICT Division (DTE-ICT), Rome, Italy; E-mail: marta.chinnici@enea.it*

Manish Lamichhane, *PERCCOM Masters Student (Previous), Software Developer (Current), Germany; E-mail: manishlamichhane@gmail.com*

Manuel Schappacher, *Institute of Reliable Embedded Systems and Communication Electronics (ivESK), Offenburg University of Applied Sciences, Offenburg, Germany; E-mail: manuel.schappacher@hs-offenburg.de*

Marina Derkach, *Volodymyr Dahl East Ukrainian National University, Severodonetsk, Ukraine; E-mail: gln459@gmail.com*

Maryna Kolisnyk, *1. Department of Computer Systems, Networks and Cybersecurity, National Aerospace University "KhAI", Kharkiv, Ukraine; 2. Automation and Control in Technical Systems Department, National Technical University "Kharkiv Politechnical Institute", Kharkiv, Ukraine; E-mail: kolisnyk.maryna.al@gmail.com*

Muhammad Adeel Pasha, *Department of Electrical Engineering, Syed Babar Ali School of Science and Engineering, Lahore University of Management Sciences, Lahore, Pakistan; E-mail: adeel.pasha@lums.edu.pk*

Mykhailo Poliakov, *Zaporizhzhia National Technical University, Zhukovskogo, Zaporizhzhia, Ukraine; E-mail: polyakov@zntu.edu.ua*

Nataliia Kunanets, *Lviv Polytechnic National University, Lviv, Ukraine; E-mail: nek.lviv@gmail.com*

Nikolaos Doukas, *Department of Computer Engineering, Hellenic Military Academy, Vari, Greece; E-mail: nd@ieee.org*

Nikolaos G. Bardis, *Department of Computer Engineering, Hellenic Military Academy, Vari, Greece; E-mail: bardis@ieee.org*

Oleg Illiashenko, *Department of Computer Systems, Networks and Cybersecurity, National Aerospace University "KhAI", Kharkiv, Ukraine; E-mail: o.illiashenko@csn.khai.edu*

Oleg Sadov, *Leader at NauLinux & Linux Cyrillic Edition projects, Russia; E-mail: oleg.sadov@gmail.com*

Oleksandr Gerasin, *Admiral Makarov National University of Shipbuilding, Mykolaiv, Ukraine; E-mail: oleksandr.gerasin@nuos.edu.ua*

Oleksandr Matsiuk, *Ternopil Ivan Puluj National Technical University, Ternopil, Ukraine; E-mail: oleksandr.matsiuk@gmail.com*

Oleksandr Solovyov, *Department of Computer Systems, Networks and Cybersecurity, National Aerospace University "KhAI", Kharkiv, Ukraine; E-mail: extsand@gmail.com*

Oleksii Duda, *Ternopil Ivan Puluj National Technical University, Ternopil, Ukraine; E-mail: oleksij.duda@gmail.com*

Oleksii Kravchenko, *Software Tools Department, Zaporizhzhia National Technical University, Zaporizhzhia, Ukraine; E-mail: akravchenko312@gmail.com*

Oleksiy Korobko, *Admiral Makarov National University of Shipbuilding, Mykolaiv, Ukraine; E-mail: oleksii.korobko@nuos.edu.ua*

Oleksiy Kozlov, *Admiral Makarov National University of Shipbuilding, Mykolaiv, Ukraine; E-mail: oleksiy.kozlov@nuos.edu.ua*

Olga Gladkova, *Software Tools Department, Zaporizhzhia National Technical University, Zaporizhzhia, Ukraine; E-mail: gladolechka@gmail.com*

Peter Arras, *Faculty of Engineering Technology, KU Leuven, Belgium; E-mail: peter.arras@kuleuven.be*

Raymond J. Garbos, *eCollaborative Ventures, Inc. New Hampshire, USA; E-mail: rgarbos@yahoo.com*

Reiner Hormann, *University of Applied Sciences and Arts Dortmund, Emil-Figge-Strasse 42, 44227 Dortmund, Germany; E-mail: reiner.hormann@fh-dortmund.de*

Robert Hiromoto, *1. University of Idaho, Moscow, ID, United States; 2. Center for Advanced Energy Studies, Idaho Falls, ID, United States; E-mail: hiromoto@uidaho.edu*

Saverio De Vito, *ENEA, R.C. Portici, Energy Technologies Department, Photovoltaic and Smart Network Division (DTE-FSN), Portici, Italy; E-mail: saverio.devito@enea.it*

Shahid Masud, *Department of Electrical Engineering, Syed Babar Ali School of Science and Engineering, Lahore University of Management Sciences, Lahore, Pakistan; E-mail: smasud@lums.edu.pk*

Svitlana Yaremchuk, *Department of General Scientific Disciplines, Danube Institute, National University "Odessa Maritime Academy", Odessa, Ukraine; E-mail: svetlana397@yandex.ru*

Ted Kochanski, *Sensors Signals Systems, Lexington, MA, United States; E-mail: tedpk@alum.mit.edu*

Torben Lippmann, *Westfälische Hochschule, Gelsenkirchen, Germany; E-mail: Torben.Lippmann@w-hs.de*

Uwe Grossmann, *University of Applied Sciences and Arts Dortmund, Emil-Figge-Strasse 44, 44227 Dortmund, Germany; E-mail: uwe.grossmann@fh-dortmund.de*

Uwe Jahn, *Fachhochschule Dortmund, Dortmund, Germany; E-mail: uwe.jahn@fh-dortmund.de*

Volodymyr Kochan, *Ternopil National Economic University, Ternopil, Ukraine; E-mail: vk@tneu.edu.ua*

Volodymyr Pasichnyk, *Lviv Polytechnic National University, Lviv, Ukraine; E-mail: vpasichnyk@gmail.com*

Volodymyr Tkachenko, *National Technical University "KhPI," Kharkiv, Ukraine; E-mail: proftva@ukr.net*

Vladimir Sklyar, *Department of Computer Systems, Networks and Cybersecurity, National Aerospace University "KhAI", Kharkiv, Ukraine; E-mail: v.sklyar@csn.khai.edu*

Vyacheslav Kharchenko, *1. Department of Computer Systems, Networks and Cybersecurity, National Aerospace University "KhAI", Kharkiv, Ukraine; 2. Centre for Safety Infrastructure Oriented Research and Analysis, Kharkiv, Ukraine; E-mail: v.kharchenko@csn.khai.edu*

Yuriy Kondratenko, *Petro Mohyla Black Sea National University, Mykolaiv, Ukraine; E-mail: yuriy.kondratenko@chmnu.edu.ua*

List of Figures

Figure 1.1	Simplified embedded system with computational electronics, several sensors, and one actuator. . . .	18
Figure 1.2	Simplified view of an automobile ABS based on the simplified embedded system shown in Figure 1.1. Sensors and an actuator specific and relevant to the function of the ABS are shown.	19
Figure 1.3	Simplified networked embedded system with computational electronics, sensors, actuators, a local network, a global interface, and the "Cloud."	22
Figure 1.4	The design for globalization process: beginning with the need of the "Customer" (labeled 1) and progressing clockwise through various design and test stages (e.g., Design Tools – 5, Virtual Instrumentation – 8) mediated by the I-GEMS Repository (labeled 15) operating in the Cloud. The final tested and verified design is delivered to the GNODE Target Platform (labeled 17) through local tools running on the Development PC (labeled 16).	33
Figure 1.5	The vital electronics design process: beginning with the local subject matter expertise and enabled by the "*In situ* Hardware Development Environment" resulting in a tested and verified "vital electronics" qualified computer based on a PSOC – safe to connect to the Internet.	34
Figure 3.1	WSN system for water flow monitoring in canals. .	65
Figure 3.2	Network node structure in (a) LEACH and (b) M-LEACH protocols.	68
Figure 3.3	Flowchart for EESNR CH selection process.	71
Figure 3.4	Flowchart for modified advertisement process proposed in EESNR.	72

Figure 3.5 Comparison of network lifespan among EESNR and other protocols. 76
Figure 3.6 Comparison of average number of successful bits transmitted. 77
Figure 3.7 Comparison of network lifespan among EESNR (with fading) and other protocols (without fading). . 77
Figure 3.8 Nodes' die-out pattern for the I-LEACH protocol. . 78
Figure 3.9 Nodes' die-out pattern in the proposed EESNR protocol. 79
Figure 4.1 Internet of Things. 85
Figure 4.2 Categories of objects. 88
Figure 4.3 Agent-based model of peers in competitive mode. . 92
Figure 4.4 Agent-based model of peers in cooperative mode. . 93
Figure 4.5 Network types. 95
Figure 4.6 Status of peers at iteration 850 in case of 250 agents and Competitive Mode. Agents in light gray are in offline mode (status 0). Agents in dark gray are in idle mode (status 1). Agents in red are in search mode (status 3). Status 2 is a transit mode where a service is assigned. Agents in green are in request mode (status 4). Statuses 5 and 6 are transit states again. The arrow originates from the requester and ends at a receiver. 96
Figure 4.7 Status of peers at iteration 850 in case of 250 agents and Cooperative Mode. Agents in light gray are in offline mode (status 0). Agents in dark gray are in idle mode (status 1). Agents in red are in search mode (status 3). Status 2 is a transit mode where a service is assigned. Agents in green are in request mode (status 4). Statuses 5 and 6 are transit states again. The arrow originates from the requester and ends at a receiver. 96
Figure 4.8 Comparative analysis of three sets of four cases in competitive mode. 98
Figure 4.9 Comparative analysis of three sets of four cases in cooperative mode. 99
Figure 5.1 Main components of the healthcare IoT infrastructure: devices with a reader, the Cloud, and a healthcare organization. 108

Figure 5.2 Detailing of components of the healthcare IoT infrastructure. 109
Figure 5.3 Communications and functions of the main components of the IoT infrastructure. 110
Figure 5.4 The structure of FTA. 112
Figure 5.5 The tree fragment of possible scenarios in which critical values of the indicators may not be transmitted to the Cloud. 112
Figure 5.6 The queueing system for the healthcare IoT system. 114
Figure 5.7 A scheme of "birth–death" process for the considered case. 115
Figure 5.8 A Markov model for the considered case of a successful attack at one stage of the service request processing (with a halt) case. 116
Figure 5.9 The stationary distribution for the considered case of a successful attack at one stage of the service request processing (with a halt). 117
Figure 5.10 A Markov model for the considered case of a successful attack at one stage of the service request processing (without halts). 117
Figure 5.11 The stationary distribution for the considered case of a successful attack at one stage of the service request processing (without halts). 118
Figure 5.12 A Markov model for the considered case of a successful attack at one stage of the service request processing (with halts and eliminating of one vulnerability). 118
Figure 5.13 The stationary distribution for the considered case of a successful attack at one stage of the service request processing (with halts and eliminating of one vulnerability). 119
Figure 5.14 A Markov model for the considered case of a successful attack at one stage of the service request processing (with halts and system has two vulnerabilities and one vulnerability is eliminated). 119
Figure 5.15 The stationary distribution for the considered case of a successful attack at one stage of the service request processing (with halts and system has two vulnerabilities and one vulnerability is eliminated). 120

Figure 5.16 A Markov model for the considered case of a successful attack at one stage of the service request processing (with halts and with halts and system has two vulnerabilities and two vulnerabilities are eliminated). 120
Figure 5.17 The stationary distribution for the considered case of a successful attack at one stage of the service request processing (with halts and with halts and system has two vulnerabilities and two vulnerabilities are eliminated). 121
Figure 5.18 Combined plot of availability functions for all considered cases. 122
Figure 6.1 General view of the structural and hierarchical scheme of the PSS of the infrastructure of the region. 130
Figure 6.2 An example of practical implementation of the structural–hierarchical scheme for the PSS of the RI facility. 131
Figure 6.3 Functional diagram of the device “motion detection/ intrusion detection subsystems.” 132
Figure 6.4 IDEF0 diagram of effects on PSMECA components. 133
Figure 6.5 Scheme of research and development of models and methods for the risk analysis of PSSs. 134
Figure 6.6 The projection of the hierarchical structure of failures on the table of the main structural elements in the PSS. 139
Figure 7.1 The results of detection of unequal one-type properties. 159
Figure 7.2 A weighted multi-graph of relations of event type properties. 163
Figure 7.3 A weighted multi-graph of relations between event types. 165
Figure 8.1 IoT four-layered architecture (based on ITU-T standard). 171
Figure 8.2 The Markov model of the router states. 186

Figure 8.3 Graphic dependencies of the AC on changing the rate values of the transitions $\lambda 731$, $\lambda 3031$, and $\lambda 1731$ in the event without the impact of attacks on the router's components and in the event of the PFC security card failure and successful attack on the router's components. 188
Figure 8.4 Graphic dependencies of the AC on changing the rate values of the transitions $\lambda 731$, $\lambda 3031$, and $\lambda 1731$ in the event with the impact of attacks on the router's components. 188
Figure 8.5 Graphic dependencies of the AC on changing the rate values of the transitions $\lambda 731$, $\lambda 3031$, and $\lambda 1731$ in the event without attacks on the router's components. 189
Figure 8.6 Graphic dependencies of the AC on changing the rate values of the transitions $\lambda 3331$, $\lambda 231$, $\lambda 331$, $\lambda 631$, $\lambda 1131$, $\lambda 1331$, $\lambda 1931$, $\lambda 2631$, $\lambda 2731$, $\lambda 2831$, and $\lambda 2931$ in the event of the PFC security card failure and successful attack on the router's components. 189
Figure 8.7 Graphic dependencies of the AC on changing the rate values of the transitions $\lambda 231$, $\lambda 331$, $\lambda 631$, $\lambda 1131$, $\lambda 1331$, and $\lambda 2631$ in the event of the PFC security card failure and successful attack on the router's components. 190
Figure 9.1 Structure of MPSAMS. 203
Figure 9.2 Simplified structure of MPSAMS. 205
Figure 9.3 RBD for the wired network subsystem. 206
Figure 9.4 RBD for the Wi-Fi subsystem. 207
Figure 9.5 RBD for the Li-Fi subsystem. 208
Figure 9.6 RBD for the Wi-Fi drone-based subsystem. 208
Figure 9.7 RBD for MPSAMS based on: (a) the wired network subsystem (SubG) and the Wi-Fi subsystem (SubW), (b) the wired network subsystem (SubG) and the Li-Fi subsystem (SubL), and (c) the wired network subsystem (SubG) and the Wi-Fi drone-based subsystem (SubD). 210

Figure 9.8 RBD for MPSAMS based on: (a) the wired network subsystem (SubG), the Wi-Fi subsystem (SubW), and the Li-Fi subsystem (SubL), (b) the wired network subsystem (SubG), the Wi-Fi subsystem (SubW), and the Wi-Fi drone-based subsystem (SubD), and (c) the wired network subsystem (SubG), the Li-Fi subsystem (SubL), and the Wi-Fi drone-based subsystem (SubD). 211
Figure 9.9 RBD for MPSAMS based on the wired network subsystem (SubG), the Wi-Fi subsystem (SubW), the Li-Fi subsystem (SubL) and the Wi-Fi drone-based subsystem (SubD). 211
Figure 9.10 Various variants of RBD for MPSAMS based on the wired network subsystem (SubG), the Wi-Fi subsystem (SubW), the Li-Fi subsystem (SubL), the Wi-Fi drone-based subsystem (SubD), and the IoT-based subsystem (SubIoT). 211
Figure 9.11 Dependence on PFFO for MPSAMS based on the wired network subsystem and the Wi-Fi subsystem on the number of the redundant drones for the Wi-Fi drone-based communication Section 9.2. 213
Figure 9.12 Dependence on PFFO for MPSAMS based on the wired network subsystem and the Wi-Fi subsystem on the number of redundant drones for the Wi-Fi drone-based communication Section 9.2 at different values of PFFO per each redundant drone. 213
Figure 9.13 Dependence on PFFO for various variants of MPSAMS on the number of the redundant drones for the Wi-Fi drone-based communication Section 9.1. 214
Figure 10.1 Overview of the virtual testbed architecture. 228
Figure 10.2 Integration of the TTCN-3 framework into VTENN. 231
Figure 10.3 Graphical overview of the VTENN implementation including TE, NM, VNs, and VCs. 233
Figure 10.4 Test topology created by a sample TC using three nodes and three channels, whereas the path between NodeA and NodeB will be removed during the TC. 236

Figure 10.5 Attachment process of the example TC. At first, nodes and channels are created. After checking for available ports at the nodes and channels (*AVailable* request), the connections will be established using the *COnnect* commands. 236
Figure 10.6 Debug output after the successful establishment of the communication channels. Both nodes and the channels print debug messages for received data. The received data at a channel port will be forwarded to the respective connected node. 237
Figure 10.7 Debug output of the detachment process of the NodeA, NodeB, and CH1. After the detachment, no communication will take place anymore. 237
Figure 11.1 Sketch of deep urban environment. 250
Figure 11.2 Some IoT applications in urban context. 251
Figure 11.3 Libelium[1] smart world: IoT use case. 252
Figure 11.4 Edge tier: architecture' scale. 254
Figure 11.5 Future air quality monitoring network will rely on different and pervasive sources of information (a) ranging from wearable system to high-accuracy conventional analyzers and fixed multisensors meeting regulatory data quality objectives (b). . . . 258
Figure 11.6 TinyNose on-board estimation of ethanol concentration by neural networks. 260
Figure 11.7 (a) The Monica cloud interface where information coming from multiple deployed multisensors is collected and fused together. (b) The smartphone real-time mapping utility allows for immediate and locally compute feedback on personal pollutant exposure. 261
Figure 11.8 The Monica architecture and its calibration update mechanism. Whenever a Monica device becomes connected to a smartphone/cloudlet/gateway, the latter asks the cloud-based backend for updated calibration parameters to maintain the calibration components. In this way, raw data are always processed with the most accurate calibration available. 264

Figure 12.1 The generalized architecture of the FPV Auto project. 277
Figure 12.2 The detailed architecture of the FPV Auto project. . 278
Figure 12.3 The main and the calibration windows of the program. 279
Figure 12.4 MSU brightness flicking. 282
Figure 12.5 MSU noise estimation. 283
Figure 12.6 The result of object recognition on the image with good lighting. 285
Figure 12.7 The result of object recognition on the image taken during rain. 285
Figure 12.8 The example of the environment with the narrow passage. 288
Figure 12.9 The example of the environment with the sharp turns. 288
Figure 13.1 Software architecture for a smart city. 295
Figure 13.2 Software architecture as the backbone of the smart city. 296
Figure 13.3 Smart urban modeling. 297
Figure 13.4 Major destination and possible attack point in a smart city. 298
Figure 13.5 Security architecture for a smart city. 299
Figure 13.6 Attack analysis. 300
Figure 14.1 ACDD collaboration chart. 318
Figure 14.2 General framework for ACDD. 319
Figure 14.3 ER diagram of logical interconnections of the requirement with other components. 324
Figure 15.1 Original German version (left) of the technology stack according to Ropohl [14] and own English translation (right). 334
Figure 15.2 Generic three-layer architecture model for metropolitan energy and mobility systems based on the architecture model of Ropohl. 335
Figure 15.3 Technology stack of the operator-controller-module (OCM) [16, 17], generalized structure of intelligent technical systems. 337
Figure 15.4 Technology stack of the ORC turbine control system, matched to the three layers of the OCM. . . 339

Figure 15.5 Demonstrator setup of the ORC turbine (left, background: stacked two containers with HT TuGen and LT TuGen, middle: direct vaporizer, right: liquefier module). 340
Figure 15.6 Thermodynamic circuits of the four modules of the ORC turbine. 341
Figure 15.7 MPC approach for the ORC turbine. 342
Figure 15.8 Coupling of state machines and controllers via SVI shared memory. 343
Figure 15.9 Nine-stage "smartification process" based on the OCM. 345
Figure 16.1 TAG architecture. 357
Figure 16.2 MQTT topic hierarchy. 362
Figure 17.1 The functional structure of the generalized SPC's IIoT system. 373
Figure 17.2 The functional structure of the generalized SPC IIoT system. 377
Figure 17.3 Control unit of the specialized pyrolysis complex MCP-5 IIoT system. 379
Figure 17.4 Main screen of the HMI of the remote SCADA system for the specialized pyrolysis complex MCP-5. 382
Figure 17.5 Graph screen of the temperature changes in the main control points of the MCP-5 reactor. 382
Figure 18.1 Functional information-communication structure of the "smart city." 392
Figure 18.2 Three-level structure of the "smart city" in the information-technological context. 395
Figure 18.3 Categories of cloud services in the "smart cities." . 399
Figure 18.4 Architecture of the "smart city" analytical data processing center. 400
Figure 18.5 Cloud service analytics architecture. 401
Figure 18.6 Computing architecture of providing context-sensitive services in the "smart cities." 403
Figure 19.1 Real-time public transport information service infrastructure. 414
Figure 19.2 A set of control points for route assignment. 422
Figure 19.3 Algorithm for assigning route number to the trolleybus. 423

Figure 19.4 The flowchart of data acquisition and information processing strategies. 424
Figure 19.5 Test road segmentation. 426
Figure 19.6 Trolleybus arrival time prediction obtained with the different techniques. 428
Figure 20.1 General architecture view. 437
Figure 20.2 A high-level diagram of the implemented system set-up. 439
Figure 20.3 A diagram displaying the high-level algorithm of semi-autonomous and autonomous modes of operation. 440
Figure 20.4 A screenshot of the varying views and representation of numerical data collected. 441
Figure 20.5 A screenshot of the web view of the user interface and configuration window. 443
Figure 20.6 A diagram displaying the structure of the HDF5 data format. 445
Figure 20.7 A high-level diagram outlining the hierarchy of the system. 446
Figure 20.8 Experimental setup for local testing of the prototype system. 448
Figure 20.9 Latency test results for the two cities where the tests were conducted. 448
Figure 21.1 Top-level system and software architecture of SSP. 460
Figure 21.2 SOLION Assistenz – radio-based assistance system (inHaus GmbH). 463
Figure 21.3 Assistance system components in a living environment (left: motion-based night light; middle: motion sensor bath (ceiling); and right: door opening contact). 463
Figure 21.4 Covibo – Communication structure. 466
Figure 21.5 The BMC. 466
Figure 21.6 AAL-scenario of partners, activities, and services used within project SSP. 467
Figure 21.7 Corporate business models and collaborative system business model. 469
Figure 22.1 Components of IoT architecture. 486
Figure 22.2 Components of WoT architecture. 489

Figure 22.3 A screenshot of the application's operation Web Thing for the direct connectivity integration template. 494
Figure 22.4 A screenshot displaying the values of the temperature sensor in the Web browser. 494
Figure 23.1 Structure of the cognitive management system. . . 506
Figure 23.2 Hierarchy of subsystems of the cognitive control system. 507
Figure 23.3 Physical models of the GOLDi lab. 509
Figure 23.4 Behavioral model of a control system. 509
Figure 23.5 Fragment of the target AG. 510
Figure 23.6 Fragment of the scenario AG. 510
Figure 24.1 Remote labs: from idea to experiment. 519
Figure 24.2 ISRT architecture. 520
Figure 24.3 Face detection example. 521
Figure 24.4 Structure of CALM, integrated courses, and labs on material science. 522
Figure 24.5 Two-point bending setup. 522
Figure 24.6 Remote lab, screenshot with view of two cameras, virtual lab. 523
Figure 24.7 New COPTURN interface. 527
Figure 24.8 Embedded factory at campus De Nayer. 529
Figure 25.1 Publication list in APM. 538
Figure 25.2 The explosive growth of IoT-devoted publications by years. 539
Figure 25.3 The diagram of the distribution of conferences by year. 541
Figure 25.4 The diagram of the distribution of authors by year. . 542
Figure 25.5 Examples of options for setting the plotter control (a and b) and the final result (c). 554
Figure 25.6 G-file for the variant of the task [Figure 25.5(b)]. . 554
Figure 25.7 Block diagram of a mini plotter. 555
Figure 25.8 Type of mini plotter (Homemade DIY CNC). . . . 556
Figure 25.9 LED ribbon control module (a) and coloring options (b). 557

List of Tables

Table 4.1 TU-based peer specifications . . . 90
Table 4.2 Services-based peer specifications . . . 90
Table 4.3 Operations-based peer specifications . . . 91
Table 4.4 Agent states . . . 91
Table 4.5 Simulation setup . . . 94
Table 6.1 FMECA table for the case of CCTV subsystem functioning in normal operation mode . . . 140
Table 6.2 PSMECA table for the case of CCTV subsystem functioning in normal operation mode . . . 141
Table 7.1 Direct and indirect one-type relations between event types . . . 164
Table 8.1 Technologies and features to create IoT systems . . . 173
Table 8.2 Examples of spyware for OS Android and Windows 181
Table 8.3 Values of transition rates, attack rates, and recovery rates . . . 183
Table 9.1 Equations for calculation of PFFO for various variants of MPSAMS in accordance with RBDs presented in Figures 9.7–9.10 . . . 212
Table 10.1 Comparison between typical specialized, generic, and nominal simulators . . . 226
Table 11.1 A tradeoff comparison among accuracy and complexity of machine learning techniques for smart air quality calibration scenarios . . . 262
Table 12.1 The comparison of CS implementations . . . 281
Table 12.2 The summary of the reliability testing . . . 283
Table 12.3 The probability of physical class failures . . . 283
Table 16.1 MQTT topic table . . . 362
Table 19.1 Characteristics of three different segments . . . 426
Table 19.2 Experimental results of trolleybus arrival time for three road segments . . . 427
Table 22.1 Curriculum structure . . . 491

Table 23.1 Values of elements of the cognitive control system . 508
Table 25.1 Number of IoT-devoted publications by years 539
Table 25.2 Number of types of IoT-devoted publications by years 540
Table 25.3 Distribution of conferences with IoT presentations by year 540
Table 25.4 Course characteristic for Master Study (wireless embedded systems) 551
Table 25.5 Course characteristic for Master Study (embedded systems) 551
Table 25.6 Metric for simulation of IoT and IoE-based systems' course 552
Table 25.7 Metric for IoT for Smart building and city course . . 553

List of Abbreviations

2nd Life	Second Life
3GPP	Third Generation Partnership Project
5G	Fifth generation wireless system
6LoWPAN	IPv6 over Low power Wireless Personal Area Networks
6LoWPAN	IPv6 Low Power Wireless Personal Area Networks
AAL	Ambient Assisted Living
ABS	Antilock Braking System
AC	Availability Function
ACDD	Assurance Case Driven Design
ACK	Acknowledgement
AD	Absolute Deviation
ADI	A Digital Identity
AES OFB	Output Feedback Mode of Operation for a Block Cipher
AES	Advanced Encryption Standard
AFR	Africa Region
AI	Artificial Intelligence
AJAX	Asynchronous JavaScript and XML
ALIOT	Internet of Things: Emerging Curriculum for Industry and Human Applications
AMQP	Advanced Message Queuing Protocol
AMS	Autonomous Mechatronics Systems
ANSI	American National Standards Institute
APE	Absolute Percentage Error
API	Application Programming Interface
ARPA	Advanced Research Projects Agency
ASHRAE	American Society of Heating, Refrigerating and Air-Conditioning Engineers
ATS	Average Travel Speed
AV	Anti-Virus
BACnet	Building Automation and Control network
BD	Big Data
BLE	Bluetooth Low Energy Technology

BLE	Bluetooth Low Energy
BMC	Business Model Canvas
BS	Base Station
BSON	Binary JavaScript Object Notation
CA	Certificate Authorities
CAE	Claim-argument-evidence
CAEC	Claim-argument-evidence-criteria
CAFFE	Convolution architecture for feature extraction
CALM	Computer-Aided Learning Module
CAN RTLS	Can Real-Time Location Systems or Radio Frequency Identification
CBD	Component-Based Development
CBSE	Component-Based Software Engineering
CCTV	Closed-Circuit Television
CDIO	Conceive Design Implement Operate
CDMA	Code Division Multiple Access
CE	European Conformity
CEN	European Committee for Standartization
CERT-US	United States Computer Emergency Readiness Team
CG	wired network communication section
CH	Case Handling Paradigm
CH	Cluster head
CHF	Multiple components (operators) which are difficult to formalize
CHW	Multiple Hardware Components
CHWH	Subset of hardware (secondary) components
CHWS	Subset of hardware (primary) components
CIDLab	Critical Infrastructure Dependability Laboratory
CIoT	Consumer Internet of Things
CMS	Content Management System
CoAP	Constrained Application Protocol
CPS	Cyber-Physical Systems
CPU	Central Processing Unit
CrS	Crisis center
CSS	Cascading Style Sheets
CSW	Multiple Software Components
cURL	Client Uniform Resource Locator
DAO	Decentralized Autonomous Organization
DB	Database

DBMS	Database Management System
DC	Data Center
DDS	Data Distribution Service
DES	Data Encryption Standard
DESL	DES Lightweight Extension
DF	Drone Fleet
df	Project Fault
DfG	Design for Globalization
DH22	Temperature-Humidity Sensor
DL	Light fidelity drone-based communication section
DLP	Data Loss Prevention
DMS	Decision-Making System
DMSS	Decision-Making Support System
DOM	Document Object Model
DoT S	Drones-of-Things Subsystem
DS	Wi-Fi drone-based sensor section
DSS	Decision Support System
DTLS	Datagram Transport Layer Security
DW	Wi-Fi drone-based communication section
EAP	East Asia and Pacific
EAS	Aggressive Environment State
ECA	Eastern and Central Asia
ECC	Elliptic-Curve Cryptography
ECDH	Elliptic-Curve Diffie–Hellman Public Key
ECE	Electrical and Computer Engineering
eCV	eCollaborative Ventures
EESNR	Energy-efficient sensor network routing
EIA	Electronic Industries Alliance
EIAS	Environment Information Components (Artificial)
EIB/KNX	European Installation Bus/Communication Bus
EINS	Environment Information Components (Natural)
EIS	Environment Information Components
enIoD	energy neutral Internet of Drones
EnOcean	Energy Harvesting Wireless Technology
ENS	Event Notification Service
ENS	Normal Environment State
EPAS	Environment Physical Components (Artificial)
EPC	Electronic Product Code
EPNS	Environment Physical Components (Natural)

EPS	Evolved Packet System
EPS	Environment Physical Components
ER	Entity Relationships
ESR	Embedded Services Router
ETA/ETD	Estimated Time of Arrival/Estimated Time of Departure
ETH	Ether
ETSI	European Telecommunications Standards Institute
EU	European Union
EU GDPR	General Data Protection Regulation
EXI	Efficient XML Interchange format
FDA	Food and Drug Administration
FMECA	Failure Modes, Effects And Criticality Analysis
FPGAs	Field-Programmable Gates Arrays
FPGE	Front-Panel Gigabit Ethernet
FTA	Fault Tree Analysis
GAIN	Global Ambient Intelligence Network
GB	Giga Bytes
GHG	Green House Gases
GIS	Geographic Information System
GISE	Globally integrated security environment
GNAT	Global Network Academic Test Initiative
GNODE	GNAT Node
GPRS	General Packet Radio Service
GPS	Global Position System
GSM	Global System for Mobile Communications
GSN	Goal Structuring Notation
GUI	Graphical User Interface
HAC	Human-Agent Collective
HART	Highway Addressable Remote Transducer
HDF5	Hierarchical Data Format - version 5
HEI	High Educational Institute
HEV	Hybrid Electric Vehicle
HF	Human Factor
hf	Operator (human) Fault
hif	Information Violations
HMI	Human–machine interface
hpf	Physical Defects
HSGB	Head Smart Garbage Bin
HTTP	Hypertext Transfer Protocol

HTTPS	HyperText Transfer Protocol Secure
HVAC	Heating, ventilation, and air conditioning system
HW	Hardware
I/O	Input/Output
ICT	Information and Communication Technology
ID	Identification
IDS (IPS)	Intrusion Detection (prevention) System
IEC	International Electrotechnical Commission
IEEE	Institute of Electrical and Electronics Engineers
IERC	European Research Cluster on the Internet of Things
IETF	Internet Engineering Task Force
if	Interaction Fault
IIC	Industrial Internet Consortium
IIoT	Industrial Internet of Things
IIMECA	Information Intrusion Modes, and Criticality Analysis
IMEI	International Mobile Equipment Identity
IMSI	International Mobile Subscriber Identity
IoCs	Indicators of Compromise
IoD	Internet of Drones
IoE	Internet of Everything
IoP	Internet of People
iOS	iPhone OS
IoT	Internet of Things
IoT S	Internet-of-Things system
IoT-A	Internet of Things Architecture
IoT-I	Internet of Things Initiative
IoTS	Internet-of-Things sensors
IoVT	Internet of Vital Things
IP	Internet Protocol
IP address	Internet Protocol address
IP	Intellectual Property
IP	Internet Protocol
IPSec	Internet Protocol Security
IPv6	Internet Protocol version 6
ISC	Internal Services Card
ISO	International Standard Organization
ISO/TR	International Organization for Standardization/Technical Reports
ISRT	Informational Systems on Reliability Tasks

ITS	Intelligent transport system
ITU	International Telecommunications Union
ITU-T	International Telecommunication Union
JMS	Java Message Service
JS	JavaScript
JSON	Javascript Standard Object Notation
JSON	JavaScript Object Notation
JSON-LD	JavaScript Object Notation for Linked Data
JTC 1	Joint Technical Committee 1
KhAI	Kharkiv Aviation Institute
KhPI	Kharkiv Polytechnic Institute
LAC	Latin America and the Caribbean
LAN	Local Area Network
LEACH	Low-Energy Adaptive Clustering Hierarchy
LED	Light-Emitting Diode
Li-Fi	Light Fidelity
Li-FiS	Light Fidelity Sensors
LiTRA	Long Term Evolution Integrated Trunked Radio
LLL	Lisp Like Language
LonWorks	Local Operating Network
LoRa	Long Range
LoWPAN	Low-Power Wireless Personal Area Networks
LPWAN	Low-Power Wide-area Network
LTE	Long-Term Evolution
LTE-A	Long-Term Evolution Advanced
M2M	Machine-to-Machine
MAE	Mean Absolute Error
MAPE	Mean Absolute Percentage Error
MB	Mega Bytes
MCB	Microcontroller Board
MCP	Multiloop Circulating Pyrolysis
MCUs	Microprocessor Units
MEAN	MongoDB, Express.js, Angular.js, Node.js
MENA	Middle East and North Africa
MongoDB	Humongous Database
MOSIS	Metal Oxide Semiconductor Implementation Service
MPC	Model Predictive Control
MPSAMS	Multi-version post-severe accident monitoring system
MPW	Municipal Polymeric Waste

MQTT	Message Queue Telemetry Transport
MQTTS	MQTT Security
MS	Metasystem
NAU	National Aerospace University
NB-IoT	Narrow Band IoT
NFC	Near Field Communication
NIM	Network Interface Module
NMS	Networked Mechatronic Systems
NM-X	Extended Single-Wide Network Modules
NoSQL	Not Only SQL
NPP	Nuclear Power Plant
NS3	Network Simulator 3
NTU	National Technical University
NVRAM	Non-Volatile Random Access Memory
OCF	Open Connectivity Foundation
OCM	Operator-Controller-Module
OECD	Organisation for Economic Co-operation and Development
OGC	Open Geospatial Consortium
OLAP	Online Analytical Processing
OMG	Object Management Group
OMS	Outage Management System
ORC	Organic Rankine Cycle
OS	Operational System
OWL	Web Ontology Language
P2P	Peer-to-Peer
PaaS	Platform as a Service
PAMS	Post-Accident Monitoring System
PCB	Printed Circuit Board
PDF	Probability Density Function
PETRAS	Privacy, Ethics, Trust, Reliability, Acceptability and Security
pf	Physical Fault
PFC	Policy Feature Card
PFFO	Probability of Failure-Free Operation
PIMECA	Physical Intrusion Modes, Effects, and Criticality Analysis
PKI	Public Key Infrastructure
PLC	Power Line Communication

PLC	Programmable Logic Controller
PoE	Power over Ethernet
PoS	Proof of Stake
PoW	Proof of Work
PSMECA	Physical Security Modes, Effects and Criticality Analysis
PSMECA	Physical Security Modes And Effect Analysis
PSOC	Programmable System On a Chip
PSS	Physical Security Systems
PTC	Parametric Technology Corporation
Pub/Sub	Publisher/Subscriber
QoS	Quality of Service
RAM	Random Access Memory
RBD	Reliability Block Diagram
RCM	Requirement Complexity Metric
RDF	Resource Description Framework
REST	Representational State Transfer
RF	Radio Frequency
RFID	Radio Frequency Identification
RFID/AutoID	Radio Frequency Identification/Auto Identification
RI	Regional Infrastructure
RL	Remote Laboratory
RMSE	Root-Mean-Square Error
ROM	Read-Only Memory
RPC	Remote Procedure Call
RPM	Red Hat Package Manager
RPMA	Random Phase Multiple Access
RSA	Public-Key Ciphers Rivest–Shamir–Adleman
RSA	Rivest Shamir Adleman
RSS	Rich Site Summary
SaaS	Software as a Service
SAGE	Semi-Automatic Ground Environment
SAR	South Asia Region
SBC	Smart Business Center
SCADA	Supervisory Control And Data Acquisition
SDN	Software-Defined Networking
SEREIN	Modernization of postgraduate studies on SEcurity and REsilience for human and Industry related domains
SG	Wired network section
SGB	Smart Garbage Bin

SHA2	Secure Hash Algorithm
SIEM	Security Information and Event Management
SL	Light fidelity sensor section
SME	Small and Medium-sized Enterprises
SMS	Short Message Service
SN	Sensor Network
SNR	Signal-to-Noise Ratio
SOA	Service-Oriented Architecture
SOAP	Simple Object Access Protocol
SOC	System-on-a-Chip
SPARQL	Protocol and Resource Description Framework Query Language
SPC	Specialized Pyrolysis Complex
SPI	Smart Personnel Interface
SQL	Structured Query Language
SRM	Software Reliability Mechanism
SS (VS)	Security (vulnerability) Scanner
SSE	Server Side Events
SSH	Secure Shell Script
SSL	Secure Sockets Layer
SSP	Smart Service Power
STARC	Methodology of SusTAinable Development and InfoRmation Technologies of Green Computing and Communication
SubD	Wi-Fi drone-based subsystem
SubG	Wired network
SubIoT	Internet-of-Things subsystem
SubL	Light fidelity subsystem
SubW	Wi-Fi subsystem
SW	Software
SW	Wi-Fi sensor section
SWMS	Smart Waste Management System
SWS	Software tools and software systems
TANS	Trolleybus arrival notification system
TCL	Tool Command Language
TCP	Transfer Control Protocol
TCP	Transmission Control Protocol
TLS	Transport Layer Security Protocol
TLS	Transport Layer Security

TV	Television
UAV	Unmanned Aerial Vehicle
uCode	Unicode
UDP	User Datagram Protocol
UE	User Equipment
UHF	Ultra High Frequency
UMTS	Universal Mobile Telecommunications System
UN	United Nations
UPnP	Universal Plug And Play
UPOE	Universal power over Ethernet
URI	Uniform Resource Identifier
URL	Uniform Resource Locator
USA	United States of America
US Tag	United States Technical Advisory Groups
UWB	Ultra Wide Band
VCS	Version Control System
VHF	Very High Frequency
VITA	VMEbus International Trade Association
VPN	Virtual Private Network
W3C	World Wide Web Consortium
WADL	Web Application Description Language
WAF	Web Application Firewall
WAN	Wide Area Network
Web	World Wide Web
WebRTC	Web Real-Time Communications
WHO	World Health Organization
Wi-Fi	Wireless Fidelity
WiFiDirect	Wi-Fi Peer-to-Peer
Wi-FiS	Wi-Fi sensors
Wikis	Knowledge management resources
WireS	Wired Sensors
WLAN	Wireless Local Area Network
WoT	Web of Things
WPAN	Wireless Personal Area Network
WSDL	Web Services Description Language
WSNs	Wireless Sensor Networks
WSS	WebSocket Security
WWW	World Wide Web
XML	eXtensible Markup Language
XMPP	Extensible Messaging and Presence Protocol

Dependable IoT for Human and Industry: Introduction and Book Scope

1. Internet of Important Things

There are numerous publications which introduce and discuss the Internet of Things (IoT). In the midst of these, this work has several unique characteristics which should change the reader's perspective, and in particular, provide a more profound understanding of the impact of the IoT on society. These salient points may be summarized as follows:

- IoT is characterized as a disruptive innovation and/or technology as defined by Christensen.
- IoT is one of the drivers of the high-tech market with an emphasis on the significance of the IoT market prediction.
- IoT has a global range; the bulk of chapters in this publication have been derived from the IDAACS 2017 conference which has become an East–West Catalyst for IoT-based innovation.
- IoT is one of the disruptive technology milestones in a technology development roadmap: beginning with the first transistor, on through era of VLSI and vital electronics, 5G, and Grand Challenges.

The history of information and communication technology "ICT" starting with the development of the first transistor is sufficiently rich to formulate general laws of technology development. The well-known and popular Moore's law is based on empirical observation over many technological generations. However, it is too simplistic to become the scientific base of logology in the ICT sphere.

In contrast, the editors of this publication recognize the existence of parallel dual worlds: one is biology based and the other one is technology based. In general, there is an urgent need for the study of the interaction between the two. The urgency is driven in part by the observation that the technology-based, virtual sphere of: smart phones, the Internet, and software applications "Apps" is no longer controllable. On a daily basis, several disparate digital media allow several billion individuals to interact in various unknown and unknowable ways. Marshall McLuhan's Global village really exists.

This realization generates fundamental challenges facing humanity. A classic example is restricting the access of a teenager to the Internet and a smart phone. This implies that we not only have to establish the existence rules between the real and virtual worlds but also need to re-establish the supremacy of humans over the emerging "Cyber-World" of robots and personal assistants. The vision outlined in the 40s of the 20th Century by Norbert Wiener has materialized.

Moore's law is a special case of a more generalized, but still empirical observation, i.e., a "generalized Moore's law," which notes that the development of ICT is ruled by total quality management style cycles. Christensen at Harvard Business School identified how a cycle of innovation originates with a disruptive innovation, initially only accessible to scientific, governmental, or business elites. This is followed by the contractual phase where the disruptive technology becomes available to society generally.

The overall development process is governed by the growing complexity of microelectronic systems and continuous integration of the technology world on all scales. The latter represents a heterogeneous ecosystem of ICT entities with virtual components targeting specific humans as users. Virtual components are databases, social media such as Facebook, search tools such as Google, community knowledge bases such as Wikipedia, personal assistants, and so on.

The integration process was initiated by the invention of a single transistor, which was followed by the first integrated device, the first embedded system, the first integrated system, the first network, the first constellation of networks, and so on. There is no upper boundary in the open-ended development process at this point. Each disruptive innovation has been associated with an application such as VLSI, the personal computer, WWW, the IoT, and so on. This process has made microelectronics globally available, with fully fledged computers as small as a grain of salt and so inexpensive as to be truly disposable. As a result of this development, an exponential function which models Moore's law in Cartesian coordinates can be replaced by an evolving spiral representation in polar coordinates.

The anticipated development of the 5G generation wireless technology, the next phase of IoT, can be viewed as an evolutionary stage of ICT technology following the generalized Moore's law. However, the fundamental difference is that 5G is "a priori technology," and needs to be designed before it can disrupt. The design process will be multifaceted and will affect both the real and virtual worlds introduced above, in ways both profound and unpredictable.

Children of the "α – generation cyber society" who will live their entire adult life in this new era have been already been born. A member of this new global society is going to experience multiple disruptive innovation revolutions during his or her life time with profound impact on redefining the professional personal and social aspects of his and/or her life.

Coming back to the IoT "State of the Union," according to the Research Nester:

> *Global IoT market reached USD 598.2 Billion in 2015 and the market is expected to reach USD 724.2 Billion at a CAGR of 13.2% by 2023.*

One of the pioneering predecessors of the IoT revolution has been the concept of Vital Electronics introduced by Dr. Ted Kochanski. His approach illustrates the ubiquitous character of IoT extremely well. According to T. Kochanski:

> *Vital Electronics is the study and use of electrical components, circuits, networks, and systems to achieve a design goal of protecting, saving, and improving critical infrastructure, and hence the quality of life. Vital Electronics' domain is a heterogeneous computing environment derived from sensors networks, embedded systems, and ambient intelligence with intelligent, robust, and trustworthy nodes capable of building Application-Centric Embedded Computers from "off-the-shelf" virtual computational and networking parts.*
>
> *Vital Electronics makes Embedded Computers more capable, reliable, energy-efficient, and optimized to their tasks. These Embedded Computers inhabit our critical infrastructure and other key applications, at increasingly low levels, and with increasing interconnectedness with their peers. At the same time, Vital Electronics enhances the ease and speed of the design of reliable Embedded Computers, and their associated Embedded Systems through the reuse of proven and certified "design elements," and other "virtual components."*
>
> *Vital Electronics is founded on the synergistic interaction between Moore's law, Metcalf's law, High-Level System Design Tools, and*

MEMS Sensors and Actuators. The increasingly capable Programmable Systems on a Chip (PSoC) such as Cypress Semiconductor's PSOC family with its companion PSoC Creator tools are the key building blocks of Vital Electronics.

However, while the original Vital Electronics was an academic international conference topic for a few years – it never had the critical mass to make a major impact on society. Thusly – Vital Electronics has been revived in the context of the on-rushing IoT tide to perhaps shape the impact:

New Vital Electronics should be synonymous with ***the Internet of Important Things*** *[IOIT] applied to realms such as:*

Health,
Housing,
Transportation,
Utility Infrastructure, etc.,

previously only peripherally and superficially affected by electronics.

The "forever" problem has been that in general, the "electrotechnical community" with the exception of people and organizations devoted to a particular market, or specialized field of endeavor didn't have the "subject matter expertise" to know where to contribute to solving "important problems," of a local or global extent.

Meanwhile, the "Subject Matter Experts" who knew what was needed didn't have the knowledge of the specialized electrotechnologies which could provide the core of a solution to a challenging problem.

Attempts to bridge the gap have typically failed due to lack of a common terminology, vernacular, or even a common context with which to discuss the issues.

Basic premise of the new Vital Electronics is to build a "Technical Ethos" to support the application of modern electro-technology to important problems at the "grass roots level" – i.e., to provide the tools [hard and soft tools] to enable the people at the subject matter expertise and "problem facing" level to define specific tasks which would make life:

Safer
More Secure
Healthier
More Efficient

The New Vital Electronics is premised on the ability to take maximum advantage of:

Tremendous recent advances in core electronics technologies driven by consumer and other high volume products, such as: smart cell phones, tablets, watches, cars, robots, etc.

Explosion of "Open Source" hardware and software driven by the renaissance in "hobbyist" hardware such as Arduino and Raspberry Pi

Revolutionary enhancements in remote learning fostered and disseminated by initiatives such as MIT Open Courseware, IEEE online courses, etc.

Democratization and global spread of supporting technologies such as: modeling and simulation, computer-based design of a myriad of things, sophisticated visualization and Augmented Reality, and the ability to "print stuff" often again spread by hobbyist /amateur interests such as video gaming

Democratization of technical knowledge at the "grass roots" level enhanced by Wikipedia and similar online sources of relatively reliable knowledge

Global spread of technical infrastructure such as cell phone networks located in the middle of African hinterlands

Global spread of package delivery on prompt basis driven by Amazon and similar online suppliers, distributors, and expediters

Key improvements in electronics technologies: absolute performance, cost performance, reliability, size, weight, and efficiency (i.e., the generalized Moore's law) for:

- *Processing*
  - *Signals*
  - *Databases*
  - *AI*
- *Sensing*
- *Communications*
  - *Wireless – e.g., 5G*
  - *Free-Space and Guided Optical*
- *Information Display*
  - *Augmented Reality*
  - *Compact Multivariable*
  - *Multimedia*
- *Power Supply Technologies*
  - *Rechargeable and Disposable Batteries*
  - *Wireless*
  - *Energy Harvesting*
- *Control of Physical Objects*
  - *MEMS*
  - *Biomorphic and Biofunctional manipulators and actuators*

Incredible improvement in availability of high performance and high function hardware and software (i.e., supercomputers and super bandwidth communications) which had traditionally been restricted to major corporate, big universities, and Federal-level governmental entities.

Hobbyists and startups are today building mechatronic systems based on the above concepts with an investment comparable to buying an SUV which in all ways outperforms systems funded by leading nation states a decade ago at the level of billions. On a larger scale, SpaceX has delivered a functioning "Falcon Heavy," a heavy-lift booster, for a fraction of NASA's budget allocation for similar performance (SLS Block-1, smaller version of Space Launch System) and at a pace inconceivable by NASA. Falcon Heavy also lands vertically and is reusable.

2. Internet of Things and Collaboratory

The disruptive character of IoT may also be illustrated by yet another example related to a new way of collaborating among researchers and ecosystem developers. The concept of service science and collaboratory are well suited to foster and accelerate the IoT application domain. According to Ray Garbos, the founder and the scientific lead of eCV collaboratory (published as "Disruptive Innovation in the Era of Global Cyber-Society: With Focus on Smart City Efforts" at IDAACS 2017), IoT, and its overseeing emerging 5G generation wireless system, is an expanding space for researchers, students, and professionals. The development of 5G IoT applications, especially as applied to Smart City applications, is hindered by the lack of available design tools, methodologies, and interoperability. The creation of a standards-based, open source hardware and software platform paired with full documentation and educational materials greatly aids in the accelerated development of IoT applications and provides a fertile ground for Science Technology Engineering and Math (STEM) educational opportunities. The developed collaboratory is an IoT-based open source, service ecosystem which can be used as both research and education development platform. The philosophy pursued by the collaboratory has been developed thanks to several innovation visionaries: Professor Miller, the President of the Olin College (who proposed the rebirth of the collaboratory initially proposed by Wulf concept in education), Dr. Jim Spohrer from IBM, who developed the theory of service science, and Prof. Christensen, Harvard University (the father of disruptive innovation). According to Wikipedia,

...Collaboratory, as defined by William Wulf in 1989, is a "center without walls, in which the nation's researchers can perform their research without regard to physical location, interacting with colleagues, accessing instrumentation, sharing data and computational resources, [and] accessing information in digital libraries ...

The IoT-based collaboratory developed at the University of New Hampshire (UNH) with paradigms and technologies including Service Science, Internet Architecture, M2M, and MQTT has been used as part of the T-shaped multidisciplinary curriculum aimed at the establishment of student startups. The universality of the presented concept allows it to address Grand Challenges in education as well as in other disciplines such as eHealth.

The concept of Grand Challenges was introduced to counterpart the Japanese program of the fifth generation of computing and since then the definition has gained popularity and recognition in many branches of human activities. Inspection of different grand challenges indicates the absence of a common consensus and standard taxonomies. However, many grand challenge solutions are enabled by and include in designing for the IoT. Thus, the IoT becomes a grand challenge fabric from hard computer engineering point of view. The IoT impact is so profound and hard to estimate today that this new computer technology may be categorized as "disruptive innovation." Yet another observation can be made related to the lack of commonly recognized and accepted collaboration schemata. One new but pragmatic collaboration approach is based on the theory of service science [5]. Service science assumes, among other things, so-called value co-creation, a truly disruptive and somewhat utopian vision of collaboration.

Based on the above observation, it is proposed to consider the following hypothesis: the collaboratory concept, which is relying upon service science principles, becomes feasible because of the IoT. Thus, service science and the 5G/IoT are serving as an enabler to address Grand Challenges. In other words, the Cartesian product of the IoT and service science is the key concept presented in this paper.

eCollaborative Ventures Inc. (eCV) was incorporated by Cheryl and Raymond Garbos, on June 30, 2015, as a B (Benefits) Corporation in New Hampshire, United States. As a Benefit Corporation, eCV is focusing on researching and developing solutions to Grand Challengers facing society and that will provide a benefit to individuals, societies, and/or the environment. eCV is organized and will operate as a Collaboratory of International Members. The eCV Collaboratory is bringing together a variety of capabilities and expertise to pursue Grand Challengers such as ecosystem development for smart/intelligent cities/regions/states. eCV invites individuals or entities to join the Collaboratory. A member signs a Collaborative Agreement that has three parts. Part 1 is a statement of the goals and objectives of the Collaboratory and that a member may choose to participate or not in any activity, the second is a non-disclosure agreement, and the third states that the individual(s) who create Intellectual Property (IP) own the IP not the eCV Corporation or other member(s).

Collaboratory members are individuals or other entities. eCV currently has 18 members (16 individuals and two small businesses). Some examples of the eCV members include Doctor/Professor Zelalem Temesgen, Mayo Clinic, Doctors Joshua and Elizabeth Van Pelt, Pharmacists; Dr. Jeff Brody, a retired IBM researcher; Dr. John Apostolos, a physicist with over 100 patents who has been a Scientific Fellow in industry since 1982; and Raymond Garbos, co-founder, CTO, and Engineering Fellow since 1984 at several large (ex. Lockheed Martin and small businesses). Our group also includes academia including Jason Jeffords, an ICT (Innovation and Communication Technologies)/IoT expert and Adjunct Professor at the UNH and the Southern New Hampshire University (SNHU) with 11 patents and Dr. Andrzej Rucinski, Professor Emeritus from UNH who was born in and has homes in both the United States and Poland. Since a focus of eCV also involves making ethical and cultural contributions, our members include Dr. Robert Straitt, a Priest/Anthropologist/Business Efficiently Expert, and Cheryl Garbos, co-founder and President eCV and a Minister/Educator/Business leader. Our small businesses specialize in software and ICT/IoT development. We have international members, Mariusz Sawinski with over 20 years of international experience bridging between technology, business, and politics from Poland, and Dr. Sumit Chowdhury, the IT major player of the 100 Smart City Project in India. In addition, each member has evolved their own network of resources that may be called upon to expand our Collaboratory capabilities as needed.

The eCV Collaboratory has formalized our current Grand Challenge ideas into several service systems-related e-categories: Health as a Service (HaaS), Learning as a Service (LaaS), Business as a Service (BaaS), and Government as a Service (GaaS). Each has a role in Smart City ecosystem evolution/revolution leveraging ICT/5G/IoT. Our concept is to focus on the individual and his/hers security in each of these areas such as patient/caregiver for HaaS, student/parent for LaaS, employee/customer for BaaS, and citizen/taxpayer/visitor for GaaS. By leveraging advances in ICT 5G and beyond and IoT, eCV is pursuing the development of a Common Open Standards-Based International Innovation Digital Infrastructure for collaborative research and development. eCV is also developing a disruptive innovative approach on how individuals of all ages from around the world (patient/caregiver, student/parent, employee/customer, and citizen/taxpayer/visitor) can safely and easily utilize and benefit from the Digital Ecosystem World IoT of the future to benefit and advanced themselves.

eCV is implementing a holistic methodology to global cyber-society that balances the technological ICT/IoT/Cloud Computing, Big Data, AI, and Analytics solutions in a secure and trusted way but we are incorporating ethical, social, cultural, and spiritual considerations. Our solutions may include policy recommendations/changes that may be needed to protect entities/individuals from rapidly changing technologies.

Dr. Sumit Chowdhury, an eCV Collaboratory Member, is a key ICT driver behind the India 100 Smart City Program and founder and CEO of Gaia Smart Cities. The Gaia Smart Cities' effort integrates IoT, industrial automation, and digitalization solutions for enterprises and smart cities. Gaia's suite of ICT applications brings together sensors, hardware, software, and analytics on a cloud-based platform and allows businesses and cities to automate processes, track metrics, and improve the performance.

The India 100 Smart City program is an innovative initiative by the Government of India. The India Smart Cities' mission is to improve the quality of life of people by harnessing technology as a means to create a smart ecosystem for their citizens. These align with the eCV Collaboratory goals and objectives presented in Section 2. The 100 Smart City objective is to promote diverse cities around the country (large, small, villages, coast line to interior) and for them to provide a core infrastructure and give a decent quality of life to its citizens, a clean and sustainable environment with the application of "Smart" solutions. A strong cooperative development is required between the government, civic groups, industries, and the citizens.

The India 100 Smart City program has defined eight critical pillars – Smart Governance, Smart Energy, Smart Environment, Smart Transportation, Smart IT and Communications, Smart Buildings, Smart Health Facilities, and Smart Education.

Pan City is an example of one city with development using three overriding frameworks:

1. Smart integration
2. Shared data
3. Shared networks.

These have three inter-related components:

1. Integrated city management
2. Security systems' management
3. Energy and environment management.

Pan City is currently focusing these in five major areas with examples of some subareas being considered:

1. Smart Energy – Smart grid, metering, renewables, and flexible response
2. Smart Water – Distribution and storm management, maintenance, and health
3. Smart Mobility – Public transit, real-time information, traffic management, and EV stations
4. Smart Public Services – Public safety, lighting, emergency management, internet availability, health, and learning

5. Smart Structures – Energy management, safety, connectivity, efficiency, and maintainability

The up-to-date eCV activities have been evolved from the IoT Lab (http://www.iotrdl.org) informally associated with the α – loft innovative organization and (http://alphaloft.org/) associated with several institutions in the state of New Hampshire, including the UNH. α – loft provides facilities, business experience, and IOT infrastructure for its participants with a goal to encourage and support startups.

The IoT laboratory has developed an open source IoT-based design and application environment enabling a seamless integration of user-defined applications into the system. The ecosystem has been prototyped and verified at the UNH using a variety of research projects, senior projects, and courses including the T-Shape Curriculum, ECE 583 – Designing with Programmable Logic, ECE 711/811 – Digital Systems, and ECE 777/877 – Collaborative Engineering. The activities have been supported by DARPA, IBM, Cypress Semiconductor, and ACM presented in several forums.

eCV is leveraging these activities and evaluate the impact as 5G and IoT evolves to address Global Grand Challenges including Smart Cities' technologies. To assure global impact of the eCV collaboratory, the team has been active outside the United States, particularly in Europe, Africa, and India. The activities in India have been conducted with a great assistance of the Indo–US Coalition for Engineering Education (http://iucee.com).

Another element of eCV plan includes contributing toward the rebirth of the renaissance engineer/architect, a professional who provides deep expertise in a selected set of engineering disciplines and a broad outlook in important social and global issues. This is consistent with the T-shape concept. It is also important as Smart City projects expand around the world and in diversity and complexity by leveraging advances in networks (5G), low-cost sensors, smart (autonomous) platforms, ICT architectures, cloud computing, Big Data, AI, deep learning, new materials, 3D printing, etc.

However, these advances also require a new equilibrium between humanity and information technologies with much greater emphasize on ethics and cultural diversity. This, in turn, requires the development of a new generation of standards. As a result, one may expect a collaborative symbiosis not only among participating members but also a balance between a human and his or her representation in the Digital Ecosystem. This noble goal is expected to be achieved through the growing activities of the eCV collaboratory.

3. Main Topics and Scope

This book comprises 25 chapters authored by 78 professionals, academics, and researchers from 14 countries in the world. There are five parts in this book. Part 1 "Internet of Vital and Trust Things" discusses the aspects of disruptive innovation in vital embedded systems and the Internet of Vital Things (Chapter 1) and an example of eCV which could provide support for key architects of complex Digital IoT Ecosystems (Chapter 2).

Part 2 is the "Modeling and Assessment" that encompasses a large-scale distributed irrigation network monitoring using a novel hierarchical routing protocol (Chapter 3) and an agent-based model of social IoT interactions in the context of P2P resource sharing (Chapter 4). Chapter 5 addresses the availability assessment of dynamism, multi-componentness, and multi-levelness of mobile health IoT systems while Chapter 6 discusses the results of the assessment of IoT-based physical security systems using the physical security modes and effect analysis technique. The results of a structural analysis of security event correlation in the next generation Security Information and Event Management systems are covered in Chapter 7. Different types of attacks on Smart Business Center are described in Chapter 8 followed by developing a Markov model to investigate on the impact of such attacks on the availability of the system. Chapter 9 addresses a general structure and underlying principles for developing an Internet of Drones-based monitoring system for post-severe NPP accidents.

Part 3 of this book is entitled "Architecting and Development". A Virtualized Testbed for Embedded Networking Nodes has been designed and implemented using 6LoWPAN communication stack (Chapter 10) to facilitate quick deployment, integration, and flexibility for debugging and testing of a protocol under development. Chapter 11 proposes a smart city concept called "Deep Urban Environment" which integrates IoT with edge computing, Big Data, and data analytics to facilitate intelligent pervasive multi-device applications. Chapter 12 discusses hybrid remote and automated control system for mobiles objects (e.g., intelligent mobile robots, intelligent transport system, etc.). Chapter 13 builds a secured software architecture to support smart solutions and services for a smart city. Chapter 14 explores the ways to increase the dependability (note: attributes are reliability, safety, survivability, and security) of IoT systems through the incorporation of additional assurance and assessment measures.

Part 4 of this book covers "Implementation and Industry Cases." Chapter 15 describes a holistic systems engineering approach which is an amalgam of innovation management and structured process approaches. This integrated approach is deployable for future energy (and mobility) systems in metropolitan regions (e.g., the German Ruhr Valley which is a use case for this study). Chapter 16 proposes an IoT and blockchain-based smart waste management system which could handle micropayments and Smart Contracts. Chapter 17 discusses an industrial IoT system designed for monitoring and automation of the control system in specialized complexes for municipal polymeric waste utilization. Chapter 18 describes a cloud-based infrastructure for a multi-layered smart city information system while Chapter 19 highlights the development of a real-time public transport information service infrastructure for the intelligent transport system. This system involves the acquisition of real-time data and prediction model for the trolleybus arrival time. The research addressed in Chapter 20 involves the development of a network architecture which employs Power over Ethernet for power as well as data transfer, and neural networks for data analytics, processing, as well as deep learning. Chapter 21 discusses three Ambient Assisted Living service-oriented projects for elderly people: Smart Service Power, SOLION, and COVIBO.

Part 5 is the concluding part of the book and it is entitled "Education and Training." Chapter 22 presents the results of a review conducted on Internet (or Web) of Things (IoT/WoT) technologies. It also discusses the structure of the curriculum entitled "Technology and development tools WoT applications" delivered in the National Aerospace University "Kharkiv Aviation Institute" (KhAI) and National Technical University "Kharkiv Polytechnical Institute." Chapter 23 discusses a remote laboratory, GOLDi, and proposes how it could be upgraded for the study of cognitive systems. Chapter 24 encompasses the conceive–design–implement–operate pedagogic approach which incorporates active learning tools/environments (e.g., remote laboratory, IoT projects, embedded factory, etc.) to better equip engineering students with problem solving, project management, and technical skills. Finally, Chapter 25 discusses the Erasmus$^+$ funded program, ALIOT, which develops innovative IoT curricula for M.Sc., Ph.D., and engineering programs in Ukraine and European Union. The topics covered are IoT applications in health systems, intelligent transport systems, ecology, and industry 4.0, smart grids, and smart buildings and cities.

PART I

Internet of Vital and Trust Things

1

Disruptive Innovation in Vital Embedded Systems and the Internet of Vital Things

Ted Kochanski

Sensors Signals Systems, Lexington, MA, United States
E-mail: tedpk@alum.mit.edu

The Internet of Things covers a wide gamut from sensors and associated processing embedded within critical infrastructure to Ping a prototype hoodie designed to alert the wearer to changes in a Facebook page. Within this extensive range of systems, some are vital to solving critical problems facing society. The challenge is how to enable local individuals with subject matter expertise to be able to contribute to the design and implementation of sophisticated electronics. A number of approaches have been developed over the years to address the challenge of designing a reliable and robust Internet of Vital Things (IoVT). While several of these approaches never had the impact, which they could have had – they can still be considered disruptive innovations. Ultimately, the concept of the IoVT, combined with a "Cloud-based" Development Framework, "empowering the subject matter experts," to design and virtually test prototype solutions, i.e., to Vital-iSolve has the potential to meet the Grand Challenges facing humanity.

1.1 Introduction and Brief History

1.1.1 Embedded Systems

Embedded systems have been a staple of electronics texts for quite a few years [1–4]. The standard definition is structural: an embedded system is a combination of computer hardware, software, and perhaps additional mechanical or other parts. It is designed to perform a dedicated function and may be a component of a larger system or product.

The other aspect of the definition is the functional: that an embedded system is an engineering artefact involving computation that is subject to physical constraints such that it must deal even-handedly with:

- Computational and physical constraints
- Software and hardware
- Abstract machines and transfer functions
- Non determinism and probabilities
- Functional and performance requirements
- Qualitative and quantitative analysis
- Boolean and real values.

Figure 1.1 shows a simplified block diagram of an embedded system consistent with the classical definitions. The system interacts with the "real world" through one or more sensors and one or more actuators. The sensors monitor some physical variables and then based on the signals and some internal models (which may be adaptive, etc.), the system acts on the "real world" through the actuator. Since by definition, embedded systems relate to the real world of moving objects, and forces, they must react to their "observations" in "real time."

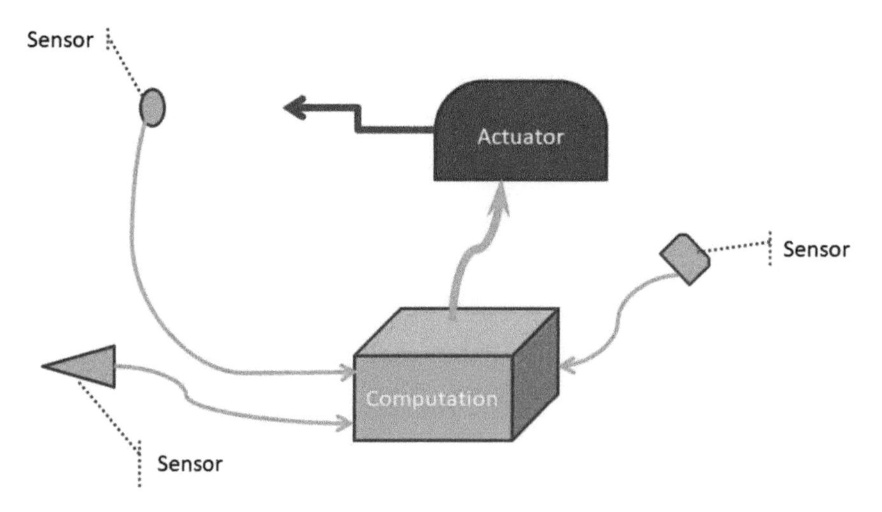

Figure 1.1 Simplified embedded system with computational electronics, several sensors, and one actuator.

Real-time operation means that the system must finish all relevant internal operations by deadlines. "Hard real time" translates into missing deadline causing system failure. "Soft real time" translates into missing deadline resulting in degraded system performance. Many embedded systems are multi-rate which translates into the requirement to handle operations at widely varying rates depending on the specific task and external factors.

For example, an embedded system tasked with maintaining a constant vehicle speed, i.e., Automotive "cruise control" must operate over a wide range of vehicle speeds, engine loads depending on terrain, and vehicle loading – hence multi-rate.

An automobile ABS, as shown in Figure 1.2, is a classic example of a practical and widely employed embedded system. This system combines several specialized sensors with a dedicated processor and an actuator which controls the brake pressure. To reduce the danger of skidding due to wheel lock-up – sensors monitor: the brake pedal, the motion of the vehicle's

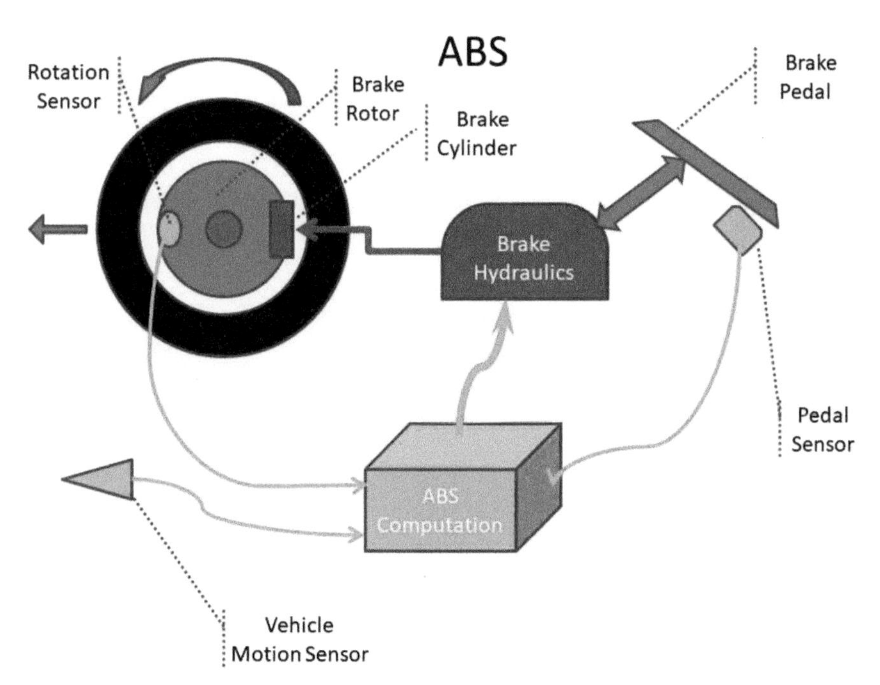

Figure 1.2 Simplified view of an automobile ABS based on the simplified embedded system shown in Figure 1.1. Sensors and an actuator specific and relevant to the function of the ABS are shown.

rotating components (e.g., brake rotor), and in some cases, the motion of the vehicle as a whole (using accelerometers and in some cases gyros). The computational electronics: processes the incoming signals, computes the vehicle dynamics, and then through an interface (actuator) modulates the brake hydraulic pressure as required. The result is that the ABS assists the driver in controlling the motion of the vehicle under "real world" slippery conditions.

1.1.2 Critical Embedded Systems

Some embedded systems are more "important" than others and necessarily demand more reliable performance than the "ordinary" embedded systems. Systems in this category are commonly referred to as "critical embedded systems" [5].

Critical embedded systems are commonly defined as those systems: which are life-critical, mission-critical, or safety-critical. The failure or malfunction of the said system may result in:

- Death or serious injury – e.g., Railroad Crossing Gate
- Direct national security impact – e.g., Anti-Ballistic Missile System
- Loss or severe damage to expensive, hard-to-replace equipment – e.g., Deep Space Probe
- Large, non-recoverable financial losses – e.g., Airline Booking System
- Serious environmental damage. – e.g., Sewerage Treatment Plant overflow release valve

VITA [5] went further in their work on defining critical embedded systems in 2005 – restricting these to high-performance, distributed computing systems that:

- Manage high-bandwidth I/O
- Involve real-time processing
- Are environmentally constrained to size, weight, and power.

Ray Alderman, Executive Director of VITA, described several requirements that a critical embedded system must meet to fit within this description.

"These are systems that must survive in harsh environments: severe shock and vibration, extreme temperatures from low to high, and contamination from dust, dirt, oil, salt spray, corrosive gases, and many other contaminants."

The boards and boxes within critical embedded systems must be designed for long life-cycle applications; these are systems that are often not replaced or updated for many years. They do not become obsolete every few months like personal computers and consumer goods. Revision management is extremely critical to a product life cycle that cannot deviate from one production lot to the next. Therefore, they cannot use components or software that are constantly changing every few months. According to Alderman, if they do, they cannot possibly be called "critical embedded systems" [5].

Once again consider the ABS – now use it to control only the brakes of a tractor trailer carrying a load of liquid chlorine, heading down a hill toward a sharp bend in the road in the winter. Clearly, under such conditions, the ABS becomes a critical embedded system.

1.1.3 The Internet of Things in Context

Many of these embedded systems are quite useful as stand-alone elements of a vehicle, building, bridge, etc. For example, a 2018 luxury sedan might have all or at least most of the following stand-alone embedded systems: air bags, ABS, event data recorder (or informally the "Black Box"), emission control, electronic ignition, electronic fuel injection, exhaust gas recirculation control, automatic traction control, automatic parking, in-vehicle entertainment, adaptive cruise control, heads-up display, satellite radio, e.g., Sirius XM, night vision headlight auto-dimming, back-up collision sensor, back-up assist video, tire pressure monitor, rain-sensing wipers, active suspension, electric power steering (PAS), electronic transmission control, cabin climate control, navigation systems, heated seats, and telematics, e.g., OnStar. Coming soon, if not already at a high-end dealer near you are drive by wire, anti-collision automatic braking, lane keeping, and automated stop-start in traffic.

However, while these all can function perfectly well as stand-alone systems, today many embedded systems are interconnected through local networks and often even globally.

Figure 1.3 shows a simplified block diagram of a networked embedded system where the generic system of Figure 1.1 has been replicated with the individual "atomic-scale" systems interconnected through a local network. In the figure, the local network is connected globally to the "Cloud" (a term of commerce and increasingly accepted even academically to refer to an amorphous, ubiquitous network such as the Internet) [6].

In such a case, the networked embedded systems are collectively referred to by the term the "Internet of Things" (IoT).

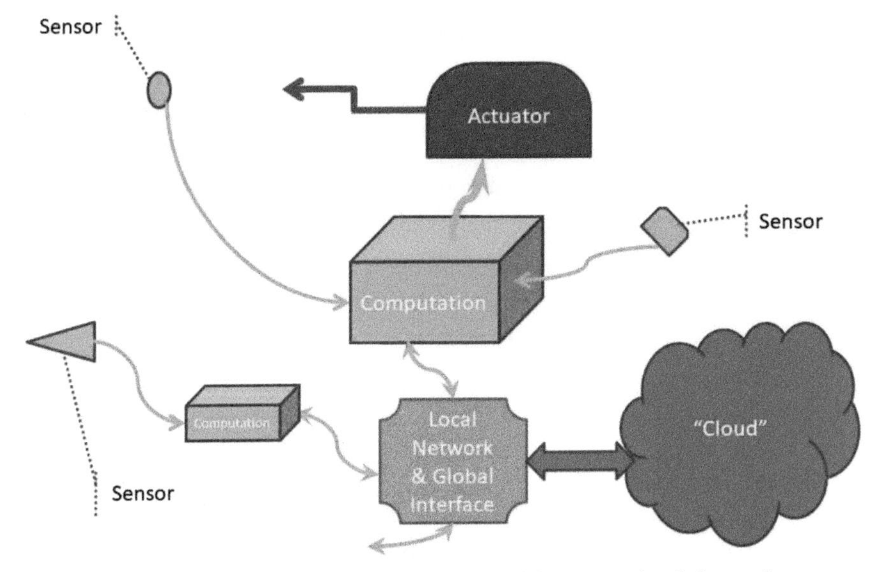

Figure 1.3 Simplified networked embedded system with computational electronics, sensors, actuators, a local network, a global interface, and the "Cloud."

A well-accepted definition of the IoT is: "the inter-networking of devices in the physical world by fitting them with sensors and network-connected devices" [6].

In its IEEE Special report on IoT issued March 2014, the IEEE described the phrase "Internet of Things" as: "A network of items – each embedded with sensors – which are connected to the Internet" [7].

The IEEE has an extensive process underway within the IEEE Standards Development Working Group on IEEE P2413 Standard for an architectural framework for the IoT [8].

The International Telecommunications Union (ITU), whose professed interest is in communications, defines the IoT as: "A global infrastructure for the information society, enabling advanced services by interconnecting (physical and virtual) things based on existing and evolving interoperable information and communication technologies" [9].

Kevin Ashton, Founder of MIT's AutoID Lab, which developed commercial RFID, is credited with coining the term, IoT in 1999.

Ten years later, in 2009, in an article for RFID Journal, he wrote, "We're physical, and so is our environment. Our economy, society and survival aren't based on ideas or information – they're based on things. You can't eat bits, burn them to stay warm or put them in your gas tank. Ideas and information are important, but things matter much more" [10].

A year later, in the spring of 2010, Margery Conner, then a senior editor for EDN Magazine, wrote an article entitled, "Sensors empower the Internet of Things" in which she wrote: "The goals for the IoT are, first, to instrument and interconnect all things and, second, to ensure that all those things are intelligent" [11].

She provided some then-cutting-edge examples including: "Ping," a hoodie with a wireless interface, from a start-up called Electricfoxy – it alerts the wearer when a Facebook page has changed, and allows quick gestural Facebook responses; Ford's Tool Link system incorporates sensors into a vehicle such as the Transit Van, so that the dashboard displays an inventory of all tools currently on-board; Disney Cruise Lines installed "virtual portholes" (round flat-panel screens), into its less-desirable windowless staterooms – The porthole displays a live video of the outside environment and can sense when people are present so that interactive Disney characters can be superimposed – resulting in rooms with a premium price.

One can argue about the societal value of the Facebook-linked Hoodie, and the Cartoon Characters on the Virtual Porthole. However, these applications are significant in themselves as they illustrate the important concept of local subject matter experts creating the solutions. The local expert, familiar with a specific challenge, such as hard-to-market windowless rooms, with access to the appropriate tech tools can develop successful solutions to "real-world problems." In this case not just a virtual porthole displaying a becalmed sea – but one with an interactive character from "Finding Nemo" to keep the kids occupied.

Ashton reprised his RFID Journal remarks in an interview with Smithsonian Magazine in 2015 where he said:

In the twentieth century, computers were brains without senses – they only knew what we told them. That was a huge limitation: there is many billion times more information in the world than people could possibly type in through a keyboard or scan with a barcode. In the twenty-first century, because of the IoT, computers can sense things for themselves. It's only been a few years, but we already take networked sensors for granted. One example is GPS-based location sensing. Civilian GPS was first authorized by congress in 2000, and the GPS systems in cellphones were not tested until 2004. Yet it's already hard to imagine a world without GPS: it helps us find our way around [12].

For a poetic-bureaucratic definition of the IoT – try: Gérald Santucci in 2010, "The Internet of Things links the objects of the real world with the virtual world, thus enabling anytime, anyplace connectivity for anything and not only for anyone. It refers to a world where physical objects and beings, as well as virtual data and environments, all interact with each other in the same space and time" [13].

Two questions arise from the Santucci statement:

First: do we need everything to interconnect and interact with everything else? For example, does a system scheduling transit vehicles in Barcelona need to know about the level of pressurization of my left rear tire on my car as I drive in Boston?

Certainly, I want to know about any significant excursion in tire pressure unrelated to environmental conditions; probably my dealer who services my car, should know if I'm in need of service on the tire. In the aggregate, the tire manufacturer and the auto manufacturer might be interested in premature failure, etc. It is even conceivable that the City of Boston might want to know where and when tires start failing due to the potholes in the roads.

However, too much connectivity introduces potential problems:

- It is costly in terms of resources to connect everything via something similar to IP addressing.
- It makes the entire universe of devices more vulnerable through the "weak link in the chain."

The second question is a bit more subtle – do we have the inherent expertise to design a system both which effectively uses all the data which we are able to collect, and from which we as a society can effectively benefit.

1.1.4 Some Observations of the Status Quo and the Near Term

So what if there are potentially Trillions of Connected Devices as Margery Conner speculated in her EDN article (based on thousands of connected devices per person and a few Billions of people) [11]. What we need are tools to build systems utilizing local expertise to solve problems using these connectable/connected networked sensing/actuating devices. We also need to ensure that these trillions of interconnected devices are as protected as we can make them, against non-local effects of failure (e.g., cascading collapse) and also hardened against malice (i.e., the weakest link in the chain effect).

There is already a proliferation of subsets of the IoT attempting to focus on major and generally "high value" environments such as the Industrial IoT coined by GE Digital in 2012 vis a vis connected to big machines [14].

One perspective is to think of the Industrial Internet as connecting machines and devices in industries such as oil and gas, power generation, and healthcare, where there is more at stake or where system failures and unplanned downtime can result in life-threatening or high-risk situations. On the other hand, the IoT *tend to include consumer-level devices such as heart monitoring fitness bands or smart home appliances. They are functional and can provide conveniences but do not typically create emergency situations if downtime were to occur.*

The IIoT, also known as the Industrial Internet, brings together brilliant machines, advanced analytics, and people at work. It's the network of a multitude of devices connected by communications technologies that results in systems that can monitor, collect, exchange, analyze, and deliver valuable new insights like never before. These insights can then help drive smarter, faster business decisions for industrial companies [14].

1.2 Internet of Vital Things (IoVT)

1.2.1 Conception of IoVT

Following the structure of the definition of critical embedded systems, here is a working definition of the IoVT – sometimes referred to as VitalNet:

The Interconnection of elements each of which can contain sensors/actuators, storage, and computation (including signal processing, "business logic," and "artificial intelligence").

Globally, the IoVT needs communications and a global infrastructure to support "Big Data," including machine-aided decision making and augmented reality/virtual reality for the human interface (to comply with Santucci) [13].

Since the failure or malfunction of the said system may result in:

- Death or serious injury
- Direct national security impact
- Loss or severe damage to expensive, hard-to-replace equipment
- Large, non-recoverable financial losses
- Serious environmental damage.

Such a system then must also be "reliable to the extreme," and robust to partial failures to insure that some of it is always working.

1.2.2 Historic Example: SAGE

An example of such a system, in the historic context, is the SAGE researched and developed by MIT Lincoln Laboratory. SAGE was developed because "Stalin had got the Bomb" and the United States was no longer a big island protected by its oceans. System engineering was by the MITRE Corporation, with deployment managed by the Airforce Electronics Systems Command at Hanscom Air Force Base in Bedford, MA [15, 16].

SAGE was developed for the US Air Force from 1950 to 1957 by the Massachusetts Institute of Technology's Digital Computer Laboratory, the Air Force Cambridge Research Laboratory, and MIT's Lincoln Laboratory. The work required scientific research in many different fields: computer hardware and software, radar, communications, and so on. The contract for manufacturing the AN/FSQ-7 computers was awarded to IBM. Western Electric Company provided buildings and internal power supply and communications. Phone lines were provided by the Bell System. System Development Corporation was responsible for the software which consisted of 500,000 lines of assembly language.

Note that for the history buffs, in 2012, SAGE was celebrated by the IEEE Boston Section as an IEEE Engineering Milestone. On 27 June, 2012, the IEEE Boston Section presented MIT Lincoln Laboratory with an engraved plaque honoring MIT Lincoln Laboratory's role in developing the "SAGE." SAGE has been designated an IEEE Milestone because it revolutionized air defense and also contributed significantly to advances in the computer industry and air traffic control. The plaque that is to be installed permanently at the Laboratory reads:

"SAGE-Semi-Automatic Ground Environment 1951–1958

In 1951, the Massachusetts Institute of Technology undertook the development of a continental air defense system for North America. The centerpiece of this defense system was a large digital computer originally developed at MIT. The MIT Lincoln Laboratory was formed to carry out the initial development of this system and the first of some 23 SAGE control centers was completed in 1958. SAGE was the forerunner of today's digital computer networks."

1.3 The SAGE Air Defense System [16]

The scope of the SAGE Air Defense System, as it evolved from its inception in 1951 to its full deployment in 1963, was enormous. The cost of the project, both in funding and the number of military, civilian, and contractor personnel involved, exceeded that of the Manhattan Project. The project name evolved over time from Project Lincoln, its original 1951 designator, to the Lincoln Transition System, and finally settled out as the Semi-Automatic Ground Environment, or SAGE. The basic SAGE architecture was cleanly summarized in the preface to a seminal 1953 technical memo written by George Valley and Jay Forrester: Lincoln Laboratory Technical Memorandum No. 20, "A Proposal for Air Defense System Evolution: The Transition Phase."

Briefly, the... system will consist of: (1) a net of radars and other data sources and (2) digital computers that (a) receive the radar and other information to detect and tract aircraft, (b) process the track data to form a complete air situation, and (c) guide weapons to destroy enemy aircraft.

The SAGE system, by the time of its full deployment, consisted of hundreds of radars, 24 direction centers, and three combat centers spread throughout the United States. The direction centers were connected to hundreds of airfields and surface-to-air missile sites, providing a multilayered engagement capability. Each direction center housed a dual-redundant AN/FSQ-7 computer, evolved from MIT's experimental Whirlwind computer of the 1950s. These computers hosted programs that consisted of over 500,000 lines of code and executed over 25,000 instructions – by far the largest computer programs ever written at that time. The direction centers automatically processed data from multiple remote radars, provided control information to intercepting aircraft and surface-to-air missile sites, and provided command and control and situational awareness displays to over 100 operator stations at each center. It was far and away the most grandiose systems engineering effort – and the largest electronic system-of-systems "ever contemplated."

1.4 Evolution of Disruptive Innovation in the Design of Microelectronic Systems and the IoVT

"Disruptive Innovation" has been defined by Clayton Christensen of Harvard Business School as what happens to a market when a "Disruptive Technology" radically transforms the traditional order [17].

More specifically, he makes the point that it is not based on a gentle transformation or even dramatic improvements to an established paradigm [18].

Disruptive Innovation describes a process by which a product or service initially takes root in simple applications at the bottom of a market – typically by being less expensive and more accessible – and then relentlessly moves upmarket, eventually displacing established competitors.

The remainder of this article attempts to chronicle a family of system engineering process disruptive innovations focused on designing and implementing critical embedded systems and now the IoVT. This work has been pioneered by Andrzej Rucinski and Thaddeus Kochanski, and a number of associates, generally connected with the University of New Hampshire, in the United States, and the Gdańsk University of Technology in Poland (Politechnika Gdańska in Polish).

1.4.1 Creation of CIDLab

In reaction to the events of September 11, 2001, the University of New Hampshire under the leadership of Prof. Andrzej Rucinski launched the Critical Infrastructure Dependability Laboratory [CIDLab] under the auspices of the Department of Electrical and Computer Engineering. CIDLab also had as Transatlantic component CIDLab-EU, led by Professor Andrzej Stepnowski, Faculty of Informatics, Gdańsk University of Technology.

The CIDLab Mission:

- Design, experimentation, implementation, verification, and test of computer systems, ranging from microelectronics to constellations of sensor networks.
- Support interdisciplinary research and education, standard compliance in critical infrastructures, cooperation with industry, governments, and non-profit organizations.
- Promote innovative collaboration methods, especially between Europe and the United States.

CIDLab began to pursue the idea of interconnected computational nodes each of which might have one or more sensors/actuators – previously investigated by Kochanski in the form of a hypothetical "Millennium Chip," conceived in the context of Telecommunications [19].

1.4.2 Concepts of GAIN and Global Systems Engineering Education

CIDLab's initial contributions were to introduce the GISE (Globally Integrated Security Environment) as an overall framework, and then to define networks of distributed sensors and actuators: GAIN (global ambient intelligence network) – the end product of the engineering process; and the global education microelectronic systems (GEMS) network – the tools to enable the creation of GAIN.

GAIN was founded on a "Near-universal Sensor Network Building Block" (GNODE), enabling the facile creation of the global ambient intelligence network. GAIN and the constituent GNODEs met the following requirements:

- Modular
- Plug in sensors/actuators
- Plug in processing and interface IP
- Reconfigurable
- Node and network configuration derived from high-level language and/or GUI tools (e.g., Matlab with Simulink)
- Node and network can adapt on-the-fly as circumstances demand
- Implementable by application engineers and end users
- Initial demonstration prototype based on "soft programmable" FPGA hardware with programming using Matlab with Simulink.

This work was described in papers presented in a number of conferences and workshops in the United States and Europe [20, 21].

CIDLab used the following definitions:

- Critical infrastructure – It is used by governments to describe material assets essential for the functioning of a society and economy.
- Dependability – In computer science, dependability is the trustworthiness of a computing system which allows reliance to be justifiably placed on the service it delivers.

Dependability includes the following attributes of a system:

- Availability: readiness for correct service;
- Reliability: continuity of correct service;
- Safety: the absence of catastrophic consequences on user(s) and the environment;
- Security: the concurrent existence of: (a) availability for authorized users only, (b) confidentiality, and (c) integrity.

At some meetings sponsored by the IEEE, the CIDLab team began to promote the concept of a Global Repository for IP components, to enable the efficient and reliable development of GAIN and similar systems.

1.4.3 I-GEMS and the Virtual Design Universe

The [I-GEMS] IEEE Global Education for Microelectronics Systems was launched on August 14, 2007 under the auspices of the IEEE to promote and facilitate the global design of microelectronic systems:

The initial project for I-GEMS was to establish a web-based, globally accessible repository of high-quality, reusable testbenches for trusted virtual components or intellectual property (IP). The repository was intended to facilitate the development of sophisticated high-reliability integrated circuits for mission-critical applications by a team of designers who may be geographically dispersed.

To utilize the repository, a supplier uploads both a testbench and a corresponding virtual component to the IEEE secure website which was to be hosted by the IEEE Boston Section (http://ieeeboston.org) and supplies information about the testbench, component (language), simulation (tool, version, library), and authors/owners. The simulation is then repeated on the IEEE webserver and the results reported initially just to the user. A charge is assessed for this self-test. Once the responses on the IEEE webserver match those in the testbench, the user may elect to receive a certificate and publish the results.

If the public option is selected, the testbench is placed in an open repository as well as a description of the component. A certificate is produced for the user with a checksum for the testbench-component tested. A charge is assessed for this certification. The user may then retain the component for proprietary reasons or contribute it to the IEEE website for others to use at no charge. The repository is modeled on MOSIS (http://www.mosis.org), with the website self-sustainable via user fees and with substantial discounts for educational users to encourage university innovation [22, 23].

A key aspect of the I-GEMS was the **Virtual Design Universe** in which:

- Everything is virtual or real
    - Nothing has to be anywhere in particular – fully virtual
    - Virtual and real are interchangeable

- Everything is mediated by Web 2.0 Services
    - SOAP, etc., tools for distributed, service-oriented systems
    - "2nd Life-like" virtual world user interface
    - Wikis, RSS, and Blogs
- Everything costs and is paid for when used:
    - Digital Cash
    - Subscriptions
    - Third party
    - Subsidy for university students
- Education and production are just variations on the same theme
    - MOSIS as the business model
    - Priority for ECE Courses.

1.4.4 Design for Globalization

However, there was a fundamental flaw in the I-GEMS concept – it took a student or a professional working in the discipline of VLSI design to create something. What was needed was a simple way for someone with knowledge about a local or global challenge (subject matter expert), but not a computer engineer, to capitalize on the cornucopia of computation resulting from "the Moore's law revolution." [24]. The following is from Intel's look-back at 50 years of Moore's law [24]:

In 1965, Gordon Moore made a prediction that would set the pace for our modern digital revolution. From careful observation of an emerging trend, Moore extrapolated that computing would dramatically increase in power, and decrease in relative cost, at an exponential pace.

The insight, known as Moore's law, became the golden rule for the electronics industry, and a springboard for innovation. As a co-founder, Gordon paved the path for Intel to make the ever faster, smaller, more affordable transistors that drive our modern tools and toys. Even 50 years later, the lasting impact and benefits are felt in many ways.

Economic Impact

Performance – aka power – and cost are two key drivers of technological development. As more transistors fit into smaller spaces, processing power increased and energy efficiency improved, all at a lower cost for the end user.

This development not only enhanced existing industries and increased productivity, but it has spawned whole new industries empowered by cheap and powerful computing.

Technological Impact

Moore's observation transformed computing from a rare and expensive venture into a pervasive and affordable necessity. All of the modern computing technology we know and enjoy sprang from the foundation laid by Moore's law. From the Internet itself, to social media and modern data analytics, all these innovations stem directly from Moore and his findings.

Societal Impact

The inexpensive, ubiquitous computing rapidly expanding all around us is fundamentally changing the way we work, play, and communicate. The foundational force of Moore's law has driven breakthroughs in modern cities, transportation, healthcare, education, and energy production. In fact, it's quite difficult to envision what our modern world might be like without Moore's law.

The Road Ahead

Like a metronome of the modern world, for 50 years, Gordon's prediction has set the pace for innovation and development. This foresight laid a fertile foundation from which all modern technology could spring, including the broad rise of digitization and personal electronics.

Moving forward, Moore's law and related innovations are shifting toward the seamless integration of computing within our daily lives. This vision of an endlessly empowered and interconnected future brings clear challenges and benefits. Privacy and security are persistent and growing concerns. But the benefits of ever smarter, ubiquitous computing technology, learning to anticipate our needs, can keep us healthier, safer, and more productive in the long run [25].

The next iteration under GISE was the concept of DfG, as shown in Figure 1.4, which expands on the concept of GAIN and I-GEMS. The assumption for DfG was that the subject matter experts interfaced with the technology experts in a way that the process could converge to the desired solution rapidly and reliably [24].

So the inevitable discussion arose, why not carry the idea to the logical conclusion. Make the tools and targets simple enough that after some brief

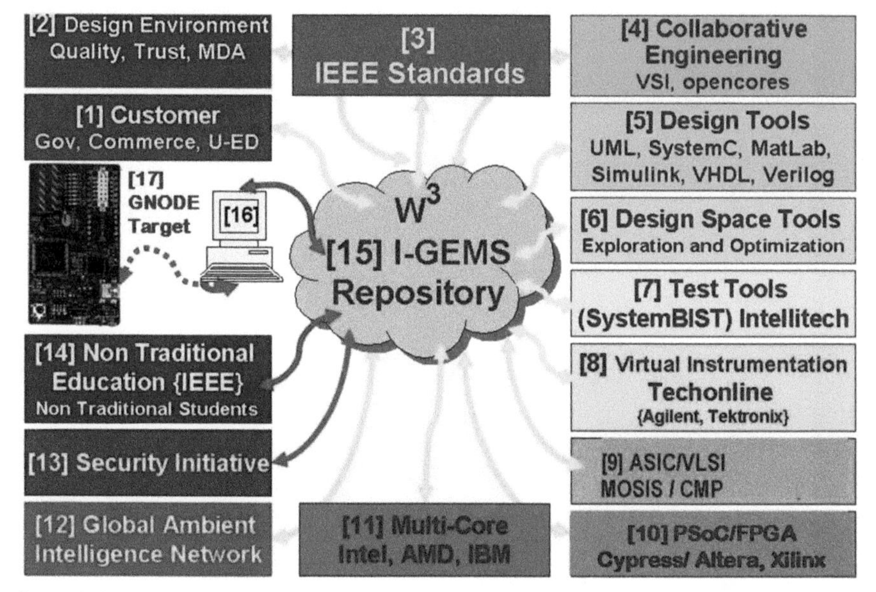

Figure 1.4 The design for globalization process: beginning with the need of the "Customer" (labeled 1) and progressing clockwise through various design and test stages (e.g., Design Tools – 5, Virtual Instrumentation – 8) mediated by the I-GEMS Repository (labeled 15) operating in the Cloud. The final tested and verified design is delivered to the GNODE Target Platform (labeled 17) through local tools running on the Development PC (labeled 16).

coaching the "subject matter expert" could go a long way to delivering a prototype of a functioning system without becoming a computer engineer.

1.4.5 Vital Electronics

The result of this evolution was the concept of **Vital Electronics**, a disruptive innovation which has been disseminated via Workshops from the United States, to Poland, Kazakhstan, India, and globally via conferences in Germany, Hungary, and a Global Engineering Colloquium in Singapore [25–30].

Vital electronics is the study and use of electrical components, circuits, networks, and systems to achieve a design goal of protecting, saving, and improving critical infrastructure, and hence the quality of life. Vital electronics' domain is a heterogeneous computing environment derived from sensors networks, embedded systems, and ambient intelligence with intelligent, robust, and trustworthy nodes capable of building application-centric embedded computers from "off-the-shelf" virtual computational and networking parts.

Vital electronics makes embedded computers more capable, reliable, energy efficient and optimized to their tasks. These embedded computers inhabit our critical infrastructure and other key applications, at increasingly low levels, and with increasing interconnectedness with their peers. At the same time, vital electronics enhance the ease and speed of the design of reliable embedded computers, and their associated embedded systems through the reuse of proven and certified "design elements," and other "virtual components."

Vital electronics is founded on the synergistic interaction between Moore's law [24, 31, 32], Metcalf's law [31], high-level system design tools [32], and MEMS sensors and actuators [33]. The increasingly capable PSoC such as Cypress Semiconductor's PSOC family with its companion PSoC Creator tools [34] are the key building blocks of vital electronics.

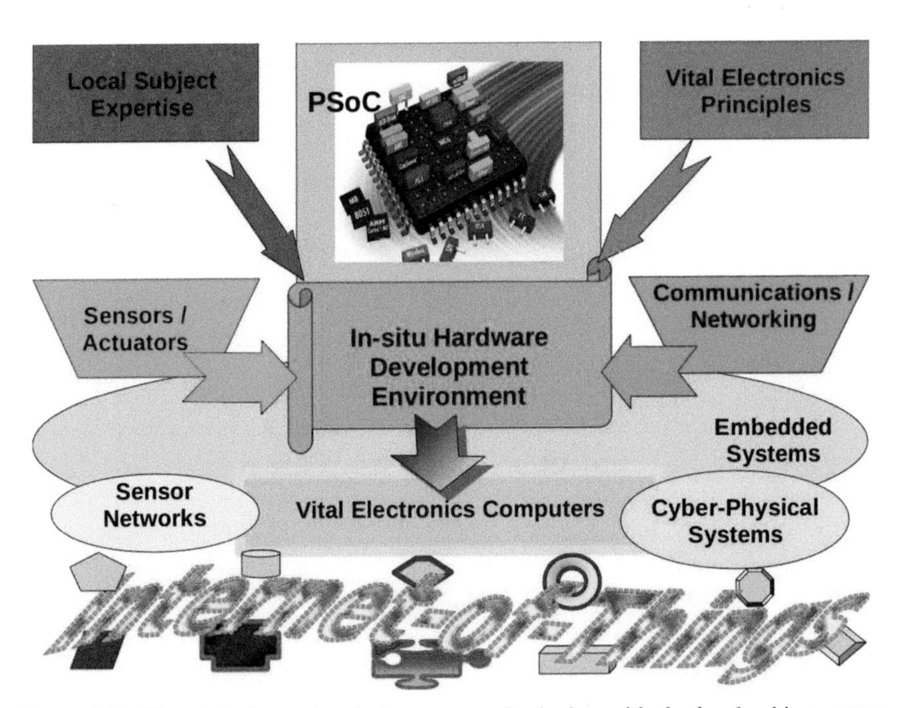

Figure 1.5 The vital electronics design process: beginning with the local subject matter expertise and enabled by the "*In situ* Hardware Development Environment" resulting in a tested and verified "vital electronics" qualified computer based on a PSOC – safe to connect to the Internet.

In Figure 1.5, the design process implemented within the context of vital electronics – in this case implemented using the Cypress PSOC family of chips and design tools [34]. This figure illustrates the key elements of PSOC-centric implementation of vital electronics as it was taught to "PSOC neophytes" (potentially "subject matter experts") in "global workshops."

From the standpoint of proving the concept that you could "make the tools and targets simple enough that after some brief coaching, the "subject matter expert" could go a long way to delivering a prototype of a functioning system without becoming a computer engineer" – vital electronics was a success. The Vital Electronics Institute with support from Cypress Semiconductor conducted one-day to one-week duration workshops in the United States, Europe, and Asia. In the most dramatic case, young women at a Women's Engineering College in Andhra Pradesh, India, who were good at programming (some were junior faculty in Computer Science), but utterly unfamiliar with computer hardware, were able to design and implement a working PSOC-based system in a few hours of "hands-on" time [30].

1.5 Vital-iSolve and the Internet of Vital Things (IOVT)

However, while the original vital electronics was an academic international conference topic for a few years [26, 28, 30], it never had the critical mass to make a major impact on society. Thusly, vital electronics has now been revived in the context of the on-rushing "IoT tide" (perhaps a Tsunami?), to help shape the impact of the IoT in general and the IoVT in particular.

New vital electronics – **Vital-iSolve** should be synonymous with the IoVT applied to realms such as:

- Health
- Housing
- Transportation
- Utility infrastructure
- National Security.

All the above areas, except for National Security were previously only peripherally and superficially affected by electronics, although today they are rapidly becoming "Moore-ified" [24].

The "forever problem" limiting the widespread impact of electronics has been – that in general the:

- "Electro-Technical Community" (e.g., the IEEE) with the exception of some groups of people and organizations devoted to a few particular markets, or specialized fields of endeavor – didn't have the "subject matter expertise" to know where and how to contribute to solving "important problems," of a local, or global extent.

- "subject matter experts" who knew what was needed didn't have the knowledge of the specialized electro-technologies which could provide the core of a solution to a challenging problem.

Attempts to bridge the gap have typically failed due to lack of a common terminology, vernacular, or even a common context with which to discuss the issues.

Thus, the basic premise of the new Vital-iSolve is to build a "Technical Ethos" to support the application of modern electro-technology to important problems at the "grass roots level" – i.e., to provide the tools [hard and soft tools] to enable the people at the subject matter expertise and "problem facing" level to define specific tasks which would "make life:"

- Safer
- More secure
- Healthier
- More efficient.

And then to deliver concept demonstrations for pilots which can be easily implemented – disruptively!

1.5.1 Vital-iSolve Fundamentals

Vital-iSolve is premised on the ability to take maximum advantage of:

- Explosive and disruptive advances in core electronics technologies driven by consumer and other high-volume products, such as: smart cell phones, tablets, watches, cars, robots, etc.
- Explosion of "Open Source" hardware and software driven by the renaissance in "hobbyist" hardware such as Arduino [35] and Raspberry Pi [36] and Semiconductor "Game Changers" such as: I386 performance, at the scale of a grain of salt, available for a dime announced by IBM [37]; and "Everest," Xilinx's real-time dynamically reconfigurable APCAP Adaptive Compute Acceleration Platform executed in 7 nm Silicon [38].
- Revolutionary enhancements in remote learning fostered and disseminated by initiatives such as MIT Open Courseware [39], IEEE online courses [40], etc.
- Democratization and global spread of supporting technologies such as: modeling and simulation, computer-based design of a myriad of things,

sophisticated visualization and augmented reality, and the ability to "print actual stuff," often again spread by hobbyist/amateur interests such as video gaming and "artist groups" [41].
- Democratization of technical knowledge at the "grass roots" level enhanced by "Wikis" (e.g., Wikipedia) and similar online sources of relatively reliable knowledge.
- Global spread of technical infrastructure such as cell phone networks located in the middle of African hinterlands (e.g., "camps" in the bush in Kenya).
- Global spread of package delivery on prompt basis, driven by Amazon and similar online suppliers, distributors, and expediters.

Much of even the Wiki's is in turn due to generalized Moore's law improvements in electronics technologies: absolute performance, cost performance, reliability, size, weight, and efficiency for:

- Processing
    - Signals
    - Databases
    - AI
- Sensing
    - MEMS
    - Nano
    - Bio
- Communications
    - Wireless – e.g., 5G [42]
    - Free space and guided optical
- Information display and non-visual presentation
    - Augmented reality
    - Compact multivariable
    - Multimedia
- Power supply technologies
    - Rechargeable and disposable batteries
    - Wireless
    - Energy harvesting
- Control of physical objects (i.e., actuators)
    - MEMS
    - Biomorphic and biofunctional manipulators and actuators [43].

The "hard drivers" are the incredible improvement in availability of high-performance and high function hardware and software (i.e., supercomputers and super storage in the "Cloud" (e.g., petabytes and even exabytes) [44] and super bandwidth communications) [45] which had traditionally been restricted to major corporate, big universities, and Federal-level governmental entities.

Hobbyists and start-ups are today building mechatronic systems based on the above hard and soft components, with an investment comparable to buying an SUV, and which in all ways outperform systems funded by leading nation states a decade ago at the level of Billions. A local company (Boston Dynamics) has produced a family of robots who have become media stars: Atlas a humanoid can back flip, run and even perform parkour at the expert level; SpotMini a "caninoid" can open doors, descend stairs backward, and run autonomously in mixed human robot packs. Both have been seen by millions of individual views [46]. SpotMini is projected to be in production within 12 months.

On a larger scale, SpaceX has successfully launched "Falcon Heavy," [47] a heavy-lift booster, developed and manufactured for a fraction of NASA's budget allocation for similar performance (SLS Block-1, smaller version of Space Launch System) [48], and at a pace inconceivable by NASA. Falcon Heavy also lands vertically and is reusable [49].

1.5.2 Vital-iSolve Ingredients

To Vital-iSolve a problem, you need to start with the following:

- Local functional components:
    - Sensors
    - Actuators/display/control
    - Signal acquisition
    - Signal generation
    - Local storage
    - Local computation
- 5G Communications and advanced "NextGen" networks [42]
- The "Cloud" – data services [50]
    - Archival storage
    - Big Data tools – AI
    - Block chain technologies including methods to pay for network services

- Cloud-based Development Platform (with Virtual Target) – iSolver – AI empowered to adapt the "iSolver" to needs of subject matter experts
    - o HW design
    - o Embedded SW
    - o Cloud SW
    - o Mechanical including Design for 3D Printed Electronics and Packaging
    - o Integration and test
- Local development target hardware
- Local subject matter expertise
- Education and practice in use of tools.

1.5.3 Example From an E-health Ambulatory Sensor Application: Heart Sensor:

- Integrated electrical, magnetic, and motion sensing
- 3D Printed electronic packaging [51] for high reliability, ease of use, low cost
- Global Secure and high integrity Distributed Data Archive and Grand AI toolkit [52] for "Big Data" to identify patterns to search for in individual patient data
- Local and global subject matter expertise in "ectopic heart rhythms", etc. [53]
- Local specialized AI tools to find the critical pattern [54] for an individual and generate appropriate alarm
- 5G network [42] to inform EMS and Critical Care Medical Professionals on alarm
- Specialized AR/VR display [55] to prep the Medical Team in advance of patient arrival.

1.6 Conclusion

1.6.1 Summing up the Recent Disruptive Innovation in Microelectronics Systems Education

Phases:

The process has evolved though the following phases over the past decade:

- Gen 0 – IGEMS
- Gen 1 – Vital Electronics PSOC Workshops India [56]
- Gen 2 – Vital-iSolve.

"Vital-iSolve" offers a new perspective on computer-aided systems engineering:

Founded on the Internet of IoVT and taking advantage of disruptive innovations in:

- Core electronics technologies driven by consumer and other high-volume products such as smart cell phones, tablets, watches, cars, robots, etc.
- Explosion of "Open Source" hardware and software driven by the renaissance in "hobbyist" hardware such as Arduino and Raspberry Pi, and "professional-lite" platforms such as "Beagle Bones"
- Revolutionary enhancements in remote learning fostered and disseminated by initiatives such as MIT Open Courseware and IEEE online courses
- Democratization and global spread of supporting technologies such as: modeling and simulation, computer-based design of a myriad of things, sophisticated visualization and augmented reality, and the ability to "print stuff" often again spread by hobbyist/amateur interests such as video gaming and "artists"
- Democratization of technical knowledge at the "grass roots" level enhanced by Wikipedia and similar online sources of relatively reliable knowledge
- Global spread of technical infrastructure such as cell phone networks located in the middle of African hinterlands
- Global spread of package delivery on prompt basis driven by Amazon and similar online suppliers, distributors, and expediters.

Building on the successful demonstration workshops using PSOC in India – Vital-iSolve enables local subject matter experts to address problems directly at the functional hardware/software prototype level.

1.6.2 The Big Question Which Needs to be Addressed

Can Flexible, highly-capable, but yet simple to implement technology such as PSOC, when combined with local subject matter expertise and local skills be used to deliver practical and viable solutions optimized to local problems which are vexing humanity on various scales?

For example, what might be needed in a system implemented in a village in Equatorial Africa may contain common elements with something solving a problem in an urban dis-developed enclave in the European Union or the United States.

Overall, IoT has already spawned an enormous amount of "Buzz-Expressions" and "Breathless Anticipatory Hype" by people who are ignorant of the true potential and challenges.

The potential of a "Global Ocean of Smartness" (or perhaps an "Atmosphere of Smartness," as the late Prof. Dertouzos of MIT would have called it in his **Project Oxygen** context) [56] – sensing and actuating, and connecting as required – is truly vast – both for doing good – but also for doing not-so-good (either by omission or commission).

Implementing such – so that the "net" on the beneficial scale is well over toward Prometheus – is also daunting as there are more than a "whole lot" of ways to "mess-up" by accident. There are as well, "folks as yet-unborn," who will take advantage of any "loop holes," or "weak links in the chain" to make life worse for a smaller or larger subset of humanity.

As a result of the above analysis, local solutions to local problems should be dominant, with global connectivity only if absolutely required. This is to be preferred to the "Big Tech Brother is watching" and will take care of all of us.

We do have a choice, at this stage of development in how the IoT is deployed, as the underlying laws, Technical Standards and regulatory framework are still only in their infancy.

Acknowledgments

Emeritus Professor Andrzej Rucinski for invaluable support and encouragement and Cypress semiconductor for technical and material support, particularly for the Indian PSOC workshops, organized by Prof. Krishna Vedula of IUCEE.

References

[1] Brown, J. (1994). *Embedded Systems Programming in C and Assembly.* New York: Van Nostrand Reinhold.

[2] Ball, S. (1998). *Debugging Embedded Microprocessor Systems.* Boston: Newnes Elsevier, p. v.

[3] Vahid, F., and Civargis, T. (2002). *Embedded System Design.* New York: Wiley.

[4] Silage, D. (2008). *Embedded Design Using Programmable Gate Arrays.* Gilroy, CA: Bookstand Publishing, pp. xi, 44, 96, 197–289.

[5] Gipper, J. (2005). *"Critical embedded systems defined," VITA Technologies*. Available at: http://www.vmecritical.com/articles/id/?5586 http://vita.mil-embedded.com/articles/critical-embedded-systems-defined/ [accessed June 05, 2018].

[6] Massachusetts Institute of Technology: Sloan School of Management. (2018). *Internet of Things: Business Implications and Opportunities, Online Short Course*. Available at: https://executive.mit.edu/openenrollment/program/internet-of-things-business-implications-and-opportunities/#.WxVz-EgvyUk [accessed June 04, 2018].

[7] IEEE. (2015). *Towards a Definition of the IoT, IEEE Special report on Internet of Things*. Available at: https://iot.ieee.org/definition.html [accessed June 05, 2018].

[8] *IoT Architecture – Internet of Things IoT Architecture*. Available at: http://grouper.ieee.org/groups/2413/April15_meeting_report-final.pdf [accessed May 15, 2018].

[9] ITU-T SG 13 (Study Period 2017) Temporary Document 244-WP1 Meeting report of Question 22/13 "Upcoming network technologies for IMT-2020 and Future Networks" (Geneva, 16–27 July 2018) Available at: https://www.itu.int/md/T17-SG13-180716-TD-WP1-0244 [accessed July 17, 2018].

[10] Ashton, K. (2009). "That 'Internet of Things' Thing; In the real world, things matter more than ideas," *RFID Journal*. Available at: http://www.rfidjournal.com/article/view/4986 [accessed June 05, 2018].

[11] Conner, M. (2010). *Sensors Empower the "Internet of Things*," EDN, pp. 32–38. Available at: https://www.edn.com/design/sensors/4363366/Sensors-empower-the-quot-Internet-of-Things-quot-[accessed June 03, 2018].

[12] Gabbai, A. (2015). Kevin Ashton Describes "the Internet of Things: The innovator weighs in on what human life will be like a century from now," Smithsonian Magazine January 2015. Available at: https://www.smithsonianmag.com/innovation/kevin-ashton-describes-the-internet-of-things-180953749/#CCK32kDiIlxVuB1j.99 [accessed June 05, 2018].

[13] Santucci, G. (2010). *"The Internet of Things: Between the Revolution of the Internet and the Metamorphosis of Objects*," 2010. Available at: http://ec.europa.eu/information_society/policy/rfid/documents/iotrevolution.pdf [accessed June 01, 2018].

[14] GE Digital. (2018). *Everything You Need to Know About the Industrial Internet of Things*. Available at: https://www.ge.com/digital/blog/

everything-you-need-know-about-industrial-internet-things [accessed May 23, 2018].

[15] MIT Lincoln Laboratory News. (2012). *SAGE named an IEEE Milestone: MIT Lincoln Laboratory's role in developing SAGE is honored.* Available at: https://www.ll.mit.edu/news/ieee-milestone-SAGE.html [accessed May 25, 2018].

[16] MIT Lincoln Laboratory History (2017). *The SAGE Air Defense System.* Available at:https://www.ll.mit.edu/about/History/SAGEairdefense system.html [accessed May 24, 2018].

[17] Christensen, C. (2003). *The Innovator's Dilemma: The Revolutionary Book that Will Change the Way You Do Business (Collins Business Essentials).* New York: Harper Collins Paperback, pp xxi-xxxi.

[18] Christensen, C. (2018). *Disruptive Innovation, Christensen Institute.* Available: https://www.christenseninstitute.org/disruptive-innovations/ [accessed May 29, 2018].

[19] Kochanski, T. (1996). "Telecommunications Applications of Giga-Scale Integration," in *2nd Conference on Telecommunications R&D in Massachusetts, U Mass Lowell, Mass Telecom Council.*

[20] Spaanenburg, H., Rucinski, A., Chamberlin, K., et al. (2007). "Globally-Collaborative 'Homeland' Security System Design," in *2007 International Conference on Microelectronic System Education*, San Diego.

[21] CIDLab Team. (2007). "CIDlab and GAIN," in *2007 IEEE Conference on Technologies for Homeland Security*, Woburn, Massachusetts.

[22] Mikelinich, K., Drozd, A., Izraylo, L., et al. (2008). "Globalization for Design: A Market for a CMP/IP Repository," in *7th European Workshop on Microelectronics Education, (EWME2008)*, Budapest, Hungary. Available at: http://www.eet.bme.hu/new/index.php?option=com_ content&task=view&id=233&Itemid=237 [accessed May 29, 2018].

[23] Kochanski, T., Rucinski, A. (2008). Invited paper, "I-GEMS and Global Security," in *1st International Conference on Information Technology*, Gdansk, pp. 515–518.

[24] Intel. *50 Years of Moore's Law: Fueling Innovation We Love and Depend On.* Available at: https://www.intel.com/content/www/us/en/silicon-innovations/moores-law-technology.html [accessed May 29, 2018].

[25] Kochanski, T. (2009). "Does it Matter If Moore's Law is Ending: Using Microelectronics to build a safer, more secure, and more prosperous world," *The luncheon keynote at 2009 IEEE International Conference on Microelectronic Systems Education (MSE 2009)*, San Francisco, IEEE.

[26] Gulley, J., Holl, F., Kochanski, T. P., Rucinski, A. (2010). "A Pilot Course in Vital Electronics," in *8th European Workshop on Microelectronics Education, (EWME2010)*, Darmstadt, Germany. Available at: http://www.mes.tu-darmstadt.de/conf/ewme2010/proceedings/proceedings.en.jsp [accessed May 29, 2018].

[27] Bikonis, K., Rucinski, A., Kochanski, T. (2010). "PSOC Embedded Systems and its Potential Application in Wireless Sensors Network," in *Proc. of the Special Session on Vital Electronics of the 2nd International Conference on Information Technology*, Gdańsk University of Technology, Faculty of ETI Annals, Information Technologies, eds A. Konczakowska and L. Hasse, Wydzial ETI Politechniki Gdańskiej, Gdańsk, Poland, vol. 18, pp. 249.

[28] Bikonis, K., Charache, H., Doran, et al. (2010). "Vital Electronics: Application-Centric approach to Computer Engineering and Computer Science," in *Proceedings of the Special Session on Vital Electronics of the 2nd International Conference on Information Technology*, Gdansk University of Technology Faculty of ETI Annals, Information Technologies, eds A. Konczakowska and L. Hasse, Wydzial ETI Politechniki Gdańskiej, Gdańsk, Poland, vol. 18, pp. 255.

[29] Kochanski, T. (2011). "Technologies and Tools to build the Internet of Things," in *IEEE Electron Devices, Solid State Circuits and Computer Societies and GBC/ACM*, Wilmington, MA: Analog Devices, 01887.

[30] Doboli, A., Kochanski, T., Panetta, K., et al. (2010). "Responding to Global Engineering Education Challenges through Vital Electronics," in *2010 ASEE Global Engineering Colloquium, Marina Bay Sands*, Singapore, ASEE.

[31] Metcalfe, B. (2013). Metcalfe's Law after 40 Years of Ethernet. *Computer*, 46(12), 26–31.

[32] Pellerin, A. (2017). Easing embedded software development with EDA tools. *Embedded*. Available at: https://www.embedded.com/design/programming-languages-and-tools/4443336/Easing-embedded-software-development-with-EDA-tools [accessed May 29, 2018].

[33] Bogue, R. (2013). Recent developments in MEMS sensors: a review of applications, markets and technologies. *Sensor Review*, 33, 4. Available at: https://www.emeraldinsight.com/doi/abs/10.1108/SR-05-2013-678 [accessed May 29, 2018].

[34] PSoC® 4. (2017). *PSoC 4000 Family Datasheet, Cypress Semiconductor, 2017*. Available at: http://www.cypress.com/file/138646/download [accessed May 29, 2018].

[35] *What Is Arduino?* Available at: https://www.arduino.cc/ [accessed May 25, 2018].

[36] *Raspberry Pi 3B Plus Family Photo Update March 2018*. Available: http://raspi.tv/2018/raspberry-pi-3b-plus-family-photo-update-march-2018 [accessed May 26, 2018].

[37] Chin, M. (2012). *IBM unveils 'world's smallest computer' (literally smaller than a grain of salt) with blockchain at Think 2018*. Available at: https://mashable.com/2018/03/19/ibm-worlds-smallest-computer/#oSy8ztxvigq7 [accessed June 02, 2018].

[38] Meyers, W. (2018). *Xilinx's ACAP Could Be A Computing Game-Changer*. Available at: https://seekingalpha.com/article/4166877-xilinxs-acap-computing-game-changer [accessed May 29, 2018].

[39] Computer System Engineering. (2009). *MIT Open Courseware, Computer Science, 6-033*. Available at: https://ocw.mit.edu/courses/electrical-engineering-and-computer-science/6-033-computer-system-engineering-spring-2009/ [accessed June 01, 2018].

[40] *The IEEE eLearning Library*. Available at: www.ieee.org/publications/subscriptions/products/elearning-overview.html [accessed May 29, 2018].

[41] *Artisan's Asylum is a non-profit maker space devoted to the teaching, learning and practice of fabrication*. Available at: https://artisansasylum.com/ [accessed May 29, 2018].

[42] *5G in Five Minutes 5G Video: Skyworks CTO Peter Gammel explains the critical components for 5G and their importance in the evolution of 5G technologies, Skyworks*. Available at: www.skyworksinc.com/5G [accessed May 29, 2018].

[43] Bogue, R., *Artificial muscles and soft gripping: a review of technologies and applications*. Available at: https://www.emeraldinsight.com/doi/abs/10.1108/01439911211268642 [accessed May 24, 2018].

[44] *The fiftieth TOP500 list of the fastest supercomputers in the world*. Available: https://www.top500.org/ [accessed June 02, 2018].

[45] Morgado, A., Huq, K. S., Mumtaz, S., et al. (2018). *Digital Communications and Networks: A survey of 5G technologies: regulatory, standardization and industrial perspectives*. Available at: https://www.sciencedirect.com/science/article/pii/S2352864817302584 [accessed May 29, 2018].

[46] Boston Dynamics. (2018). *SpotMini autonomously navigates a specified route through an office and lab facility*. Available at: https://youtu.be/Ve9kWX_KXus [accessed May 29, 2018].

[47] Spacex. *Falcon Heavy: the most powerful operational rocket in the world by a factor of two.* Available at: http://www.spacex.com/falcon-heavy [accessed May 27, 2018].

[48] NASA. *SLS Block 1 configuration Space Launch System Building America's New Rocket for Deep Space.* Available at: https://www.nasa.gov/sites/default/files/atoms/files/sls_fact_sheet_final_10112017.pdf [accessed May 29, 2018].

[49] Etherington, D. (2018). *SpaceX landed two of its three Falcon Heavy first-stage boosters.* Available at: https://techcrunch.com/ [accessed May 29, 2018].

[50] Wayner, P. (2018). *AWS Lambda tutorial: Get started with serverless computing: 12 steps to building an application in the Amazon cloud, without breaking a sweat or a budget, InfoWorld.* Available at: https://www.infoworld.com/article/3263908/cloud-computing/aws-lambda-tutorial-get-started-with-serverless-computing.html [accessed June 02, 2018].

[51] Bailey, C., Stoyanov, S., Tilford, T., et al. (2016). "3D-printing and electronic packaging – current status and future challenges," in *18th IEEE Electronics Packaging Technology Conference (EPTC).*

[52] *Ectopic heartbeat.* Available: https://medlineplus.gov/ency/article/001100.htm [accessed May 29, 2018].

[53] *PRTools User Guide: PRTools is a Matlab toolbox for pattern recognition.* Available at: http://37steps.com/ [accessed May 29, 2018].

[54] *Welcome to the AI Toolkit for Azure IoT Edge: The integration of Azure Machine Learning and Azure IoT Edge.* Available at: https://github.com/Azure/ai-toolkit-iot-edge [accessed May 22, 2018].

[55] Greig, J. (2018). *Google and LG's high-resolution OLED display sets stage for future of enterprise AR and VR.* Available at: https://www.techrepublic.com/article/google-and-lgs-high-resolution-oled-display-sets-stage-for-future-of-enterprise-ar-and-vr/ [accessed June 02, 2018].

[56] Kochanski, T., Rucinski, A., (2012) "A ubiquitous roadmap towards global computer engineering and science, in: Contemporary Issues in Engineering Education, Vol. 3, R. Natarajan and N.R. Shetty, Editors, the Indian Society for Technical Education, Macmillan Publishers India Ltd., Bangalore, India, 2012 pp. 226–234, ISBN: 978-935-059-058-4.

[57] Brown, E. (2001). Project Oxygen's New Wind: Look out PCs., MIT's ubiquitous computing effort is taking technology out of the box. *MIT Technology Review.* Available at: https://www.technologyreview.com/s/401308/project-oxygens-new-wind/K [accessed May 29, 2018].

2

How to Support Creativity in the Complex IoT with Ethics and Trust for Users

Raymond J. Garbos

eCollaborative Ventures, Inc. New Hampshire, USA
E-mail: rgarbos@yahoo.com

The World Wide Web and Internet of Things are revolutionizing how individuals interact with each other and the environment around them. Using natural language, users can interface with other people and devices which will learn and "understand" the user traits. Although this may simplify how an individual or group uses the digital IoT ecosystem, the actual complexity of the underlying system-of-systems digital IoT ecosystem is still exponentially increasing. This complexity requires us to contemplate how we organize, learn, and develop the wisdom and tools to architect the future in a dependable and trusted environment to protect and benefit individuals and society. This is particularly important when inventions lead to new applications not complemented during the initial development: example ARPA Net (Advanced Research Projects Agency Network) to wireless fifth-generation (5G) RF and optical networks and IoT "devices" for Smart – individuals, homes, buildings, infrastructures, and cities. This is a revolution in connectivity resulting in a new global system-of-systems digital ecosystem. Since its start, it has been in an ever-growing state of change due to the throbbing beat of evolution. This chapter focuses on three areas: 1) who and how to support some of the key architects of the complex digital IoT ecosystem, 2) the need to develop complex IoT system-of-systems analysis and optimization tools to aid the architects, and 3) how to help the architects become aware and design a trusted and ethical interface for individual users, especially those who use but do not fully understand the complexity and security of the digital IoT ecosystem.

2.1 Introduction

From the outset, the WWW ("web") created a series of ***"now I can"*** wows. Today, the average person knows ***"I'm connected"*** to virtually everywhere. However, some now notice ***"now I can't"*** or ***"I don't want to."*** Most people are asking, ***"what can you do for me now?"*** and some ***"how am I protected?"*** Although we live in this new complex system-of-systems digital IoT ecosystem where people, information, and "things" are connected, we must also be aware of the responsibility to consider ethical and cultural impacts on individuals not aware of the web pitfalls. So how do we innovate to the benefit of the users and to protect them in this complex, ever expanding 5G [1] system-of-systems IoT digital ecosystem?

One approach is to look back in history to understand how a few innovators architected very complex solutions in their time – from the Pyramids in Egypt, to the Great Cathedrals in Europe, personal computers/video games, the WWW/social media, and autonomous systems/automation, etc. Whether these were motivated by religion, politics, profit, accident, or desire to benefit society in some way, they represent only a small subset of solutions that when we look back at them, we realize they had a great impact on their current and future societies.

I like to think of inventors, innovators, and idea people as the architects of their future. These architects create the key innovative concept that would generally then be implemented by a team consisting of the master craftspersons, craftspersons, apprentices, workers, and then de facto or by design the "users" who benefitted from the solutions and provided feedback. Then via training, education, and mentoring, some apprentices become craftspersons and some craftspersons become master craftspersons, etc. As people understand the innovation, some became "wiser" and improved on the original innovation with many additions that were never contemplated by the original architect, or the knowledge gained from the initial innovation triggered a new unrelated innovation, like the US military developing the GPS satellite network now used for personal and autonomous navigation.

As time went by, various master craftspersons (today called system engineers/architects) and their "team" of craftspersons formed guilds where they could interact with their "peers" collaborate with other guilds, share ideas, and discuss and develop the approaches and solutions to their particular "modern" problems, the Grand Challenges of the Day. Master craftspersons within these guilds gained great respect and influence becoming the "grand masters" in their world. The Institute of Electrical and Electronics Engineers

(IEEE) [2] is an example of a modern guild with some being recognized as IEEE Fellows, some of our grand masters, and lead system engineers of today.

About two decades ago, to formalize and document the process of developing a complex system, the United States government, industry, and academia formalized the system engineering design process. I believe this was in part due to the complexity of systems being developed, the diversity of skills needed, and the time needed to develop a product/solution.

Prior to this formal process, the lead system engineer or system architect was a recognized "expert"; their approach was accepted and then eventually verified during some simulation, test, and demonstration. However, specific decisions/trade studies made by the system architects were not always documented so if a problem occurred during test or operation, one was not always able to retrace the decision process leading to delays and cost overruns. The formal system engineering design process, now in common practice, was designed to document and track all the processes that lead to a solution. This formal process was not designed to necessary lead to the best or optional solution, but a solution that worked within the constraints of the project (e.g., resources, performance, cost, and schedule). In my experience, the formal process also assumed that the problem itself is defined and the requirements have been reasonability established. So, it is more about the implementation process for a specific system design once a concept is developed than how the problem is uncovered or defined, or how the creativity or innovative ideas were initially conceptualized. One was able to get a post-graduate degree is system engineering without ever being involved in a system design.

As the world's interconnectivity with people and things (IoT) continues to exponentially increase, partially enabled by computing power, reduce size and power devices, software/algorithm advances, and faster networks, we need to think more globally. Therefore, I believe that we will need to continually evolve our approaches and processes to innovate and solve complex problems. With advances in passive/active bi-metrics, AI, deep learning, analytics, quantum computing, (processing/sensors minimization and capability), networks, Cloud Computing, etc., society will be able to do more dependable and trusted real-time or near real-time situational awareness and decision making. With the new wireless 5G networks coming online, RF wireless-based communication will enable full bandwidth at low latency and optical communications will be able to provide 10–100 times more useable bandwidth than over the improved RF networks. These rates will enable telemedicine, telepresence, smart homes/cities/regions/states, autonomous system concepts, etc., to become an everyday reality.

2.2 Architecting the Future

The complexities, processes, and high bandwidth interconnectivity of the Web require us to consider better ways to approach our future challenges. I believe that there are several approaches that should be considered in architecting our future. I will discuss three major concepts that should be addressed to successfully target the increased complexity of our evolving digital IoT ecosystem:

- Recognizing and nurturing conceptual architects and key innovators in this evolution, with the encouragement to form, participate in, and lead Digital eGuilds.
- Formalizing IoT system-of-systems processes, solutions, and optimization to solve Grand Challenges to aid the conceptual architects and craftspersons.
- Help the conceptual architects and craftspersons recognize and develop dependable, trusted, and safe systems, with ethical and cultural awareness, and including protections to benefit the individual users and the society as a whole.

2.2.1 Conceptual Architects

In 1984, I was selected to become an Engineering Fellow at a major aerospace company in the USA. Years later, I was selected by the Fellows and Senior Management to manage the company's Scientific and Engineering Fellows Program. These Fellows were the innovators and master craftspersons of their particular disciplines. During discussions about the future of engineering, one of the Fellows proposed the idea of a conceptual engineer or conceptual architect. The conceptual architect is often the "idea" or "concept" person, an innovator who sees some unmet opportunity, the person who can understand that there is a problem or opportunity, usually before it is identified by others, and is able to define the problem/opportunity, and develop some concepts and approaches to solve the problem. Sometimes, this person lacks specific technical knowledge and may not be considered a master craftsperson in that field but has a ***surprising and crucial intuition of what can be done***. This kind of unique mind is often not recognized or appreciated but we should seek them out and enable them. Many times, rewards go to the developers of the solution or product itself, rather than the person who developed the initial concept.

One of the Engineering Fellows I worked with was Ralph Baer [3]. He invented the device, circa 1969, that attached to the input RF connector on a TV where the device could now control the display with an initial game program called "Ping Pong." He realized that this could create a new home gaming industry as well as a low-cost simulation market. He is now recognized as the "Father of Video Gaming" and a true American Inventor. For me, he was a mentor and is an example of a conceptual architect.

More recently, Mark Zuckerberg [4], chairman and chief executive officer of Facebook and Social Media forerunner is another example of a conceptual architect; using the Internet and technological advancements to connect us more and more tangibly to the Web in ways we never envisioned (for good or ill). Facebook says, "Welcome, I'm a growing part of the Internet of Things."

Just as the inventors of the ARPA Net [5], circa 1969, did not envision the WWW of today or the explosion of smart IoT devices, we cannot fully understand what and how the inventions of tomorrow will revolutionize our future. Since the complexity of both problems and solutions will increase, perhaps exponentially, in our time, no one individual will understand all the technologies which underlie these innovations.

However, I believe that conceptual architects will be key to this evolution and sometimes create disruptive advances not thought of at the time. Also, such innovators are likely to form their own virtual networks or "eGuilds." The eCollaborative Ventures (eCV), Inc. [6] Collaboratory [7], ***a New Hampshire, USA Benefits Corporation*** [8], which I co-founded with my wife Cheryl (President), is an eGuild. We are a multi-disciplinary group of both conceptual architects and grand masters trying to solve Grand Challenges to benefit society (eHealth, eLearning, eGovernment, eBusiness, Smart City [9], and structures).

The UN [10] and the WHO [11] have both defined various Grand Challenges facing our world. As the world gets more interconnected and complex, conceptual architects will be addressing the world's current and future Grand Challenges. Many of these challenges, like Smart Cities, must be considered as examples of complex system-of-systems and not as a collection of individual, unrelated systems.

Generally, the path to becoming a master craftsperson or expert (Fellow) involves both formal education, experience (apprenticeship), and mentoring by a master. However, as the complexity increases, individual guild members need to have both depth in several specific areas and breadth in many areas *and* the ability to communicate with experts in their field and others in other fields. This can be referred to as "T-shaped" [12] skills. The leg of the "T" includes the specific areas of expertise and the top of the

"T" represents the breadth of experience required. Society needs to develop more multi-disciplined T-shaped education and apprenticeship opportunities.

However, the formal path to becoming a conceptual architect is less defined. In many ways, the conceptual architect is one who thinks and sometimes operates ***"outside the box."*** This is a person who can look at the world a little differently and tends to see barriers as challenges or opportunities. Society needs to find ways to early identify these traits in individuals and provide encouragement and support in non-traditional ways especially as level of complexity within system-of-systems and expansion of IoT. We will need more conceptual architects.

2.2.2 System-of-Systems

In an individual system, the inter-relation with the different components/parameters is generally well understood by today's engineers and scientists and tradeoffs like performance, size, mass, power, cost, reliability, etc., can be implemented using the current system engineering processes previously discussed. However, designing a system-of-systems is much more complex. It is like the difference between designing a new airplane and designing an international fleet of aircraft to serve a worldwide transportation market. An airplane may have thousands of parameters that must be considered in a tradeoff among all the critical factors needed to create a workable, practical, and efficient design, but the fleet will have several orders of magnitude more parameters, with interactions between all possible parameters which are less understood and should be considered.

If we apply this idea to health, we can think of a heart as a single system or even a person, but we will miss a lot of information since we also know that family, environment, and many other factors have major impacts on health, and thus an individual's health should be considered more holistically and treated as a system-of-systems. With advances in computing technology (hardware/software) such as IBM Watson [13], we are now able to analyze many more parameters or interactions as more health data are collected and processed not just from an individual and his or her family but from a local area, region, and around the world. This will be important as telemedicine becomes more commonplace around the world. With the diversity of medications, the interactions between medicines and possible side effects will be able to be analyzed, tracked, and their impact on the overall wellness better understood. As eHealth system-of-systems solutions evolve due to technology like Watson and faster networks with more environment and personal data being collected in real time via IoT (sensors, processing, and algorithms), the UN and WHO Grand Challenge for all individuals can be slowly realized and our eHealth system-of-systems will evolve and become more optimized.

There has been recent interest around the world in Smart City technology. A Smart City should be considered a system-of-systems problem and not approached by looking at individual parts like energy, waste management, etc. All elements of a city (government, citizens, health, transportation, energy, emergency responses, education, business, infrastructure, financial, etc.) need to be included in any analysis. For example, a few years ago, India initiated a 100 Smart City Program [14]. They selected 90 winning proposals around the country with different ecosystem issues to research and implement various solutions that will benefit citizens and the city overall. They developed a best practices approach, so results will be shared within the country and around the world. The key to a city evolving to become a Smart City is the digital ecosystem using WWW and IoT and advances in computing hardware/software. However, this approach is somewhat trial and error, and the results may or may not turn out be the most efficient or cost effective. Although benefits will likely result, they are the optimum. Since funding is always a constraint, criteria for evaluation should include: ***"is the city getting the best for each dollar"*** and ***"are the solutions optimized."*** The problem is so complex that no individual can analyze all the possibilities or "what if" situations.

One of the engineering challenges now is to develop more formal system-of-systems processes. This would allow the architects to analyze the multi-millions of parameter interactions that would be involved in a complex environment such as a Smart City implementation. Fortunately, there is research in developing tools that would allow one to analyze complex system-of-systems. The challenge will be to define all the parameters of a complex system-of-systems and identify the analytical relationship between each parameter and all the others. However, once done, the optimization tools can narrow down the multi-million combinations and reduce them to a reasonable number that could lead to a more optimum solution.

As our wireless network increases in capability and IoT expands with wearable sensors for health and local sensors for smart homes and infrastructures, our world will still need to develop and recognize conceptual architects to innovate within the system-of-systems complex digital ecosystems using new evolving system-of-systems optimization tools and analytics. This will enable decision processes to be documented providing the most cost-effective and most beneficial solutions to society as a whole. Smart Cities with limited budgets, legacy systems, and infrastructure will be able to create the most optimum solution for their goals with these new tools. What may be optimal for one city may not be for another but collecting each data base will help with the analytics, benefiting all. However, in this optimization how do we provide a trusted environment for the individual?

2.2.3 Trusted Interfaces

With more individuals connected to the WWW and IoT collecting environmental and personal data 24/7 and in most locations, we must also protect the individual from information being misused. An individual should be able to personalize information and protect themselves. Trust must be incorporated into the digital IoT ecosystem. An individual should be able to trust that specific information is protected and entities collecting and requesting information are ethical and aware of cultural differences, whether the data itself are accurate, etc. This is even more important as the underground or dark web expands.

One concern is that how are ethical and cultural concerns and individual protections incorporated in the digital ecosystem and IoT evolution. Many companies, services, browsers apps, etc., do little to protect the individual. Many have not simply included sufficient protection for the user. As "things" get connected, lack of built in protections may allow data on devices to be accessible by third parties for activities not approved by the users. A couple of examples: in many software apps, an individual must approve certain "permissions" to use that app. For example, Google's Arts and Culture app uses your "selfie" taken on your smartphone, compares your face to museum portraits, and selects the one closest artwork that "matches" your face. Sounds like fun, but now they have a key bio-identifier of you, for non-secure purposes that could be missed by them or a third party. You also must allow the app to have access to the camera data, storage, etc., on your device, which is also common to many apps that have no apparent reason to need that information. You are usually required to give them access to ALL your photos and many "free" apps require that you give them permission to use, even publish any or all of your personal information.

New voice activated household assistants like Google Home and Amazon Alexa are usually listening all the time and require that you allowed them to process all "nearby" voices. Therefore, you may not be able to preclude them from responding to children's voices or protect you or the children from things they may reveal, guests, service people, without their awareness or approval. On their side, most such systems and many, many other apps rarely, if ever, include an agreement about what they can or cannot or will or will not do with any of your data that they collect, unless buried in the small print. Some may list all the requirements, some may not, but rarely can you refuse to give them permission and still use the app. Also, whole computer services; browsers; email, photo taking apps; shopping apps; and some social networks, games, etc., without which many users would feel disconnected from their friends and family and society, require such blanket permissions.

In some cases, you can protect yourself. For example, many computers and mobile devices including tablets, have multiple choices in their system settings about what you want to share and what you want them to share with you. However, they are often preset when you buy them in their favor. Some may sound good but open you to sharing a lot of your system's information. Searching the Internet for settings on your system will often reveal sites and videos, by others who are not part of your computer company, like industry magazines, which will explain settings, tell you why you might or might not want them, and then show you how to change them to what you want. None of this is easy and protecting yourself is time consuming or may require outside help, but a few important ones are well worth the effort. Unfortunately, the average user will not take the time or "care" many because they are not aware of the negative impacts the information may cause them in the future.

In some cases, the IoT may exacerbate these problems, because the "smart" devices want to be "affordable" and therefore developers may not want to include protections and leave it up to the buyer – or the ***buyer beware*** approach.

This book talks about dependable systems. Typically, dependable systems refer to systems that are available, reliable, durable, and maintainable and in some cases refer to safe and secure consideration. It is the latter two that we are concerned about in this chapter. Does safe and secure refer to the encryption of data between one destination and another and how the Cloud protects data and identity as it moves around the network? Does the dependable system include ethical and cultural processes? Can it be trusted by the user? eCV believes that these areas, especially as they relate to individuals, are not sufficiently considered or developed by the conceptual architects et al. or included in system-of-systems development tools.

As interconnectivity and complexity increase, the IoT developers or governments need to include user protection policies. It may even require legislation. Developers must specifically identify what and where any information is used and that reasonable efforts are made to protect that information.

We must also have developers respect cultural (including religious) differences and develop alternative approaches when necessary so certain individuals are not excluded from technical advances. This becomes more important as we address the UN and WHO Grand Challenges involving people from around the world. We may also need to develop policies where individuals in some public, but many non-public situations should be notified when information is being collected. For example, if a home voice control system hears a "new" voice possibly from a guest or worker, it stops collecting and asks for permission to continue. In the USA, there are some laws about recording without permission, so it may be illegal for a homeowner to have the system activated when a non-family member is in the house. Some of the wearable smart fitness devices tract the location of the person, so a person's routine can be determined which may put the person at risk. Data collected by any smart device are now being used as evidence in legal cases. Recently, Apple announced that their iPhone running the iOS 11 update will allow storage of an individual health records. These records are sensitive and if released somehow can have very negative effects on the individual. Unfortunately, many users not aware of the negative aspects of technology much like not being aware of side effects from drugs. As IoT evolves, we must be aware of possible misuse (intentional or unintentional) of the technology not necessarily by the initial party but third parties, dark wed, etc.

Our eCV eGuild is researching approaches to help an individual interact with this new digital ecosystem in a more trusted manner. One thought is to create personalized adaptive algorithms that can act as the individual's interface and buffer to the digital ecosystem – the individual's ***"Guardian Angel."*** An individual interfaces, via natural language and mannerisms, to a "Smart" (AI based) personnel interface (SPI) on their smartphone, personal computer, etc. The SPI then communicates to the digital world. This SPI uses passive and active bio-metrics to provide a first stage trusted interface that the person using the system is the person whose data are being generated or accessed. The SPI buffers and protects the individual while interfacing to the WWW as well as the dark web much like a parent or personal assistant for a manager and act as their personal digital ecosystem ***"Guardian Angel."*** Using AI and deep learning, the SPI continuously "adapts" to the individual's interest, needs, and traits. The SPI then uses this knowledge for emails, messages, searches, and data protection.

The SPI is the person's Digital Avatar helping the individual use IoT and the digital ecosystem in a trusted way for their benefit especially as it relates to eHealth, eLearning, eBusiness, eGovernment, and Smart Infrastructures.

A few examples – for eHealth, the SPI will be able to monitor via wearable technologies and IoT and advise on the health maintenance plans and provide warnings associated with their medical conditions and local environment. For eLearning, the SPI will be able to customize the learning experience by finding lesson plans and problems along the interest of the user and at a pace equal to the individual's capabilities. For eBusiness, the SPI will be able to customize a training and retraining program and find policy and legal information, etc. For eGovernment, the SPI will be able to quickly find specific on-line government data, tax information, customized to the user, etc.

In some ways, companies like Google, Apple, Amazon, and Facebook are also trying to learn the user traits, but they are doing it for marketing and paid advertisers. Recently, I searched the web for information on a specific disease affecting one of my relatives. The initial search responses were all pharmaceutical companies and ads. It took me several additional steps to get to the specific disease information I was looking for. The SPI would be able to "understand" I was looking for specific disease information and filter the data and continue the search finding the specific data of interest for me. I also recently brought an item online and almost immediately started to receive pop-up ads from that company. This was a unique item that I would not tend to buy again, but still were receiving ads from that company months later. SPI would be able to filter these as well. The filter parameters within the SPI will be able to be set or overridden by the user.

When information is requested over the WWW or a link shows up in an email, SPI will be able to search and investigate the source to determine the "risk" of providing that information or selecting that link. Again, SPI can have user set thresholds for different data like social security number, health data set at very high risk, and information like birth place at low risk and age, sex, race, and religion somewhere in the middle depending on the user wishes. Many of these ideas are in various specific applications already, but the SPI will integrate them into one trusted application. The eCV eGuild wants to use advanced technology to benefit and protect the individual as the digital ecosystems becomes more intrusive in everyday life.

2.3 Conclusion

The WWW and the IoT will revolutionize how individuals interact with the world around them. With the complexity and interconnectivity of the underlying digital ecosystem exponentially increasing, we need to respect and provide support to the conceptual architects and their eGuilds as they innovate the future. At the same time, we need to develop formalized system-of-systems analysis tools so that more optimize complex solutions can be developed faster and at lower cost and risk. We must consider the ethics and cultural differences when developing IoT devices and software. Although much of the digital ecosystem will benefit individuals, we must also protect individuals using trusted personalized adaptive technology to "understand" the user interest, needs, and traits to make it easier for the user to efficiently operate within the digital ecosystem while protecting the individual and their personal information. New ways of operating like how the eCV International Collaboratory, a new eGuild, will use the web to form a virtual high-tech group to address some of the Grand Challenges facing society. We want to ensure that dependable systems include ethical and cultural considerations and that the everyday user can trust that technology is also being used to protect them as the IoT evolves and propagates on the body (wearable devices) and within the home, workplace, and public/private spaces.

Acknowledgments

I would like to acknowledge my wife Cheryl Garbos, President and COO of eCollaborative Ventures, Inc. (eCV), and Dr. Andrzej Rucinski, a charter member of the eCV Collaboratory, for reviewing, commenting, and editing this chapter. I would also like to thank the other individuals who have joined our international collaboratory (our eGuild) and have offered to help address several Grand Challengers to benefit society.

References

[1] Segan, S. (2018). *5G – 5th Generation Wireless Network – "What Is 5G? AT&T, Verizon, and other carriers will start to launch 5G networks this year. But what exactly is 5G, and how fast is it compared with 4G? Here's what we know so far."* Available at: https://www.pcmag.com/article/345387/what-is-5g [accessed February 1, 2018].

[2] Institute of Electrical and Electronics Engineers (IEEE). Available at: https://ieeeusa.org

[3] Baer, R. *The Father of the Video Game: The Ralph Baer Prototypes and Electronic Games*. Available at: http://americanhistory.si.edu/

collections/object-groups/the-father-of-the-video-game-the-ralph-baer-prototypes-and-electronic-games

[4] Zuckerberg, M. *Facebook Founder*. Available at: https://astrumpeople.com/mark-zuckerberg-biography-success-story-of-facebook-founder-and-ceo/

[5] Advanced Research Projects Agency (ARPA) Net. Available at: https://www.darpa.mil/about-us/timeline/arpanet

[6] eCollaborative Ventures, Inc. (eCV) (2015). Available at: www.ecollaborativeventures.com

[7] Collaboratory. Available at: https://en.wikipedia.org/wiki/Collaboratory

[8] New Hampshire Benefit Corporation. (2014). *"Benefit corporations: coming soon to New Hampshire - A new state law gives socially minded businesses an alternative corporate option", Colleen Lyons*. Available at: http://www.nhbr.com/November-28-2014/Benefit-corporations-coming-soon-to-New-Hampshire/

[9] Smart City. Available at: https://www.ibm.com/smarterplanet/us/en/smarter_cities/overview/

[10] United Nations (UN). Available at: http://www.un.org/en/sections/what-we-do/index.html

[11] World Health Organization (WHO). Available at: http://www.who.int/about/en/

[12] Guest, D. (1991). *T-Shaped*. Available at: https://www.ptc.com/en/product-lifecycle-report/why-engineers-need-to-develop-t-shaped-skills

[13] IBM Watson. Available at: https://www.ibm.com/watson/

[14] India 100 Smart City Program. Available at: www.smartcities.gov.in

PART II

Modelling and Assessment

3

Design and Simulation of an Energy-efficient Sensor Network Routing Protocol for Large-scale Distributed Environmental Monitoring Systems

Awais Ahmad[1], Muhammad Adeel Pasha[1,*], Shahid Masud[1] and Axel Sikora[2]

[1]Department of Electrical Engineering, Syed Babar Ali School of Science and Engineering, Lahore University of Management Sciences, Lahore, Pakistan
[2]Institute of Reliable Embedded Systems and Communication Electronics, Hochschule Offenburg, Offenburg, Germany
[*]Corresponding Author: adeel.pasha@lums.edu.pk

Due to climate change and scarcity of water reservoirs, monitoring and control of large-scale irrigation systems is now becoming a major focal area for researchers in cyber-physical systems (CPS) domain. It is a unique yet daunting task to design an energy-efficient distributed system – for large-scale environmental monitoring – that lasts for decades. However, the evolution in digital electronics and microelectromechanical systems has enabled the researchers in CPS domain to use technologies like Internet of Things in automating such large-scale systems. Wireless sensor networks (WSNs) are also rapidly finding their way in the field of irrigation and play a key role as data gathering technology in CPS domain. They are efficient for reliable monitoring and early climate change detection, giving farmers an edge to take precautionary measures. However, designing an energy-efficient WSN system requires a cross-layer effort and energy-aware routing protocols play a vital role in the overall energy optimization of a WSN. Reliable and energy-efficient transmission of data in energy-constrained WSN is the primary challenge to achieve a next-level efficiency control and monitoring system. In this book chapter, we are proposing a new hierarchical routing protocol that is suitable for large-area environmental monitoring such as distributed monitoring of the large-scale irrigation network existing in the Punjab province of Pakistan. The proposed protocol resolves the issues faced by traditional hierarchical routing protocols such as low-energy adaptive clustering hierarchy (LEACH), M-LEACH, and I-LEACH, and enhances the lifespan of sensor nodes that, in turn, results in an increased lifespan of the whole WSN system. Increasing the lifespan and improving the quality of service also increase the reliability of the WSN. However, it has the inverse relation with the power consumption. Retransmissions can reduce the loss in data packets but it will increase the power consumption of WSN; but periodic transmissions reduce the data packet loss and hence increase the reliability of the WSN and also reduce the power consumption of the overall WSN. We used the open-source NS3 simulator for simulation purposes and results show that our proposed protocol results in an average gain of 27.8% or higher in lifespan of the overall network when compared against the protocols reported in the literature.

3.1 Introduction

3.1.1 Context and Motivation

Water is a critical resource for sustainable economic development of Pakistan. Irrigation System of Pakistan's largest province by population, Punjab, consists of about 23,184 miles in length of canals with cultivable commanded area of about 21 million acres [1]. There is a need for an automated system for monitoring of water distribution and flow of water in these canals and associated tributaries. Wireless sensor networks (WSNs) are now widely being used for low-power distributed monitoring [2] in various cyber-physical systems (CPS) like precision agriculture, forest fire monitoring, as well as monitoring of infrastructural health of buildings and bridges [3–5]. Designing of a low-cost embedded system for monitoring and control of waterways is a challenging task, but we have already developed a hardware design and performance evaluation of smart water metering solution, which a GSM-based powerful tool for water monitoring [6]. In the current contribution, we are using a distributed WSN system for automatic monitoring and control of large-scale waterways in Punjab, Pakistan. Figure 3.1 shows an example of such a distributed system.

A WSN consists of a large number of wireless nodes where each node consists of a micro-controller, a sensing or actuating element, an analog-to-digital converter, a radio transceiver, an optional power amplifier, and an energy source (battery). Given the size and cost constraints of a WSN node, it cannot be equipped with a very large battery source. Moreover, replacing batteries is also not feasible as the nodes are deployed in remote locations with difficult access. To extend the lifetime of a WSN, we need both a node-level as well as network-level design effort as it is important to distribute the energy dissipated among the nodes evenly in the whole network to have a uniform/stable coverage. In this book chapter, we are proposing an energy-efficient routing protocol that improves the overall lifespan of the WSN where nodes are placed in a random fashion to simulate a dense network of water tributaries present at the far-end of an irrigation network.

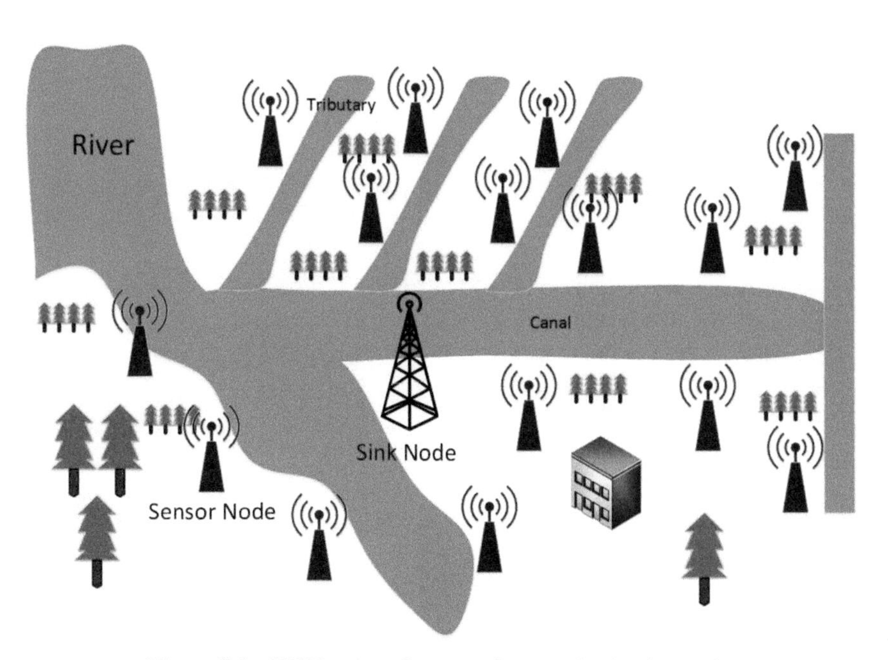

Figure 3.1 WSN system for water flow monitoring in canals.

Routing topology plays an important and key role in extending the lifespan of a WSN by optimizing the overall energy consumption of nodes. WSN nodes can be organized hierarchically by creating clusters that are groups of neighboring nodes. Hierarchical routing protocols save individual node's energy by cutting down longer transmission distances to smaller chunks.

As far as the reliability of the network is concerned, it can be directly ensured through the acknowledgment and retransmissions in the case when the acknowledgment is not received from the receiver. Retransmission is a widely used strategy in a resource-constrained WSN for recovering the lost packets [7]. For the data-critical applications where data packet loss is not affordable, retransmission of the whole data packet is necessary. But for the applications where periodic data are needed and missing a data packet is a trivial issue (e.g., water-level sensing in canals and tributaries), periodic transmissions can cover the gap of data packet losses. In our proposed framework, we are considering the second type of application and using network lifespan as an indirect mean of increase in the reliability of the network health.

Low-energy adaptive clustering hierarchy (LEACH) [8] and its variants [9, 10] are the most famous hierarchical routing protocols used in energy-efficient WSN systems where nodes divide themselves into clusters and one of the nodes is randomly selected as cluster head (CH), giving equal probability to each node of becoming CH. Each node forwards its sensed data to its own CH which, in turn, aggregates the data for whole cluster and transmits it to a base station (BS).

I-LEACH [11] is a recently developed hierarchical protocol that offers further advancement over LEACH by adding different energy-efficient features like multi-hopping from M-LEACH [12] and data-thresholding from TEEN [13] protocols. Furthermore, it checks the remaining battery life of the node during the process of CH selection and saves the node from dying by denying it to become a CH if its remaining energy is less than a desired threshold. This makes I-LEACH protocol more energy-efficient than other state-of-the-art protocols. However, all the stated hierarchical protocols have the advertisement phase where as soon as a node is selected as CH, it broadcasts a message to all other nodes to join it to form a cluster. Each CH must advertise in every round, so the number of advertisements is directly proportional to the number of rounds during the network lifespan. Hence, reducing the number of advertisements can, in turn, increase the number of transmission rounds and the whole network lifespan will be increased.

3.1.2 Contributions

Our major contribution lies in addressing the issue of number of advertisement packets in the existing hierarchical protocols. In this work, we propose a new routing protocol with a lesser number of advertisement phases by elongating a single round. Our proposed protocol incorporates the feature of hopping data to neighboring CHs in the direction of the BS to reduce the transmit distance making it more energy-efficient than existing M-LEACH protocol. Data-thresholding feature of TEEN and remaining battery-life check of I-LEACH are also incorporated in our protocol. The symbolic metric of energy efficiency in our framework is the network lifespan. This is a logical choice as the higher the energy efficiency of the overall network will be, the longer it will last. We are comparing the number of rounds lasted (i.e., the network lifespan) resulted in our proposed protocol against those in other existing protocols while keeping the initial energy of the nodes, transmission energy, reception energy, and all other factors constant. We call our proposed protocol more energy-efficient by comparing the lifespan of our proposed protocol against the existing energy-efficient routing protocols reported in the literature. We use the open-source network simulator, NS3, to simulate a network with nodes having individual protocol stacks. Just to clarify, we are assuming the nodes to be stationary and not mobile. The simulator allows to create nodes with the network layer, data-link layer, and transport layer to interact among themselves. Simulation results show that our proposed protocol gives an average gain of 27.8% or higher in network lifespan by cutting short of the advertisement phases when compared with the state-of-the-art protocols such as I-LEACH, LEACH, and M-LEACH.

The second contribution of this work is to add realistic channel fading at different hop levels to mimic real-life application scenarios. Our simulation results show that even in the presence of fading, our proposed protocol is better in performance than other existing protocols reported with ideal conditions (i.e., without fading).

3.2 Related Work

LEACH [7] is a cluster-organized self-adaptive energy-efficient TDMA-based routing protocol. It consisted of the features like random CH selection and distributed cluster formation, and it had been proven to be more energy-efficient than non-hierarchical routing protocols. In LEACH, nodes in the network organize themselves in clusters and one of them randomly becomes CH as shown in Figure 3.2(a). Energy is uniformly distributed to each node

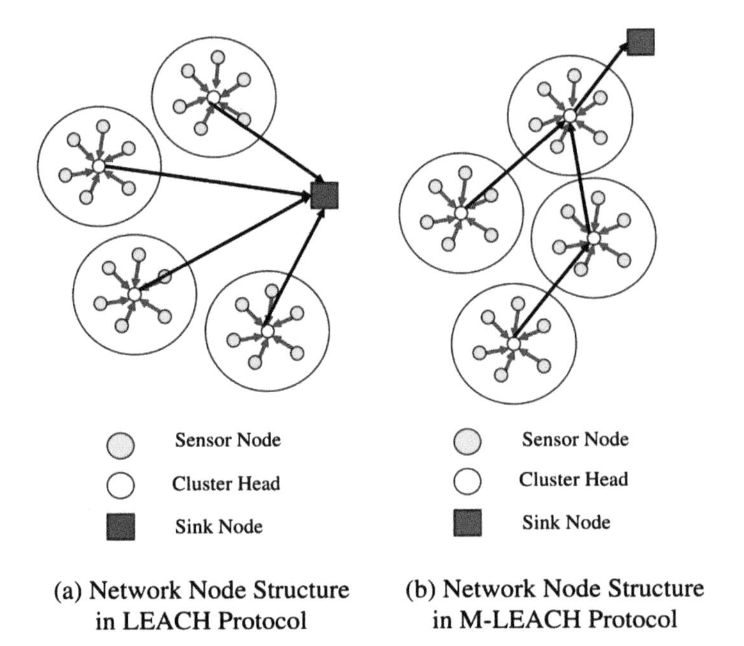

Figure 3.2 Network node structure in (a) LEACH and (b) M-LEACH protocols.

and energy load on each node is also identical as LEACH gives equal probability to each node of becoming CH. The CH selection threshold criteria for any node is given in

$$p_r = \frac{P}{1 - P(r \,\%\, \frac{1}{P})} \tag{3.1}$$

where r is the recent round number, p_r is the probability of a node to become CH in the rth round, P is the probability of CH forming, and $\%$ is the modulo operator. One set of synchronous data transmissions from each alive node to the BS is called a "round."

There are two phases within each round:

- **Cluster Formation or Setup Phase:** In the setup phase, nodes dynamically divide themselves and form a cluster. Some of the nodes randomly become CH and each CH broadcasts an advertisement message to all neighboring nodes to invite them to join him to form a cluster. After the cluster is formed, the CH broadcasts the TDMA schedule to each node in their clusters.

- **Transmission Phase:** In this phase, each node in a cluster sends its data to its CH according to the assigned TDMA slot. The CH of each cluster aggregates the data for its own cluster and transmits all data including its own directly to the BS.

The biggest advantage of LEACH protocol was that it cuts down the individual direct transmissions between each node and the BS. In fact, a CH accumulates the data from its members and performs a single long-hop communication that results in saving up a lot of energy when compared with individual long-hop communications. However, if a cluster is far away from the BS, then this single long-hop (and an overall two-hop) communication between the CH and BS channels would be significantly high according to Friis's free space equation [14] where transmission energy is directly proportional to square of the communication distance. This was a significant shortcoming of the basic LEACH protocol.

The M-LEACH [12] routing protocol tackled the issue of single long-hop transmission in LEACH as it cut down the distance of each CH to BS communication by hopping the data from originating CH to one of the neighboring CHs in the direction of the BS. It could be a single- or multi-hop communication depending upon the position of originating CH and the BS. If there was no CH between the originating CH and the BS, then there would be a direct transmission like the original LEACH protocol. Reduction in the transmission distance significantly reduced the amount of energy consumed by any CH that resulted in enhancing the lifetime of the whole WSN. Figure 3.2(b) shows the overall node structure of the M-LEACH protocol.

WSN nodes have sensors to sense different attributes of a system and it would be a smart choice if some pre-processing is done at node level to see if the newly sensed value of an attribute is different than the previously sensed and transmitted value. The TEEN [13] protocol added this local processing to a WSN node and adjusted the thresholds for sensed value to qualify for transmission. By analyzing the sensed data, TEEN reduced the number of transmissions which resulted in saving energy without loss in critical data values.

I-LEACH [11] was also a hierarchical protocol with random CH selection. It incorporated the energy-efficient features of TEEN (i.e., data thresholding), random CH selection (LEACH), and multi-hop transmission (M-LEACH) to propose a more energy-efficient protocol than the individual protocols themselves. It also introduced an additional constraint of remaining battery life during the cluster formation phase which kept the selection process random but the nodes with only a sufficient amount of battery life could participate in becoming a CH. The modified CH selection criterion is given in

$$p_r = \frac{\alpha P}{1 - P(r \,\%\, \frac{1}{P})} \tag{3.2}$$

where α is the remaining energy fraction in the battery of a node and all other variables are already defined in Equation (3.1).

3.3 Proposed Protocol EESNR

All the hierarchical routing protocols stated above have an identical advertisement phase where the CH, once randomly selected, must broadcast its captaincy by asking other nodes to join it and form a cluster. Once the clusters are formed, the CH must transmit a TDMA schedule to all cluster members to avoid data collisions. These two control messages are broadcasted in every transmission round that contributes to a significant amount of transmission energy. We tackle this inherent weakness of the LEACH-based protocol through the energy-efficient sensor network routing (EESNR) protocol. It is also hierarchical but outperforms the existing hierarchical protocols by reducing the control message overhead of advertisement and scheduling phases.

3.3.1 Network Topology Model for EESNR

Being a hierarchical protocol, EESNR follows a tree structure with the BS as root, CHs as children to the root node but parents for leaf nodes. So, a route will be from the node to CH to the BS. The total number of CHs in a network depends upon the number of total nodes in WSN used for distributed monitoring and all the nodes take part in the CH selection process but by observing the constraint on battery life that was proposed in the I-LEACH protocol. CH selection process is totally random and any node that fulfils the criterion, given in Equation (3.3), can become a CH for that round.

$$p_r = \frac{\propto P}{1 - P(\frac{r}{\beta} \,\%\, \frac{1}{P})} \tag{3.3}$$

where $\propto$, P, and r are the same as defined in Equation (3.1) and β is the round elongation factor that ensures to delay the CH selection process until the previously selected CHs have completed the elongation period. By elongating a transmission round, we save a significant energy through the reduction in control overhead resulting from broadcast messages for advertisement and TDMA scheduling processes.

Here are the new phases defined in the EESNR protocol:

1. **CH Selection and Advertisement:** In the setup phase, nodes decide to become CHs for current round based upon the percentage of CHs for network and how many times that node has already been a CH in the past. P is the desired percentage of CHs which is normally set as 0.05 (5% of total nodes). Once P the percentage of nodes has become CHs, they seek their cluster members through advertisement messages to all neighboring nodes. In EESNR, by keeping the same cluster for a longer duration though, the control overhead is reduced. After the elongation period expires, new CH nodes are selected on the same random basis using Equation (3.3) and they again start a new advertisement phase. By elongation, a CH is asked to carry out the whole process with the same cluster formation which in turn reduces the advertisement, schedule transmissions, and cluster joining requests packets from other nodes. Figures 3.3 and 3.4 show the working of CH Selection process and the modified advertisement process, respectively, during the first phase.

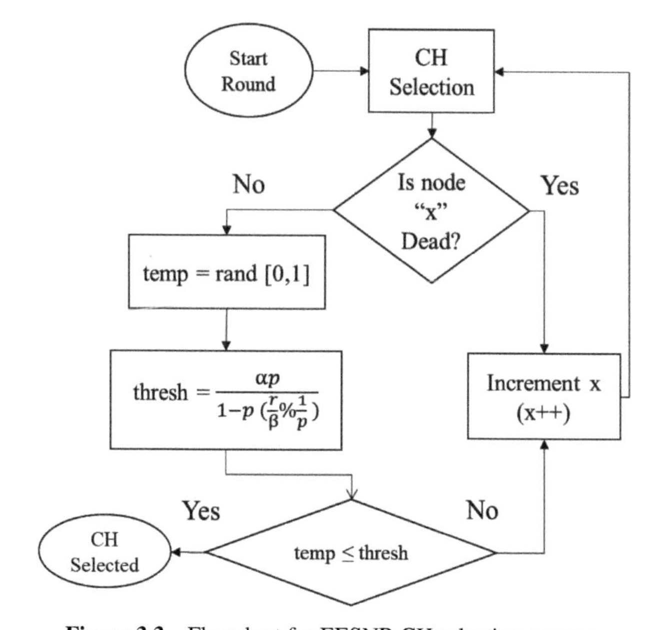

Figure 3.3 Flowchart for EESNR CH selection process.

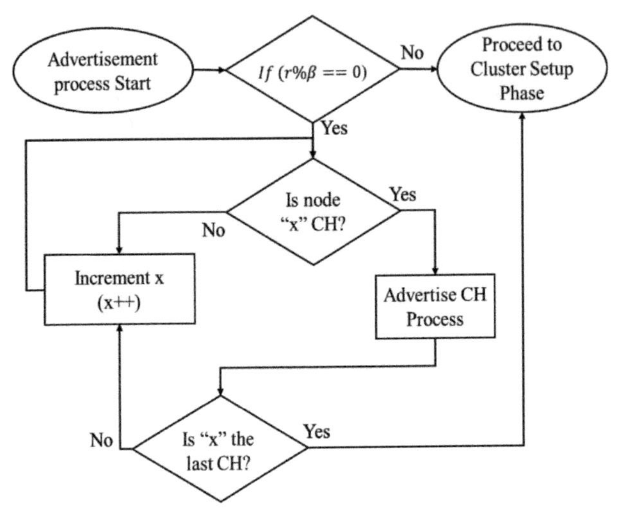

Figure 3.4 Flowchart for modified advertisement process proposed in EESNR.

2. **Cluster Setup and Schedule:** After the advertisement is sent from all CHs to non-CH nodes, the non-CH nodes decide to join clusters based upon the distance to the CHs. The process is totally random and member nodes dynamically decide it during each round by sending an acknowledgment message to the respective CH they want to be associated with. Following the concept of the original LEACH-based protocols, currently, the thresholding of cluster size during the cluster formation phase is not tackled in this work and it can be an interesting future research direction. When the clusters are formed, CHs assign a data transmission slot to each member according to a TDMA scheme to avoid data collisions. This is called the scheduling process.
3. **Data transmission:** After the schedule has been sent to all the members by respective CHs, they start sending their data to CHs in their respective time slots. When each CH receives the data from its member nodes, it decides whether to transmit data directly to the BS or to some other intermediate CH based on the distance between the CH and the BS. If a CH (say CH1) decides to send data to an intermediate CH (say CH2), then the intermediate or receiving CH, i.e., CH2 adds CH1 to its cluster members and assigns it a TDMA time slot to avoid data collision with other cluster members. Through this process, a tree-based data-link is established among all nodes to the BS through intermediate CHs and

data starts to flow in this tree. A transmission round is completed when the last CH neighboring to the BS sends its accumulative data to the BS. At this moment, the average energy of the whole network is calculated to check whether the network has a sufficient amount of energy left to carry on with another transmission round.

3.3.2 Path Loss/Fading

In all the above-discussed proposed protocols, the data-link quality is assumed to be ideal, i.e., to have to no fading at all along the propagation path in routing. None of the data packets is lost due to fading in channel as no noise is introduced in the communication link. This assumption, however, can be quite contradictory to practice (i.e., in case of real-life networks). Hence, to make EESNR more realistic, we introduced Rayleigh fading, with zero-mean Gaussian noise [9] in different communication links, i.e., between cluster members to CH, one CH to another CH, and the last CH to the BS. Rayleigh fading model [9] assumes to have varied signal strength when it has been transmitted through such a fading medium. We assume the noise to have zero mean that is evenly phase distributed between 0 to 2π radians. The probability density function (PDF) of such a random variable "R" is given by

$$PDF_r(d) = \frac{2d}{E(R^2)} e^{\frac{-d^2}{E(R^2)}} \tag{3.4}$$

3.3.3 Radio and Data Transmission Model

Energy is consumed in transmitting and receiving data by each node. Being a hierarchical multi-hop protocol, in EESNR, only CHs transmit data to relatively larger distances, i.e., to either the BS or a neighboring CH. All other nodes communicate to their respective CHs which results in a shorter distance and low-energy communication as energy consumption is linearly related to the amount of data being transmitted while squarely related to the transmission distance. The transmission and reception energies can be modeled using Equations (3.5) and (3.6), respectively.

$$E_{trans} = \zeta_{elec} \times DataLen_{trans} + \sigma_{amp} \times Dist^2 \tag{3.5}$$

$$E_{receive} = \zeta_{elec} \times DataLen_{receive} \tag{3.6}$$

where E_{trans} is the energy (in joules) consumed by a node in transmitting data of packet size $DataLen_{trans}$ (in bits) on distance $Dist$ (in meters) and

$E_{receive}$ is the energy (in joules) consumed by a node in receiving data of packet size $DataLen_{receive}$ (in bits). ζ_{elec} is the energy per bit and σ_{amp} is the energy per bit per square meter required by the power amplifier to transmit the data over a certain distance in meters.

3.4 Simulation Setup and Results

3.4.1 Simulation Setup

To compare our proposed protocol against the other existing protocols, we created a test-bench in Network Simulator 3 (NS3) which is an open-source Linux-based simulation platform to simulate different networking protocols. Our experimental setup has a data structure called "node" which stores different parameters for a node such as its location, battery energy, CH count, cluster member count, nearest CH, current sensor reading, last transmitted sensor reading, etc. A distributed WSN comprises many of such nodes. Just like other existing protocols, we run our simulation until the average energy of the whole network, calculated as the end of each transmission round, reaches a certain threshold. Algorithm 3.1 describes the CH selection criterion used in EESNR.

Algorithm 3.1 CH Selection in EESNR

```
Initialization;
while (Network Energy > u) do
        for (each node x ∈ G where G is network, round = r) do
              if x.head ≠ dead then
                    temp = [∅–1]; R number
                    thresh = αp/(1-p((r/β) mod(l/p)));
                    if (temp ≤ thresh) then
                          x becomes cluster head;
                          x.head++;
                    else
                          x++;
                    end
              else
                    x++;
              end
        end
end
```

The structure also has a variable named node.c member which represents the number of nodes joined to a particular CH. It has NULL in it if the node is not CH in current round. node.bETotal and node.bERem represent the total battery energy in a node and remaining battery energy in a node, respectively. They are used in the calculation of factor α which is used in the CH selection process.

To achieve an acceptable signal-to-noise ratio (SNR), we chose ζ_{elec} to be 50 nJ/bit and σ_{amp} to be 100 pJ/bit/m^2. The values for ζ_{elec} and σ_{amp} are consistent with other existing protocols [5] and are defined in Section 3.5. The transmission channel is assumed to be symmetric, i.e., the amount of energy consumed in transmitting data from node X to Y or Y to X is the same.

Algorithm 3.2 Multi-hop routing process in EESNR

```
Initialization;
for each node x ∈ G where G is network, round = r;
        temp = ∅;
        dist2dest = ∅;
        smallest = 999999;
        CHIndex = -9;
        for each node y ∈ G where G is network, round = r;
                srcD = dist(x, BS);
                relayD = dist(y, B5);
                temp = dist(x,y);
                if ((relayD<srcD) && (temp<smallest) && (temp<srcD))
                then
                        smallest = temp;
                        CHIndex = y;
                else
                        x.nearestCH = CHIndex;
                end
        end
end
```

As we are working with hierarchical protocol, data transmitted by any cluster member to a CH will be of the same size (i.e., of 16 bits) but the data transmitted by the CH will be an accumulation of all the data values from its cluster members and the overall packet size will depend upon the number of nodes in that cluster. We assume that the frequency of data transmissions is much higher than the frequency of the changes appeared in physical phenomenon (e.g., water level or flow-rate in a canal). Hence, if any packet is lost, it will be compensated in the next round.

3.4.2 Simulation Results

We tested our proposed approach with a simulation run for 200–600 nodes with a coverage area of 500 × 500 square units and with the BS placed at 250 × 250 (i.e., at the center). Each simulation was repeated and averaged over 200–1000 to observe the law of large numbers.

Figure 3.5 shows the results for the number of rounds lasted (a measure of network lifespan) of our proposed protocol, EESNR, against other existing protocols with the random nodes' deployment. The average network lifespan increases with an increase in the number of nodes as more nodes mean shorter distances in multi-hop communication. It can be clearly seen that thanks to our proposed modifications to the advertisement and scheduling phases, the reduction in energy consumption has resulted in an average 27.8% increase in the number of transmission rounds the network has lasted with the same initial battery powers when compared against I-LEACH protocol (the best of the existing protocols in comparison).

Figure 3.6 shows the results for the difference in the average number of successful data bits transmitted by EESNR and I-LEACH. Just to clarify, it is the gross traffic, i.e., the information packets as well as the control packets. However, since in EESNR, we have reduced the number of control packets, the overall increase in number of successful data bits means an increase in net traffic (i.e., information packets).

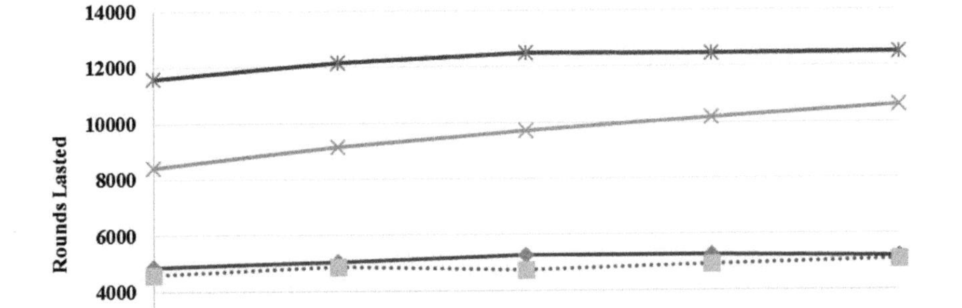

Figure 3.5 Comparison of network lifespan among EESNR and other protocols.

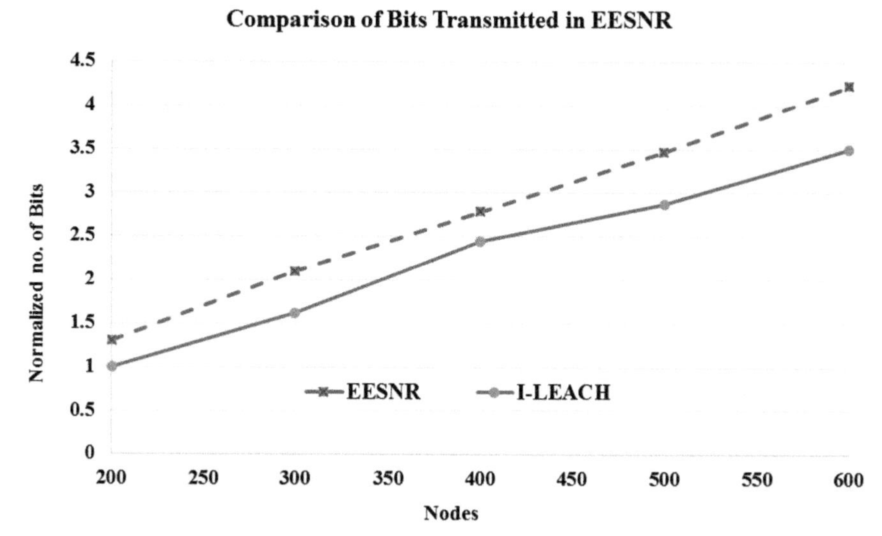

Figure 3.6 Comparison of average number of successful bits transmitted.

Figure 3.7 shows the results after adding Rayleigh fading in the communication channel of the EESNR protocol. If the fading value is high, then the node must transmit data with more power, to achieve a reasonable

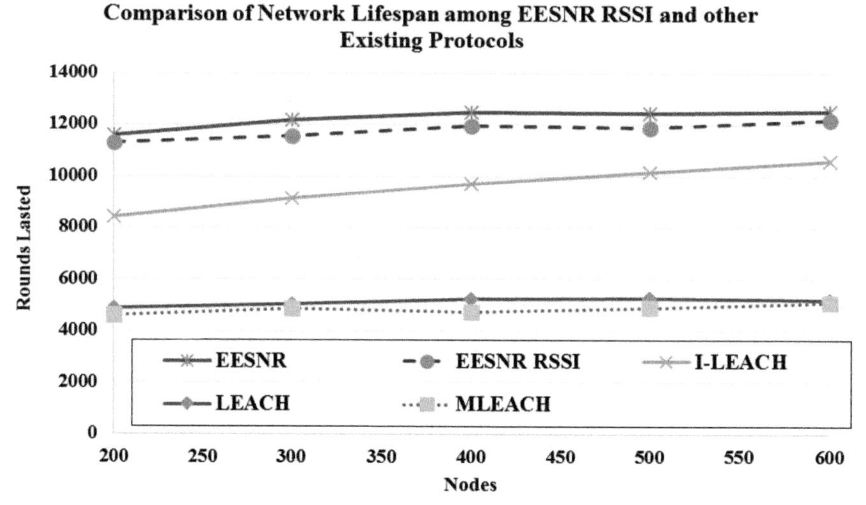

Figure 3.7 Comparison of network lifespan among EESNR (with fading) and other protocols (without fading).

SNR to achieve a successful transmission. Naturally, transmitting data with more power drains more energy and hence graph shows a decrease in the overall network lifespan as shown in the second curve (EESNR-RSSI) when compared against the top curve, EENSR without fading. However, it can still be seen that even with the realistic assumption of channel fading, EESNR is still significantly better than other existing energy-efficient protocols proposed in the literature such as I-LEACH, LEACH, and M-LEACH that are implemented without considering channel fading.

Finally, we tested the proposed EESNR protocol for die-out patterns of the nodes. To ensure a stable coverage throughout the network lifespan from the network reliability perspective, the nodes must die-out in a random fashion over the geographical area. This feature ensures that no region in the area covered by the distributed WSN will have a significantly higher concentration of early died nodes and this stable coverage, in turn, can be used as a quality of service (QoS) and reliability parameter [15]. This was the case with direct transmissions and the initial LEACH [8] protocol where the nodes that were far away from the BS used to die out earlier than the rest of the nodes due to higher energy consumption of long-hop communications. Randomness in die-out pattern was an important feature proposed in the I-LEACH protocol as can be seen in Figure 3.8 that we cannot identify a

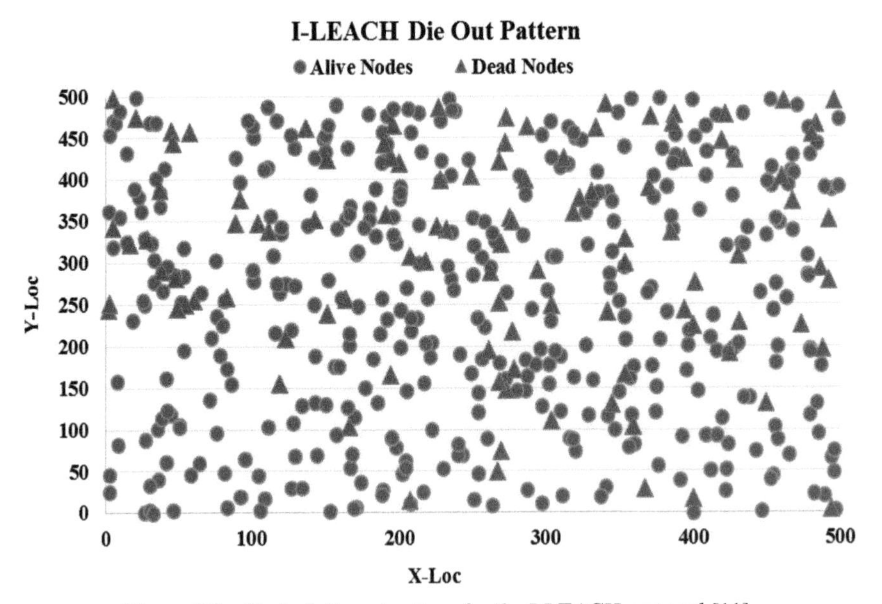

Figure 3.8 Nodes' die-out pattern for the I-LEACH protocol [11].

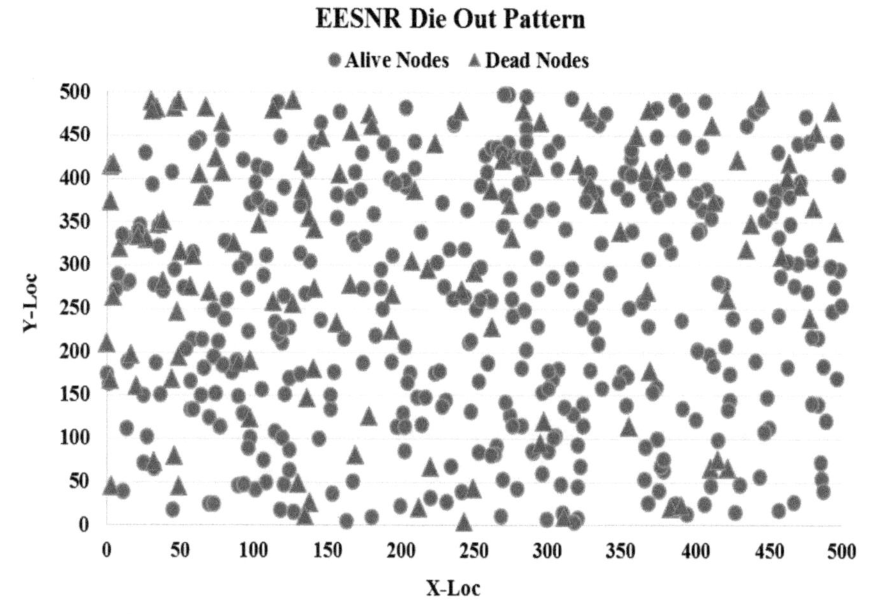

Figure 3.9 Nodes' die-out pattern in the proposed EESNR protocol.

region where there is a higher concentration of nodes that have died out. Figure 3.9 shows the node die-out pattern of our proposed EESNR protocol and it can clearly be seen that EESNR protocol also holds a similar randomness in the die-out pattern. The disperse pattern of dead nodes is also an indication of the reliability of the WSN. As mentioned earlier, if the dead nodes are concentrated in one region, that area will be cut off from the rest of the network to result in a highly unreliable network for that region.

3.5 Conclusions

Due to climate change and scarcity of water reservoirs, monitoring and control of irrigation systems has become a major focal area for researchers in CPS domain. The evolution in digital electronics and MEM technology has resulted in the advent of Internet of Things domain that includes WSN systems as well. WSN has a great potential for monitoring and control of large-scale waterways but typical WSN nodes are not equipped with large energy sources like bulky batteries given the size and cost constraints. Hence, there has been a continuous cross-layer effort to improve the energy efficiency of a WSN system both at node and system levels. Energy-aware routing protocols play a vital role in improving the lifespan of the overall WSN system. In this book chapter, we proposed an energy-efficient protocol, EESNR, which can be used in any large-scale distributed environmental monitoring WSN application. We took the example of irrigation system monitoring in the Punjab province of Pakistan which is one of the World's largest man-made canal network with around 23,184 miles in length of canals. The proposed protocol was simulated in the open-source NS3 simulator and the results show that it provides an average gain of 27.8% or higher in network lifespan over the state-of-the-art hierarchical routing protocols such as I-LEACH, LEACH, and M-LEACH. This increased lifespan can be interpreted as an indirect measure of increase in energy efficiency as starting with the same energy, the nodes in our proposed protocol lasted longer than the other protocols reported in the literature. Moreover, EESNR also resulted in a random die-out pattern of the nodes in the network that resulted in a more stable coverage of the network throughout the lifespan. This stable coverage or randomness in die-out patterns can be translated into a better QoS and increased reliability of the network. Another contribution of this book chapter was to add realistic channel fading at different hop levels to mimic real-life applications. It is important to note that even in the presence of fading, our proposed protocol is better in performance than other existing protocols reported with ideal conditions (i.e., without fading).

Acknowledgments

This work was funded by German Academic Exchange Service (DAAD) through its grant number SRG-352 awarded to Department of Electrical Engineering, LUMS, and Hochschule Offenburg. The authors are grateful for this support.

References

[1] Govt. of Punjab. (2018). "*Introduction and Overview of PMIU*", Tech Report. Available at: http://irrigation.punjab.gov.pk/introduction.aspx [accessed May, 2018].

[2] Phuong, N. M., Schappacher, M., Sikora, A., Ahmad, Z., and Muhammad, A. (2015). "Real-time water level monitoring using low-power wireless sensor network," in *Embedded World Conference*.

[3] Ahmad, Z., Pasha, M. A., Ahmad, A., Muhammad, A., Masud, S., Schappacher, M., and Sikora, A. (2017). "Performance evaluation of IEEE 802.15.4-compliant smart water meters for automating large-scale waterways," in *The 9th IEEE International Conference on Intelligent Data Acquisition and Advanced Computing Systems: Technology and Applications (IDAACS-2017)*, Bucharest. 746–751.

[4] Gutierrez, J., Villa-Medina, J. F., Nieto-Garibay, A., and Porta-Gandara, M. A. (2014). "Automated irrigation system using a wireless sensor network and GPRS module," in *IEEE Transactions on Instrumentation and Measurement*, 63(1), 166–17.

[5] Kim, Y., Evans, R. G., and Iversen, W. M. (2008). "Remote sensing and control of an irrigation system using a distributed wireless sensor network," *in IEEE Transactions on Instrumentation and Measurement*, 57(7), 1379–1387.

[6] Ahmad, Z., Asad, E. U., Muhammad, A., Ahmad, W., and Anwar, A. (2013). "Development of a low-power smart water meter for discharges in indus basin irrigation networks," in *Wireless Sensor Networks for Developing Countries*, Springer, Berlin Heidelberg, 1–13.

[7] Mahmood, M. A., Seah, W. K. G., Welch, I. (2015). "Reliability in wireless sensor networks: a survey and challenges ahead. *Computer Networks*, 79.

[8] Heinzelman, W. R., Chandrakasan, A., and Balakrishnan, H. (2000). "Energy-efficient communication protocol for wireless microsensor networks," in *Proceedings of the 33rd Annual Hawaii International Conference on System Sciences (HICSS '00)*, 08, 8020.

[9] Subramanian, G., Ahmed, Z., Okelola, N. and Murugan, A. (2015). "LEACH protocol based design for effective energy utilization in wireless Sensor networks," in *International Conference on Science and Technology (TICST)*, Pathum Thani, 385–389.

[10] Ahlawat, A., and Malik, V. (2013). "An extended vice-cluster selection approach to improve V leach protocol in WSN," in *Third International Conference on Advanced Computing and Communication Technologies (ACCT)*, Rohtak, 236–240.

[11] Pasha, M. A., Khan, J. H., and Masud, S. (2015) "I-LEACH: energy-efficient routing protocol for monitoring of irrigation canals," in *SIMULATION: Transactions of The Society for Modeling and Simulation International*, 91(8), 750–764. doi: 10.1177/0037549715602957

[12] Kodali, R. K., Venkata Sai Kiran, A., Bhandari, S. and Boppana, L. (2015). "Energy efficient m-level LEACH protocol." *International Conference on Advances in Computing, Communications and Informatics (ICACCI)*. Kochi, pp. 973–979.

[13] Manjeshwar, A. and Agrawal, D. P. (2001). "TEEN: a routing protocol for enhanced efficiency in wireless sensor networks." *Proceedings of the 15th International Parallel and Distributed Processing Symposium,* San Francisco, CA, 2009–2015.

[14] Friis, H. T. (1946). "A note on a simple transmission Formula," in *Proceedings of the IRE*, 34(5), 254–256. doi: 10.1109/JRPROC.1946.234568

[15] Kumar, S., and Lobiyal, D. K. (2013). Sensing coverage prediction for wireless sensor networks in shadowed and multipath environment. *The Scientific World Journal,* 2013:565419. doi: 10.1155/2013/565419

4

Modeling and Assessment of Resource-sharing Efficiency in Social Internet of Things

Kashif Zia*, Arshad Muhammad and Dinesh Kumar Saini

Faculty of Computing and Information Technology, Sohar University, Sohar, Oman
*Corresponding Author: kzia@su.edu.om

In addition to seamless connectivity and smartness, the objects in Internet of Things are expected to have social capabilities. In the literature, these objects are named "social objects." In this chapter, an intuitive paradigm of social interactions between these objects is argued and modeled. The impact of social behavior on the interaction pattern of social objects is studied taking P2P resource sharing as an example application. The model proposed in the chapter studies the implications of competitive vs. cooperative social paradigm while peers try to acquire shared resources/services. These contradictory behaviors are modeled through an agent-based model. The simulation results reveal that the social capabilities of peers impart a significant increase in the quality of interactions between social objects. Typically, the nature of underlying structure of network connectivity has a profound impact. The small-world network is not only closer to the social paradigm that surrounds us but also competes with an unbounded network, such as a mesh network.

4.1 Introduction

Computer networks offer amazing possibilities to exchange data and deal with files [1]. Old communication systems placed resources on a centrally managed server, which can then be accessed by client machines that connect to the central server (a "client–server" relationship). In contrast, peer-to-peer (P2P) technology allows client machines on a network to share their own resources (file store, processing power, and peripherals.) with other network connected machines with little or no involvement of any central server [2]. Thus, it enables machines to act as clients and servers at the same time. This powerful technology makes it possible for the contents to be distributed widely without the need for the central provision of large resources in terms of computing power, storage, or in particular, network bandwidth [3].

P2P technology enabled a revolution in machine-to-machine communication. While the machines are increasing exponentially, there is a desire to reduce their cost almost to nothing to enable communication between even very small and ordinary things around us. This endeavor has resulted in technologies and systems like wireless sensor network (WSN), Internet of Things (IoT), cyber-physical systems, and human-agent collectives (HAC) (sociotechnical systems) [4]. IoT has gained a tremendous interest not only from the researchers but also from the industry. Interest groups have been formed to define frameworks and standards for the IoT. Major IT companies have introduced a number of products and services based on IoT and invested substantial amount such as Nest by Google and SmartThings by Samsung [5]. According to [6], major takeover of Nest by Google for $3.2 billion followed by acquisitions of Dropcam by Nest. A number of leading ICT organizations have introduced IoT solutions such as Amazon Web Services, Ericsson, Huawei, IBM, and others.

According to the CISCO [7], 25 billion devices connected to the Internet in 2015; it will increase to 50 billion by 2020. Referring to Figure 4.1, Cisco estimated that IoT was born in between 2008 and 2009. Going back in past, in 2003, 6.3 billion people in the world and 500 million devices were connected; there were 0.08 device for every person even though smart devices such as smart phones were introduced in the market. According to [8] that by 2020, revenues for the IoT vendors could exceed $470 billion and IoT market size will grow to $3.7 billion; [9] predicted that 75.4 billion devices will be connect to the Internet in 2025, which is a huge number of devices.

The claim that IoT [10] is the future of computing [11] can definitely be debated. But no one can deny its profound impact on the society of the future. Although the technological advancements empowering and enhancing IoT infrastructure are cherished with enthusiasm and fervor on a daily basis, the social dimensions are often overlooked. This can raise many concerns in the society of the future. For example, the effect of blindly developing

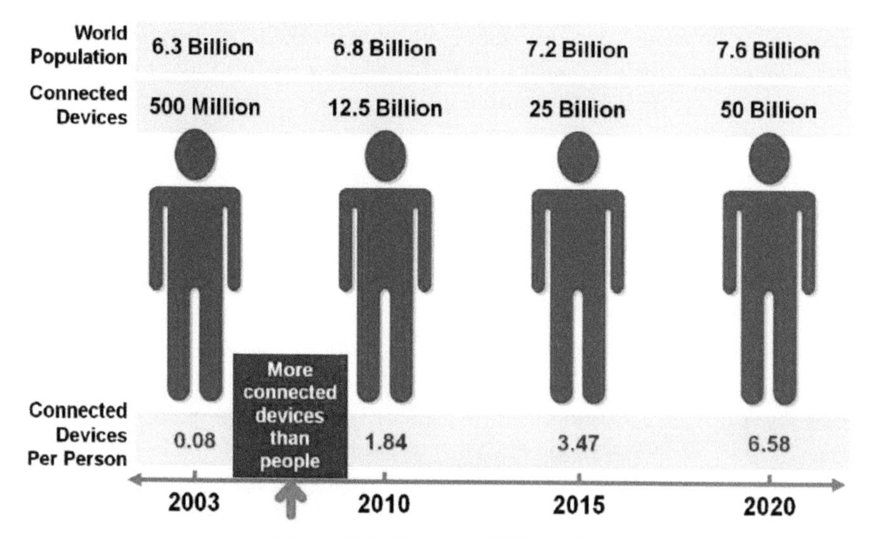

Figure 4.1 Internet of Things [7].

and adopting Internet-enabled mobile computing and social networks has raised many questions in societal context [12]. The notion of "social objects" provides opportunity to explore this dimension. Social IoT is an overlapping of IoT and HAC in which "coordinating human collectives" in addition to autonomous human agents [4] are enabled by autonomous interacting things.

The authors in [13, 14] investigate the impact of IoT technologies and application on human values. In particular, the importance of trust on technology is highlighted to improve the person-to-person communication using the IoT medium. Another model evaluating honesty is presented in [15, 16]. The overall context of users owning IoT devices is implemented as a collaboration module in [17]. An IoT-based study to verify users based on the pattern of their activities is presented in [18], thus emphasizing the importance of social aspects.

The relevance of imparting social capabilities in the objects of IoT is evidenced in the growth of IoT itself. The concept about things is now moving from smart objects to the acting objects. But, the future is of social objects. By "smart objects," researchers refer to objects capable of communicating with human social networks, whereas by "acting objects," an agent as a representative of a human being performs a pseudo-social behavior, influencing and being influenced by the environment. There are already a number of applications in the above two domains [19]. However, the real challenge is the futuristic "social objects," which would have the capability to build their own social network. If not carefully modeled, objects interacting, acting, and influencing their self-managed networks may turn out to be counter-productive or even harmful for the society [19]. The purpose of the model is to analyze the impact of social-like capabilities on the interaction pattern of social objects in a P2P resource sharing scenario. Adopting an agent-based modeling method and simulation, a comparative study of competitive vs. cooperative behavior of peers is demonstrated.

This chapter is organized as follows. In Section 4.2, we present IoT-related work, followed by motivation in Section 4.3. Section 4.4 presents the proposed model of P2P resource sharing, followed by simulation and results in Section 4.5. In Section 4.6, the authors conclude this chapter.

4.2 Related Work

Industry 4.0 [20] is a European initiative promising to transform industrial system of the future by integrating new technologies toward sustainability, efficiency, and safety. According to [21], collaboration issues are one of the most demanding aspects of this movement. Many social aspects are identified such as strategic decision-making, behavioral and trust modeling, collaborated group achievement and optimization, and evolving network dynamics. Without any doubt, IoT as the underlying network of Industry 4.0 needs to be *at least* socially aware thus acting as an enabler to address these challenges.

The study of the overall working mechanism of a social system *being competitive, cooperative, or mixed-type* has been a topic of interest [22]. Cooperation with peers is found in many natural systems [23]. Human societies, trained to be competitive [24], are learning to appreciate cooperation as the winning strategy [25]. Hence, in many research contributions, the conditions of transforming a population from competitive to cooperative mode are studied.

For example, the influence of an activity like migration for outbreak of cooperation is studied [26]. A simplistic interaction situation of iterated Prisoner's dilemma is implemented. However, the model is population based and does not focus on individual characteristics of interacting entities which is necessary to investigate emergence of cooperation. The specifics of formation of reciprocal appreciation in small groups are studied using an agent-based model [27]. Although the intuition of the model is social interaction based, it does not provide functional specification of activity performed through inter-agent interactions.

A recent work [28] presents a model of interaction in a P2P network using a multi-agent-based system. The authors presented a comparison between interaction efficiency whereby the model of interaction is influenced by social dimensions of competitiveness vs. cooperation. While sharing a common resource, according to the model, the probability of interaction increases with an increase in difference between the sizes of the agents. The size of an agent is directly proportional to the portion of common resource acquired by an agent. Hence, a peer will cooperate with another peer if one can offer and the other can receive. However, the model is proposed for a symmetric setting in terms of agents' function, services, and network configurations. Motivated from the above model, in this chapter, an agent-based model influenced by the principles of social interactions in an asymmetric setting is proposed.

One of the main aims of SIoT services to make autonomous decisions without or less human involvement is the vital factor to provide seamless connectivity and secure and reliable services [29]. Alongside security and privacy, reliability is a factor concerning proper function of the system to achieve dependability of the system for the delivery of secure and reliable services. Reliability concerns that the device can perform the desired job. In SIoT, security becomes a major challenge and it increases uncertainty among the users to consume on IoT services and applications [30, 31]. Trust management needs to be enhanced in order to overcome users' fear in SIoT usage and consumption. The concept of trust has been studied in numerous disciplines such as psychology, P2P network [32], WSNs [33], IoT [34], etc. A trust relationship involves two or more entities that rely on each other for a common benefit, i.e., devices/services being used. Trust management is different in SIoT from social networks due to the various requirements and constraints such as enormous number of entities/devices involved, highly dynamic because of large number of nodes joining and leaving the network at any point of time, energy consumption, etc. A trust can be classified based on the five design dimensions such as trust composition, trust propagation, trust aggregation, trust update, and trust formation [35]. Social trust, which comes under trust composition, derives from the social relationship among the owners of the IoT devices and it can be measured by intimacy, honesty, privacy, centrality, and connectivity. Social trust is prevailing in SIoT systems, where devices must be evaluated not only by QoS trust but also on social trust, i.e., devices' willingness to perform the requested service(s). In SIoT, it is recommended a distributed trust propagation scheme for the SIoT due to the distributed nature of IoT [36], i.e., each IoT device stores the trust values of other IoT devices [34].

4.3 Motivation

A guiding outline about the capabilities of a social object is presented in [19], with the following features:

1. Autonomous nature of social objects takes decentralized decisions using knowledge available at the local level.
2. Social objects broadcast their presence and availability to provide a specific service.
3. Inter-object interaction is supported and social objects trust their peers.

The model presented in this chapter is structured around these features. As featured above, the social objects must be autonomous and self-organized. Having an underpinning on the concept of agency [37], these must possess a "self-organizing collective behavior not resulting from the existence of a 'central controller,' but due to their own interactions with the other agents" [28]. The interactions require awareness of the proximity. Hence, the first two points are indigenous to the agent-based modeling paradigm. As of [19], P2P computing [2, 38] is taken into consideration as a typical application of such a setting, assuming that interacting peers have absolute trust on each other. Figure 4.2 shows different categories of objects such as smart, active, and social objects. In addition to the proposed model adhering to the notion of "social objects" [19], it supports asymmetric peers in terms of their capabilities and services configurations, and realization of a more realistic real-world networking. This particular feature is an extension of the research we have already reported [39, 40].

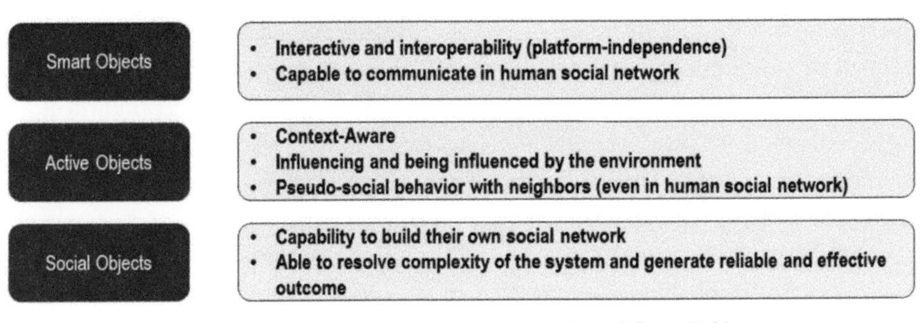

Figure 4.2 Categories of objects, adapted from [19].

4.4 The Proposed Model

The model adheres to the specifications of social objects. Each peer/agent acts as a decision-making entity using only the knowledge it possesses without the assistance of a central coordinator. Agents are not mobile, but the nature of interaction is dynamic due to situation-based peer selection. All agents are available and their operations/statuses are transparent to others. For simplicity, it is assumed that all agents in a specific range (neighborhood) are equally accessible and trust each other unconditionally.

The models are tightly entangled with services specifications and agents' basic interactions' capabilities. A detailed description of these specifications is given in Section 4.4.1. The agent-based models of competition and cooperation follow in Sections 4.4.2 and 4.4.3.

The services provided and requested by the peers are disparate, so is the network configuration. The configurations supported by the network are of different types. By default, the peers can communicate with the peers in the neighborhood. However, a "small world network" configuration [41], in which a few long distance connections augment an otherwise "regular network," is used as a realistic social possibility. Hence, the services available from distant peers are supported through a few long distant peers.

4.4.1 P2P Resource Sharing Specifications

It is assumed that four services are provided by the network. Each service requires a time effort to complete it. The first service, *Serv*1, requires an effort equal to 25 service completion units (SCU), *Serv*2 requires 50 SCU, *Serv*3 requires 75 SCU, and *Serv*4 requires 100 SCU. The fifth service, *Serv*0, is a different service, requiring no time resources (0 SCU). It indicates that an agent is serving other agents rather than being served for one of the four services mentioned above. The timing in the model is a hypothetical time unit (TU) which corresponds to SCU. Roughly, it is assumed that a TU maps to a minute in reality with an initial value, TU = 0, which increases with progression of the simulation (in timestamps), and reaches to 1440 TU (a day consisting of 24 h). After the completion of one day, the second day starts, and so on.

The peer specifications dependent on TUs are given in Table 4.1. The peer specifications related to services are given in Table 4.2. The peer specifications related to operations are given in Table 4.3. Based on these specifications, an agent would either perform its action in competitive or

Table 4.1 TU-based peer specifications

Time Unit	Specifications
up-duration	The duration for which a peer remains available on the network every day (a random value from the series 0, 120, 240, . . . , 1440 is assigned).
up-time	The time at which a peer becomes part of the network every day (a random value from the series 0, 60, 120, . . . , 1380 is assigned). An extreme case would be when a peer's *up-duration* is 1440 and the peer remains *up* for the full day.
down-time	The time at which a peer becomes unavailable on the network every day. This is equal to *up-time+up-duration*.

Table 4.2 Services-based peer specifications

Services	Specifications
service0-perc	The percentage of time a peer is willing to *give* to other peers. A random value from the series 0, 25, 50, . . . , 100 is assigned to the peer. However, being a 100% *giver* does not mean a fully devoted peer for others. Rather this is the value of willingness of a peer which relates to being fair and not to be a free rider.
current-service	The service currently being executed by the peer; one out of *Serv*0, *Serv*1, *Serv*2, *Serv*3, and *Serv*4. A peer after completing a service can decide to request for a new service after an *idle-time*.
idle-time	The duration of time for which a peer would remain idle (not executing any service from service 1 to 4). This would be a random value between 0 and 60 TUs and weighted with *service0-perc*. The idle time is applicable uniformly between two service requests or right after *up-time*.
units-completed	The TUs of the *current-service* which have been completed. If the *units-completed* are equal to SCU of the *current-service*, the service is completed and the peer will enter into the idle state.

cooperative mode. There are six states of an agent on which the model of competitive and cooperative behavior is based, as stated in Table 4.4. The *duration-incurrent-status* would track for how long a peer is in a specified status and helps in transiting from one state to another if a certain condition is met.

4.4.2 Agent-Based Model of Peers in Competitive Mode

Figure 4.3 represents the model of the peers in competitive mode.

- From being in the "off" state, an agent would transit into the "idle" state, if the current time (CT) of simulation is equal to *up-time* of the agent. Otherwise, the *duration-in-current-status* would be incremented.

Table 4.3 Operations-based peer specifications

Operations	Specifications
consistency	Represents how consistent the peer is in terms of continuity of the connectivity. The possibility of connectivity being disrupted is captured through this parameter. A random value between the specified ranges is assigned to a peer.
recent-services-completed	A list of services completed by a peer in the recent past indicated by last *l* services.
possible-service-providers	A list of service providers available for *current-service* being executed. These are "eligible" peers which have recently completed the desired service.
corresponding-peer (CP)	The peer chosen by a peer from the possible list of service providers.
Status	The state in which a peer is in [out of possible six or seven states depending on the mode of interaction (competing or cooperating) being used].
duration-in-current-status	The duration for which a peer has been in a current state.

Table 4.4 Agent states

Status	Conditions
status 0	When an agent is not on the network and it is physically in the "off" state
status 1	When an agent is in the "idle" state representing no service required or delivered
status 2	When an agent in the "assign" state that is ready to go from an idle to active state
status 3	When an agent is in the "search" state that is searching for *possible-service-providers* for the execution of the required service
status 4	When an agent is in the "request" state that is requesting for resources from the searched peer
status 5	When an agent is in the "proceed" state that is having completed the current service and is ready to execute another process cycle
status 6	When an agent would be in the "serve" state that is just serving another agent while being idle for *x* TUs, where *x* is equal to *serviceOperc* times the original *idle-time*.

- If a peer is in the "idle" state, it would transit to the "assign" state, if the duration it had to be in the idle state is complete. Otherwise, the *duration-in-current-status* would be incremented.
- In the "assign" state, a random service is assigned to the agent and it transits to the "search" state.

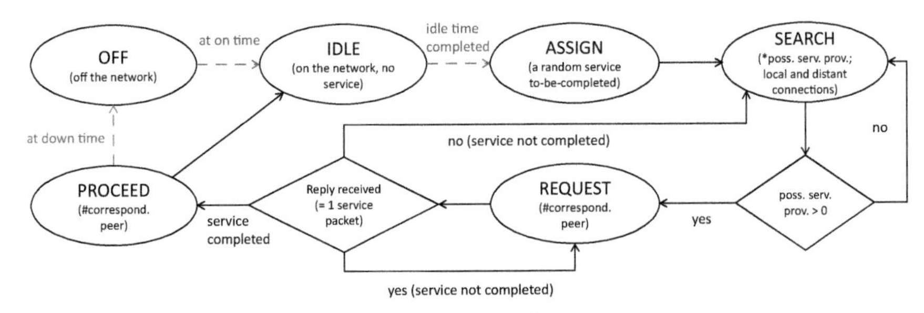

Figure 4.3 Agent-based model of peers in competitive mode.

*poss. serv. prov = possible service providers (peers completing the desired service recently).
#correspond. peer = a provider in IDLE state & consistent (not in OFF state often).

- In the "search" state, an agent searches for *possible-service-providers*, both in neighborhood as well at a distance. These are agents who have already completed the desired service recently (after last status 0 to status 1 transition). If the search is successful, the agent transits to the "request" state. Otherwise, the *duration-in-current-status* would be incremented.
- In the "request" state, the chosen *corresponding peer* would reply against the request if its state is idle and the value of *consistency* allows it. If such a reply is received, it is considered that one TU of the current requested service has been completed. Otherwise, the agent transits back to the "search" state. If *units-completed* are greater than or equal to SCU of the current service, the service is completed. The agent would record the service competed in *recent-services-completed* structure and would transit to the "proceed" state.
- In the "proceed" state, a peer either transits back to state "off" or state "idle" depending on its *down-time*. All service related parameters are also reset.

4.4.3 Agent-Based Model of Peers in Cooperative Mode

When compared with competitive mode, the major refinement occurs in status 3, and a new state is introduced, namely, "serve" state with status = 6. Figure 4.4 shows peers in cooperative mode.

- In the "search" state, an agent searches for *possible-service-providers*, both in neighborhood and at a distance. If there are some possible service providers, and one of them is in status 1 or 6, the agent can readily start utilizing the service by entering into status 4. If there is no service provider available, the peer will keep searching. In case of two or

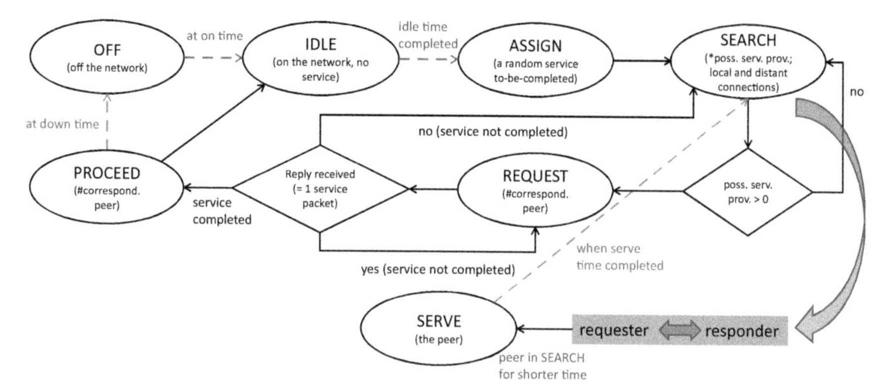

Figure 4.4 Agent-based model of peers in cooperative mode.

*poss. serv. prov = possible service providers (peers completing the desired service recently).
#correspond. peer = a provider in IDLE or SERVE state & consistent.

more peers searching for each other, the conflict resolution mechanism would be invoked. The conflict would be resolved in favor of one of the interacting peers based on the *duration-in-current-status*. If the *duration-incurrent-status* of the requester is greater than the *duration-in-current-status* of the respondent, the status of the requester would be set to 6, which is the requester would transit into the *serve* state. The respondent status would be set to 4, and thus ready to utilize the service provider by the requester. An entirely opposite procedure would be followed if the *duration-incurrent-status* condition is false.

- In the “serve” state, a peer would just serve the other peer being idle for x TUs, where x is equal to *service0perc* multiplied with the original *idle-time*.

4.5 Simulation and Results

4.5.1 Simulation Setup

Simulation is implemented in NetLogo, an open-source agent-based modeling simulator [42]. Two scenarios are generated: scenario 1 simulated all agents/peers in the competitive model, whereas scenario 2 simulated all agents/peers in the cooperative model. There are three population sizes: 100 agents, 250 agents, and 500 agents. All agents are static and do not move. Each of the above situations (of total 6) is simulated in four interaction modes: (i) a mesh network, in which an agent interacts with any other agent (neighbor or distant), (ii) a regular network, in which an agent only interacts with its direct neighbors, (iii) a small world network with a beta value of 0.1

Table 4.5 Simulation setup

Case	Scenario	Population	Network Type
Case 1	Competitive	100	Mesh
Case 2	Competitive	100	Regular
Case 3	Competitive	100	Small world (beta = 0.1)
Case 4	Competitive	100	Small world (beta = 0.2)
Case 5	Competitive	250	Mesh
Case 6	Competitive	250	Regular
Case 7	Competitive	250	Small world (beta = 0.1)
Case 8	Competitive	250	Small world (beta = 0.2)
Case 9	Competitive	500	Mesh
Case 10	Competitive	500	Regular
Case 11	Competitive	500	Small world (beta = 0.1)
Case 12	Competitive	500	Small world (beta = 0.2)
Case 13	Cooperative	100	Mesh
Case 14	Cooperative	100	Regular
Case 15	Cooperative	100	Small world (beta = 0.1)
Case 16	Cooperative	100	Small world (beta = 0.2)
Case 17	Cooperative	250	Mesh
Case 18	Cooperative	250	Regular
Case 19	Cooperative	250	Small world (beta = 0.1)
Case 20	Cooperative	250	Small world (beta = 0.2)
Case 21	Cooperative	500	Mesh
Case 22	Cooperative	500	Regular
Case 23	Cooperative	500	Small world (beta = 0.1)
Case 24	Cooperative	500	Small world (beta = 0.2)

(only 10% connections are distant), and (iv) a small world network with a beta value of 0.2 (only 20% connections are distant). The neighborhood range is a static value equal to 5 cell radius. Table 4.5 enumerates these cases for further discussion.

There are four possible services (shared resources). A service is either denied or completed in each iteration (TU). For a service, after been assigned to an agent would either accumulate the denied or completed value based on what occurs in that TU. How efficiently a service is progressing is dependent on ratio of denials and completions. More formally:

1. SCU denials: The SCUs denied so far (an accumulated value at the system level).
2. SCU completed: The SCUs completed so far (an accumulated value at the system level).
3. SCU served: The serving efficiency at the system level, calculated as: %SCU served = SCU denials / (SCU denials + SCU completed).

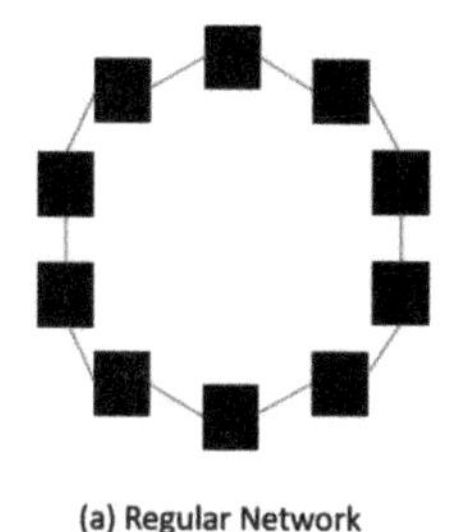

(a) Regular Network

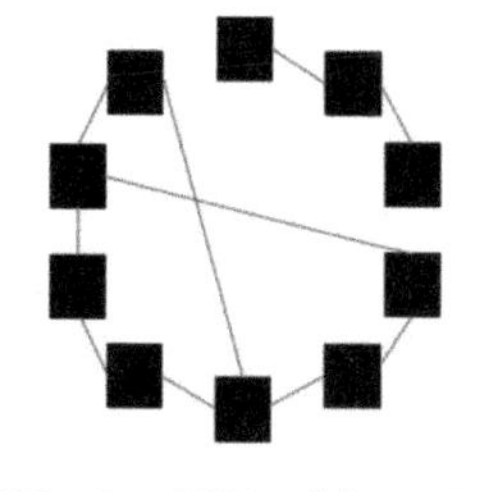

(b) Small-world Network (beta = 0.2)

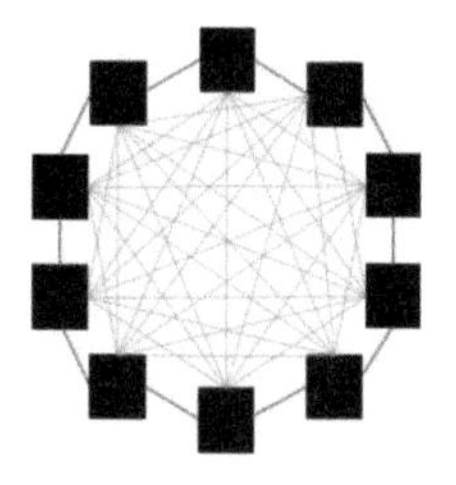

(c) Mesh Network

Figure 4.5 Network types.

Three network types are considered (shown in Figure 4.5). A regular network is a network in which all nodes are connected with each other with a regularity (e.g., for all nodes in a network, a neighborhood of equal sized radius). A regular network of degree 2 is shown in Figure 4.5(a). A regular network becomes a small network when few edges form long distant connections rather than regular connections. Figure 4.5(b) shows a small-world network of beta = 0.2 of the original regular network [Figure 4.5(a)]. Finally, a mesh network is a network in which any node in the network is connected to any other node [shown in Figure 4.5(c)]. Each case is run 100 times, each time with different random configurations. The results of 100 runs are averaged to present a normalized picture.

4.5.2 Simulation Results

The simulation starts at day 0, at midnight. The next midnight completes day 1 (1,440 min). The simulation ends after 10 days (14,400 min) or 14,400 iterations (timestamps). Figure 4.6 presents screenshots of simulation after 14 h and 15 min (850 iterations). Three cases are shown representing 250 agents in competitive mode: (a) mesh network, (b) regular network, and (c) small-world network. In a small-world network, with this density of agents, few distant contacts (shown with lengthy gray edges) are able to elevate % SCU completed (accumulated till this timestamp) to 0.121 from 0.088 (when compared with regular network having only neighborhood contacts). However, a mesh network outperforms with % SCU completed equal to 0.590.

Figure 4.7 also presents screenshots of simulation after 14 h and 15 min (850 iterations). Three cases are shown representing 250 agents in cooperative mode: (a) mesh network, (b) regular network, and (c) small-world network. Incidentally, at this point, the regular network and small-world

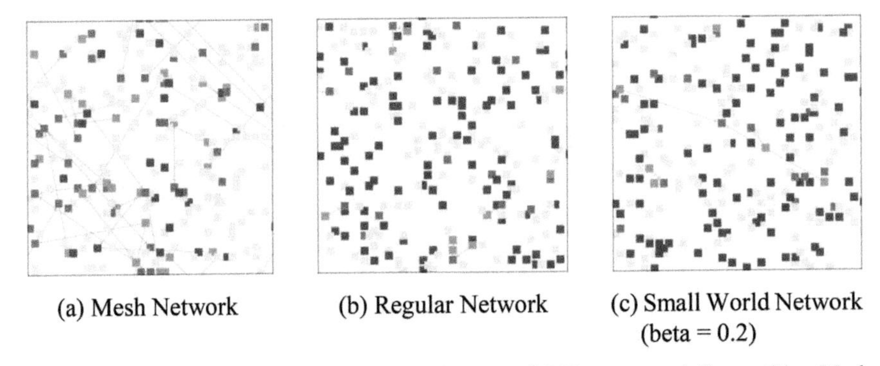

Figure 4.6 Status of peers at iteration 850 in case of 250 agents and Competitive Mode. Agents in light gray are in offline mode (status 0). Agents in dark gray are in idle mode (status 1). Agents in red are in search mode (status 3). Status 2 is a transit mode where a service is assigned. Agents in green are in request mode (status 4). Statuses 5 and 6 are transit states again. The arrow originates from the requester and ends at a receiver.

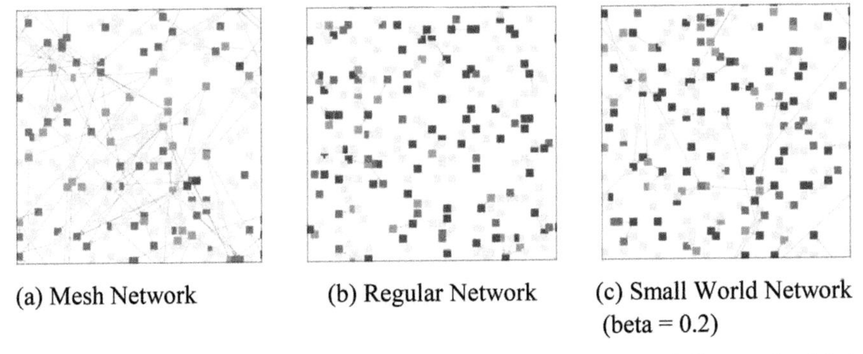

Figure 4.7 Status of peers at iteration 850 in case of 250 agents and Cooperative Mode. Agents in light gray are in offline mode (status 0). Agents in dark gray are in idle mode (status 1). Agents in red are in search mode (status 3). Status 2 is a transit mode where a service is assigned. Agents in green are in request mode (status 4). Statuses 5 and 6 are transit states again. The arrow originates from the requester and ends at a receiver.

network perform equally well (at about % SCU completed equal to 0.81). This value does not lag behind by much when compared with mesh network (% SCU completed equal to 0.859). However, accessing and maintaining a mesh network require a lot of resources when compared with a small-world network. A regular network requires the least resources. Although not evident in this situation, a regular network, alone, generally does not perform well. Hence, the real comparison is between the mesh and the small-world network.

Figure 4.8 shows graphs representing comparative analysis of three populations: 100 (a), 250 (b), and 500 (c) agents of scenario 1 (competitive). Each graph plots % SCU completed against simulation iteration, for four cases [mesh, regular, small-world (beta = 0.1), and small-world (beta = 0.2) network]. Similarly, Figure 4.9, shows graphs representing comparative analysis of three populations: 100 (a), 250 (b), and 500 (c) agents of scenario 2 (cooperative). Each graph plots % SCU completed against simulation iteration, for four cases [mesh, regular, small-world (beta = 0.1), and small-world (beta = 0.2) network].

The following conclusions are drawn from the simulation results:

1. The cooperation-based strategy in a small-world network matches with the cooperation-based strategy in a mesh network, typically when the population of agents is sufficiently large [see Figures 4.9(b) and (c)]. A visual analysis of it is possible if Figures 4.7(a) and (c) are compared with each other. In the mesh network, a large number of agents are in the request state, but apparently, most requests are denied. In the small-world network, the requests made have high probability of completion. Hence, overall, the effect is the same, but the mesh network imposes a lot of loads on the network in the form of requests which are denied.
2. In the competition-based strategy, the small-world network (with beta = 0.2) mostly outperforms the mesh network, typically when the population of agents is high. This is evident from Figure 4.8(c). A visual analysis of it is possible if Figures 4.6(a) and (c) are compared with each other. Similar to the above justification, it can be inferred that in the mesh network, a large number of requests are denied. This is not the case with the small-world network.
3. Overall, when comparing Figures 4.8 and 4.9, it is concluded that the cooperative strategy is much more efficient than the competitive strategy.

In summary, the simulation results reveal that a small-world network is not only quite efficient but also socially compatible. Further, with an increase in density of the objects, the beta value of the small-world network may be reduced while keeping the standards of service provisioning at the same level.

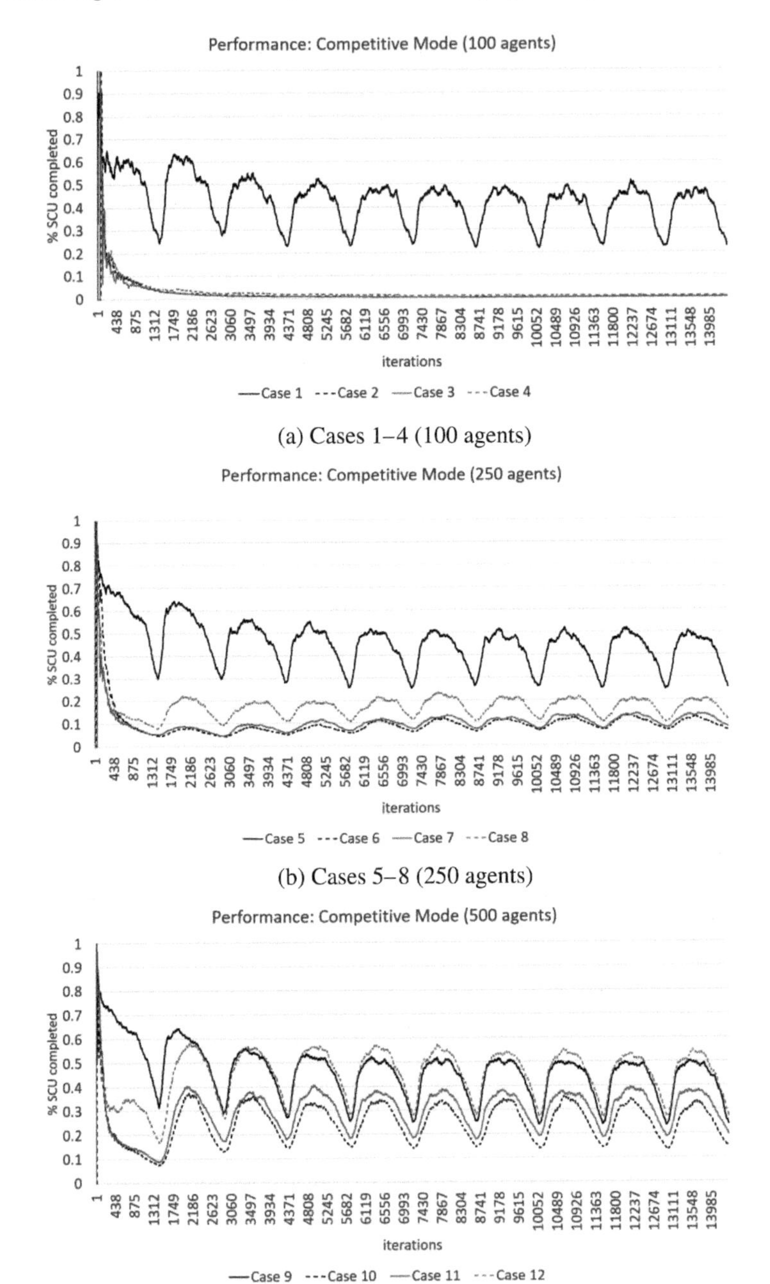

(a) Cases 1–4 (100 agents)

(b) Cases 5–8 (250 agents)

(c) Cases 9–12 (500 agents)

Figure 4.8 Comparative analysis of three sets of four cases in competitive mode.

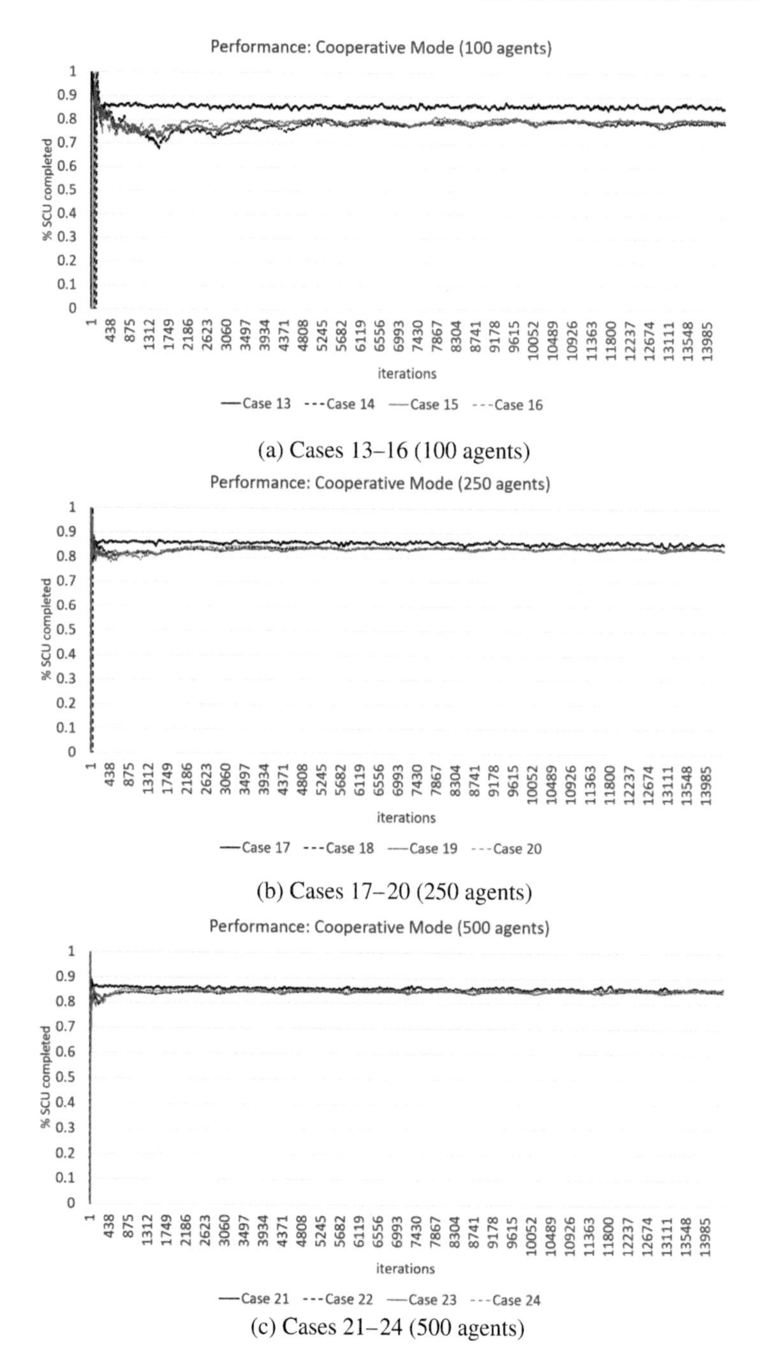

(a) Cases 13–16 (100 agents)

(b) Cases 17–20 (250 agents)

(c) Cases 21–24 (500 agents)

Figure 4.9 Comparative analysis of three sets of four cases in cooperative mode.

4.6 Conclusions

A P2P resource sharing scenario is taken up to analyze the potential of social web of things. While accepting competitive behavior as a default behavior to model a resource-sharing scenario, a model of cooperative behavior is also proposed. It is observed that cooperation is important for futuristic web of things constituted by social objects. Through an agent-based simulation study, it is proved that the cooperative strategy is more efficient than the competitive strategy. The simulation results show that peer communicating in a mesh network achieves the best results. However, accessing and maintaining a mesh network would require a lot of resources when compared with a regular or a small-world network. It is observed that a small-world network performs equally well with substantially less number of connections required. A natural extension of the model would be to use a trust model when making a choice between peers, resulting in favor of the more trustworthy peer based on the satisfaction history of the interaction.

References

[1] Fayez, G. (2015). *Analysis of Computer Networks*. Springer Publishing Company, Incorporated, 2nd edition.

[2] Adriana, I, Trunfio, P. Ledlie, L. and Schintke, F. (2010). "Peer-to-peer computing," in *European Conference on Parallel Processing*, 444–445. Springer.

[3] Nima, J. N., and Milani, F. S. (2015). A comprehensive study of the resource discovery techniques in peer-to-peer networks. *Peer-to-Peer Networking and Applications*, 8(3):474–492.

[4] Marina, P., Podobnik, V., and Jezic, G. (2016). Beyond the internet of things: The social networking of machines. *International Journal of Distributed Sensor Networks*, 12(6):1–15.

[5] Tibken, S. (2015). Samsung, smartthings and the open door to the smart home. *cnet CES*.

[6] Felix, W., and Fluchter, K. (2015). Internet of things. *Business and Information Systems Engineering*, 57(3):221–224.

[7] Dave, E. (2011). The internet of things: How the next evolution of the internet is changing everything. *CISCO White Paper*, 1(2011):1–11.

[8] Louis, C. (2015). Roundup of internet of things forecasts and market estimates. *Forbes*.

[9] Sam, L. (2016). Energy savings from nest. *IHS Technology White Paper*, 2016.

[10] Jayavardhana, G., Buyya, R., Marusic, S., and Palaniswami, M. (2013). Internet of things (iot): A vision, architectural elements, and future directions. *Future Generation Computer Systems*, 29(7):1645–1660.

[11] Hyuncheol, P., Kim, H., Joo, H., and Song, J. (2016). Recent advancements in the internet-of-things related standards: A onem2m perspective. *ICT Express*, 2(3):126–129.

[12] Sebastian, V., Park, N., and Kee, K. F. (2009). Is there social capital in a social network site? Facebook use and college students' life satisfaction, trust, and participation. *Journal of Computer-Mediated Communication*, 14(4):875–901.

[13] Maria, C., Ramani, R., Worthy, P., Weigel, J., Viller, S., and Matthews, B. (2016). Could the inherent nature of the internet of things inhibit person-to-person connection? in *Proceedings of the 2016 ACM Conference Companion Publication on Designing Interactive Systems*, 177–180, ACM.

[14] Peter, W., Matthews, B., and Viller, S. (2016). "Trust me: doubts and concerns living with the internet of things," in *Proceedings of the 2016 ACM Conference on Designing Interactive Systems*, 427–434. ACM.

[15] Katayoun, F., and Zia, K. (2017) Trust reality-mining: evidencing the role of friendship for trust diffusion. *Human-centric Computing and Information Sciences*, 7(1):1–16.

[16] Upul, J., and Lee, G. M. (2017). "A computational model to evaluate honesty in social internet of things," in *Proceedings of the Symposium on Applied Computing*, 1830–1835. ACM.

[17] Seung, W. K., Lim, T. B., and Jong Il Park. (2015). Design and implementation of iot collaboration module supporting user context management. *IEMEK Journal of Embedded Systems and Applications*, 10(3):129–137.

[18] Fazel, A., Aloqaily, M, Kantarci, B., Erol-Kantarci, M., and Schuckers, S. (2017). Social behaviometrics for personalized devices in the internet of things era. *IEEE Access*, 5:12199–12213.

[19] Luigi, A., Iera, A., and Morabito, G. (2014). From "smart objects" to "social objects": The next evolutionary step of the internet of things. *IEEE Communications Magazine*, 52(1):97–105.

[20] Christoph Jan, B. (2017). "The concept industry 4.0," in *The Concept Industry 4.0*, Springer, 27–50.

[21] Camarinha-Matos, L. M., Fornasiero, R., and Afsarmanesh, H. (2017) "Collaborative networks as a core enabler of industry 4.0.," in *Working Conference on Virtual Enterprises*, Springer, 3–17.

[22] Christoph, H. (2006). Cooperation, collectives formation and specialization. *Advances in Complex Systems*, 9(4):315–335.

[23] Marco, D., Birattari, M., and Stutzle, T. (2006). "Ant colony optimization," in *IEEE Computational Intelligence Magazine*, 1(4):28–39.

[24] Kucuksenel, S., (2012). A theory of fairness, competition and cooperation. *Journal of Public Economic Theory*, 14(5):767–789, 2012.

[25] Andre Barreira da Silva, R., and Laruelle, A. (2013). Evolution of cooperation in the snowdrift game with heterogeneous population. *Advances in Complex Systems*, 16(08):1350036.

[26] Frank, S., and Behera, L. (2012). Optimal migration promotes the outbreak of cooperation in heterogeneous populations. *Advances in Complex Systems*, 15(supp01):1250059.

[27] Koponen, I. T., and Nousiainen, M. (2016). Formation of reciprocal appreciation patterns in small groups: an agent-based model. *Complex Adaptive Systems Modeling*, 4(1):24.

[28] Leonidas, F. C., Caiafa, C. F., Ausloos, M., and Proto, A. N. (2015). Cooperative peer-to-peer multiagent-based systems. *Physical Review E*, 92(2):022805.

[29] Truong, N. B., Lee, H. Askwith, B., and Lee, G. M. (2017). Toward a trust evaluation mechanism in the social internet of things. *Sensors*, 17(6):1346.

[30] Wafa, A., Amel Zayani, C., Amous, I., and Sedes, F. (2016). "Trust management in social internet of things: a survey," in *Conference on e-Business, e-Services and e-Society*, Springer, 430–441.

[31] Kowshalya, A. M., and Valarmathi, M. L. (2017). Trust management for reliable decision making among social objects in the social internet of things. *IET Networks*, 6(4):75–80.

[32] Arshad, M., Arabo, A., Merabti, M., Shi, Q., and Askwith, B. (2010). "A secure gateway service for accessing networked appliances," in *5th International Conference on Systems and Networks Communications (ICSNC)*, 183–188. IEEE.

[33] Grover, J., and Sharma, S. (2016). "Security issues in wireless sensor network - a review," in *5th International Conference on Reliability, Infocom Technologies and Optimization (Trends and Future Directions) (ICRITO)*, 397–404. IEEE.

[34] Michele, N., Girau, R., and Atzori, L., (2014). "Trustworthiness management in the social internet of things," *IEEE Transactions on knowledge and data engineering*, 26(5):1253–1266.

[35] Jia, G., and Chen, R. (2015). "A classification of trust computation models for service-oriented internet of things systems," in *12th IEEE International Conference on Services Computing (SCC)*, 324–331. IEEE.

[36] Ray, C., Guo, J., and Bao, F. (2016). "Trust management for soa-based iot and its application to service composition," in *IEEE Transactions on Services Computing*, 9(3):482–495.

[37] Dirk, H. (2012). Agent-based modeling. *Social Self-Organization*, 25–70.

[38] Arshad, M., Fergus, P., Merabti, M., and Askwith, B. (2010). "Peer-to-peer overlay gateway services for home automation and management," in *Advanced Information Networking and Applications Workshops (WAINA), 2010 IEEE 24th International Conference on*, 880–886. IEEE.

[39] Kashif, Z., Muhammad, A., and Saini, D. K. (2017). "Socially aware peers for futuristic web of things," in *9th IEEE International Conference on Intelligent Data Acquisition and Advanced Computing Systems: Technology and Applications (IDAACS)*, 2, 579–583. IEEE.

[40] Kashif, Z., Saini, D. K., Farooq, U., and Ferscha, F. (2017). "Web of social things: Socially-influenced interaction modeling," in *Proceedings of the15th International Conference on Advances in Mobile Computing and Multimedia, MoMM2017*, New York, NY, 123–130. ACM.

[41] Carsten, G., Grosskinsky, S., Kurths, J., and Timme, M. (2015). Collective relaxation dynamics of small-world networks. *Physical Review E*, 91(5):052815.

[42] Uri, W., and Rand, W. (2015). *An introduction to agent-based modeling: modeling natural, social, and engineered complex systems with NetLogo*. MIT Press.

5

Modeling and Availability Assessment of Mobile Healthcare IoT Using Tree Analysis and Queueing Theory

Anastasiia Strielkina, Dmytro Uzun, Vyacheslav Kharchenko and Artem Tetskyi

Department of Computer Systems, Networks and Cybersecurity, National Aerospace University "KhAI", Kharkiv, Ukraine
E-mail: a.strielkina@csn.khai.edu; d.uzun@csn.khai.edu; v.kharchenko@csn.khai.edu; a.tetskiy@csn.khai.edu

In the field of healthcare, networked healthcare devices are closely intertwined in the structure of the Internet of Things. Such extensive application of IoT gadgets and IoT applications offers many opportunities for monitoring patients, tracking their location and condition, obtaining, analyzing, and sharing patient health data.

A healthcare IoT infrastructure with a brief description of each component is presented. These components are a device with a reader, the Cloud, the healthcare provider, and a communication channel. Networked healthcare devices sense electrical, thermal, chemical, and other signals from the patient's body. They directly sense and collect biomedical signals, that is, information about the physical and mental states of health. Thus, it is clear that such devices are safety critical because a human's life depends on its performance. The application of failure/attack trees to identify security problems of the IoT infrastructure is considered. An abstract example of a fault tree for the IoT system is given. The paper presents simple cases with a few models of healthcare IoT system based on the queueing theory. The models describe streams of the requests and attacks on vulnerabilities and procedure of recovery by restart and eliminating of one and/or two vulnerabilities. In addition, using the submitted models, it is possible to calculate an availability function.

5.1 Introduction

5.1.1 Motivation

The Internet of Things (IoT) is a new step in technological progress. The IoT allows people and "things" to connect anytime, anywhere using a variety of communication networks. According to the preliminary forecasts [1,2], about 50 billion devices will be connected to the Internet and the IoT market will reach about $1.7 trillion by 2020.

The IoT represents new, exciting opportunities for almost every area of our life. Moreover, of course, a healthcare is not an exception. The IoT can significantly improve the existing healthcare system. Modern healthcare has risen to an unattainable level earlier over the past decade. Today, the healthcare sector is a high-tech industry, where all areas of healthcare are successfully developing that can save lives of previously hopeless patients. The technical equipment of healthcare institutions has significantly been improved; it has become possible to diagnose the disease at the earliest stage and to quickly restore the working capacity of patients.

According to the forecasts of the researchers [3], the market of healthcare IoT gadgets and IoT applications will grow to $136.8 billion in 2021. The IoT in healthcare provides opportunities:

- To obtain, analyze, and share patients' health data;
- For personalized treatment that can be given at a more detailed level (e.g., access to real-time information, predictive diagnostics);
- For interaction between individual devices with the entire healthcare system (e.g., management of patient experience while gaining cost efficiencies), etc.

The IoT will allow to monitor patients, track their location and condition (control of temperature, pressure, and other physical indicators), as well as monitor the healthcare institution itself, its internal microclimate.

One of the most widely used IoT devices is an insulin pump. According to the statistics of the World Healthcare Organization presented in 2016 [4], about 8.5% of the world population had diabetes in 2014. The diabetes epidemic has serious health and socio-economic consequences. The prevalence of diabetes is increasing more rapidly in middle- and low-income countries.

This confirms the relevance of the applicability of such devices. The insulin pump can be placed in an inconspicuous place under the clothes, so a patient can carry out and control the injection of insulin with a special console or smartphone.

However, with all the advantages of using such devices, the security and safety risks are increasing. Thus, the assessment of such systems is a complex process. Specific features of such systems' availability assessment include dynamism, multi-componentness, and multi-levelness that lead to a large number of failure factors. In turn, these cause reduction of availability. Changing the parameters of failure flows caused by software faults and attacks on vulnerabilities leads to an increase in model complexity. One of the ways to reduce this complexity is the fragmentedness of the models being developed.

5.1.2 State of the Art

The mainstreaming of this paper is confirmed by existing standards and regulatory acts in the field of healthcare provision [5–7]. These documents provide different approaches for risk management to determine safety and security of healthcare devices.

In [8], the metric-based approach based on the analysis of Markov chain's stiffness, decomposability, sparsity, and fragmentedness was shown. According to that paper, the fragmentedness can be achieved dividing the model into N fragments (they have the same structure but different parameters). Such an approach may help to present the interconnection between different parts of a system.

An approach to assess security and availability of a building automation system using the attack tree analysis (for investigating any intrusion into the system) and Markov models (for calculating system availability) was presented in [9]. However, techniques of modeling and quantitative assessment of IoT health system availability considering cyberattacks on vulnerabilities, failures caused by hardware and software components, and recovery procedures are often not well known.

As a result of the foregoing, the task of such system availability modeling is in demand and relevant.

5.1.3 Aim and Objectives

In this paper, the authors presented the application of failure/attack trees to identify safety/security problems of the IoT infrastructure and a simple model using the queueing theory of the healthcare IoT system. There are many solutions that allow using smart sensors working with clouds, different types of devices. In this case, it is proposed to develop a complex system that allows taking into account the specificity of end user devices, communication channels, and technologies of data flows. With all the variety of existing techniques, mathematical models of such solutions are extremely rare in the literature. It is planned to achieve the result of the availability assessment by means of the results' implementation of the simulation on the developed models.

The remainder of this paper is structured as follows. Section 5.2 presents a healthcare IoT infrastructure with a brief description of each component. Section 5.3 presents fault tree analysis for failure occurrence nature of healthcare IoT infrastructure, a justification of applicability possibilities of the queueing theory, and presents a case study for considered system followed by concluding remarks.

5.2 Healthcare IoT Infrastructure

A typical healthcare IoT system generally consists of such components as depicted in Figure 5.1.

Thus, the constituent comVponents of the healthcare IoT infrastructure are: (1) Cloud, (2) a device (with a reader, also often called a "base station"), (3) healthcare providers, and (4) communication channels between the device and the Cloud, and the healthcare provider and the Cloud (Figure 5.2).

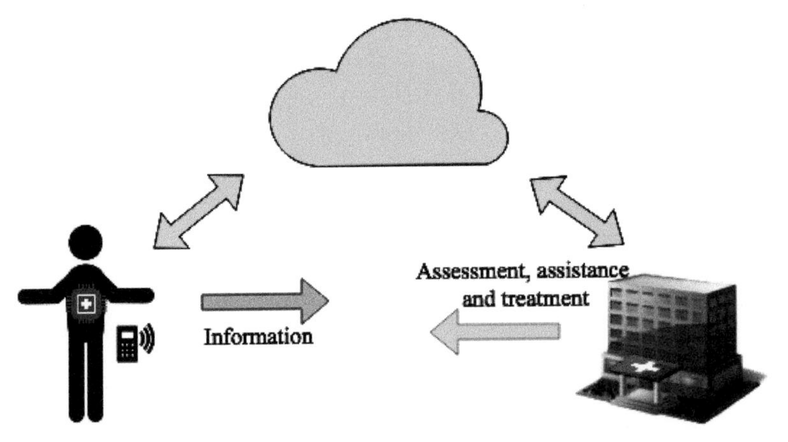

Figure 5.1 Main components of the healthcare IoT infrastructure: devices with a reader, the Cloud, and a healthcare organization.

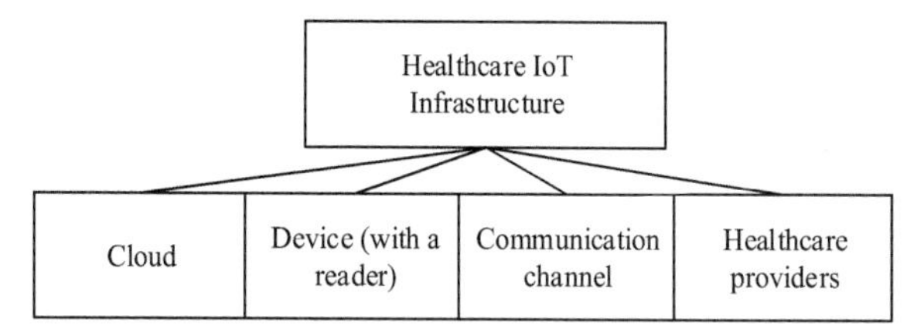

Figure 5.2 Detailing of components of the healthcare IoT infrastructure.

The number and variety of devices have increased significantly. According to [10], healthcare devices can be classified as:

- Consumer products for health monitoring;
- Wearable, external devices;
- Internally embedded devices;
- Stationary devices.

Such devices sense electrical, thermal, chemical, and other signals from the user's body. They directly sense and collect biomedical signals, that is, information about the physical and mental states of human's health.

Therefore, FDA provides risk-based classification [11, 12] of the considered devices:

- Class I – low risk – general controls (least amount of regulatory control). This class includes elastic bandages, crutches, thermometers, skull plate anvils, which are used for skull reconstruction, and the battery-powered artificial larynx.
- Class II – medium risk – general and special controls (assure safety and effectiveness). This class includes the tests for sickle cell disease, stationary and mobile X-ray systems, and pneumatic surgical drills, air handlers for surgical operating rooms, oxygenators used during cardiopulmonary bypass surgery, and dura substitute devices.
- Class III – high risk – general controls and premarket approval (support or sustain human life). This class includes the life-sustaining, life-supporting and/or devices of substantial importance in preventing impairment of human health.

It is understandable and logical that the insulin pump refers to the Class III because a human's life depends on its performance.

Main limitations of devices are small storage, limited energy, and computational capabilities. For communication with the Cloud, devices frequently use external devices that are referred as the readers (i.e., laptops, smartphones, etc.). Thus, the data can be sent to the Cloud directly or indirectly

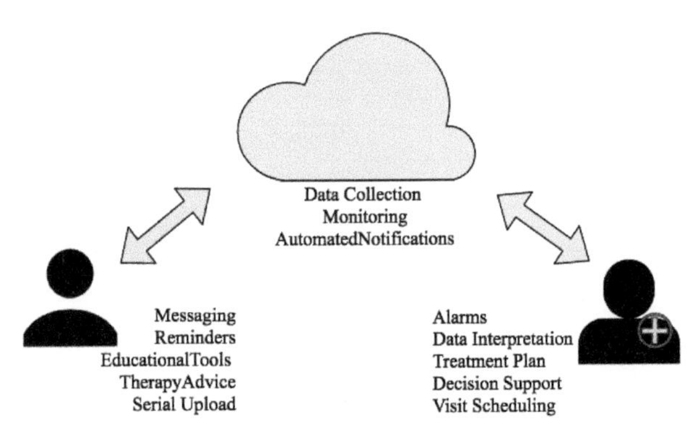

Figure 5.3 Communications and functions of the main components of the IoT infrastructure.

(through the reader). The reader is less severe because it has its own resource constraints. Many communication technologies are well known and used in the IoT such as WiFi, Bluetooth, ZigBee, and 2G/3G/4G cellular. Inasmuch as devices and readers are resource-constrained, received data from sensors are usually sent to the Cloud servers. These data are processing and storage for a long term there. Authorized users (healthcare providers) have access to these data at any time and in any place. Healthcare providers include healthcare professionals, organizations and their business associates, insurance companies, etc. As before, specialists – doctors and healthcare personnel – will occupy the central place in the treatment. The IoT will help them to work more effectively. Explanation of the functions, activities, and inter-relationships of the main components is depicted in Figure 5.3.

Thus, it is possible to describe the healthcare IoT system. The healthcare device communicates wirelessly with a reader. This reader through the access point loads the taken indications for service in the Cloud. Hospitals, emergency doctors, and health care providers can access this service. Detailed descriptions of the integration of data from the healthcare device to the Cloud are presented in [13, 14].

5.3 Applicable Approaches and Methods for Modeling and Simulation of Healthcare IoT

5.3.1 Fault Tree Analysis for Failure Occurrence Nature of Healthcare IoT

To analyze the security and/or safety problems of the IoT system, a fault or attack tree analysis method can be applied. Guidance for Industry and FDA

Staff [5, 6] mentions the use of this method when creating healthcare devices to identify and classify hazards. The faults/attacks trees are a flexible tool that is used to model various undesirable events in different spheres of human activities. For example, in [15], trees were used to analyze hacked email victimization scenarios. In [16], FTA was used to calculate the medication error. The scope of failure trees is not limited to information technology; this method is also used in such areas as the aviation industry, the military industry, the nuclear industry, etc. The fault tree aggregates the possible ways of achieving the main event (component failure, subsystem failure, and successful attack). To build the fault tree, it is necessary to analyze possible attacks on the IoT infrastructure or its individual components.

In the analysis of security, the cyber assets of the system should be identified – this is something that has value and importance for the system owner or for the intruder who can attack the system. Understanding what consequences of the failure or the successful attack can be allows a sober assessment of the criticality of potential security problems. Since the IoT infrastructure consists of several components and communication channels, it may be advisable to consider the security problems of each component and communication channels separately. In this case, several trees will be built as a decomposition of one large tree.

Consider an abstract example of a fault tree for the IoT system. As mentioned above, first we need to identify the main event. Examples of such an event may be the fault of a certain sensor, violation of the integrity, or confidentiality of the transmitted information. For healthcare systems, the issues of violation of the information privacy can be especially relevant. For critical decision-making systems, the main event can be the damage to the patient's health.

After determining the main event, it is necessary to determine the most frequent options for achieving this event. Assume that the failure of at least one of the components – cloud storage, information gathering device, communication channel, or healthcare provider – can lead to the occurrence of the main event "IoT infrastructure failure." The failure of the cloud storage can occur if all previous events $A_1, \ldots, A_k$ occur. The failure of the information-gathering device may arise under the condition that all previous events $B_1, \ldots, B_l$ occur. The failure of the communication channel can occur if all previous events $C_1, \ldots, C_m$ occur. The failure of a healthcare provider can arise under the condition that all previous events $D_1, \ldots, D_n$ occur. In this case, k, l, m, and n are the numbers of events preceding the corresponding event. The described tree structure is shown in Figure 5.4.

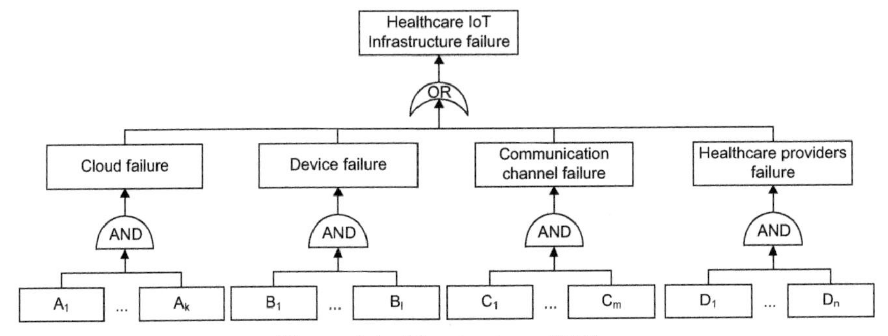

Figure 5.4 The structure of FTA.

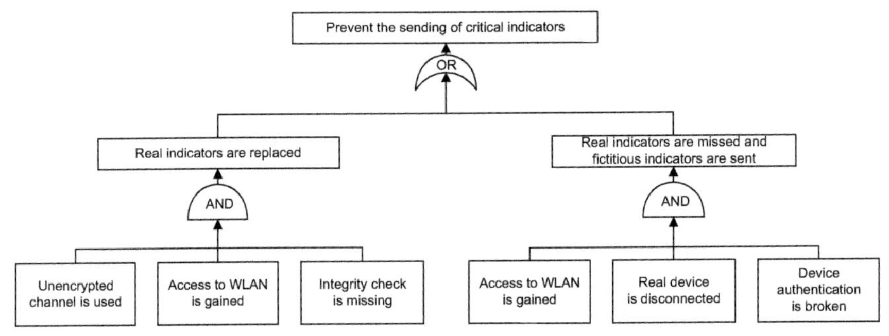

Figure 5.5 The tree fragment of possible scenarios in which critical values of the indicators may not be transmitted to the Cloud.

Let us consider the fragment of the fault tree in the subsystem of the device worn by the patient. Calculation of the amount of insulin administered occurs on the local device, and the current blood sugar level is transferred to the Cloud for monitoring by a healthcare professional. The critical values of the blood sugar level can be obtained if the measurement is incorrect or when incorrect insulin doses are administered. Information about such an event should be noted in the controlling healthcare organization. The fault tree fragment shown in Figure 5.5 shows possible scenarios in which critical values of the indicators may not be transmitted to the Cloud.

Using this model probability of system failure (prevent the sending of critical indicators), P_{sf} can be calculated using the following formula (5.1)

$$P_{sf} = P_1 P_2 P_3 + P_1 P_4 P_5 \tag{5.1}$$

where P_1–P_5 are probabilities of the following events correspondingly: access to WLAN is gained; unencrypted channel is used; integrity check is missed; real device is disconnected; and device identification is broken.

It is assumed that sending prevention of the critical values of the indicators in the cloud can be achieved in two ways. The first method is shown on the left side of the tree; in this case, the sensor data have been replaced due to transmission over the unencrypted channel and lack of integrity check. The second method is shown on the right side of the tree; it replaces the actual device with a dummy device, which sends to the cloud indicators that are certainly not critical.

In practice, attack or fault trees have a more complex structure, combining various options for using the elements AND/OR. The structure depends on the most likely scenarios of attacks considered in the process of analyzing the security problems of the healthcare IoT system. In this case, different scenarios can be complex and have intersections in the form of identical events.

To calculate the probability of the main event occurrence, it is necessary to parameterize the constructed tree. By parameterization we mean the definition of the probability of occurrence of events at the lower level. Data for parameterization can be obtained by collecting statistics of attacks and component failures. In the absence of statistics and, as a result, the impossibility of determining quantitative indicators, it is possible to use the method of expert assessments and the use of variables of fuzzy logic (for example, low, medium, and high probability).

The advantages of this method are universality, flexibility, visibility, and the availability of tools that allow modeling. Using the considered trees allows to graphically represent possible scenarios for reaching the main event, assess the likelihood of the main event, and identify the weaknesses in the security or safety of the infrastructure. In addition, the fault tree is convenient to use when choosing countermeasures to counter possible failures and attacks.

Trees are a universal method that can be applied to both security and reliability assessment. In our case, failures in the system can arise due to external actions (attacks), or without them (failures caused by processes occurring within the components of the system). The purpose of the tree is to determine the probability of the main event. The tree in 5.5 can be viewed as part of the tree in 5.4, which is used to evaluate the reliability of the system.

The disadvantage of fault trees is the impossibility of determining all possible scenarios for the occurrence of the main event. There is also the problem of detailing each scenario, since excessive detail can reduce the visibility of the overall security state of the infrastructure in question. Besides, in case of implementing maintenance procedures, the development of trees becomes very complex and their parameters such as recovery rates or middle time of repair cannot be taken into account as a whole.

5.3.2 Justification of Applicability of the Queueing Theory

In this paper, the queueing systems are those systems at which random service requests from IoT device (customer) are received at random times, while incoming requests are serviced by means of the available service channels [17,18]. Under the flow of service is understood the flow of requests, serviced one after another by one continuously occupied channel (for example, Cloud). Figure 5.6 shows the queueing theory model for the considered case of the healthcare IoT system.

When considering a single serviced device, the proceeding processes can be represented by a Markov stochastic process (chain) with discrete states and discrete time. However, when considering the IoT infrastructure (due to the huge number of devices and the difference in their characteristics), it should be noted that in this case, all flows are simplest, and the process occurring in the IoT system is a Markov stochastic process (chain) with discrete states and continuous time. The assumptions of the Markov stochastic process (chain) using are that the intensity of failures λ and repairs μ is assumed to be constant, events are independent of each other, and the probability of occurrence of two or more events during a short time interval is much less than the probability of occurrence of one event during the same time period.

Obviously, there is a stationary state in this process. It is not necessary to formulate the Kolmogorov equation since the structure is regular; the necessary formulas are given in the reference books [17, 18].

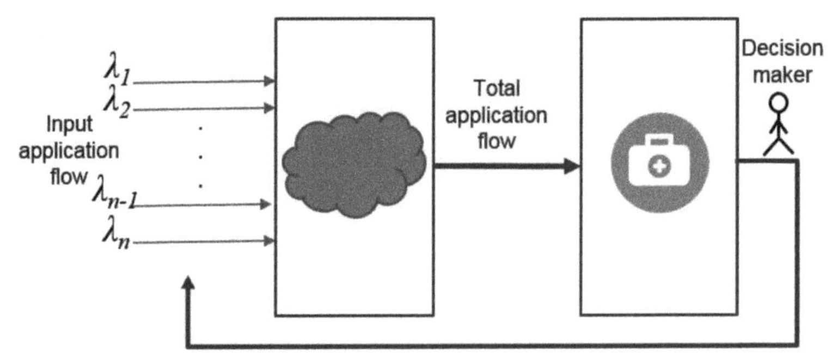

Figure 5.6 The queueing system for the healthcare IoT system.

5.4 Case Study: Modeling of Healthcare IoT Using Queueing Theory

5.4.1 Initial Model "Birth–Death"

Let us describe a hypothetical situation. The hospital, which is part of the IoT infrastructure, receives three service requests from users of the networked insulin pump (for example, results of sugar level planned measurement, request an appointment, and statistics on the number of injections). The flow of all requests is the simplest. The average time of receiving a new request from an insulin pump user is 30 min (determined by subjective characteristics), t. When the request is received, the healthcare staff begins to process it. The processing time for one request is distributed according to the exponential law and on average is 10 min, t_{pr}. At the initial time, there are no requests in the system. It is required to determine the reliability characteristics of this system [19, 20].

In this way, there are four possible states in this system:

- S_0 – There are no service requests in the system.
- S_1 – There is one service request in the system.
- S_2 – There are two service requests in the system.
- S_3 – There are three service requests in the system.

We will assume that the processes of receipt and processing of requests are homogeneous Markov; simultaneous receipt of several service requests and simultaneous processing are practically impossible. Since all applications are equivalent, from the point of view of the reliability, it does not matter which service request is in the state S_3, it is important that only one.

With this in mind, the situation is modeled by the "birth–death" process (Figure 5.7).

According to Figure 5.7, λ_{01}, λ_{12}, and λ_{23} are the intensities of service requests flow, and μ_{10}, μ_{21}, and μ_{32} are the intensities of processing flow.

The intensity of receipt of one service request is equal to $\lambda = 1/t$, and the intensity of processing of one request is equal to $\mu_{10} = 1/t_{pr}$.

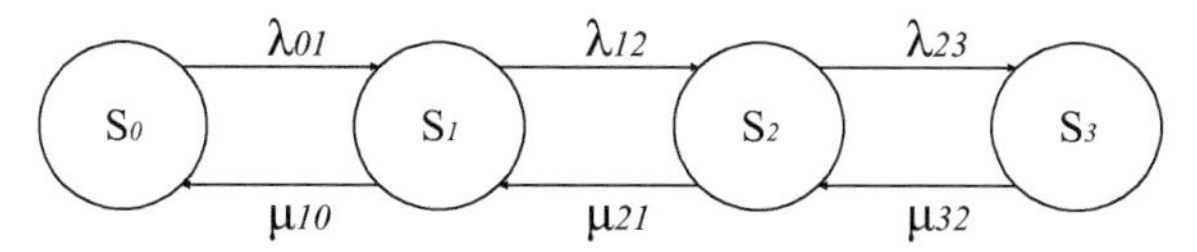

Figure 5.7 A scheme of "birth–death" process for the considered case.

There are no requests in the state S_0; consequently $\lambda_{01} = 3\lambda$; in the state S_1, one request was received – $\lambda_{12} = 2\lambda$, and in the state S_2, two requests were received – $\lambda_{23} = \lambda$. In the state S_3, two requests are processing, so $\mu_{10} = 3\mu$, for the state S_2 $\mu_{21} = 2\mu$, and for S_1 $\mu_{32} = \mu$.

According to [11], the probability of a state when there are no requests in the system:

$$P_0 = \frac{1}{1 + \frac{\lambda_{01}}{\mu_{10}} + \frac{\lambda_{01}\lambda_{12}}{\mu_{10}\mu_{21}} + \frac{\lambda_{01}\lambda_{12}\lambda_{23}}{\mu_{10}\mu_{21}\mu_{32}}} \approx 0.4219.$$

Similarly, the remaining probabilities are calculated, which are equal to P_1=0.4219, P_2=0.14, and P_3=0.016.

5.4.2 The Model Considering Attacks on Vulnerabilities

In case of a successful cyberattack on a single vulnerability in the healthcare IoT system, the developed "birth–death" model is modified as follows, as shown in Figure 5.8.

Let us suppose that the successful attack on the healthcare IoT system occurs with a probability of 20% (every fifth attack is successful), and μ_r=0.9. Let us suppose that at the initial instant of time with a probability 100% is in state S_0. Figure 5.9 shows the convergence on the stationary distribution (iterating for 50 steps). The model was simulated using R package "markovchain" [21].

If the healthcare IoT system continues to perform its functions without halts during the attacks, in this case, an appropriate Markov model is shown in Figure 5.10. Figure 5.11 shows the convergence on the stationary distribution.

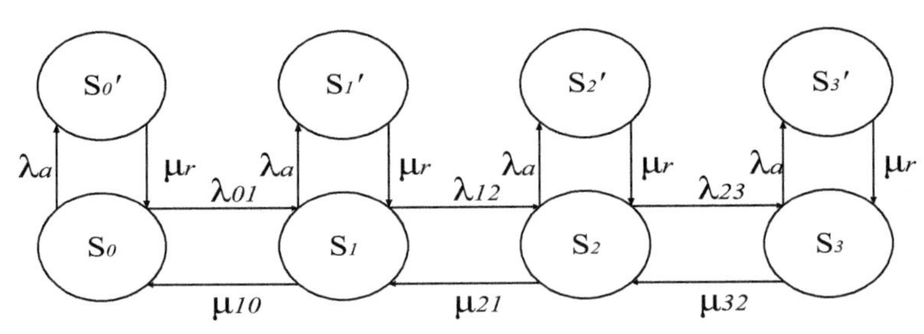

Figure 5.8 A Markov model for the considered case of a successful attack at one stage of the service request processing (with a halt) case.

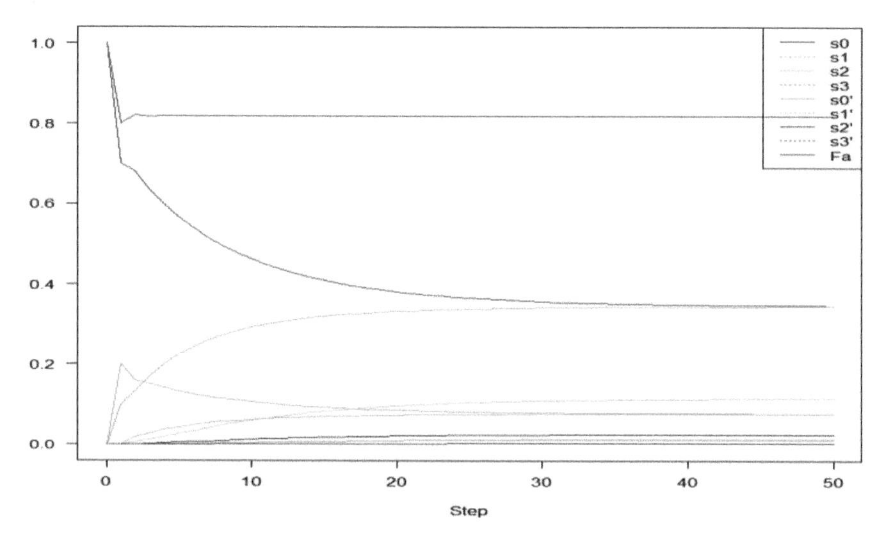

Figure 5.9 The stationary distribution for the considered case of a successful attack at one stage of the service request processing (with a halt).

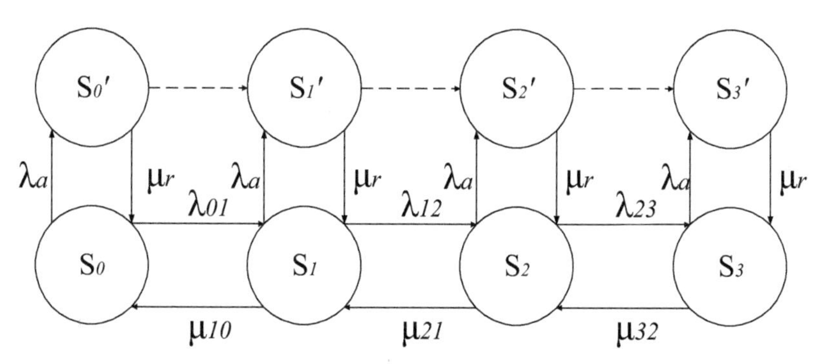

Figure 5.10 A Markov model for the considered case of a successful attack at one stage of the service request processing (without halts).

5.4.3 The Model Considering Elimination of Vulnerabilities

In the cases discussed above, the vulnerabilities are not eliminated, and the system just restarts and continues to function in the same way. Figure 5.12 illustrates a case that when the healthcare IoT system has one vulnerability is eliminated (with $\mu_r' = K_\mu \cdot \mu_r$, $K_\mu < 1$, $t_r' > t_r$). Figure 5.13 shows the convergence on the stationary distribution.

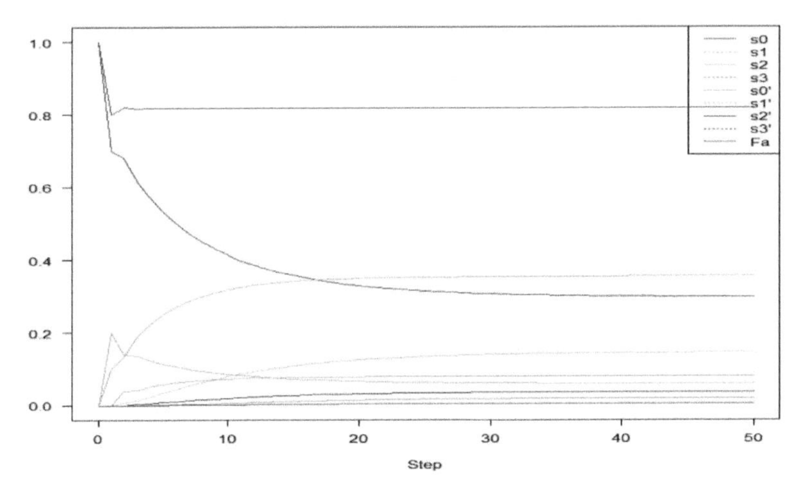

Figure 5.11 The stationary distribution for the considered case of a successful attack at one stage of the service request processing (without halts).

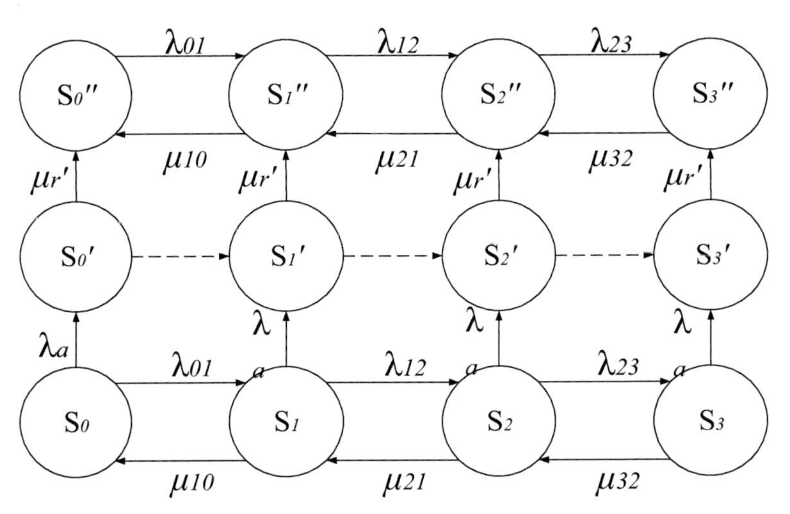

Figure 5.12 A Markov model for the considered case of a successful attack at one stage of the service request processing (with halts and eliminating of one vulnerability).

In the previous cases, a system with one vulnerability was considered. Further models with two vulnerabilities are proposed (Figures 5.14 and 5.16). In the first case (Figure 5.14), as in the cases depicted in Figures 5.8 and 5.10, the second vulnerability is not eliminated and the system just restarts and continues to function in the same way (with $\lambda_a' = K_\lambda \cdot \lambda_a$, $K_\mu < 1$). Figure 5.15 shows the convergence on the stationary distribution.

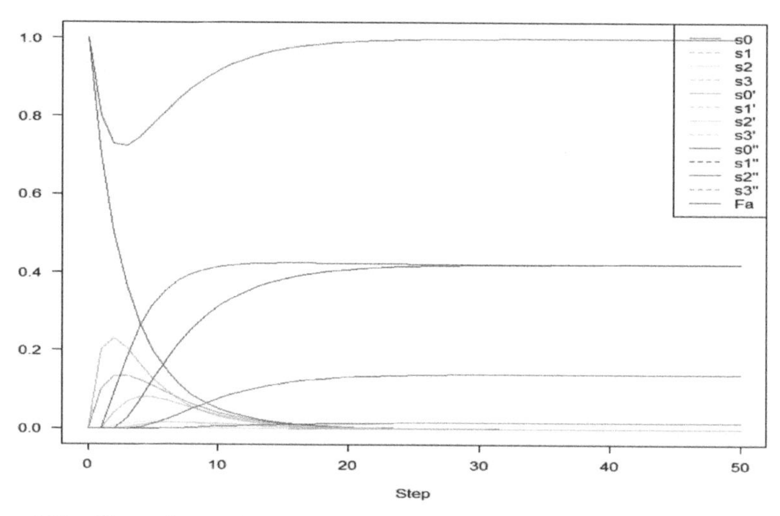

Figure 5.13 The stationary distribution for the considered case of a successful attack at one stage of the service request processing (with halts and eliminating of one vulnerability).

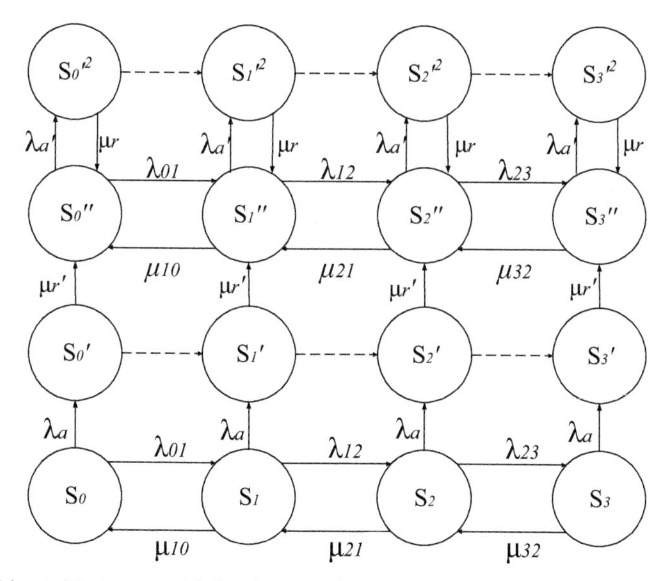

Figure 5.14 A Markov model for the considered case of a successful attack at one stage of the service request processing (with halts and system has two vulnerabilities and one vulnerability is eliminated).

In the second case (Figure 5.16) as in the case depicted in Figure 5.12, the second vulnerability is eliminated. Figure 5.17 shows the convergence on the stationary distribution for the considered case.

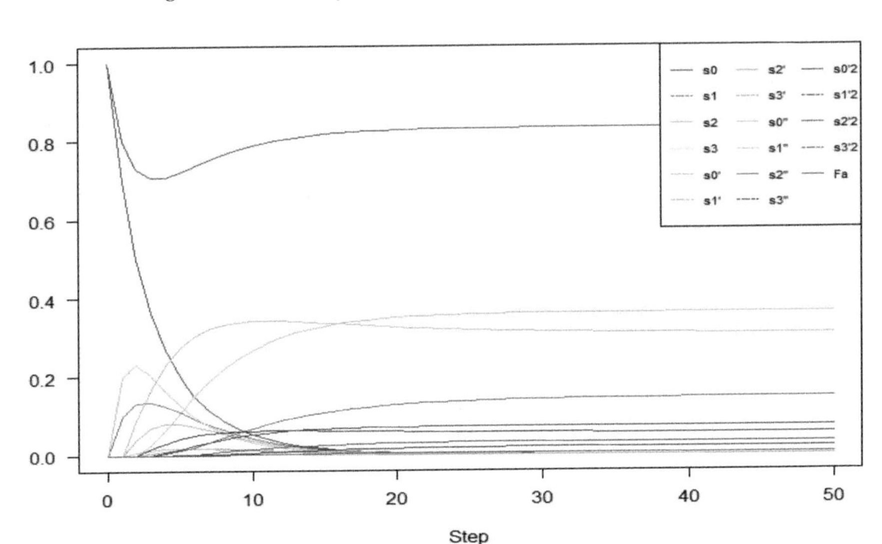

Figure 5.15 The stationary distribution for the considered case of a successful attack at one stage of the service request processing (with halts and system has two vulnerabilities and one vulnerability is eliminated).

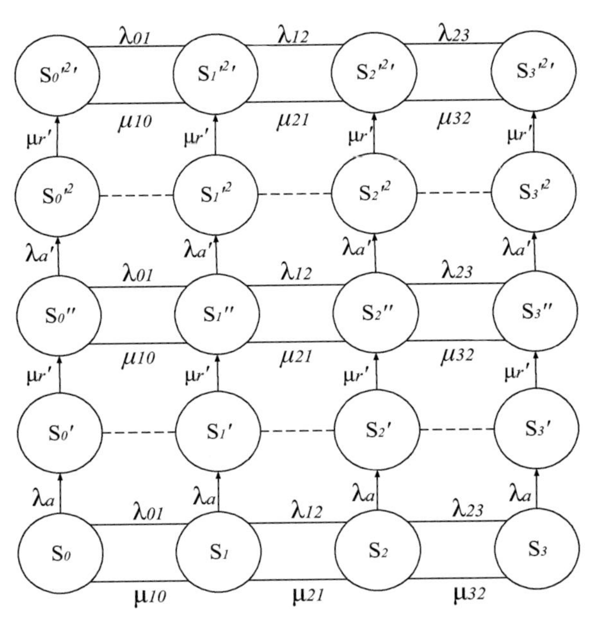

Figure 5.16 A Markov model for the considered case of a successful attack at one stage of the service request processing (with halts and with halts and system has two vulnerabilities and two vulnerabilities are eliminated).

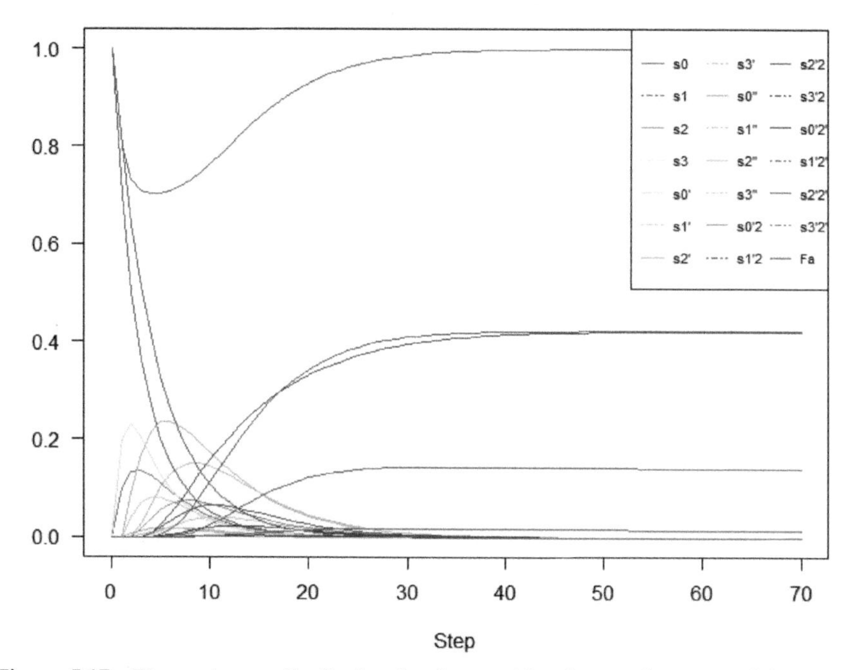

Figure 5.17 The stationary distribution for the considered case of a successful attack at one stage of the service request processing (with halts and with halts and system has two vulnerabilities and two vulnerabilities are eliminated).

For the cases depicted in Figures 5.8 and 5.10, the availability functions are calculated as:

$$F_a(t) = P_{s0}(t) + P_{s_1}(t) + P_{s_2}(t) + P_{s_3}(t), \tag{5.2}$$

and for Figures 5.12 and 5.14 as:

$$\begin{aligned} F_a(t) = {} & P_{s0}(t) + P_{s_1}(t) + P_{s_2}(t) + P_{s_3}(t) + P''_{s0}(t) + P''_{s_1}(t) \\ & + P''_{s_2}(t) + P''_{s_3}(t) \end{aligned} \tag{5.3}$$

and for Figure 5.16 as:

$$\begin{aligned} F_a(t) = {} & P_{s0}(t) + P_{s_1}(t) + P_{s_2}(t) + P_{s_3}(t) + P''_{s0}(t) + P''_{s_1}(t) + P''_{s_2}(t) \\ & + P''_{s_3}(t) + P'^{2}_{s0}(t) + P'^{2}_{s_1}(t) + P'^{2}_{s_2}(t) + P'^{2}_{s_3}(t) + P'^{2\prime}_{s0}(t) \\ & + P'^{2\prime}_{s_1}(t) + P'^{2\prime}_{s_2}(t) + P'^{2'(t)}_{s_3} \end{aligned} \tag{5.4}$$

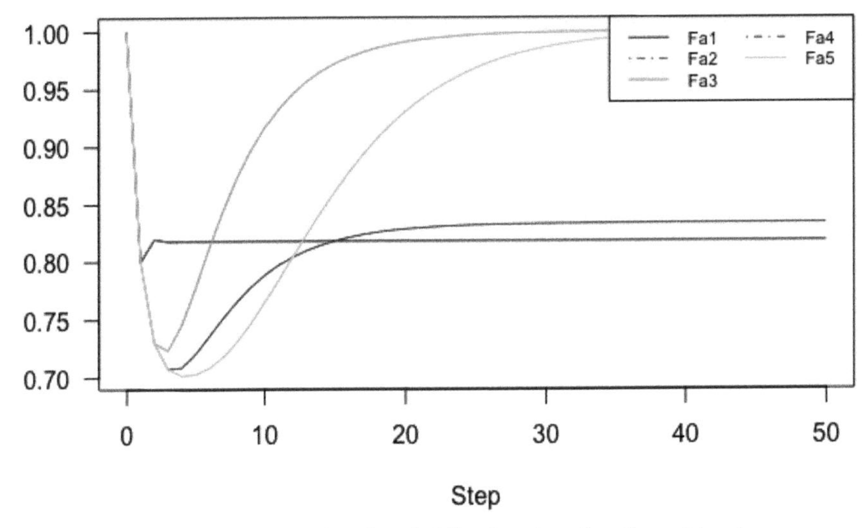

Figure 5.18 Combined plot of availability functions for all considered cases.

5.4.4 Discussion of the Simulation Results

The analysis of the obtained results shows that when one vulnerability and for other case two vulnerabilities are eliminated, the healthcare IoT system has a higher probability to be operational than just restarting; however, the t'_r value affects the duration of the availability function transition period to a stationary mode. Figure 5.18 shows the combined plot of availability functions for all considered cases discussed above.

For the cases depicted as Fa3 (Figure 5.13), Fa4 (Figure 5.15), and Fa5 (Figure 5.17), the availability functions fall below due to the loss of availability for removing vulnerabilities ($t'_r > t_r$) and then increase to a steady value (after removing vulnerabilities, the healthcare IoT system becomes more secure and correspondingly more reliable) and it takes more time to get to a steady value for a case depicted as Fa5 (Figure 5.17) due to the fact that the resources spent on the removal of the second vulnerability. System availability enhancing can be achieved by increasing the parameters μ_r and μ'_r that is speeding up the recovery uptime.

5.5 Conclusions

In this paper, we discussed opportunities and prospects for the IoT application in the field of healthcare providing. We then analyzed the existing healthcare IoT infrastructure with a description of each component. To determine the security and/or safety problems of the healthcare IoT infrastructure, a method for analyzing fault/attack trees was examined, which makes it possible to estimate the probability of failure/attack on the healthcare IoT system. This paper includes the justification of applicability possibilities of the queueing theory and presents the case study for modeling the considered IoT system. The considered models describe streams of the requests and attacks on vulnerabilities and procedure of recovery by restart and eliminating of one and/or two vulnerabilities. As a result, using the developed models, it is possible to calculate availability function at the stationary and non-stationary states. After analyzing the obtained results, we can conclude that the procedure for removing vulnerabilities is advisable because the health-care IoT system becomes more secure and reliable due to the fact that the availability of the system will increase. The presented trees can be applied as a technique for visualizing possible attacks, failures, and their scenarios for the further construction of the Markov model, which describes the behavior of the healthcare IoT system.

In summary, we observe that the rapid development of a large of portable healthcare devices in the context of IoT, like insulin pumps, is enabling a new step of care in which remote monitoring and care become a key feature for enforcing safety.

Next steps of our research will be dedicated to expansion and refine-ment of fault tree with different and various attack scenarios and failures, and developing new more sophisticated Markov models associated with influence of different kinds of vulnerabilities (both hardware and software vulnerabilities) of the healthcare IoT system (including Cloud, networked healthcare device, communication channels, and human factor, i.e., integrate unhealthy status of a patient as a system failure) and increasing bondability and dependency of tree analysis approach and Markov models.

Acknowledgments

This paper results from the Erasmus+ program educational project ALIOT "IoT: Emerging Curriculum for Industry and Human Applications" (reference number 573818-EPP-1-2016-1-UK-EPPKA2-CBHE-JP, website http://aliot.eu.org) in which the appropriate course is developed (ITM4 – IoT for health systems) within its framework, and we have developed mod-ules related to IoT systems' modeling. The authors would like to thank colleagues on this project, within the framework of which the results of this work were discussed. The authors would also like to show their deep gratitude to colleagues from the Department of Computer Systems and Networks, National Aerospace University "KhAI" for their patient guidance, enthusiastic encouragement, and useful critiques of this paper.

References

[1] *Reality Check: 50B IoT devices connected by 2020 – beyond the hype and into reality*. Available at: www.rcrwireless.com/20160628/opinion/reality-check-50b-iot-devices-connected-2020-beyond-hype-reality-tag 10 [accessed December 13, 2017].

[2] Lund, D., MacGillivray, C., Turner, V., and Morales, M. (2014). *Worldwide and Regional Internet of Things (IoT) 2014–2020 Forecast: A Virtuous Circle of Proven Value and Demand.* Framingham, MA: IDC.

[3] Malik, M. A. (2016). "Internet of things (IoT) healthcare market by component (implantable sensor devices, wearable sensor devices, system and software), application (patient monitoring, clinical operation and workflow optimization, clinical imaging, fitness and wellness measurement) – global opportunity analysis and Industry forecast, 2014–2021," *Alied Market Research*, 124.

[4] World Healthcare Organization. (2016) *Global Report on Diabetes.* Available at: http://www.who.int/iris/handle/10665/204871

[5] U.S. Department of Health and Human Services, Food and Drug Administration, Center for Devices and Radiological Health, Office of Device Evaluation. (2016). *Applying Human Factors and Usability Engineering to Medical Devices: Guidance for Industry and Food and Drug Administration Staff.*

[6] U.S. Department of Health and Human Services, Food and Drug Administration, Center for Drug Evaluation and Research Center for Biologics Evaluation and Research. (2006). *Guidance for Industry: Q9 Quality Risk Management.*

[7] ISO 14971. (2007). *Medical devices – Application of risk management to medical devices.*

[8] Butenko, V., Kharchenko, V., Odarushenko, E., and Butenko, D. (2015). "Metric-based approach and tool for modeling the I&C system using Markov chains," in *Proceedings of ICONE-23 23rd International Conference on Nuclear Engineering*, Chiba, Japan, 1–9.

[9] Qahtan, M. A.-S., A., and Kharchenko, V. (2016). "Availability and security assessment of smart building automation systems: combining

of attack tree analysis and markov models," in *Proceedings of Third International Conference on Mathematics and Computers in Sciences and in Industry,* Chania, Greece, 302–307.

[10] Healey, J., Pollard, N., and Woods B. (2015). "The healthcare internet of things: rewards and risks," in *Atlantic Council*, 17.

[11] U.S. Food and Drug Administration. *Classify your device*. Available at: www.fda.gov/MedicalDevices/DeviceRegulationandGuidance/Overview/ClassifyYourDevice/ucm2005371.htm [accessed December 18, 2017].

[12] Sutton, W. M. (2015). "Classification overview," *U.S. Food and Drug Administration.*

[13] Mohan, A. (2014). "Cyber security for personal medical devices internet of things," in *Proceedings of the IEEE International Conference on Distributed Computing in Sensor Systems*, 372–374.

[14] Mohan, A., Bauer, D., Blough, D., Ahamad, M., Bamba, B., Krishnan, R., et al., (2009). "A patient-centric, attribute-based, source-verifiable framework for health record sharing. GIT CERCS technical report GIT-CERCS-09-11," *Georgia Institute of Technology*, 10.

[15] Nagaraju, V., Fiondella, L., Wandji, T. (2017). "A survey of fault and attack tree modeling and analysis for cyber risk management," in *Proceedings of the IEEE International Symposium on Technologies for Homeland Security (HST)*, 1–6.

[16] Lyons, M., Adams, S., Woloshynowych, M., and Vincent, C. (2004). "Human reliability analysis in healthcare: a review of techniques," in *International Journal of Risk and Safety in Medicine*, v. 16, 223–237.

[17] Lakatos, L., Szeidl, L. and Telek. M. (2012). *Markovian Queueing Systems. Introduction to Queueing Systems with Telecommunication Applications*. New York, NY: Springer, 199–224.

[18] Sztrik, J. (2012). *Basic Queueing Theory*. Available at: irh.inf.unideb.hu/~jsztrik/education/16/SOR_Main_Angol.pdf [accessed December 18, 2017].

[19] Insulin Pumps Consumer Guide. (2017). Available at: http://main.diabetes.org/dforg/pdfs/2017/2017-cg-insulin-pumps.pdf [accessed May 10, 2018].

[20] Medtronics. *The Basics of Insulin Pump Therapy*. Available at: https://www.medtronicdiabetes.com/sites/default/files/library/download-library/workbooks/BasicsofInsulinPumpTherapy.pdf [accessed May 10, 2018].

[21] Spedicato, G. A., Kang, T. S., Bhargav, S., and Yadav, D. (2017). *The Markovchain Package: A Package for Easily Handling Discrete Markov Chains in R*. Available at: https://cran.r-project.org/web/packages/markovchain/vignettes/an_introduction_to_markovchain_package.pdf [accessed December 13, 2017].

6

PSMECA Analysis of IoT-based Physical Security Systems

Al-Khafaji Ahmed Waleed, Vyacheslav Kharchenko, Dmytro Uzun, Oleg Illiashenko and Oleksandr Solovyov

Department of Computer Systems, Networks and Cybersecurity,
National Aerospace University "KhAI", Kharkiv, Ukraine
E-mail: eng_ahmed.waleed@yahoo.com; v.kharchenko@csn.khai.edu; d.uzun@csn.khai.edu; o.illiashenko@csn.khai.edu; extsand@gmail.com

This paper presents the results of research of the physical security systems (PSSs) and its assessment using the physical security modes and effect analysis (PSMECA) technique considering the infrastructure facility of a region. The following results have been obtained: structural and functional decomposition of the PSS for the regional infrastructure (RI) has been performed; engineering solutions for the implementation of the standard functions for the subsystems of the research object have been developed; and a set-theoretic model of the failure occurrence in the context of the study of the PSS has been developed. An example of PSMECA application is discussed.

6.1 Introduction

6.1.1 Motivation

The modern qualitative (intensive) and quantitative (extensive) growth of the achievements of science and technology has served as the driving force for creating a multitude of scientific and practical developments. The importance of such developments is difficult to overestimate, because the average person is the carrier and/or implementer of many different ideas and technical solutions. In addition, it is necessary to take into account such aspects of socialization as culture and traditions, politics, religion, which often are catalysts to the acceleration of the diffusion of scientific and technical solutions and modern society.

Given that the objective existence of a set of positive scenarios for the application of science and technology achievements is undeniable, it is also necessary to take into account potential destructive actions and/or their scenarios. One of the systems, on which such destructive actions can be directed, is the physical security systems (PSSs) of the sophisticated objects related to state buildings, buildings of infrastructure objects, buildings of educational units (universities, schools, etc.), buildings cultural fund buildings, and so on.

The analysis of information from the world-known, generally accepted open sources [1–5] allows drawing a conclusion about a large number of terrorist acts on state, infrastructure facilities, and objects of cult of cultural heritage in Iraq. The reasons for such attacks are obvious – inadequate security of objects of social significance.

The main points in the security of physical security of this kind of objects include four types of events: physical security (guards), frames – metal detectors, closed-circuit television (CCTV), and electronic access cards. However, the real problem is that the operators of the control room are exposed to the type of blackmail, intimidation, etc.; as a result, they often become involuntary accomplices in crimes. Therefore, the problem arises of automating the functions of operators. On the other hand, there are multi-vector attacks, such as malicious disconnection of electricity, which disables video surveillance and access control systems or provocations to distract attention with the aim of enabling the penetration of intruders into the protected territory.

Another aspect that should be underlined is that the global physical security market size was valued at USD 133.94 billion in 2016, registering a compound annual growth rate of 9.1% over the forecast period [6].

Taking into account all of the above, the need of physical security to an environment aimed to mitigate or reduce terrorists acts, crime, or vandalism through theft, burglaries, and fire is anticipated to be the key trends driving the market and society.

6.1.2 The Objectives, Approach and Structure

Work-related analysis shows that authors do not provide a holistic approach of analysis of intrusion modes, their effects, and further risk assessment by ranging of its criticality. Threat assessment and response for intrusions applied to power substations are presented in [7]. The same researchers describe the physical security monitoring system with the use of multi-agent system [8] which can be applied for CCTV. The process of designing, analyzing, and selecting an exterior PSS is studied in [9]. The importance of system vulnerability assessment as an outcome of analysis and evaluation is underlined in [10]. The importance of determination of vulnerabilities and threats is considered as one of the most critical considerations for physical systems in [11]. Although the assessment guide of PSSs, developed by US Department of Energy [12] describes assessment methods and outlines their use, it contains only the overall picture without addressing the technologies that have been used for providing security systems on the market.

There are various definitions of PSS and approaches to ensure dependability and resilience of complex systems. Some of them are reviewed in [13–15]. In the context of the research interest, the PSS of the RI object is, on the one hand, a subsystem of the RI, and on the other hand, it includes a deterministic (finite) set of subsystems (or components), which it consists of. Each subsystem (or component) can be represented structurally in the form of separate elements and connections among them.

The formulation of the task directly includes the research of the functioning of PSS. There are various formulations for the definition of PSS and approaches to ensure it [16–18]. In general, PSS can be represented by an appropriate subsystem within the boundaries of the enterprise's integrated security system (facility or region). The object of research and analysis is the PSS of the facility belonging to the Ministry of Education and Science of Iraq (as an infrastructure object of the region), and the territory of compact residence of students and employees.

In addition, it is necessary to gather the following information:

- the types of failures which can occur in the system;
- the ways of distribution of the possible failures over the subsystems (components) and its elements;
- the likelihood of failures' occurrence;
- estimation of the risk of a successful attack on the protected object;
- the time needed to restore the normal functioning mode of the corrupted subsystem (or its component);
- the criticality of each specific type of attack, which can be a set (vector) of one and/or more failures (failure scenarios) provided their natural or artificial occurrence;

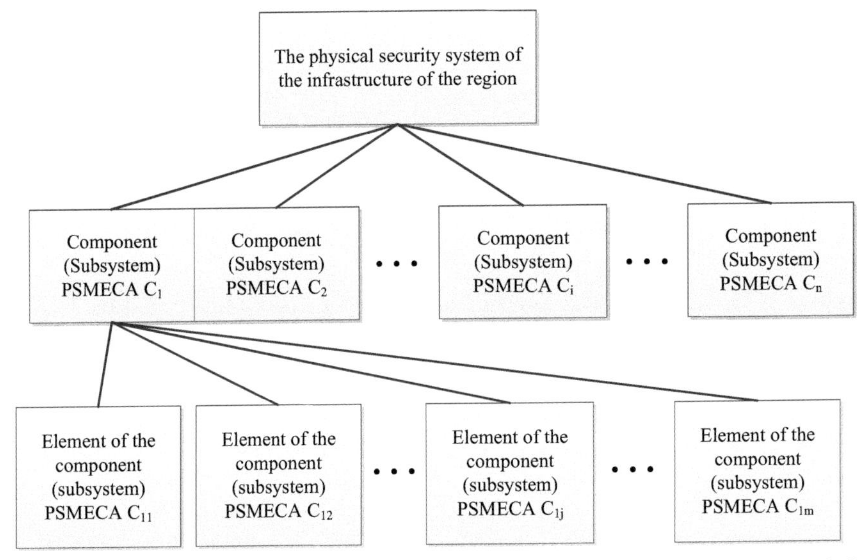

Figure 6.1 General view of the structural and hierarchical scheme of the PSS of the infrastructure of the region.

- determination of both sufficient and cost-effective countermeasures in order either to eliminate identified (or even possible in future) attacks, vulnerabilities, and threats or make them difficult (or even impossible) to exploit by an attacker [19].

The actual decomposition of the real PSS of a specific infrastructure facility of the region can be described by the filling of the components and elements in accordance with the specifics of the technical implementation (see Figure 6.1). In the figure, for each element (subsystem), the proposed method of security analysis, which will be discussed in more detail later, is indicated.

After the designing of the structural–hierarchical scheme, it is necessary to research and analyze the behavior of the system and the individual elements and the interactions between them during the time.

Thus, the objectives of this paper are the following:

- analysis of IoT-based PSS (Section 6.2);
- development of the scheme of research and development of models and methods of risk analysis of PSS, model of functions and components of PSS, and fault models of PSS;
- discussion of analysis results and of the occurrence of failures in PSS (Section 6.3).

6.2 IoT-based Physical Security System

An example of the practical implementation of the structural–hierarchical scheme for the PSS of the RI can be represented by a set of subsystems, e.g., motion/intrusion detection subsystem and access control subsystem; 24/7 monitoring and signaling/alerting subsystems; CCTV subsystem; lighting subsystem; subsystem of communications, and others.

The general view of the structural and hierarchical scheme of the PSS of the RI must be filled by the above subsystems (shown in Figure 6.2). Based on the example of the practical implementation of the structural–hierarchical system for the physical security of the RI facility, shown in Figure 6.2, we will consider modeling a prototype system using Raspberry Pi [20, 21] as the main module. The Raspberry Pi microcomputer was chosen as the main control module due to the advantages in low power consumption, which allow creating an autonomous workstation for performing the tasks of automation. Due to the functional deployment using remote access and full-fledged graphic interfaces, this system is completely "friendly" for the operator and the end user, which is not unimportant in the processing of information data [22]. Technical capabilities allow simulating the behavior of devices as connected directly through analog interfaces and remotely via wireless systems. As the analogs of the microcomputer Raspberry Pi, the less expensive version which was studied is Banana Pi [23]. This is a hardware-software complex that allows performing operations like Raspberry Pi,

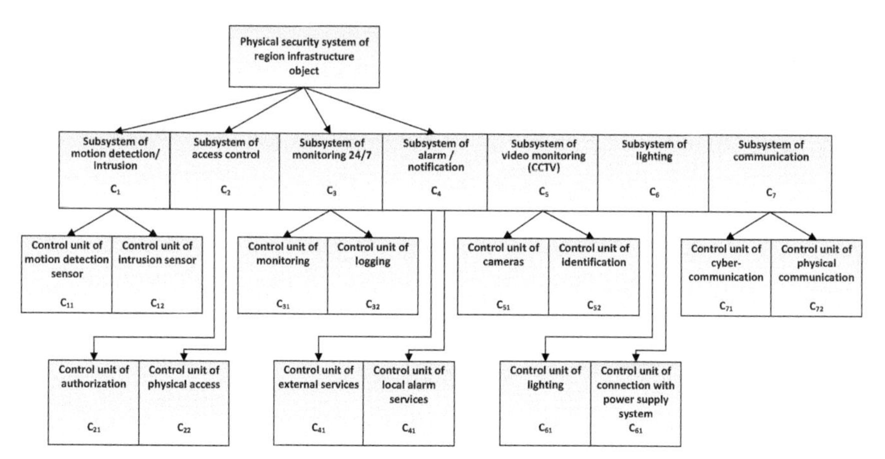

Figure 6.2 An example of practical implementation of the structural–hierarchical scheme for the PSS of the RI facility.

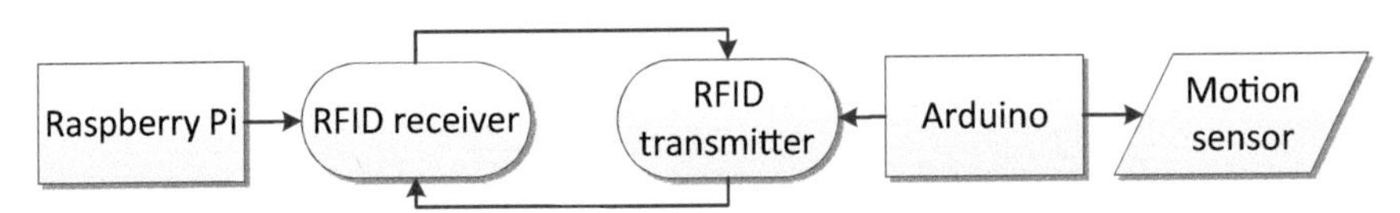

Figure 6.3 Functional diagram of the device "motion detection/intrusion detection subsystems."

but with some hardware deviations and reduced processing power. And a more expensive analog existed on the market which was reviewed during the research is CubieBoard4 [24].

Based on the device behavior pattern in the context of the common system, according to Figure 6.3, the purpose of the final product imposes a certain format of interaction between the modules.

The basic scheme of the prototype functioning of the "motion detection/intrusion detection" device combines a complex of hardware and software components that allow identifying the problem zones in the area of the PSS (see Figure 6.3).

The "motion detection/intrusion detection" allows identifying problem zones in the area of the physical security of specific objects and taking measures to restructure security protocols including the development of a full complex of the automated security system, which implies both an objective binding and global access to a single server, which is responsible for response in case of unauthorized access. Functionally, the prototype is made of a low-power Raspberry Pi microcomputer connected to the network of the inspected object and a set of external sensors combined via technology of radio-frequency identification (RFID) using long-range identification capabilities. This approach allows solving a wider range of problems at a distance of up to 50 m. During the process of implementing the prototype, the data channel through RFID technology raises the current issue of security of transmitted signals, which, in case of interception and substitution, will allow an attacker to gain control both over the server side of the system and directly over the sensors. As a sample for a crypto-component, the DESL algorithm is taken as the basis, which is a modified version of the DES algorithm for use in conditions of insufficient number of software and hardware resources [25].

Despite the shortcomings of DESL in the form of a small key size of 56 bits, this method will be hacked for several months, which is enough to test the necessary functions to automate the object within the prototype, with further replacement with more powerful security systems with a high level of cryptographic security. This approach allows one to check the basic functionality of the system and the possibility of unauthorized access, interception, substitution, or jamming activities with the most rational resource consumption.

The capabilities of the prototype which is based on Raspberry Pi in connection with Arduino chips allow in the shortest possible time with less costs for components, energy resources, and development to assemble a finished product aimed at the format of work at a certain position facility, check the relevance of the system from the point of view of safety, and give a list of recommendations, procedures, and planned works to create a real fully functional sample based on more protected (expensive) software/hardware components.

For the description of attack scenarios (intrusion) or cascade failure of subsystems/elements, the CASE tool with the ability to describe the processes occurring in the system can be applied. To provide clarity, the scenario of power outage (accidental or intentional) in the interconnection of lighting and video surveillance subsystems, described in IDEF0 notation, is presented in Figure 6.4.

Thus, the next stage is to conduct the failure, modes, effects, and criticality analysis (PSMECA) according to PSS [26] which from the point of view of existing methodologies allows effectively solving the following problems:

- determination of possible types of failures of components (subsystems) of the system;

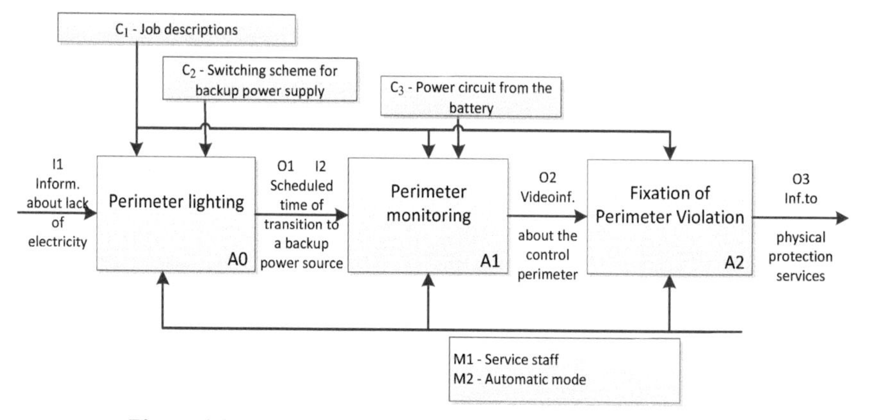

Figure 6.4 IDEF0 diagram of effects on PSMECA components.

- analysis of the impact of these failures on the functioning of the system;
- establishing the countermeasures aka the possibilities (methods) of preventing failures and/or eliminating the effect of failures on the functioning of the system.

The prototype of the object of research and analysis is the system of physical security of the student campus of one of the universities of Baghdad (Iraq).

6.3 Establishment of the Models of PSS

The process of research and development of models and methods for the risk analysis of PSSs has been carrying out in the corresponding scheme shown in Figure 6.5, where HW – hardware, SW – software, HF – human factor, PIMECA – physical intrusion modes, effects, and criticality analysis, and IIMECA – information intrusion modes, effects, and criticality analysis.

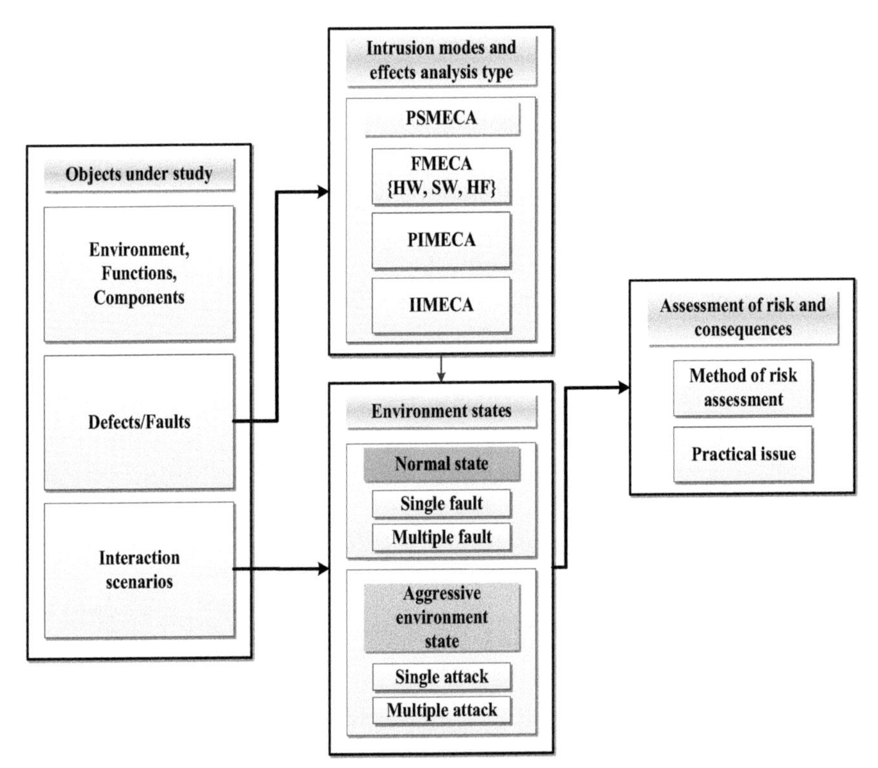

Figure 6.5 Scheme of research and development of models and methods for the risk analysis of PSSs.

PIMECA and IIMECA are both modifications of FMECA. More information about variations of FMECA-family analysis methods, which specifically concentrates on corruption of information security and cybersecurity in a form of intrusions in complex systems, could be found in [19] and [27–29]. The problem of choice of FMECA-family techniques and tools for safety analysis of critical systems is described in [30].

Objects under study represent as follows: components of the system, their interrelations and functions, as well as environment, which also plays a significant role during evaluation as well as its defects and faults. Environment states include both normal state (when single and multiple faults can occur, but their criticality and the related risk can be easily mitigated and so doesn't harm the security properties) and aggressive environment state (with indication of single and multiple attacks, which can harm the security properties of the object); assessment of risk and consequences includes the appropriate method of risk assessment and its practical issues.

6.3.1 Models of Functions and Components of PSS

This section contains the formal description of the functions and components of PSS. *PSS* is a system of physical security, which is a part of the metasystem (*MS*), which in its turn includes the environment of the system (*ES*):

$$MS = \{PSS, ES\} \tag{6.1}$$

PSS is designed to perform the following functions:

$$SFPSS = \{FVis, FVDet, F\inf\}, \tag{6.2}$$

where *FVis*, *FVDet*, and *F*inf are subsets of visualization, detection, and information processing correspondingly.

PSS consists of a set of disjoint components:

$$SCPSS = CHF \cup CHW \cup CSW, \tag{6.3}$$

where *CHF* – multiple components (operators) which are difficult to formalize, *CHW* – multiple hardware components, and *CSW* – multiple software components. In order to reveal prime reasons of failure occurrence, the intersection of hardware components and software components, human factor and hardware components, and human factor and software components is defined as null:

$$CHF \cap CHW = \varnothing, CSW \cap CHF = \varnothing, CHW \cap CSW = \varnothing. \tag{6.4}$$

In its turn

$$CHW = CHWS \cup CHWH, CHWS \cap CHWH = \varnothing, \tag{6.5}$$

where *CHWS* – a subset of hardware (primary) components – media of software (storage devices and data stores) and *CHWH* – a subset of hardware (secondary) components – video cameras, motion sensors, presence, etc.

In its turn, the dependency between system software and applications could be written as:

$$CSW = CSWS \cup CSWA, CSWS \cap CHWA = \varnothing, \tag{6.6}$$

where *CSWS* – a subset of system software (operating systems) and *CSWA* is a subset of application software (specialized software).

The environment includes physical components (*EPS)* and information components or subsystems (*EIS*). *EPS* and *EIS* subsystems are divided into natural (passive) subsystems (*EPNS* and *EINS*) and artificial (active or aggressive with respect to the system) – *EPAS* and *EIAS*.

From one side, systems' environment consists of its physical and information components

$$ES = \{EPS, EIS\}. \tag{6.7}$$

From other side, it could be represented in a form of its environment states – normal (*ENS*) or aggressive (*EAS*)

$$ES = \{ENS, EAS\}. \tag{6.8}$$

In other words, the medium can be described by the Cartesian product

$$ES=\{EPS, EIS\} \times \{ENS, EAS\}=\{EPNS, EINS, EPAS, EINS\}. \tag{6.9}$$

There is a mapping ΩEC of sets of subsystems of the environment of functions

$$SFPSS = \{FVis,\ FDet, FInf\} \tag{6.10}$$

on sets of components

$$SCPSS = CHF \cup CHW \cup CSW \tag{6.11}$$

which could be represented as

$$\Omega EC : SFPSS \to SCPSS, \tag{6.12}$$

which is described by a Boolean matrix *BFC*, such that 0, if there is no influence (dependence); 1, if there is some influence; ∅, if the nature of the indicators is different.

There is a mapping ΩEF of sets of subsystems of the environment of functions

$$SFPSS = \{FVis, FDet, FInf\} \tag{6.13}$$

on sets of components

$$SCPSS = CHF \cup CHW \cup CSW \tag{6.14}$$

which could be represented as

$$\Omega FC : SFPSS \rightarrow SCPSS, \tag{6.15}$$

which is described by a Boolean matrix *BFC* with the following values:

- 0 – in case if there is no influence (dependence);
- 1 – in case if there is influence;
- ∅ – in case if the nature of the indicators is different.

6.3.2 Fault Models of Physical Security System

In accordance with [1] and [3], the faults are divided into four types:

- physical (*pf*),
- project (*df*),
- operator (*hf*),
- interaction (*if*).

Respectively, a number of faults of the *SDPSS* of the *PSS* system consist of disjoint subspaces:

$$SDPSS = SDpf \cup SDdf \cup SDhf \cup SDif, \tag{6.16}$$

and

$$SDpf \cap SDdf = \varnothing, SDdf \cap SDif = \varnothing, SDpf \cap SDif = \varnothing, ... \tag{6.17}$$

The non-intersection of subsets of faults means that they concern different causes of their occurrence, but not consequences.

Given that

$$\begin{aligned} ES &= \{EPS, EIS\}, \\ SDPSS &= SDpf \cup SDdf \cup SDhf \cup SDif \cup SDiif. \end{aligned} \tag{6.18}$$

Errors associated with the actions of the operator can also be divided into those that cause physical defects (*hpf*) or information violations (*hif*).

In this case

$$SDPSS = SDpf \cup SDdf \cup SDhpf \cup Sdhif \cup SDipf \cup SDiif. \quad (6.19)$$

There is a mapping ΩDC of set of system faults *SDPSS* on the set of components *SCPSS*:

$$\Omega DC : SDPSS \rightarrow SCPSS, \quad (6.20)$$

which is described by a Boolean matrix *BDC*, such that 0, if there is no influence (dependence); 1, if there is some influence; ∅, if the nature of the indicators is different.

6.3.3 Investigation and Analysis of the Occurrence of Failures in PSS

At this stage, it is necessary to determine the uniqueness of the correspondence of the failures arising in the PSS (in fact – violations in the implementation of the functions specified in the system design) and the components of this system (necessary to perform the functions). Thus, taking into account the occurrence of failures of different nature (hardware, software, and human factor ones), the sought-for match is represented as a projection of the hierarchical structure on the table of the basic structural elements of the PSS.

The construction of the table is caused by the need of justification of formal confirmation (proof) of the reason for including different types of components in the generated fault matrixes. PSMECA tables imply information from both FMECA and IMECA. This construction grounds on formulas (6.1–6.20) from Sections 6.3.1 and 6.3.2 and allow to formalize different nature of failure occurrence. The implementation is presented in Tables 6.1 and 6.2.

Considering the dynamical nature of failures in the system of physical security, the necessity of defining set of scenarios is existed. The set of scenarios (*SScen*) consists of different consequences of events, which drive to failure. So, taking into account the scenarios of dynamical occurrence of failures in the system of physical security under investigation:

$$SScen = \sum SSceni, \ i = 1, ..., n, \quad (6.21)$$

taking into account the factor of time (*t*).

	PSMECA							
	FMECA				IMECA			
	pf	*df*	*hf*		*if*			
			hpf	*hif*	*ipf*		*iif*	
					ip(n)f	*ip(a)f*	*ii(n)f*	*ii(a)f*
HW	1	1	1	1	1	1	0	1
SW	0	1	0	1	0	1	1	1
OP	Ø	Ø	1	1	1	1	1	1

Figure 6.6 The projection of the hierarchical structure of failures on the table of the main structural elements in the PSS.

Thus, the developed formalization of the hierarchical structure of failures in connection with failure source nature will allow creating PSMECA tables based on set-theoretical model of the PSS components.

6.4 Conducting of PSMECA

6.4.1 An Example of PSMECA Tables for the Case of CCTV Subsystem Functioning in Normal Operation Mode

The process of creating PSMECA tables begins from developing the similar (basic or source) FMECA tables, which are modified according to developed set-theoretical model of the PSS components. The main goal of such a modification is to go deep into the structure of thc analyzed system failure sources to provide a more strictly formalized approach, based on additional structure elements and levels of hierarchy, as shown in Figure 6.6. Thus, for this example, the first step will be developing the FMECA table of video surveillance subsystem. Table 6.1 depicts the results of FMECA analysis in the form of table, where P – probability, S – severity, M – maintainability, and C – criticality. Probability, severity, and maintainability are ranged from low (L) through medium (M) to high (H) and the assessment is expert based. The resulted level of criticality (C) is indicated by the maximum range of probability, severity, and maintainability for the corresponding mode of failure. Such fuzzy values (low, medium, and high) are chosen just to

Table 6.1 FMECA table for the case of CCTV subsystem functioning in normal operation mode

Subsystem	Failure Type	Failure Mode	Failure Cause	Failure Effect	P	S	M	C
Motion/ intrusion detection subsystem	HW	Does not start	Installation error or emergency stop (interrupt)	Movement monitoring within the controlled perimeter is disabled	L	H	L	H
		Improper functioning			M	M	M	M
	SW	Does not work	Staff error or design error		L	H	M	H
		No feedback			L	M	M	M
Access control subsystem	HW	Does not start	Installation error or emergency stop (interrupt)	Unauthorized access to the secured area can be granted	L	H	L	H
		Improper function-ing			M	M	M	M
	SW	Improper function-ing	Staff error or design error		L	M	M	M

demonstrate the opportunity of implementation of the developed approach without unnecessary complication of calculations.

FMECA table for the case of CCTV subsystem functioning in normal operation mode should be modified into a similar PSMECA table according to the developed set-theoretical model of the PSS components (see Table 6.2). The assessment of probability, severity, and maintainability is also based on expert judgment. The probability for PSMECA is established as low (L), low to medium (L/M), which depends on aggressive environment conditions (e.g., in case of intensification of terrorist activities), medium (M), and high (H). For severity and maintainability, the same range (low, medium, and high) as in in previous case is used.

The developed PSMECA table can be used for setting the more detailed causal relationship between subsystems, their failure types, and PSS security risks.

Thus, based on the results of Table 6.2, obtained from Table 6.1, it is possible to determine the cause of the failure occurrence in the PSS and the value of failure criticality more accurately.

Table 6.2 PSMECA table for the case of CCTV subsystem functioning in normal operation mode

Subsystem	Failure Type				Failure Mode	Failure Cause	Failure Effect	P	S	M	C
Motion/ intrusion detection subsystem	HW	pf			Does not start	Installation error or emergency stop (interrupt)	Movement monitoring within the controlled perimeter is disabled	L	H	L	H
		df						L	H	L	H
		hf	hpf		Improper functioning			M	H	L	H
			hif					M	M	M	M
		if	ipf	ip(n)f				L	L	L	L
				ip(a)f				L/M	L	M	M
			iif	ii(a)f				L/M	H	H	H
	SW	df			Does not work	Staff error or design error		L	H	M	H
		hf	hif		No feedback			L	M	M	M
		if	ipf	ip(a)f				L/M	H	M	H
			iif	ii(n)f				L	L	M	M
				ii(a)f				L/M	H	H	H
Access control subsystem	HW	pf			Does not start	Installation error or emergency stop (interrupt)	Unauthorized access to the secured area can be granted	L	H	L	H
		df						L	H	L	H
		hf	hpf		Improper functioning			M	H	L	H
			hif					M	M	M	M
		if	ipf	ip(n)f				L	L	L	L
				ip(a)f				L/M	L	M	M
			iif	ii(a)f				L/M	H	H	H
	SW	df			Improper functioning	Staff error or design error		L	H	M	H
		hf	hif					L	M	M	M
		if	ipf	ip(a)f				L/M	H	M	H
			iif	ii(n)f				L	L	M	M
				ii(a)f				L/M	H	H	H

6.4.2 Discussion of the PSMECA

The proposed technique for the PSS security assessment called physical security modes and effects criticality analysis (PSME(C)A) combines two well-known techniques taking into account PSS particularities [2]. The first technique is failure modes, effects and criticality analysis and the second one – intrusion modes, effects, and criticality analysis (IME(C)A).

The features of the PSME(C)A technique are the following:

- The technique is based on the analysis of component and system faults according with set SDPSS considering that SDipf and SDiif are decomposed on subsets faults caused by natural reasons (n) and aggressive environment (f), i.e., *ip(n)f* and *ip(a)f*, *ii(n)f* and *ii(a)f*. Figure 6.6 describes sets of the faults for different components (hardware, software, and human factor).
- The results of analysis are represented by a set of rows describing by a vector <component, type of faults, modes, and effect of failure in point of view PSS security, probability *Prob*, severity *Sev*, and complexity (time and costs) of up-state recovery *Crec*>.

$$\text{PSS security Risk} = \text{Prob} * \text{Sev} * \text{Crec}. \tag{6.22}$$

- Taking into account the proposed PSS structures and platform (Figures 6.2 and 6.3), the hierarchical PSMECA, which consists of FMECA/IMECA (HF/IME(C)A), can be applied as shown in Tables 6.1 and 6.2.

6.5 Conclusions and Future Steps

This article describes an example of applying PSMECA analysis to a university building. Moreover, various IoT-based facilities for public use may be the object of using PSMECA analysis.

Thus, based on the research of the PSS, the following results have been obtained:

- Structural and functional decomposition of the PSS of the RI was developed.
- Engineering solutions for the implementation of the standard functions of the subsystems in the research object were proposed and some of the were reviewed in the paper.

- The set-theoretical models of the PSS components, environment, and faults, and general issues of PSMECA-based assessment have been analyzed.

This article describes the context with a static system. Before conducting the assessing in dynamics, it is necessary to consider attack scenarios. In the case of a dynamic process, *a posteriori* analysis should be performed, i.e., if there is a specific event (failure mode), it is necessary to reassess the criticality of the effect of failure on the subsystem and on the system and conduct PSMECA once again.

The PSS should be analyzed periodically to ensure that the original protection objectives remain valid. Future research can be dedicated to developing scenarios of the physical and cyber-attacks including multi-step intrusions and multiple failures and considering these circumstances during PSMECA in dynamics.

Acknowledgments

This paper is based on the experience obtained during the constant work within of the EU-funded projects, namely, in the frame of the educational projects: SEREIN project funded under Tempus program – «Modernization of postgraduate studies on SEcurity and REsilience for human and INdustry related domains» (543968-TEMPUS-1-2013-1-EE-TEMPUS-JPCP) [31] and ALIOT project funded under Erasmus+ program «Internet of Things: Emerging Curriculum for Industry and Human Applications» (573818-EPP-1-2016-1-UK-EPPKA2-CBHE-JP) [32], as well as research project STARC «Methodology of SusTAinable Development and InfoRmation Technologies of Green Computing and Communication» funded by the Ministry of Education and Science of Ukraine. The authors appreciated the scientific society of consortiums of both abovementioned projects and in particularly the Department of Computer Systems, Networks, and Cybersecurity of National Aerospace University – Zhukovsky «KhAI» for invaluable inspiration, hardworking, and creative analysis during the preparation of this paper.

References

[1] 2010 Baghdad church massacre. Available at: https://en.m.wikipedia.org/wiki/2010_Baghdad_church_massacre

[2] CNN (2009). Deadly bombings worst Iraq attack in two years. Available at: http://edition.cnn.com/2009/WORLD/meast/10/25/iraq.violence/index.html

[3] BBC News. (2010). Gunmen attack Iraqi central bank. Available at: http://www.bbc.com/news/10304652

[4] The Guardian. (2009). *Six bombs, 95 dead – carnage and despair return to Iraq*. Available at: https://www.theguardian.com/world/2009/aug/19/iraq-baghdad-bombings

[5] The New York Times. (2010). *Suicide Bomber Kills Dozens in Attack on Iraqi Army Recruits*. Available at: https://mobile.nytimes.com/2010/08/18/world/middleeast/18iraq.html

[6] Grand View Research. (2018). *Physical Security Market Size, Share, & Trends Analysis Report By Component, By Hardware, By Services, By End-use (Energy, Utility, Retail, Commercial), And Segment Forecasts, 2018–2025*, 51. Available at: https://www.grandviewresearch.com/industry-analysis/physical-security-market

[7] Jing, X., Liu, C.-C., Sforna, M., Bilek, M., and Hamza, R. (2014). "Threat assessment and response for physical security of power substations," in *Proceedings of Innovative Smart Grid Technologies Conference Europe (ISGT-Europe)*, IEEE PES, Istanbul, 1–6.

[8] Jing, X., Liu, C.-C., Sforna, M., Bilek, M., and Hamza, R. (2015). "Intelligent physical security monitoring system for power substations," in *Proceedings of 18th International Conference on Intelligent System Application to Power Systems (ISAP)*, Porto.

[9] Han, L., Burnett, D., Sheaffer, D., and Arnold, E. (2009). "Applying decision analysis process to exterior physical security system technology design and selection," in *Proceedings of Security Technology, 43rd Annual 2009 International Carnahan Conference on*, Zurich, IEEE, 312–312.

[10] Siva, R. P. (2017). *How to design effective physical security system*. Available at: https://www.linkedin.com/pulse/how-design-effective-physical-security-system-siva-rp-cpp-psp/

[11] Kline Technical Consulting. "*The 7 Most Critical Considerations for Physical Security Systems*" Whitepaper. Available at: http://www.klinenm.com/uploads/common/The_7_Most_Critical_Considerations_for_Physical_Security_Systems.pdf

[12] U.S. Department of Energy. (2016). *Physical Security Systems Assessment Guide*. Available at: https://www.energy.gov/sites/prod/files/2017/02/f34/PhysicalSecuritySystemsAssessmentGuide_Dec2016.pdf

[13] Avizienis, A., Laprie, J.-C., Randell, B., and Landwehr, C. (2004). "Basic concepts and taxonomy of dependable and secure computing," in *IEEE Transactions on Dependable and Secure Computing*, 1(1):11–33.

[14] Yastrebenetsky, M., and Kharchenko, V. (eds). (2014). *Nuclear Power Plants Instrumentation and Control Systems for Safety and Security*. Hershey, PA: IGI Global, 470.

[15] Qahtan, M. A.-S. A., and Kharchenko, V. (2016). "Availability and security assessment of smart building automation systems: combining of attack tree analysis and markov models," in *Proceedings of Third International Conference on Mathematics and Computers in Sciences and in Industry*, China, Greece, 302–307.

[16] Charlie., F., and Brayon, M. "Physical Protection Principles," *Nuclear Installation Dept*. AELB. Available at: www.aelb.gov.my

[17] Harris, S. (2013). *Physical and Environmental Security. In CISSP Exam Guide*. United States: McGraw-Hill, 6th ed., 427–502.

[18] Conrath, J., (1999). "Structural Design for Physical Security: State of the Practice," *Task Committee, Structural Engineering Institute, ASCE Reston*, 264.

[19] Kharchenko, V. S, Illiashenko, O. A, et.al. (2014) "Security informed safety assessment of NPP I&C systems: GAP-IMECA technique," in *International Conference on Nuclear Engineering, Volume 3: Next Generation Reactors and Advanced Reactors; Nuclear Safety and Security, ASME, 22nd International Conference on Nuclear Engineering ICONE.*

[20] Monk, S. (2015). *Programming the Raspberry Pi: Getting Started with Python*. McGraw Hill Professional, 192.

[21] Blum, J. (2013). *Exploring Arduino: Tools and Techniques for Engineering Wizardry*. Jonh Wiley & Sons, 384.

[22] Raspberry Pi Official page. Available at: https://www.raspberrypi.org/products/raspberry-pi-3-model-b/

[23] Banana Pi Official Page. Available at: http://www.banana-pi.org/

[24] Cubieboard Forum Page. Available at: http://cubieboard.org/model/cb4/

[25] Poschmann, A., Leander, G., Schramm, K., and Paar, C. (2007). "New light-weight crypto algorithms for RFID," in *IEEE International Symposium on Circuits and Systems*, New Orleans, LA, 1843–1846.

[26] Waleed, A. K. A., Kharchenko, V., Uzun, D., and Solovyov, O. (2017). "IoT-based physical security systems: structures and PSMECA analysis," in *9th IEEE International Conference on Intelligent Data Acquisition and Advanced Computing Systems: Technology and Applications (IDAACS)*, Bucharest, 870–873.

[27] Gorbenko, A., Kharchenko, V., Tarasyuk, O., and Furmanov, A. (2006). "F(I)MEA-technique of web services analysis and dependability ensuring," *Lecture Notes in Computer Science*, 4157, 153–167.

[28] Babeshko, E., Kharchenko, V., and Gorbenko, A. (2008). "Applying F(I)MEA-technique for SCADA-based industrial control systems dependability assessment and ensuring," in *Third International Conference on Dependability of Computer Systems DEPCOS-RELCOMEX*, 309–315.

[29] Kharchenko, V., Illiashenko, O., Kovalenko, A., Sklyar, V., and Boyarchuk, A., (2014). "Security informed safety assessment of NPP I & C systems: GAP-IMECA technique," in *Proceedings of the 22nd International Conference on Nuclear Engineering ICONE*, Prague, Czech Republic.

[30] Illiashenko, O., Babeshko, E. (2012). Choosing FMECA-based techniques and tools for safety analysis of critical systems. *Information & Security: An International Journal*, 28(2), 275–285.

[31] Tempus SEREIN Project. Available at: http://serein.eu.org/

[32] Erasmus+ ALIOT Project. Available at: http://aliot.eu.org/

7

IoT Security Event Correlation Based on the Analysis of Event Types

Andrey Fedorchenko[1,2] and Igor Kotenko[1,2]

[1]St. Petersburg Institute for Informatics and Automation of Russian Academy of Sciences, Saint Petersburg, Russia
[2]St. Petersburg National Research University of Information Technologies, Mechanics and Optics, Saint Petersburg, Russia
E-mail: fedorchenko@comsec.spb.ru; ivkote@comsec.spb.ru

This chapter is devoted to the structural analysis of security event types for event correlation in the next-generation SIEM systems. Such systems should function in various cyber-physical infrastructures with a conditionally unlimited number of information sources. This concept includes adaptive support for IoT networks integrated into IT infrastructures, as well as the application of technologies for collecting, storing, and processing of large data. This chapter deals with general issues of security event correlation, as well as current problems in this research field. We describe an approach for the preliminary analysis of source data. This approach allows us to derive numerical characteristics of relations between security events in order to establish their interrelationship. The approach essence lies in the segmentation of the general set of event instances by types and detecting direct and indirect links between them. A weighted multi-graph model is constructed based on the obtained relations between the event types. This model is used to quantify the similarity of event instances. A distinctive feature of the proposed approach is the automated preprocessing of the initial information from the conditionally indeterminate cyber-physical infrastructure. The approach is focused on self-adaptation to the target infrastructure during implementation, as well as automated configuration (tuning) of the correlation module. We demonstrate examples of the structural analysis of the Windows event log and describe the obtained results. The transition from non-numerical description of security events to numerical assessment of their similarity allows using data mining and machine learning methods.

7.1 Introduction

The current level of information technology development requires increased attention to cyber security. More and more spheres of human activity are influenced by information technologies, and therefore the relevance of security problems increases significantly. At the same time, the rapid growth of globalization and the integration of information and physical resources demand additional requirements on information protection tools. In particular, we see the great importance of guarantying security for cyber-physical infrastructures containing the large number of physical sources connected by the network, for example, Internet of Things (IoT). Proactive detection of threats in both information and cyber-physical infrastructures is very important. One of the classes of such tools is security information and event management (SIEM) systems of the next generation.

The main task of SIEM systems is to collect information from different network applications (sensors), reveal alerts on security breaches, and determine security incidents [1, 2]. To achieve this task, SIEM systems use methods of normalization, aggregation, filtering, and correlation of events. However, with the development of such threats to the security of computer infrastructures as targeted attacks and attacks on the IoT, currently applied methods and approaches are often unable to provide an adequate level of security. This trend is aggravated by increasing the amount of data, and processing of such huge data volume is becoming more difficult. The process of data correlation in SIEM systems plays a fundamental role. This process basically aims at defining causal relationships between processed events. It enables to detect malicious, attacking, and abnormal activity, and attack sources and targets. It depends on the implementation of various specific phases of the correlation process [3]. Despite the diversity of correlation methods, the most widely used is a rule-oriented approach.

In this chapter, we describe an approach for performing the correlation process based on the structural analysis of event types. In this approach, we propose to identify differences in the event structures those indicate differences in event types, and to find direct and indirect relations between event types. Then we introduce a mechanism for estimating the "distance" between instances of events, based on the principle of similarity between the values of equivalent and unequal one-type properties. The main research goal is to develop an adaptive technique for event correlation. The necessity and importance of adaptation in IoT are explained by the large number of heterogeneous security information sources, significant differences in their representation formats and data exchange, and the difficulty of manual configuration of rules for information processing. In general, the proposed technique is oriented to work in a conditionally indeterminate cyber-physical infrastructure of IoT that undergoes sufficient changes at the operational stage. The proposed approach to assess the similarity of events is focused on the semantic analysis of information through quantitative indicators to overcome the heterogeneity and uncertain semantics of the IoT source data. We assume that the suggested approach to the structural analysis of initial data allows expanding the functionalities of existing correlation methods. Besides, it is worth noting that the procedures described in this chapter are intended only for the structural analysis stage, whereas the general correlation technique also implies the stages of functional, behavioral, and evolutionary analysis. The implementation of these stages is a further research direction for the correlation process in the next-generation SIEM systems.

The proposed approach is not aimed at assessing security as a whole and has only an indirect impact on security tasks. A security event can be either a low-level message about an action that occurred (or its attempt), or an incident registered by the defense. The definition of the similarity of events by direct and indirect links will allow tracing the patterns in the origin of incidents for proactive security monitoring. In this case, we rely on the compactness hypothesis in the following interpretation: more similar events are most likely related to each other within the physical segment (log, host, subnet, and infrastructure as a whole).

This chapter is an extended version of the paper presented at the IEEE International Conference on Intelligent Data Acquisition and Advanced Computing Systems: Technology and Applications (IDAACS'2017) [4]. In the chapter, in contrast to the paper [4] and the earlier results of the authors (outlined, for example, in [1, 2, 5]), the relations of the proposed graph of event types were expanded through the analysis and identification of indirect relations between types. We take into account these extensions in the results of experiments on detection of the same unequal properties from the security event log of the Windows family. Moreover, in this chapter, we propose an approach to calculate the similarity indexes between the types of properties and the relative weight of the relation between two instances of the event. This approach allows us to introduce numerical metrics for estimating the distances between the objects of the analyzed sample. The transition from a non-numerical description of events to a numerical representation of the ratio of their instances provides the possibility of applying the methods of data analysis and machine learning.

This chapter is organized as follows. Section 7.2 describes various methods and techniques for security event correlation. The most important publications were used as sources of information for analyzing the current state. Section 7.3 is devoted to the general definition of the place and role of the correlation process in SIEM systems, as well as a formal description of the proposed approach and requirements for its application. Section 7.4 contains the results of practical implementation of particular tasks of security event correlation within the confines of the developed approach. In this section, we also analyze the results of experiments, where the source information is a set of data from the Windows NT security event log.

7.2 State of the Art

Correlation of data has been initially applied in the intrusion detection systems (IDS) for identifying relationships between network events with the purpose of their aggregation and subsequent detection of attacks (including distributed and multi-step ones) [3]. Exactly from the systems of the given class, the methods of correlation were adapted to correlation of information in SIEM systems. In general, the correlation process can be divided into several stages, including normalization, aggregation, filtering, anonymization, prioritization, and correlation [3]. Availability and supplementary decomposition of these stages in a particular solution depends on its implementation. From our point of view, each of the stages is necessary for full implementation of the correlation process.

At present, there are many methods for correlation of events and security information. All methods can be nominally divided into signature-based and heuristic (behavior analysis). These methods are based on similarity analysis, statistical analysis, data mining, etc. The difficulty of quality assessment of correlation methods is stipulated by the fact that the manufacturers of SIEM systems in order to protect intellectual property do not disclose the technological solutions used in their systems. Besides, even after buying a SIEM system, the study of correlation module is hampered by the fact that its tuning is mainly in the generation of new (additional) rules and exceptions. However, along with paid solutions of SIEM systems, there are also open source projects, as well as many scientific publications.

The most popular and easiest in implementation is the rule-based method [6–8] based on the fixed correlation of events under certain conditions. These conditions may contain logical operations on data, their properties, and calculated indicators. The main drawback of this method is the complexity and long time for computing the rules by the security administrator. The efficiency of the rule-based correlation also directly depends on the skill of the administrator. Many methods, such as the ones based on templates (scenarios) [6], graphs [9,10], finite state machines [9,11], similarity [12,13], and others, inherently have different models to represent events and their relationships, but ultimately, they can also be expressed in the form of rules. Current direction in event correlation is the application of self-learning approaches such as Bayesian networks [6,9,14], immune networks [9,14], artificial neural networks [9, 14, 16], and others. The advantage of these approaches is in the possibility of an independent (unconditional) event correlation with minimization of manual settings. However, building of learning models requires a preliminary data analysis and it is difficult to automate it. In addition, the application of intelligent approaches imposes the requirements for assessing the adequacy and quality of the models, and the original training data should be fairly complete.

The use of parametric and non-parametric correlation indices (linear, various ranks, and other), including Pearson coefficients, is described in [17–19] for the evaluation of algorithms, detecting patterns of distributed DoS attacks.

Reference [20] considers the deterministic and stochastic approaches to event correlation for network attack detection. A probabilistic event correlation model is proposed, which is based on the spatial–temporal analysis. Here event spaces are linked in a chain of sequences, and a concrete state from the set of states corresponds to each space at the current moment. The resulting chain sequences are used to calculate the probability of a specific attack scenario.

Reference [21] demonstrates the correlation process for attack detection in a cloud computing environment represented as a distributed sensor system. It is suggested to use the complex event processing technology. The proposed ontological model of attack detection includes scenarios, indicators, symptoms, attack impacts, and the intended target state.

Reference [22] supposes to use the application behavior model to identify illegitimate and abnormal activity. The initial data for constructing the model are events that reflect the system calls of all possible applications. Five levels of the model representation are selected. The upper level is the allocated functionality. Based on the initial information, a normal behavior profile is formed by processing a multi-graph where each vertex is an event.

7.3 Approach to Security Event Correlation

7.3.1 Security Correlation and Sources of Information

For the formulation of the research task, initially, we must determine the place and role of the correlation process in SIEM systems. It is believed that the correlation process is aimed at (1) determining the relationship between events and security information, (2) grouping of low-level events into higher level events, and (3) detection of incidents and security alerts. Thus, the implementation of the correlation process starts with the collection of data from different sources and ends at the security report stage. It should also be noted that the correlation process is continuous and should be designed to run in real time.

The global task is to develop methods of automated correlation of heterogeneous security information. The suggested division of the task is defined by the relevant aspects of the complexity of computer infrastructures as complex dynamical systems. In the current research, the private task is to develop the correlation approach based on the analysis of event types. The novelty of the approach is in the way to automate the search of causal relationships between disparate events. This approach would ensure a smooth addition of event types that were not previously known, but only after conversion to the normalized representation.

Various sources of external and internal information may serve as the input data of the correlation processes: (1) databases of virus signatures, exploits, attack patterns, installed and vulnerable platforms, and vulnerabilities; (2) configurations of operational systems (OS), networks, and applications; (3) network traffic and user and application behavior; (4) incident and event logs of different security systems, including anti-virus (AV), IDS and intrusion prevention system, security and vulnerability scanners, web application firewalls, data loss prevention systems, etc.

It should be noted that the first and the second types of sources contain security information with conditionally static content. In turn, the third and the last types of sources consist of data with dynamic content and characterize the current changes in the state of individual elements or the entire infrastructure as a whole. This separation is necessary because of the complexity of the correlation of information from different categories taking into account the time scale (for dynamic content). It is also worth noting that at this stage, the developed approach focuses mainly on input data with dynamic content, since any change in the conditionally static information may also be represented as an event. However, this fact does not exclude accounting of the data with conditionally static content. The scheme also includes tools for intermediate processing of input information and generating higher level events. The original data for the correlation are heterogeneous and multi-level that must be taken into account when solving the global task.

7.3.2 Events, Event Types, and Properties

The *event* is understood as a result of an action (completed, denied, and failed) or an attempt to commit the action generated by either the action source or by its processing system having a predefined description format, understandable by processing system, and also having specific properties that describe the action itself. The events of different types within a single log are the initial data for our research and can be expressed as follows:

$$\{e_1,\, e_2,\, \ldots,\, e_k\} \subset E^L, \quad \{t_1,\, t_2,\, \ldots,\, t_n\} \subset T^L, \tag{7.1}$$

where E is a set of events of the log L and T is a set of events types of the log L.

The analysis of event types is proposed to be held on real input data (events logs). In this case, we eliminate the possibility of errors associated with changes in the format of types, and in the presence of such a change, such events will be appropriately marked. On the basis of log analysis, we generate the structures of events types with certain properties:

$$\{p_1,\, p_2,\, \ldots,\, p_m\} \subset P^T, \tag{7.2}$$

where P is the set of properties of the set of types T.

Event properties can be nominally divided into the following groups: (1) *identification properties* whose values for each event are unique within a set of events of a single log (a group of logs) or the system. For example, the identifiers of event records; (2) *properties of membership*, whose values indicate the content of events in certain sets, such as the type, provider, and host; (3) *temporal properties* that reflect the time of creation, recording, start, finish, and other temporal characteristics of actions; (4) *audit properties*, determining the result of the action that describes the action as successful, forbidden, failed, etc.; and (5) *information properties* reflecting the specific characteristics of the actions described in the event (it is the most extensive group of attributes). Thus, the log analysis is aimed to identify the structures of types, and their properties can be represented in the form of mapping of a set of events to a set of types and properties of events types:

$$E \rightarrow T, P. \tag{7.3}$$

7.3.3 Correlation Method Based on Analysis of Event Types

As the identified types of the events consist of properties characterizing the actions described in the events, the relations between types of events can be formed by relations between their properties. To determine the place of structural analysis in the problem of determining the relations, we introduce the classification of the relations between the properties of event types.

The relations on equivalent and non-equivalent properties are separated.

Equivalent property p is the same property of two different event types t_1 and t_2:

$$\forall p \in P^T\colon \;\; p \in P^{t_1} \cap P^{t_2}; \;\;\; \{t_1, \, t_2\} \subset T. \tag{7.4}$$

In their turn, the relationships *on non-equivalent properties* are divided into the one-type and intertype ones.

The one-type non-equivalent properties p_1 and p_2 are properties that are equivalent in content type (according to semantic values of properties):

$$\{p_1, \, p_2\} \subset P^T\colon p_1{\sim}p_2. \tag{7.5}$$

Intertype non-equivalent properties are properties that are equivalent in content values with the apparent difference between the content types.

In addition, a single event type t can contain several one-type and intertype non-equivalent properties p in its structure:

$$\forall \{p_1, \, p_2, \, \ldots, \, p_s\} \subset P^t : p_1{\sim}p_2 \sim \cdots {\sim}\, p_s, \;\, t \, \in \, T, \tag{7.6}$$

where s is the number of the one- type or intertype properties.

On the subsequent analysis, the presence of equivalent properties for different types of events will be treated as *direct relations* between the properties of events types, and the presence of the one-type and intertype non-equivalent properties – as the *one-type and intertype indirect relations*, respectively.

For example, when comparing the structures of two types of security events of OS MS Windows “a privileged service was called” (4673) and “a process has exited” (4689), one of the equivalent properties of both types is the “ProcessId” (initiator process), which is a direct relation between these types of events.

When analyzing the structure of the event type “a new process has been created” (4688) in addition to properties “ProcessId,” there were identified properties “NewProcessId” and “Execution ProcessId.” All these three properties are one-type and characterize the process identifier, only in the first case – initiator process ("parent"), in the second – child process (the heir), and in the third – process source (agent) of the event.

This relation is an indirect one-type on the content type (the type is "processId"), and therefore it allows us to trace additional functional relations between events of different types and in this case to identify the events of working sessions of processes and their inheritance hierarchy.

Also the event type "a new process has been created" (4688) contains a property named "ProcessName." When considering the properties "ProcessId" and "ProcessName," the types of their contents are clearly different: "ProcessId" – in the first case and "location (executable file) in the file system" – in the second one. However, both properties describe the identifying characteristics of the process. In the first case, this characteristic is tied to the time scale: the identifier is assigned by the system to each created process and has a unique random value within the session process (from creation to completion). In the second case, the identification feature is more static and has no reference to time scale. As the result of calculating the frequency–time characteristics between the values specified among the non-equivalent properties of different types, an indirect intertype relation can be determined.

During the analysis of structures of event types, an undirected graph G of event type relations is formed:

$$G = (T,\ P,\ \varphi), \quad \varphi{:}P \to T \times T. \tag{7.7}$$

Further statistical analysis of the source data without additional processing is difficult, since most of the event properties are not of the numerical type (process name, type and access rights, result of action, etc.). If the set of values for an individual property is conditionally limited, then we can calculate the rank for each value based on its usage frequency. However, in practice, this operation may not yield the desired result.

As a solution for this problem, we propose to use relations between the event types to calculate the similarity of their instances. To do this, it is useful to introduce the concepts of the absolute weight of the relation between two events types and the relative weight of the relation between two event instances.

Absolute weight $w_{abs}^{t_1,t_2}$ between two event types t_1 and t_2 is the number of their comparable (directly or indirectly) properties $p_{eq}^{t_1,t_2}$ (where N – function for determining the number of comparable properties):

$$w_{abs}^{t_1,t_2} = N\left(p_{eq}^{t_1,t_2}\right) \tag{7.8}$$

In this case, the value of the absolute weight between two specific types is limited by the number of event type properties with the smallest number of properties. That is, if the event type t_1 has *m* properties, and the event type t_2 has *k* properties, then the maximum possible absolute weight of these event types is *min(m,k)*. Also it should be noted that for each relation kind (direct, indirect one-type, and indirect intertype), the absolute weight is calculated separately. Due to this separation, the formed graph of the relations of event types is transformed into a weighted multigraph.

The *relative weight* of the relation between two event instances is determined by the number of comparable properties whose values are identical. This metric is represented by the expression:

$$w_{rel}^{t_1,t_2} = \frac{N(p_i^{t_1} = p_i^{t_2})}{n}, \quad i = 1 \ldots n, \quad n = w_{abs}^{t_1,t_2}, \quad p_i^{t_1} \sim p_i^{t_2}, \qquad (7.9)$$

where $p_i^{t_1} = p_i^{t_2}$ denotes the set of comparable properties and *N* is the function for determining the number of properties whose values are identical.

The possible values of the relative weights are in the range [0, 1]. Like the absolute weight, the relative weight is also calculated for each relation kind separately.

Thus, a necessary and sufficient condition for calculating the absolute and relative weights of a certain relation kind is the availability of corresponding comparable properties between related types.

The introduction of a relative weight by type makes it possible to heuristically calculate the distance between event instances. Based on this metric, we can to construct a matrix of distances between events. Each event is described by the distances to all other events of the analyzed dataset.

The transition from the non-numerical description of the relations between events to the numerical one also allows using certain intelligent methods of data analysis. However, a limited number of methods can work with the data type in the form of distance matrices, in particular, the nearest-neighbor method, the method of the Parzen window, and the method of potential functions (particular case of the first method) [23, 24].

7.3.4 Input Data Requirements

The proposed approach has a number of constraints on the input data. It is supposed that before detecting the patterns of event types within the same model, the format of events is normalized. Normalization of structures is mainly reflected in the following condition: the structure of one event type t must not have equivalent properties p_1 and p_2:

$$\forall \{p_1, p_2\} \subset P^t : p_1 \neq p_2, \quad t \in T \tag{7.10}$$

This restriction is necessary to avoid looping on a single event in the course of using the proposed approach. However, we should adhere to the normalized (single-valued) format for records of properties corresponding to events of different types.

It should be noted that the initial data must also satisfy the necessary condition for the completeness of various types of events within the frame of the discussed model and sufficient condition for the completeness of the number of different types of events to perform the time–frequency and behavioral analyses.

In addition, in connection with the sensitivity and binding the proposed approach to real time, the time properties of events in the same model should be synchronized. Thus, for the correct application of the proposed approach, the time parameters of the events must be synchronized within the system log, segment, and infrastructure, respectively.

The definition of the time window of the event log for performing the behavioral analysis should be made taking into account the requirement of the sample representativeness. Thus, the size of the sample is proposed to be determined on the basis of: (1) the time–frequency analysis of input event types and (2) the dynamics (frequency) of the changes in the values of equivalent, unequal one-type, and unequal intertype properties.

7.4 Implementation and Experiments

Within the performed research, the security event log of OS MS Windows 8, not included in the local domain, was used as the source data. The computer with OS MS Windows 8 is an administrative host for the IoT infrastructure of a smart home system.

The experimental dataset has the following characteristics:

1. The log size was 4 GB (7 GB in the XML format).
2. The duration of the log recording was 1 month.
3. Depending on the task performed the processing of the log was from 20 min (identification of types and construction of a graph by direct relations between types) up to 2 h (identification of indirect one-type relations and refinement of the graph), the calculations were performed on the Intel Core i5-3570, 3.4 GHz, x64 in single-threaded execution.

4. The number of log events was equal to 6,700,000.
5. The number of identified types of events was 81 of 418 stated in the documentation [25] for the given OS version (the event types of the previous version were not taken into account; the number of instances of events of this version was no more than 20).
6. The number of identified properties was equal to 158, where 14 properties were common (found in all types of events), 53 unique (found only in one type of events), and 89 adjacent (occurred in more than one type of events).
7. The volume of unique values of event properties was about 200 Mb.

To *identify unequal one-type properties,* the following technique was used:

1. *Generation of values of equivalent event properties.* This operation performs an accumulating function and, when processing in real conditions, requires a sufficient amount of information about possible values of properties. We assume that initially, the values of all properties are conditionally countable (bounded). In practice, this condition is not always feasible. Therefore, further we also propose to divide the properties with a static or dynamic range of values. This fact will influence on the correlation process results in the form of different life-times of the corresponding property values. On the one hand, this will make it possible to isolate the current state of the infrastructure from possible "noises" in information recorded much earlier than the current time. On the other hand, long-term storage of property values can be used to identify the long-term trends in the internal processes of the infrastructure, and in particular, the attacks distributed over time.
2. *Calculation of the similarity index of unequal properties.* This operation is performed by the definition of the intersection of the sets of property values and subsequent normalization in accordance with the following expression:

$$Sim\,(p_1, p_2) = \frac{|V^{p_1} \cap V^{p_2}|}{|V^{p_1} \cup V^{p_2}|}, \quad \{p_1, p_2\} \subset P, \quad |V^{p_1} \cup V^{p_2}| > 0, \tag{7.11}$$

   where V^{p_1} and V^{p_2} – the sets of property values p_1 and p_2, respectively. The range of similarity index values belongs to the interval [0, 1].
3. *Combining pairs of unequal one-type properties.* This operation is performed by calculated indexes of similarity between the event properties.

The most significant results of the search for the one-type unequal properties (using the presented source data) are shown in Figure 7.1. Graph nodes denote properties of event types, and edges – weighted relations between properties (with indicating the similarity index based on the sets of analyzed values). Relations with the similarity index close to 0 are not displayed.

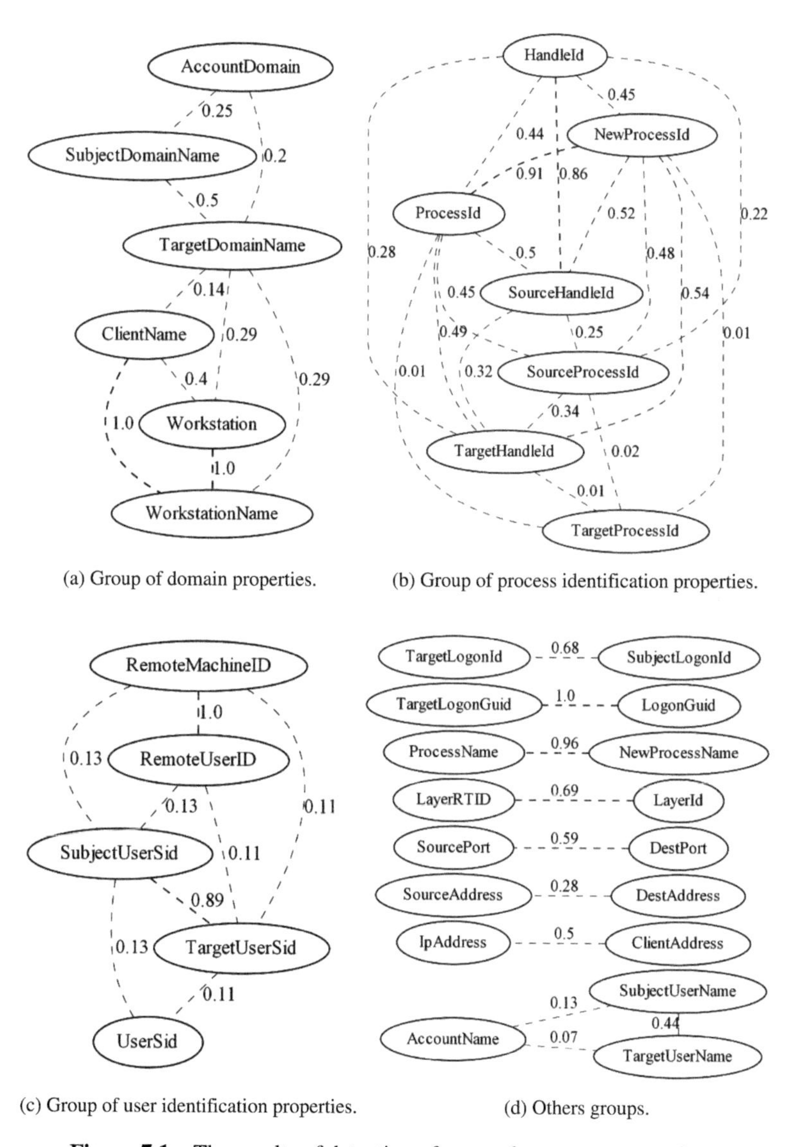

(a) Group of domain properties.

(b) Group of process identification properties.

(c) Group of user identification properties.

(d) Others groups.

Figure 7.1 The results of detection of unequal one-type properties.

In Figures 7.1(a)–(c), three main groups (sets) of the one-type properties are distinguished: the domain property group (a), the group of process identification properties (b), and the group of user identification properties (c). The remaining binary relations are presented in Figures 7.1(d); 57 properties (40% of the total, without counting the common properties) were identified as one-type with respect to one or more unequal properties. Seven properties (12% of the identified) were clearly identified as erroneous. In turn, properly identified one-type properties form 46 indirect relations (32% compared to equivalent properties) without regard to their transitivity.

In Figure 7.1, not all similarity indexes of properties have a sufficient value (>0.1) for the automated decision making about their uniformity. However, we nevertheless take such relations as correct on the basis of the expert assessment of their apparent similarity.

After performing the experiments, we formulated some remarks to the process of determining the one-type unequal properties, namely:

1. *The variety of input data is proportional to the quality and number of indirect relations found.* The increase of diversity for the indirect relations can be achieved by analyzing more initial data and collecting information from more sources. This condition is necessary when adapting the proposed approach to the target infrastructure.
2. *Using the quantitative data type to describe the properties of events* leads to the increase of one-type unequal properties which are incorrectly identified. For example, in the Windows security log, such properties as key length ("KeyLenght"), network port numbers ("IpPort," "DestPort," and "SourcePort"), protocol type ("Protocol"), and many others are described similarly. At the same time, the use of the quantitative data type to describe the values of such properties as the protocol type ("Protocol"), the change type ("ChangeType"), and others distorts the result of determining the indirect one-type relations. This distortion is also observed in the properties with the reference data type. To overcome the possible uncertainty, the properties must be represented in a string form (for example, "Protocol = udp"), and the reference values must be dereferenced. When the quantitative representation of certain properties could not be changed to non-quantitative, an additional analysis of their values should be fulfilled. This analysis is carried out by calculating such characteristics as the range of possible values, the accuracy of values, etc.

3. *The inconsistent assignment of identifiers to the objects described in the events* also increases the number of false indirect relations. Thus, the one-type properties (for example, different process identifiers) must have a unique identifier format among all sets of unequal properties.
4. *The inconsistency of the use of zero values* negatively affects the search for existing one-type unequal properties, and can also lead to incorrect determination of the relations between them. For example, in the processed data set, the zero values are represented as: "-," "null," "0," "0x0," "{00000000-0000-0000-0000-000000000000}," etc. In the analysis, we artificially excluded such values when calculated the similarity index. However, this fact complicates the implementation of automated adaptation.
5. *Using different data types for the values of one property* is a gross error, which negatively affects the accuracy of the correlation process, violating the sense of the event properties. For example, in the Windows security log, the "SourceAddress" property can have values of IP (for example, "127.0.0.1") or MAC (for example, "1234: 5678: abcd") addresses. Such use increases the time for analyzing the log by experts and significantly complicates the automated analysis of events. In particular, the intersection of different data types in values of the same property reduces its similarity index in relation to other properties.
6. *Increasing the accuracy of determining indirect one-type relations* is possible by using the recognition patterns of value types, for example, regular expressions. In this case, you can define data types such as IP address, MAC address, path in the file system, URL link, GUID identifier, and others.

Unfortunately, most of the above remarks can be eliminated only by the developers of logging subsystems. However, since most of the identified problems are related to the stage of data normalization, we can also solve it by applying the appropriate rules for processing the source information.

Figure 7.2 represents a fragment of the multi-graph of relations between the events types for the equal and unequal one-type properties (graph nodes). For example, we take the following events types: "an account was successfully logged on" (4624), "an operation was attempted on a privileged object" (4674), "the Windows Filtering Platform has permitted an application or service to listen on a port for incoming connections" (5154), and "the Windows Filtering Platform has allowed a connection" (5156). Dashed edges show direct relations between properties, and dotted edges indirect one-type relations. In this case, direct and indirect relations between types are weighted by the absolute value.

For convenience, the displayed multi-graph shows only properties directly or indirectly associated with one or more other properties. And unequal one-type properties are also associated with comparable properties within the self-type (if possible). Some properties are indicated as indirectly related based not on calculated similarity indexes, but on the sense of property names (for example, "IpPort," "SourcePort," and "DestPort"). For such relations, the similarity index is indicated by the symbol "?".

A more structured representation of Figure 7.2 is shown in Table 7.1. The left part of the table shows all directly or indirectly related properties of the event types from the presented example. The header indicates the number of the event types and in parentheses the total number of properties in the type (not including the common ones). Each cell at the intersection of a property and a type can contain one of the following values: (1) "-" – this property is not in the type; (2) "+" – this property is directly contained in the type, and (3) "$\sim n$ (m)" – the property is indirectly related to the property n (through numbering in the table), with the similarity index m. The third kind of values can be written to the cell several times. Exceptions are some properties related in meaning, but not in terms of similarity index. They are labeled "???". The result of the computed direct (solid non-directed edges) and the indirect one-type (dotted non-directed edges) absolute weights of the relations between the specified types of security events (graph nodes) is shown in Figure 7.3.

The indirect one-type relation weights within types 4624, 4674, and 5156 are indicated by directed edges. It should be noted that when calculating the absolute weights of the relations between the events types in current experiments, the transitivity of the relations of unequal one-type properties was not taken into account.

If we take them into account, the presence of the indirect one-type relation will be determined by a set of comparable properties. For example, the properties "AccountDomain" and "TargetDomainName" have the similarity index $Sim = 0.2$, and the properties "TargetDomainName" and "Workstation" – 0.29. Therefore, we can assume that the properties of "AccountDomain" and "Workstation" are also unequal one-type ones, although their similarity index $Sim = 0$. A corresponding set of comparable properties will be the domain property group [Figure 7.1(a)].

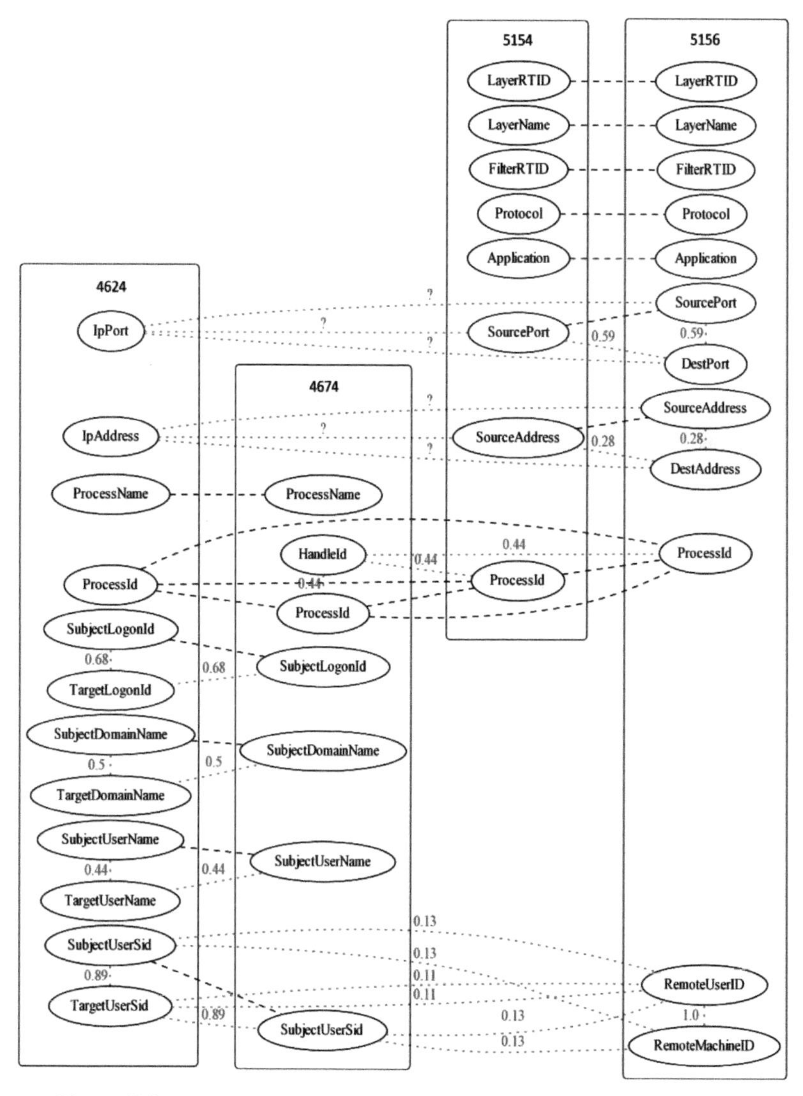

Figure 7.2 A weighted multi-graph of relations of event type properties.

The division of the initial sample of security events by type has practical advantages: (1) *verification of events formats arriving for the analysis.* It allows registering new event types and checking already registered ones; (2) *the ability of tabular data representation for storage.* It simplifies and accelerates the access to information through distributed collection

Table 7.1 Direct and indirect one-type relations between event types

N	Property Name \ Event Type	4624 (21)	4674 (12)	5154 (8)	5156 (13)
1	SubjectUserSid	+	+	–	~3 (0.13) ~4 (0.13)
2	TargetUserSid	+	~1 (0.89)	–	~3 (0.11) ~4 (0.11)
3	RemoteUserID	~1 (0.13) ~2 (0.11)	~1 (0.13)	–	+
4	RemoteMachineID	~1 (0.13) ~2 (0.11)	~1 (0.13)	–	+
5	SubjectUserName	+	+	–	–
6	TargetUserName	+	~5 (0.44)	–	–
7	SubjectDomainName	+	+	–	–
8	TargetDomainName	+	~7 (0.5)	–	–
9	SubjectLogonId	+	+	–	–
10	TargetLogonId	+	~9 (0.68)	–	–
11	ProcessId	+	+	+	+
12	HandleId	~11 (0.44)	+	~11 (0.44)	~11 (0.44)
13	ProcessName	+	+	–	–
14	IpAddress	+	–	~15 (???)	~15 (???) ~16 (???)
15	SourceAddress	~14 (???)	–	+	+
16	DestAddress	~14 (???)	–	~15 (0.28)	+
17	IpPort	+	–	~18 (???)	~18 (???) ~19 (???)
18	SourcePort	~17 (???)	–	+	+
19	DestPort	~17 (???)	–	~18 (0.59)	+
20	Application	–	–	+	+
21	Protocol	–	–	+	+
22	FilterRTID	–	–	+	+
23	LayerName	–	–	+	+
24	LayerRTID	–	–	+	+

and parallel processing, taking into account the large amount of raw data, bypassing the parsing stage; and (3) *preliminary selection of data for processing based on the identified relations between types.* First of all, the description of equivalent and unequal one-type properties between events types is analyzed.

The task of event correlation is especially important for security monitoring of IoT networks. In this case, we assume that the initial data for analysis have a large volume, and the data relationships are not known. Manual and

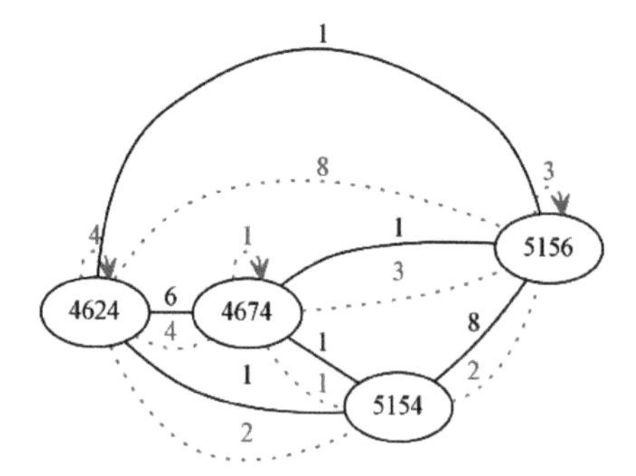

Figure 7.3 A weighted multi-graph of relations between event types.

rule-oriented processing of such information is significantly hampered, and temporary, technical, and human resources for its implementation are unlikely to justify themselves. The presented examples of the initial data analysis show that the proposed approach is able to perform adaptive information processing in conditions of uncertain infrastructure, including IoT networks.

The final results of the correlation process are aimed at a more accurate definition of the current state of the analyzed infrastructure and improving the quality of countermeasure generation. This advantage is achieved by taking into account all the initial data and calculating their semantic connectivity. We propose to use the calculated weights indicators in the future as features of events to analyze the source data. For example, using the logistic regression method, we can get the probability of an incident using a number of low-level events for a classification problem (multi-classification). The resulting assessment can be used for risk assessment, analytical attack modeling, and other monitoring and security management tasks. Evaluation of the impact of the proposed approach on the implementation of these tasks and on security as whole is the further direction of research.

7.5 Conclusion

This chapter proposed the approach for processing of events types and their properties for security event correlation. We developed the technique to implement the particular procedures of adaptive correlation, as well as pre-processing of initial data. This technique consists in isolating individual events types and in revealing direct and indirect one-type relations between them. The main advantage of the proposed approach is in the transformation of non-numerical information into numerical and normalized indicators.

It will allow applying previously unused intelligent methods to perform the process of correlation of security events in cyber-physical systems. The implementation of the proposed data analysis mechanisms takes into account the requirement for automatic or automated adaptation to a conditionally uncertain cyber-physical infrastructure. Practical results confirm the validity of the proposed approach for direct application in the tasks of security information correlation for IoT.

Further work involves the further development of the technique for the accumulation of information from heterogeneous sources and its integration in the form of a graph of relations. In particular, it is proposed to identify the groups of properties with ranking according to the level of functional impact on the macro objects of the target infrastructure. We also plan to realize the correlation procedures of behavioral and evolutionary analysis based on the determination of the most important groups of properties as features for their application in machine learning methods.

Acknowledgments

This research was partially supported by grants of RFBR (Project Nos. 16-29-09482 and 18-07-01488), by the budget (Project No. AAAA-A16-116033110102-5), and by Government of the Russian Federation (Grant 08-08).

References

[1] Kotenko, I., and Chechulin, A. (2012). "Common framework for attack modeling and security evaluation in SIEM systems," *in Proceedings of 2012 IEEE International Conference on Green Computing and Communications, Conference on Internet of Things, and Conference on Cyber, Physical and Social Computing*, Los Alamitos, CA. IEEE Computer Society, 94–101.

[2] Kotenko, I., Polubelova, O., and Saenko, I. (2012). "The ontological approach for SIEM data repository implementation," in *Proceedings of IEEE International Conference on Green Computing and Communications, Conference on Internet of Things, and Conference on Cyber, Physical and Social Computing*, 761–766.

[3] Kruegel, C., Valeur, F., and Vigna, G. (2005). Intrusion detection and correlation: challenges and solutions. *Advances in Information Security*. Vol. 14. Springer, 5–118.

[4] Fedorchenko, A., Kotenko, I., and Baz, D. E. (2017). "Correlation of security events based on the analysis of structures of event types," in *Proceedings of the 9th IEEE International Conference on Intelligent Data Acquisition and Advanced Computing Systems: Technology and Applications*, IDAACS'2017, 270–276.

[5] Kotenko, I., Chechulin, A., and Novikova, E. (2012). "Attack modelling and security evaluation for security information and event management," in *Proceedings of the International Conference on Security and Cryptography (SECRYPT)*. Rome, Italy, 391–394.

[6] Sadoddin, R., and Ghorbani, A. (2006). "Alert correlation survey: framework and techniques," in *Proceedings of the International Conference on Privacy, Security and Trust: Bridge the Gap Between PST Technologies and Business Services.*

[7] Hanemann, A., and Marcu, P. (2008). "Algorithm design and application of service-oriented event correlation," in *Proceedings of the 3rd IEEE/IFIP International Workshop on Business-Driven IT Management*, 61–70.

[8] Limmer, T., and Dressler, F. (2008). Survey of event correlation techniques for attack detection in early warning systems. Tech report, *Department of Computer Science*, University of Erlangen, Nuremberg.

[9] Muller, A. (2009). *Event Correlation Engine*. Master's Thesis, Swiss Federal Institute of Technology, Zurich.

[10] Ning, P, and Xu, D. (2008). Correlation analysis of intrusion alerts. Intrusion Detection Systems: series Advances in Information Security. *Opt. Lett*. 38, 65–92.

[11] Ghorbani, A. A., Lu, W., and Tavallaee, M. (2010). *Network intrusion detection and prevention*. Springer, 3–224.

[12] Hasan, M. A. (1991). "Conceptual framework for network management event correlation and filtering systems," in *Proceedings of the Sixth IFIP/IEEE International Symposium on Integrated Network Management*, 233–246.

[13] Zurutuza, U., and Uribeetxeberria, R. (2004). "Intrusion detection alarm correlation: a survey," in *Proceedings of the IADAT International Conference on Telecommunications and computer Networks*, 1–3.

[14] Guerer, D. W., Khan, I., Ogler, R., and Keffer, R. (1996). An artificial intelligence approach to network fault management. *SRI International*, 1–10.

[15] Tiffany, M. (2002) *A survey of event correlation techniques and related topics*. Available at: http://www.tiffman.com/netman/netman.html [accessed January 22, 2018].

[16] Elshoush, H. T., and Osman, I. M. (2011). "Alert correlation in collaborative intelligent intrusion detection systems — a survey," in *Proceedings of the Applied Soft Computing*, 4349–4365.

[17] Kou, G., Lu, Y., Peng, Y., and Shi, Y. (2012). Evaluation of classification algorithms using MCDM and rank correlation. *International Journal of Information Technology and Decision Making*, 11, 197–225.

[18] Wei, W., Chen, F., Xia, Y., and Jin, G. (2013). "A rank correlation based detection against distributed reflection Dos attacks," in *IEEE Communications Letters*, Vol. 17, 173–175.

[19] Beliakov, G., Yearwood, J., and Kelarev, A. (2012). Application of rank correlation, clustering and classification in information security. *Journal of Networks*, 7, 935–945.

[20] Jiang, G., and Cybenko, G. (2004). "Temporal and spatial distributed event correlation for network security", in *Proceedings of the American Control Conference,* IEEE, 996–1001.

[21] Ficco, M. (2011). Security event correlation approach for cloud computing. *International Journal of High Performance Computing and Networking,* 7(3), 173–185.

[22] Davis, M., Korkmaz, E., Dolgikh, A., and Skormin, V. (2017). "Resident Security system for government/industry owned computers," in *Lecture Notes in Computer Science*, Springer-Verlag, 185–194.

[23] Yang, L., and Jin, R. (2006). *Distance metric learning: a comprehensive survey*. Department of Computer Science and Engineering Michigan State University, 3–48.

[24] Yeung, D.-Y., and Chow, C. (2003). "Parzen-window network intrusion detectors," in *Proceedings of the 16th International Conference on Pattern Recognition*, 385–388.

[25] Windows security log events. Available at: https://www.ultimatewindowssecurity.com/securitylog/encyclopedia/Default.aspx/ [accessed January 22, 2018].

8

Investigation of the Smart Business Center for IoT Systems Availability Considering Attacks on the Router

Maryna Kolisnyk[1,2,*], Vyacheslav Kharchenko[1] and Iryna Piskachova[3]

[1]Department of Computer Systems, Networks and Cybersecurity, National Aerospace University "KhAI", Kharkiv, Ukraine
[2]Automation and Control in Technical Systems Department, National Technical University "Kharkiv Politechnical Institute", Kharkiv, Ukraine
[3]Department of Computer Science and Control System, Ukrainian State University of the Railway Transport, Kharkiv, Ukraine
*Corresponding Aurthor: kolisnyk.maryna.al@gmail.com

Connection to the Internet of things (IoT) can occur both wirelessly and wired. For smart office solutions (smart business center – SBC), data exchange between devices should take place without failures, in a protected mode, with confidentiality of transmitted information, data protection stored on the server, and devices and software should function with a high degree of reliability. The IoT systems can be affected by attacks both on wireless networks and attacks on network devices. This paper describes the main types of attacks on the SBC systems vulnerabilities. Markov model for an important device that is part of the IoT network – a router functioning – is developed and analyzed. The model considers the impact of various attacks, faults, and failures of the router main subsystems on the IoT systems availability.

8.1 Introduction

There are several definitions of the term Internet of Things (IoT). According to IEEE [1]: “IoT is a system consisting of networks of sensors, actuators, and intelligent objects, the purpose of which is to combine all “things” including every day and industrial objects, in such a way as to make their smart, programmable, and more able to interact with people and with each other”.

ISO/IEC AHG1 proposed the definition of IoT, which was adopted by SWG 5 [2]: “The infrastructure of interconnected objects, people, systems, and information resources, along with intelligent services, allowing them to process information and physical and virtual world and respond.” The IoT paradigm should integrate any electronic devices into the Internet environment and support new innovations and interaction between people and things [3, 4]: A global infrastructure for the information society, enabling advanced services by interconnecting (physical and virtual) things based on existing and evolving interoperable information and communication technologies. Now there are standards organizations for IoT definition [4, 5]: JTC 1, ISO, ITU-T, IEC, IoT@Work, IoT-A, IoT-I, 3GPP, CEN, GS1, OGC, OMA, OMG, IEEE, IIC, One M2M, US Tag, IETF, IERC, and ETSI.

General requirements for the IoT reference architecture [1–6] are the following: regulation compliance; autonomous functionality; auto-configuration; scalability; discoverability; heterogeneity; unique identification (names/addresses); usability; standardized interfaces; well defined components with auto configuration; network connectivity; timeliness; time-awareness; location-awareness; context-awareness; content-awareness; modularity; robust/reliability/resilient; security; confidentiality of information and privacy; legacy components; manageability; risk management; low cost; green (minimizing of power consumption), etc. The IoT layered architecture based on [4] shown in Figure 8.1.

Paradigms and principles of IoT [6, 7] are: case handling paradigm; fog computing paradigm; component-based software engineering, also known as component-based development, and end-to-end principle. IoT is estimated with related concepts: edge computing, that covers a wide range of technologies including wireless sensor networks, mobile data acquisition, mobile signature analysis, cooperative distributed peer-to-peer Ad Hoc networking and processing also classifiable as local cloud/fog computing and grid/mesh computing, dew computing, mobile edge computing, cloudlet, distributed data storage and retrieval, autonomic self-healing networks, remote cloud services, augmented reality, and Future Internet; Big Data; robotics; and semantic technologies.

<table>
<tr><td colspan="2">Management Capabilities (QoS Manager, Device Manager)</td><td colspan="2">Application layer</td><td colspan="2">Privacy and Security Capabilities</td></tr>
<tr><td rowspan="8">Generic Management Capabilities (management platforms: Firmware management, Remote Control, Device Registration, Device Provisioning)</td><td rowspan="8">Specific Management Capabilities (Thing Interaction, Hybrid Interaction, Virtual Representation of Things, Searching, Finding, Accessing Things)</td><td colspan="2">IoT applications (Smart Hospital, Smart Office, Smart home, Industrial IoT, Smart Vehicle, Smart City, Smart Grid, Smart Hospital, Smart Retail, Smart Supply Chain, applications in Insurance, Airline, Pharmaceutical, Business Services)</td><td rowspan="8">Generic Security Capabilities (Transport Layer Security (TLS), Secure Socket Layer (SSL), Internet Protocol Security (IPSec), AES-128, PKI)</td><td rowspan="8">Specific Security Capabilities (DPIA (Data Protection Impact Assessment), BLOCKCHAIN, security in the cloud and hybrid solutions, Artifical Intelligence)</td></tr>
<tr><td colspan="2">Service and Application Support Layer</td></tr>
<tr><td>Generic Support (Data Aggregation/Processing (Scribe, Apache Storm, Kafka, RapidVQ), Data Analysis, Data Storage)</td><td>Specific support (Visualization, Data Analysis, Machine Learning, Data Mining)</td></tr>
<tr><td colspan="2">Network layer</td></tr>
<tr><td>Networking Capabilities (local cloud/fog computing and grid/mesh computing, dew computing, mobile edge computing, cloudlet, distributed data storage, autonomic self-healing networks, remote cloud services, wireless sensors networks, augmented reality, Future Internet; Big Data, etc., wired and wireless technologies: GSM, CDMA, 802.15.4a, Ethernet, Bluetooth, Zigbee, Wi-Fi, Z-Wave)</td><td>Transport and Session Capabilities (RPL, 6LoWPAN, IPv4, IPv6, HTTP, COAP, XML, JSON, FTP, HTTP, SSH, Telnet, XMPP, MQTT, etc.)</td></tr>
<tr><td colspan="2">Device layer</td></tr>
<tr><td>Devices, sensors (embedded sensors and microprocessors, laptops, servers, smartphones, tablets, i-Pads, RFID tags, camera, Barcodes, actuators, GPS terminal, etc.)</td><td>Gateways (Industrial gateways with data aggregation, real-time analytics, persistent storage functions: HP Enterprise, Cisco, Dell, Fujitsu, Microsoft, IBM, redhat, Oracle, vmware, Huawei, Pivotal)</td></tr>
</table>

Figure 8.1 IoT four-layered architecture (based on ITU-T standard).

There is a specific set of requirements for an IoT application domain [8]: change to use of sensory web applications; access to sensors via the Internet; detection of sensors and data from sensors; self-description of sensors for people and software (using standard coding); monitoring the sensors in real time via the Internet; using of standard web services to access data from sensors and sensor observations; the implementation of sensory observation systems to search for phenomena of immediate interest and warnings in case of danger; the ability to respond to warnings issued by other sensors; software with the possibility of geolocation on demand and processing of observations from a newly discovered sensor without *a priori* knowledge of this sensor system; sensors with the ability to configure and perform tasks through standard, common web interfaces; and autonomous sensors and sensor networks.

The objectives of the research are the following:

- To analyze possible attacks on the software vulnerabilities of IoT wireless networks and protection methods (smart business center – SBC. SBC consists a router with the ability to connect devices to both wired and wireless networks, a server, sensors, a program-controlled switch) (Section 8.2);
- To develop and research of the Markov model of the SBC router states under the influence of various attacks on the software vulnerabilities (Section 8.3);
- To formulate the recommendations to increase dependability of the SBC router.

8.2 Security Challenges for IoT Technologies

8.2.1 Technologies and Features to Create IoT Systems

IoT systems can be organized using wired and wireless technologies, merged into local access networks and distributed networks. IoT devices make communication via radio frequency (RF) signals and Internet gateway devices; both variants of communication offer a variety of attack vectors against IoT [9]. Table 8.1 describes the technologies that can be used to create IoT networks.

Thus, IoT vulnerabilities include both interception of the radio signal and traditional methods of network attacks, as well as spyware attacks. Traditional methods of penetration into the network, such as cloud interrogation, direct connection, cloud infrastructure, and malware attacks, can also be used for IoT devices. The ability to manage IoT devices with software at a remote distance provide sample opportunities for intruders.

Table 8.1 Technologies and features to create IoT systems

IoT Elements		Samples
Identification	Naming	EPCglobal, uCode
	Addressing	IPv4, IPv6, RPL
Sensing		Smart Sensors, wearable sensing devices, embedded sensors, actuators, RFID tag, Bluetooth Low-Energy, EIB/KNX, X10, CAN RTLS and on-board sensors RFID/AutoID, RFID/WSN Wi-Fi-based T5 Sensor Tags
Communication		LAN: RFID, NFC, UWB, Bluetooth Smart (BLE), Z-Wave, IEEE 802.11a/b/g/n, WiFiDirect, IEEE 802.15.4 (ZigBee), ANSI/EIA709.1-B LonWorks, X10, EnOcean, IEEE 802.3, ISO 8000-7 DASH7, ANSI/ASHRAE 135-2016 BACnet, ISO/IEC 14908 LonTalk, Wireless HART, 6LoWPAN, UPnP WAN: 3GPP 3G/4G/5G (UMTS, LTE, LTE-A, LiTRA), 2G (GSM, GPRS), NB-IoT, LoRa, Random Phase Multiple Access (RPMA), CDMA, IEEE 802.3 10/40/100GigabitEthernet, Cat-M, SIGFOX, IEEE 802.11ah
Computation	Hardware	Microcontrollers, Microprocessors, SOCs, FPGAs, Smart Things, Arduino, Phidgets, Intel Galileo, Raspberry Pi, Gadgeteer, BeagleBone, Cubieboard, Smart Phones, NEST, Routers, UDOO, FriendlyARM, Z1, WiSense, Mulle, Xbee, T-Mote Sky, Iris Mote, Sun SPOT
	Software	OS (Contiki, TinyOs, LiteOS, Riot OS, Android, Unison OS 5.2, BlackBerry 10, iOS 5 and later, Windows Phone 8.1, Android 4.3 and later, Linux 3.4); Cloud and Big Data (Numbits, Hadoop, Scribe, Kafka, Cassandra, Apache Storm, Apache Spark, Luxune, Flume, RapidMQ, Fluentd), NS-2, LinuxMCE, NetSim, OPNET, TinyDB-TOSSIM, SWANS, Avrora
Service		Identity-related (shipping), Information Aggregation (Smart Grid), Collaborative-Aware (Smart Home), Ubiquitous (Smart City), Service Registry and Search interfaces, Service Composition Environment, Service Runtime, IoT.est Services components, IoT@Work Event Notification Service (ENS), IoT@Work Directory Service
Web services		SOAP/WSDL, REST
Semantic		Resource Description Framework (RDF), Web Ontology Language (OWL), Efficient XML Interchange (EXI) format, OpenIoT sensors directory, URI/RDF (W3C SSN Compliant), SPARQL/Semantic Schemes for RDF-compliant resources

8.2.2 Vulnerabilities and Types of Attacks in Wireless IoT Systems

Often there are reports of new software vulnerabilities of devices used in organizing IoT. Unlike traditional equipment or consumer technologies, many IoT devices do not have an interface for downloading and installing security updates (patches). If consumers themselves buy and install IoT devices, in particular, to create a smart home, they have a little knowledge of security or privacy issues, and the attacker can use them when attacking. The hacker can exploit the vulnerability of an open IP port for a variety of different attacks on Wi-Fi, cellular networks, Internet Protocol (IP) over Bluetooth, ZigBee, Z-wave, etc.

Vulnerabilities in IoT lead to attacks [10]: insecure web interfaces default accounts (XSS, SQL injection); inefficient authentication/authorization weak passwords, no two-factor authentication insecure network services ports open (universal plug and play (UPnP)), denial-of-services (DoS), distributed denial-of-services (DDoS attacks); lack of transport encryption no use of transport layer security (TLS), misconfigured TLS, custom encryption; private data unnecessary private information collected; insecure cloud interfaces default accounts, no lockout; inefficient mobile interfaces weak passwords, no two-factor authentication insufficient security configurability ports open, use of UPnP, (DoS attacks); insecure software/firmware; old device firmware, unprotected device updates; and poor physical security exposed USB ports, administrative accounts.

The use of a large number of poorly protected heterogeneous devices, most of which do not allow patching, network devices, and convergence of different technologies, makes it possible for successful attacks on IoT devices by attackers. Attackers use system vulnerabilities to cause unintentional or unforeseen behavior of the software and/or hardware of the IoT devices (cameras, network equipment, server, and digital sensors). The goals of hackers often are: unlocking a smart lock, disabling security, receiving an access to video cameras, and stealing data stored on the server [11]. There are passive attacks at installation system time in low power transmission mode, and then interception and decryption of the set key message are happened. To fix it, it is necessary to check the current key state before it is set and using cryptography and authentication.

An example of an exploit might be the use of a diagnostic port vulnerability to exploit a buffer overflow that provides access to IP networks, processor vulnerabilities [12], and the operating system. In some cases, the firmware can refer to the source code. When collecting big data, there will soon be a problem of ensuring their confidentiality.

Both wired and wireless IoT networks can be affected by various kinds of attacks, but wireless networks are more vulnerable to the influence of intruders. The ability to read RF identification tags (RFIDs) at a distance or the use of new technologies [Bluetooth (BLE) (iBeacon)] opens hackers the possibility of attacks. Hackers can control device drivers for monitors, cameras, and microphones to spy on users and collect data, images, video, and passwords [13, 14].

The purpose of the attacks on IoT [15]: information war; masking of illegal activities; political opposition; implementation of terrorist activities; intellectual and industrial competition; attempt of self-affirmation; personal dislike; and theft of the personal data of users of the IoT device for the purpose of blackmail and financial gain.

Attacks affecting IoT systems include spy attacks, DDoS attacks, special attacks on the power system, and traditional network attacks. Also in the past year [15, 16], there has been an increase in the number of DDoS attacks on the IoT infrastructure of things and using the components of the IoT. All these attacks are possible due to the presence of software vulnerabilities for managing smart devices and operating systems (OS), firewalls, servers, routers, switches, and software, and there has been an increase in the number of DDoS attacks on the IoT infrastructure of things and using the components of the IoT, hardware bookmarks, and incorrect and incomplete settings of firewalls and security policies.

8.2.3 Security Issues of Some Wireless Technologies of IoT

Requirements in wireless IoT networks [17]: authenticity; confidentiality; integrity; and availability. To protect the SBC, it is need to take appropriate measures. Currently, there are a number of methods to protect against attacks in wireless networks. Consider the security features of some wireless technologies used in the creation of the IoT, in particular, an SBC.

8.2.3.1 ZigBee technology

ZigBee technology is used to connect sensors to IoT, but it is potentially dangerous in terms of security, since the radio amplifier works with a default user name and password and is vulnerable to attacks [18].

Main attacks on ZigBee [18–20]:

- Physical attacks – physical access to the device (for example, using Bus Pirate and GoodFet interface cards) connected via ZigBee to intercept a hard-coded encryption key that is downloaded from the flash memory to the random access memory (RAM) when the device is powered on.
- Attacks on keys – uses remote tools to obtain encryption keys.
- Attacks with the device (for example, the hardware–software KillerBee), which simulates a node in the network ZigBee and collects traffic in a wireless network. Minimal session protection does not allow detecting intrusions, and the attack is almost impossible to detect.
- Repeated attacks and injections – can use key-based attacks mixed with packet attacks and/or injections to trick the ZigBee device to perform unauthorized actions.

Security services such as key establishment, encryption, frame integrity protection, and device authentication were included in the specifications of open wireless protocols such as ZigBee. Although these security services are built on top of the recognized cryptographic algorithms such as symmetric encryption message (Advanced Encryption Algorithm AES-CCM* 128 bits), authentication integrity protection (message integrity code (MIC 0–128 bits)), cipher block chaining message authentication code (CBC-MAC), and replay protection, successful attacks against them have been demonstrated that exploit the implementation vulnerabilities or insecure key management practices [18, 19].

8.2.3.2 Z-wave technology

Z-wave protocol stack includes five layers: application, security, network, transport, and physical. On security layer using anti-replay and MAC, encryption AES OFB mode, data authentication CBC-MAC, 128 cipher and MAC-keys, custom key establishment protocol, 64 bit nonce message freshness, 128 random network key [21].

Z-wave technology can have the malicious impact of hackers possible when [22]: it intercepts an encrypted key exchange and decrypts it using hard-coded. Key exchange occurs only at the initial installation of the system or reinstallation, which limits the attack time window. Z-wave devices can switch their radio receiver to low power transmission mode during the key exchange process, in order to make interception of packets during an attack much more difficult. In the application domain, a cyber-attacker can exploit a number of vulnerabilities, many of which are the result of a poor understanding of user security. Using simple passwords and passwords by default simplifies access to devices and accounts. Similarly, sending unencrypted confidential information allows an attacker to collect information by sniffing or listening. Increasingly, hackers use complex social engineering attacks to deceive unsuspecting employees or users in disclosing confidential information.

Attacks of the network domain are directed to the basic network infrastructure and access networks. Attacks can exploit vulnerabilities in the protocols (in particular, the secure sockets' layer (SSL) protocol from 2014 should be excluded in favor of the TLS protocol, as it contains vulnerabilities, and botnets of IoT devices used tunnels using this protocol for DDoS attacks) and replace the IP and MAC addresses of the device, which leads to unauthorized access to the network. Attackers can also use the error information used by engineers to debug systems, to determine what information professionals do and do not control, and to determine network topologies, device IDs, and information flows.

Devices and Equipment: Attacks are usually aimed at connected devices and equipment, such as supervisory control and data acquisition systems, sensors, or appliances. Many of these devices depend on hard-coded access keys, which makes them vulnerable to brute-force attacks and substitution.

All intruders have signatures that can be used to determine who is attacking and how. They can include anything related to the attacker's IP address, network/host artifacts and tactics, methods, and procedures (TTP). Collectively, these methods are referred to as indicators of compromise [21].

This technology uses out-of-band authentication. The device-specific key is the first 16 bytes of the 32-byte long elliptic-curve Diffie–Hellman (ECDH) public key of the joining node [22].

8.2.3.3 Long-Term Evolution/Long-Term Evolution Advanced (LTE/LTE-A) technologies

In LTE and LTE-A networks, such vulnerabilities and attacks are possible [23–27]: threats of user identification and confidentiality – an attacker can gain illegal access to access network services by forging security keys or block user access to certain services by changing IE settings; spyware attacks based on IP address tracking, which may have international mobile subscriber identity or other identifier; attacks on base stations and the process of transferring data; attacks associated with the transmission of false system information over the network, leading to a failure of broadcast or multicast alarms; DoS; risk of data manipulation risk when managing the system – changing the evolved packet system (EPS) protocol signal to include bugging devices; attacks of unauthorized access to the network; attacks to compromise node of base station (eNB) credentials, as well as physical attacks on eNB – when false configuration data, fake or cloned credentials, and data associated with remote algorithmic attacks; security attacks, which include removing fake messages in the eNB by using any protocol vulnerabilities; and attacks on the core network based on the location of the main nodes.

Network security (I) provides security and the reliability of the system by encrypting information between the mobile equipment and the universal subscriber identity module network.

Security of the domain EPC – protection from network attacks (II) allows nodes to securely exchange alarm data and user data.

Domain security (III) provides mutual authentication and secure access to mobile stations.

Security of the application area (IV) ensures the exchange of messages in a safe mode based on rules that allow the provider's domain and the user application.

Without security, 3GPP (V) defines the rules for obtaining rights to enter the EPC and protect access.

In wireless SBC networks, 3G/4G can use such cipher algorithms [23–27]: Kasumi block encryption algorithm, developed as a building block for UMTS (UEA1) encryption algorithms and integrity algorithms (UIA1) with a 128-bit key and two display functions for creating encrypted text (S-boxes); block encryption algorithm SNOW 3G is designed to prevent algebraic attacks and as the main component of both UEA2 and UIA2; block encryption algorithm Milenage has a block size and a key size of 128 bits and uses the basic form of the AES encryption algorithm as the main function; and ZUC encryption algorithm is planned as the main component for UEA3/UIA3 [25].

8.2.3.4 Low-power Wide-area Network (LoRAWAN) technology

Attacks against LoRAWAN systems [28] include ACK spoofing, replay attacks, eaves dropping attacks, bit flipping attacks, DoS and DDoS attacks, data modification attacks, and attacks on power system.

LoRAWAN technology has such protect mechanisms [29]:

- The use of a local master key that fulfills the requirements for the separation of a cryptographic network is never transmitted to the radio access network and is not very exposed;

- Implementation of mechanisms for mutual authentication of the user equipment EPS core for authenticating both the terminal and the network and encryption of information transmitted between them;
- All operations with a private eNB key used to authenticate a device must never leave the trusted environment;
- Autonomous verification, in which the network is guaranteed by successful authentication, can be performed only by a trusted device;
- DDoS-attacks' network protection by limiting the number of connections on the eNB in the network; connect to the core network only verified eNB;
- The user's privacy is ensured by secretly storing IMSI and IMEI inside the device and within the distribution environment, as well as the consistency of the signaling and user data;
- Encrypt messages and generate keys using the advanced encryption standard (AES) AES-128 algorithm. The permanent key used in the AES protocol should never be visible outside the security module;
- Authorization to connect to the core networks and to ensure the integrity of the software.

8.2.3.5 Radio Frequency IDentification (RFID) technology

RFID tags may be vulnerable to eavesdropping, traffic analysis (sniffing), spoofing or denial of service, device location tracking, spoofing attacks, re-attack attacks, physical attacks, insert attacks, DoS and DDoS attacks, and viruses.

Methods to protect RFID [30–32] include a trusted reader platform and standard public key infrastructure (PKI), access control, anti-counterfeiting, and the synchronized secrets' method.

8.2.3.6 Bluetooth Low Energy technology (BLE)

Attacks against BLE systems [33]: identity device tracking, eaves dropping, and man in the middle attack, duplicate device, network infiltration, ransomware attacks, or data theft.

Protection methods of BLE are [33,34]: security model along with secure simple pairing, security techniques (MAC address hiding), ECDH-based key exchange and LE secure connections, just work method (no key exchange between the devices and no user interaction required), capability exchange, numeric comparison, out-of-band pairing method, bonding, 128-bit AES, signed data over an unencrypted channel with trusted relationship, private address generation and resolution are implemented in the controller, static device address, using of Received Signal Strength Indicator (RSSI) as the indication of beacon signal strength, hidden the unique MAC address of the devices.

Protection methods of IoT are [3]: cryptographic ciphers such as AES, secure hash algorithm (SHA2), public-key ciphers Rivest–Shamir–Adleman, elliptic-curve cryptography, TLS protocol, SSL protocol for authentication and information encryption using the ciphers mentioned, PKI as building blocks for authentication and trust through a digital certificate standard and certificate authorities, and lightweight cryptography.

8.2.4 Spyware in IoT

Network devices in order to disable them and steal confidential data. The methods that cybercriminals use to steal important information can be such as eavesdropping and video recording without the knowledge of office employees by installing spyware and software. Table 8.2 shows examples of spyware for frequently used operating systems in the office. Functions of spyware for OS MS Windows [35, 36] include control of keystrokes of the keyboard for the purpose of collecting passwords and logins; tracking the visits to pages of websites (history, search queries, frequently visited sites, the length of visits to the site); reflection of the user's device screen via the internet from another computer or tablet; screen shots and snapshots using webcams, IP cameras, or mobile phone cameras; change the settings of the control system (on, off, idle time, connectable removable media); interception of the contents of the clipboard; monitoring and recording calls via Skype, Vyber, and the microphone of a smart device (listening to the environment); interception of data sent for printing and copying to the external media; keeping statistics of work on the computer; control of user location; and record all actions on a smart device or server (launching programs, operations with files, etc.).

Purposes of spyware in OS Android [35, 36] are recording of calls, sms; hidden recording of voice, photo, and video; keylogger; interception of correspondence VK, Kate, WhatsApp, Viber, Skype, and Instagram; false calls at a certain time; and collect information about the location of the device using WiFi/GPRS. US-CERT gave recommendations for protection against spyware [17]:

- To install special anti-spyware programs both on devices that connect to the Internet using wireless technologies and on the server;

Table 8.2 Examples of spyware for OS Android and Windows

Examples of Spyware for OS MS Windows	Examples of Spyware for OS Android
NeoSpy	Cell Tracker
Real Spy Monitor	Children Tracker
Actual Spy	Ear Spy
SpyGo	Mobile Hidden Camera
Snitch	Sneaky Cam
KGB SPY	Spy Message
StaffCop	Secret Calls
HideTrace	Secret Agent Fake Call
WebWatcher	Secret Agent
Dame Ware Mini Remote Control Server	Observer
PUNTO Switcher	
PC Pandora	
Micro Keylogger	
Expert Home	
Kickidler	
Total Spy	
SniperSpy	
Kidlogger Pro	
Remote Manipulation System	
System Surveillance Pro	
Soft Activity Keylogger	
Snooper	
The Best Keylogger	
Ardamax Keylogger	
Spytech Spy Agent	
Spyrix Personal Monitor	
All In One Keylogger	
Keylogger remote	
Elite Keylogger	
HackToolKeyGen.Win32	
Power Spy	

- If the suspicious program window appears, the device must be rebooted immediately;
- To configure the firewall access policies correctly;
- To install special anti-virus software with the function of tracking suspicious software and install special programs to search for hardware bookmarks and perform a visual inspection of devices;
- To reset the device configuration in case of detection of suspicious programs to the factory settings.

8.3 The Markov Model of the SBC Router States

In a smart office system, an important component is a router that allows devices to connect to the network and connect them to the Internet. Manufacturers offer routers for distributed services with the ability to connect devices using both wired technologies and wireless ones created specifically for the IoT. Given that the devices connected to the Internet are heterogeneous, the routers provide special methods of protection against various kinds of attacks. This can be the ability to programmatically configure access policies, to create tunnels for secure data transfer, and to install firewalls in both hardware and software. To conduct research, the Cisco 5915 embedded services router with UHF, VHF, and Wi-Fi radio modules was used [37]. Routers have such components: line cards, control interface plane, control processing plane, data plane, services plane, network interface modules (NIMs), internal services card (ISC), NM-X, front-panel gigabit Ethernets (FPGEs), multi-gigabit fabric, policy feature card (PFC), read-only memory (ROM), RAM, non-volatile RAM (NVRAM), and flash memory. The architecture of the router includes optional security feature cards (PFC) with special software. These cards have the same hardware in the form of two boards with different mechanisms of operation in the event of an attack: with a mechanism for monitoring the beginning of attacks hardware 1 and with a mechanism for monitoring the behavior of the traffic hardware 2 with one version of the software.

8.3.1 Assumptions and Initial Data for Modeling

For the developed model, the following assumptions were chosen [13, 14]:

- The flow of failures that occurs in the SBC system is a process without aftereffects, each time in the future the system's behavior depends only on the state of the system at this time and does not depend on how the system has passed to this state;
- The structure of the network includes the reservation of the server and the router, and time of transition to the reserve in the event of a failure of the main device is minimal;
- The number of DDoS attacks and the number of primary defects in the software are constant;
- The monitoring and diagnostic tools are in good working order and determine the technical state of the system with a high degree of authenticity.

The Markov model allows determining the most critical state of the router with a certain set of input data in the attack conditions, and without them. Assume that the flow of events in the subsystems of the router is ordinary, stationary, and does not have aftereffects. Then the flow of transition events from one state of the model to another state has an exponential distribution and is the simplest one. The flow of faults and failures of both software and hardware is the simplest and obeys the law of exponential distribution. The failure flow in the SBC system has the Markov property. Then the events occurring in the subsystems of the router can be described by the mathematical apparatus of Markov random processes and the flow of faults and failures of the router hardware and software has the Poisson distribution law, and hence it can be described by a Markov random process [38].

Initial data for modeling are shown in Table 8.3.

Table 8.3 Values of transition rates, attack rates, and recovery rates

Groups of Rates	Values of Transition Rates, Attack Rates, and Recovery Rates		
The attack rates for different router devices in IoT systems often reach these values	α228=10^8 1/h α238=10^8 1/h α2216=10^8 1/h α2316=10^8 1/h α3216=10^8 1/h	α3226=10^8 1/h α3227=10^8 1/h α2227=10^8 1/h α2327=10^8 1/h α2633=10^8 1/h	α2226=10^8 1/h α2326=10^8 1/h α2232=10^6 1/h α2332=10^6 1/h
The transition rates of the router subsystems failure during the successful attack increase and have been chosen their values:	λ12=10^{-6} 1/h λ13=10^{-8} 1/h λ231=10^{-6} 1/h λ331=10^{-6} 1/h λ14=10^{-6} 1/h λ45=10^{-8} 1/h λ15=10^{-5} 1/h λ56=10^{-8} 1/h λ530=10^{-5} 1/h λ16=10^{-8} 1/h λ631=10^{-6} 1/h λ17=10^{-8} 1/h λ87=10^{-2} 1/h λ731=10^{-2} 1/h λ18=10^{-5} 1/h λ19=10^{-8} 1/h λ930=10^{-5} 1/h	λ112=10^{-6} 1/h λ1213=10^{-5} 1/h λ113=10^{-7} 1/h λ1331=10^{-6} 1/h λ114=10^{-5} 1/h λ1430=10^{-4} 1/h λ115=10^{-5} 1/h λ1530=10^{-6} 1/h λ116=10^{-2} 1/h λ1617=10^{-3} 1/h λ1630=10^{-3} 1/h λ1731=10^{-6} 1/h λ118=10^{-7} 1/h λ1819=10^{-5} 1/h λ1830=10^{-6} 1/h λ119=10^{-7} 1/h λ1931=10^{-7} 1/h	λ123=10^{-6} 1/h λ124=10^{-6} 1/h λ2429=10^{-3} 1/h λ2430=10^{-3} 1/h λ125=10^{-6} 1/h λ2529=10^{-3} 1/h λ2530=10^{-3} 1/h λ2530=10^{-3} 1/h λ126=10^{-6} 1/h λ2631=10^{-2} 1/h λ127=10^{-6} 1/h λ2731=10^{-2} 1/h λ128=10^{-6} 1/h λ2831=10^{-2} 1/h λ129=10^{-6} 1/h λ2931=10^{-2} 1/h λ130=10^{-5} 1/h

(*Continued*)

Table 8.3 Continued

Groups of Rates	Values of Transition Rates, Attack Rates, and Recovery Rates		
	$\lambda110=10^{-6}$ 1/h $\lambda1011=10^{-6}$ 1/h $\lambda1030=10^{-8}$ 1/h $\lambda111=10^{-7}$ 1/h $\lambda1131=10^{-7}$ 1/h	$\lambda120=10^{-8}$ 1/h $\lambda2022=10^{-2}$ 1/h $\lambda121=10^{-7}$ 1/h $\lambda2122=10^{-2}$ 1/h $\lambda122=10^{-6}$ 1/h	$\lambda3031=10^{-2}$ 1/h $\lambda133=10^{-6}$ 1/h $\lambda3331=10^{-2}$ 1/h
The recovery rates of the router components after a fault or failure will have values	$\mu21=12$ 1/h $\mu31=10$ 1/h $\mu41=10$ 1/h $\mu51=10$ 1/h $\mu61=10$ 1/h $\mu71=10$ 1/h $\mu81=10$ 1/h $\mu91=10$ 1/h $\mu101=10$ 1/h $\mu111=10$ 1/h $\mu331=10$ 1/h	$\mu121=10$ 1/h $\mu131=12$ 1/h $\mu141=12$ 1/h $\mu151=12$ 1/h $\mu161=10$ 1/h $\mu171=12$ 1/h $\mu181=10$ 1/h $\mu191=10$ 1/h $\mu201=3$ 1/h $\mu211=3$ 1/h	$\mu221=10$ 1/h $\mu231=2$ 1/h $\mu241=5$ 1/h $\mu251=5$ 1/h $\mu261=1$ 1/h $\mu271=5$ 1/h $\mu281=5$ 1/h $\mu291=4$ 1/h $\mu301=6$ 1/h $\mu311=0.083$ 1/h
If the PFC card is operational with assumption that all types of attacks are detected and prevented, the average statistical intensities of the transitions are	$\alpha228=10^{-8}$ 1/h $\alpha238=10^{-8}$1/h $\alpha328=10^{-8}$ 1/h $\alpha2316=10^{-8}$ 1/h $\alpha3216=10^{-8}$ 1/h $\alpha3226=10^{-8}$ 1/h $\alpha3227=10^{-8}$ 1/h $\alpha2633=10^{-10}$ 1/h $\alpha2326=10^{-8}$ 1/h $\alpha2332=10^{-8}$ 1/h $\lambda12=10^{-6}$ 1/h $\lambda13=10^{-8}$ 1/h $\lambda231=10^{-8}$ 1/h $\lambda331=10^{-8}$ 1/h $\lambda14=10^{-6}$ 1/h $\lambda45=10^{-8}$ 1/h $\lambda15=10^{-8}$ 1/h $\lambda56=10^{-6}$ 1/h $\lambda530=10^{-5}$ 1/h $\lambda16=10^{-10}$ 1/h $\lambda631=10^{-5}$ 1/h $\lambda17=10^{-8}$ 1/h $\lambda87=10^{-7}$ 1/h $\lambda731=10^{-8}$ 1/h $\lambda18=10^{-8}$ 1/h $\lambda19=10^{-8}$ 1/h $\lambda930=10^{-5}$ 1/h $\lambda110=10^{-8}$ 1/h	$\lambda1230=10^{-6}$ 1/h $\lambda113=10^{-7}$ 1/h $\lambda1331=10^{-5}$ 1/h $\lambda114=10^{-8}$1/h $\lambda1430=10^{-8}$ 1/h $\lambda115=10^{-6}$ 1/h $\lambda1530=10^{-7}$ 1/h $\lambda116=10^{-5}$ 1/h $\lambda1617=10^{-7}$ 1/h $\lambda1630=10^{-7}$ 1/h $\lambda1731=10^{-6}$ 1/h $\lambda118=10^{-5}$ 1/h $\lambda1819=10^{-5}$ 1/h $\lambda1830=10^{-5}$ 1/h $\lambda119=10^{-7}$ 1/h $\lambda1931=10^{-8}$ 1/h $\lambda120=10^{-5}$ 1/h $\lambda2022=10^{-8}$1/h $\lambda121=10^{-6}$ 1/h $\lambda2122=10^{-7}$ 1/h $\lambda122=10^{-5}$ 1/h $\lambda123=10^{-6}$ 1/h $\lambda124=10^{-7}$ 1/h $\lambda2429=10^{-7}$1/h $\lambda2430=10^{-8}$ 1/h $\lambda125=10^{-7}$1/h $\lambda2529=10^{-8}$ 1/h $\lambda2530=10^{-6}$ 1/h	$\lambda2931=10^{-6}$ 1/h $\lambda130=10^{-8}$ 1/h $\lambda3031=10^{-6}$ 1/h $\lambda133=10^{-8}$ 1/h $\lambda3331=10^{-6}$ 1/h $\mu21=12$ 1/h $\mu31=10$ 1/h $\mu41=10$ 1/h $\mu51=10$ 1/h $\mu61=10$ 1/h $\mu71=10$ 1/h $\mu81=10$ 1/h $\mu91=10$ 1/h $\mu101=10$ 1/h $\mu111=10$ 1/h $\mu121=10$ 1/h $\mu131=12$ 1/h $\mu141=12$ 1/h $\mu151=12$ 1/h $\mu161=10$ 1/h $\mu171=12$ 1/h $\mu181=10$ 1/h $\mu191=10$ 1/h $\mu201=3$ 1/h $\mu211=3$ 1/h $\mu221=10$ 1/h $\mu231=2$ 1/h $\mu241=5$ 1/h

Table 8.3 Continued

	$\lambda1011=10^{-5}$ 1/h	$\lambda126=10^{-6}$ 1/h	$\mu251=5$ 1/h
	$\lambda1030=10^{-6}$ 1/h	$\lambda2631=10^{-6}$ 1/h	$\mu261=1$ 1/h
	$\lambda111=10^{-6}$ 1/h	$\lambda127=10^{-8}$ 1/h	$\mu271=5$ 1/h
	$\lambda1131=10^{-5}$ 1/h	$\lambda2731=10^{-5}$ 1/h	$\mu281=5$ 1/h
	$\lambda112=10^{-8}$ 1/h	$\lambda2831=10^{-8}$ 1/h	$\mu291=4$ 1/h
	$\lambda1213=10^{-8}$1/h	$\lambda129=10^{-5}$ 1/h	$\mu301=6$ 1/h
	$\mu311=0.083$ 1/h	$\mu331=10$ 1/h	

8.3.2 Description of the SBC Router States' Graph

The Markov model of the router availability is shown in Figure 8.2. The model has 33 states: good-working state of the router (1); failure of the router's operating system (OS) (2); failure of NVRAM (3); failure of the service plane (4); fault of the control processing plane (5); failure of the control processing plane (6); failure of the control plane interface (7); fault of the control plane interface (8); failure of the multi-gigabit fabric (9); failure of one of the line cards (10); failure of all line cards (11); failure of one NIM (12); failure of all NIMs (13); failure of ISC (14); failure of enhanced service module (SM-X) (15); failure of the one cores of the data plane (16); failure of all cores of the data plane (17); failure of one of the FPGEs (18); failure of all FPGEs (19); failure of the PFC hardware special-cases' limiter (20); failure of the PFC hardware special-cases' limiter by matching policy (21); failure of the PFC hardware (22); failure of the control-plane policing software (23); failure of the switch processor in the control plane (24); failure of the routing processor in the control plane (25); failure of the router's ROM (26); failure of the router's RAM (27); failure of the router flash memory (28); failures of the routing and switch processors in the control plane (29); fault of the router (30); failure of the router (31); failure of the PFC hardware and software (32); and failure of the power supply unit of the router (33).

Figure 8.2 presents the states of the router components and rates of the transitions from one state to another (the fail and failure rates – λij, the recovery rate of the system after a fail or failure – μij, and the attacks rate in the event that the board and the firewall software and hardware – PFC failure – αij). Red color is used for transitions, which describe possible directions of attacks in the case of router's PFC failure. In this event, attacks by intruders can be successful and incapacitate the router's hardware and software.

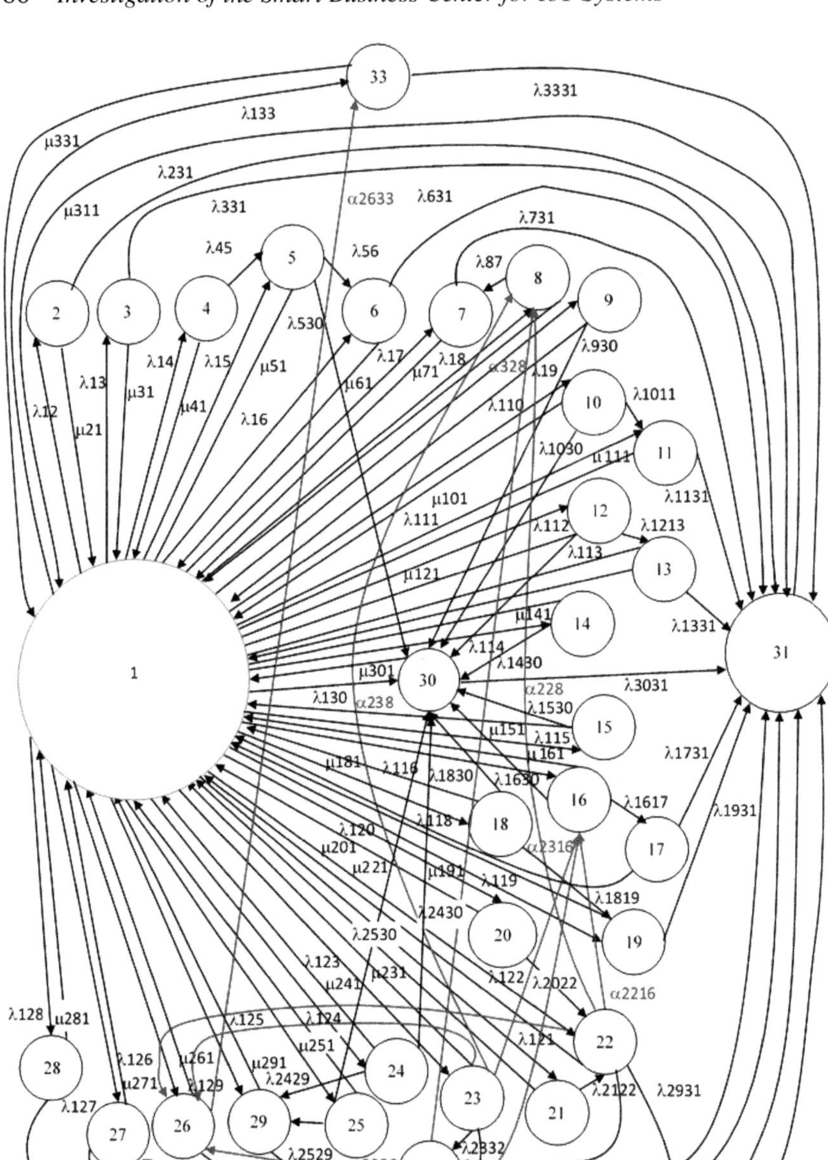

Figure 8.2 The Markov model of the router states.

To assess the SBC router availability to perform the specified functions under attacks, the stationary availability ratio AC has been selected as a complex indicator of dependability (reliability, security/integrity, and maintainability). This factor is determined by the expression:

$$\mathrm{AC} = \mathrm{P1\,(t) + P4\,(t) + P10\,(t) + P12\,(t) + P16\,(t) + P18\,(t) + P20\,(t)} \\ + \mathrm{P21\,(t)}. \qquad (8.1)$$

For the developed Markov model, a system of linear differential equations of Kolmogorov–Chapman was constructed under the initial conditions:

$$P1\,(0) = 1; \sum_{i=1}^{33} Pi(t) = 1.$$

Pi(t) – the probability of SBC router states.

8.3.3 Simulation Results

For the chosen values of the transition rates, the AC was calculated and the graphical dependencies of the AC of the SBC router on the change in transition rates under the influence of attacks were obtained (Figures 8.3–8.7). Figures 8.3–8.5 show the change in the AC for the most critical states of the router and Figures 8.6 and 8.7 show some other states.

The Markov model allows determining the most critical state of the router with a certain set of input data in the attack conditions, and without them. Analysis of the graphical dependences shown in Figure 8.3 showed that the value of the router availability ratio with the selected initial data in case of a failure of the card and the protection software is reduced by an order of magnitude compared to the value of the router availability factor without the impact of attacks. Figure 8.4 shows that in the event of PFC failure, the value AC = 0.9999869986348 for the range of values $\lambda731$ = 0 ... 0.000002 1/h changes insignificantly, and for the same range of values $\lambda3031$ and $\lambda1731$, the value of AC = 0.9999869986335.

Analysis of the graphical dependences shown in Figure 8.5 showed that the value of the availability factor of the router with the selected initial data and good working state of PFC meets the requirements for availability and is AC = 0.999995594197653. With an increase in the failure rate $\lambda731$, $\lambda1731$, and $\lambda3031$ to 0.000002 1/h, the change in the AC value is not significant: for $\lambda731$ – in 12 position of decimal places.

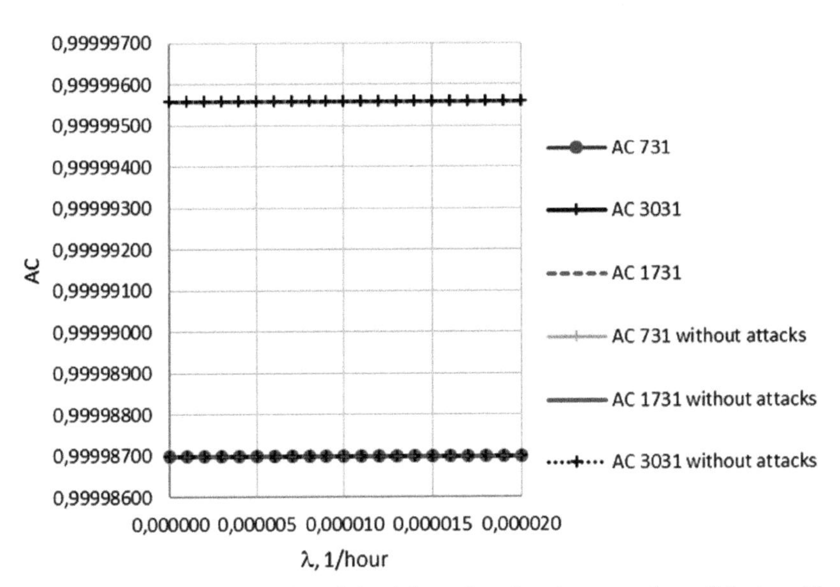

Figure 8.3 Graphic dependencies of the AC on changing the rate values of the transitions λ731, λ3031, and λ1731 in the event without the impact of attacks on the router's components and in the event of the PFC security card failure and successful attack on the router's components.

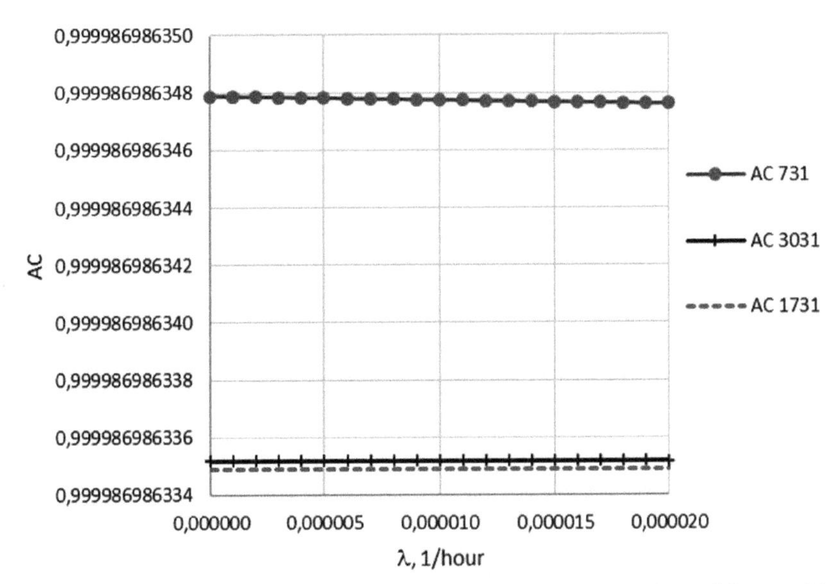

Figure 8.4 Graphic dependencies of the AC on changing the rate values of the transitions λ731, λ3031, and λ1731 in the event with the impact of attacks on the router's components.

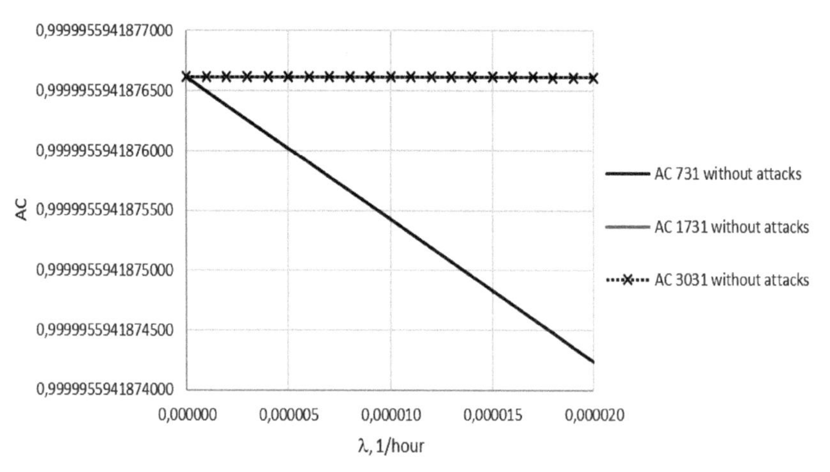

Figure 8.5 Graphic dependencies of the AC on changing the rate values of the transitions λ731, λ3031, and λ1731 in the event without attacks on the router's components.

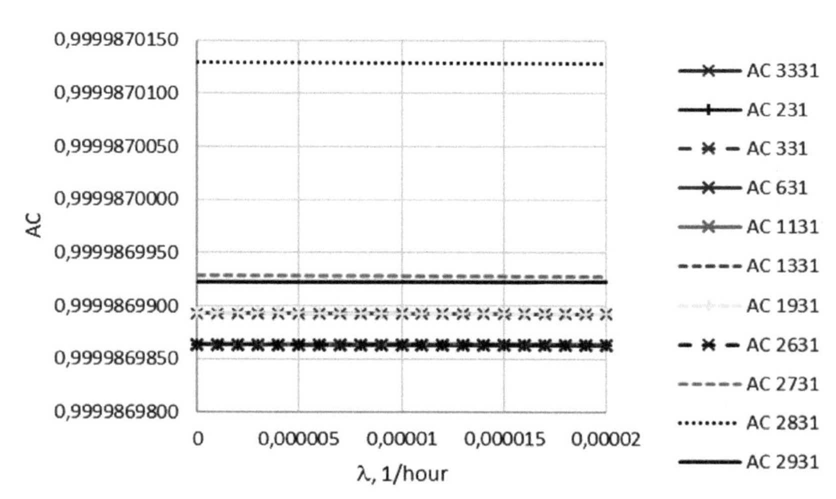

Figure 8.6 Graphic dependencies of the AC on changing the rate values of the transitions λ3331, λ231, λ331, λ631, λ1131, λ1331, λ1931, λ2631, λ2731, λ2831, and λ2931 in the event of the PFC security card failure and successful attack on the router's components.

Analysis of the graphical dependences of the transition intensities from the failure state or the failure of the individual small components of the router to its router failure state (Figure 8.6) showed that when the transitions rates λ3331, λ231, λ331, λ631, λ1131, λ1331, λ1931, λ2631, λ2731, λ2831, and λ2931 change in the range 0–0.000002 1/h, the value of the availability factor

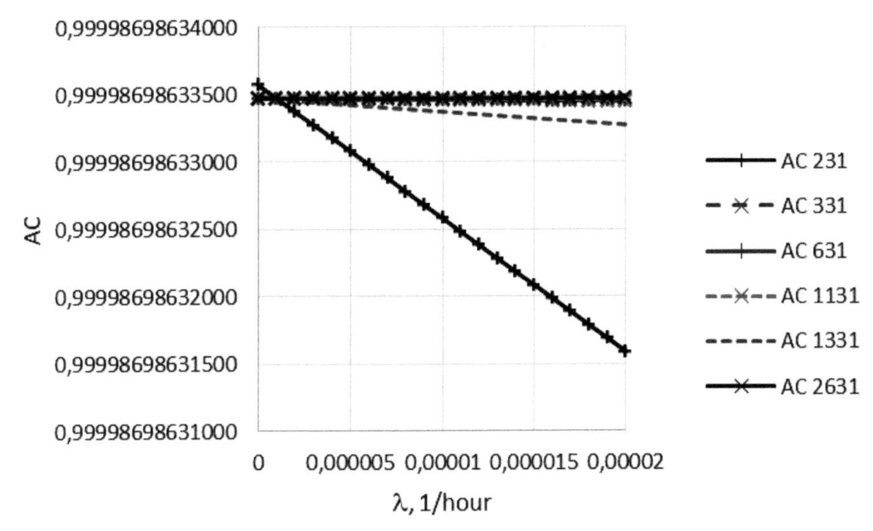

Figure 8.7 Graphic dependencies of the AC on changing the rate values of the transitions λ231, λ331, λ631, λ1131, λ1331, and λ2631 in the event of the PFC security card failure and successful attack on the router's components.

for PFC failure and the successful impact of attacks does not exceed 0.999987 (for λ2831) and 0.999986995 for the other intensities shown in Figure 8.6.

Analysis of the graphical dependences (Figure 8.7) showed that when the values of λ231, λ331, λ631, λ1131, λ1331, and λ2631 are in the range 0–0.000002 1/h change, the AC has the value 0.999986986335. Most of all, the ROM responds to the attack and its AC decreases to the value of AC = 0.9999869863165.

8.4 Conclusion

The results of the research allow substantiating ways of increasing the dependability of router operation, which is one of the main components of complex SBC systems.

The analysis of the SBC router states in cases of attacks impact is carried out. To investigate and evaluate the reliability parameter – the router's availability ratio – a Markov model is constructed that includes all possible states of a complex system router. Thanks to the developed model of the router states, it became possible to determine the most critical states in attack conditions.

The most critical states of the model are 22, 23, and 32. The considered scheme of the router is reliable. But, taking into account the model research results, it is necessary to take measures to improve both the security and reliability of RAM, processors, and PFC hardware and software.

Research has shown that attacks reduce the availability of even highly reliable router schemes. Therefore, it is necessary to patch the vulnerabilities in the router firewall in a timely manner.

The analysis showed that the AC of the router is strongly affected by the technical state of the PFC. In case of its failure, or its absence, the value of the PFC is reduced by an order of magnitude. Therefore, it is necessary to take measures to maintain the efficiency of the protection card. This can be: reservation of the PFC as a whole, multi-version of its software.

It is necessary to reduce the time of recovery of the router after faults and failures and in particularly difficult situations – reserving of the router.

Further research can be aimed at clarifying the time of attacks on SBC components, so that the router software vulnerability is closed by the patch in a timely manner to prevent the failure of SBC components.

Acknowledgments

This research is supported by the project STARC (Methodology of SusTAinable Development and InfoRmation Technologies of Green Computing and Communication) funded by the Department of Education and Science of Ukraine. Besides, the authors thank the colleagues on Erasmus + project ALIOT (Internet of Things: Emerging Curriculum for Industry and Human Applications, 573818-EPP-1-2016-1-UK-EPPKA2-CBHE-JP) for discussion during the development of the M.Sc. and Ph.D. courses related to IoT systems' dependability and security investigation and assessment. Part of this book is the result of research activities in the framework of Horizon2020 640073 R2RAM Project. The authors would like to thank acknowledge the European Commission, the Research Executive Agency (REA) and the former project officers Traian Branza (for SkyFlash) and Giuseppe Daquino (for R2RAM).

References

[1] Roberto, M., Biru, A., and Rotondi, D. (2015). *Towards a definition of the "Internet of Things (IoT),"* Revision 1, 86. Available at: https://iot.ieee.org/images/files/pdf/IEEE_IoT_Towards_Definition_Internet_of_Things_Revision1_27MAY15.pdf.

[2] ISO/IEC JTC 1 Preliminary Report. (2014). *Internet of Things (IoT)*, 17. Available at: https://www.iso.org/files/live/sites/isoorg/files/developing_standards/docs/en/internet_of_things_report-jtc1.pdf.

[3] Rajkumar, B., and Dastjerdi, A. V. (2016). "*Internet of things principles and paradigms*," Morgan Kaufmann is an imprint of ELSEVIER, 241.

[4] Telecommunication standardization sector of ITU, ITU-T Y.2060 (2012). "*Series Y: global information infrastructure, internet protocol aspects and next-generation networks. Next Generation Networks – Frameworks and functional architecture. Overview of the Internet of things*" Available at: https://www.itu.int/rec/T-REC-Y.2060-201206-I.

[5] Preliminary Report. (2014). *ISO*, Switzerland, 17. Available at: https://www.iso.org/files/live/sites/isoorg/files/developing_standards/docs/en/internet_of_things_report-jtc1.pdf.

[6] IERC AC4. (2012–2014). "*IoT Semantic Interoperability: Research Challenges, Best Practices, Solutions and Next Steps.*" IERC AC4 Manifesto – 'Present and Future'," 105.

[7] Rajkumar, B., and Dastjerdi, A. V. (2016). "*Internet of Things Principles and Paradigms*," 380. Available at: http://www.cs.newpaltz.edu/~lik/publications/books/e978012 8053959.pdf.

[8] Vitor, R., Denisczwicza, M., Dutraa, M., Ghodousb, P., da Silvab, C. F., Moayeric, N. et al., (2016). "*An application domain-based taxonomy for IoT sensors*," 248–258. Available at: https://hal.archives-ouvertes.fr/hal-01581127/file/2016-TE2016-Taxonomy-for-IoT-Sensors.pdf.

[9] Mark, D. (2017). "*A security assessment of z-wave devices and replay attack vulnerability. GIAC (GSEC) gold certification*," the SANS Institute, 29. Available at: https://www.sans.org/reading-room/whitepapers/internet/security-assessment-z-wave-devices-replay-attack-vulnerability-37242.

[10] ESET. (2018). "*Spectre and meltdown vulnerabilities discovered*," ESET Customer Advisory 2018-0001. Available at: https://support.eset.com/ca6643/

[11] Brian, B. (2017). "*How hackers contributed to the z-wave security framework.*" Available at: http://www.ioti.com/security/how-hackers-contributed-z-wave-security-frame work.

[12] Matt, L., and Parseghian, P. (2018) "Today's CPU vulnerability: what you need to know." Available at: https://security.googleblog.com/2018/01/todays-cpu-vulnerability-what-you-need.html

[13] National Highway Traffic Safety Administration. (2016). *Cybersecurity Best Practices for Modern Vehicles*. (Report No. DOT HS 812 333), Washington, DC.

[14] Semantec. (2016) "*An internet of things reference architecture*," 22. Available at: https://www.symantec.com/content/dam/symantec/docs/white-papers/iot-security-reference-architecture-en.pdf.

[15] Pennie, W. (2017). "*The risks of using portable devices*," 5. Available at: https://www.us-cert.gov/sites/default/files/publications/RisksOfPortableDevices.pdf.

[16] Semantec. (2017). "*Internet security threat report ISTR*," 22, 77. Available at: https://www.symantec.com/content/dam/symantec/docs/reports/istr-22-2017-en.pdf.

[17] Jun, Z., and Jamalipour, A. (2009). "*Wireless sensor networks: A networking perspective*," Institute of Electrical and Electronics Engineers, 489.

[18] Cognosec. (2015). "*Black hat, ZigBee exploited – The good, the bad and the ugly*," 8. Available at: https://www.blackhat.com/docs/us-15/materials/us-15-Zillner-ZigBee-Exploited-The-Good-The-Bad-And-The-Ugly-wp.pdf.

[19] Niko, V., Haataja, K., Patino-Andres, J. L., and Toivanen, P. (2013). "Security threats in zigbee-enabled systems: vulnerability evaluation, practical experiments, countermeasures, and lessons learned," *Conference Paper: Conference: System Sciences (HICSS)*. Available at: https://www.researchgate.net/publication/261457783_Security_Threats_in_ZigBee-Enabled_Systems_Vulnerability_Evaluation_Practical_Experiments_Countermeasures_and_Lessons_Learned.

[20] Fan, X., Susan, F., Long, W., and Li, S. '*Security Analysis of Zigbee*,' 18. Available at: https://courses.csail.mit.edu/6.857/2017/project/17.pdf.

[21] Behrang, F., and Ghanoun, S. (2013). *Hacking Z-Wave home automation systems*. Available at: https://www.slideshare.net/sensepost/hacking-zwave-home-automation-systems.

[22] Sigma. (2016). "*Introduction to the z-wave security ecosystem*," ABR, 13.

[23] Warda, A., Anwar, S., and Arshad, M. J. (2016). "Security architecture of 3GPP LTE and LTE-A network: a review," in *International Journal of Multidisciplinary Sciences and Engineering* [ISSN: 2045-7057], 7(1). Available at: http://www.ijmse.org.

[24] Bikos, A., and Sklavos, N. (2013). "Security problems of LTE/SAE in wireless networks 4G," in *IEEE Security and Privacy Magazine*,

11(2), 55–62. Available at: https://www.researchgate.net/publication/236117981_LTESAE_security_issues_on_4G_wireless_networks.

[25] Bikos, A., and Sklavos, N. (2013). LTE/SAE security issues in 4G wireless networks. Available at: https://www.researchgate.net/publication/236117981_LTESAE_security_issues_on_4G_wireless_networks.

[26] Schoinas, P. C. (2013) "Secure military communications on 3G, 4G and WiMAX," in *The NPS Institutional Archive DSpace Repository 2013-09*." Available at: https://calhoun.nps.edu/bitstream/handle/10945/37712/13Sep_Schoinas_Panagiotis.pdf?sequence=1.

[27] Xenakis, C., and Merakos, L. (2004). Security in third generation mobile networks. *Computer Communications*, 27, 638–650.

[28] Robert, M. (2016). "*LoRa security building a secure LoRa solution*," MWR Labs Whitepaper, 18. Available at: https://www.labs.mwrinfosecurity.com.

[29] Alliance (2017). *LoRaWAN*^{TM} *1.1 Specification,* LoRa Alliance, 101. Available at: https://lora-alliance.org/sites/default/files/2018-04/lorawantm_specification_-v1.1.pdf.

[30] Vince, S. (2003). "Pervasive computing goes the last hundred feet with RFID systems," 6. Available at: http://echo.iat.sfu.ca/library/stanford_03_pervasive Comp_ RFID. pdf.

[31] Alexander, I., Michahelles, F., Lehtonen, M., and Ostojic, D. (2009). "*Securing RFID systems by detecting tag cloning*," 18. Available at: http://www.avoine.net/rfid/download/papers/LehtonenOIM-2009-perva sive.pdf.

[32] Nimish, V., Patwardhan, A., Joshi, A., Finin, T., and Nagy, P. (2006). "*Protecting the privacy of passive RFID tags*," 5. Available at: http://ebiquity.umbc.edu/_file_directory_/papers/330.pdf.

[33] William, O., Filippoupolitis, A., and Loukas, G. (2017). "*Evaluating the impact of malicious spoofing attacks on bluetooth low energy based occupancy detection systems*," ISBN Information: INSPEC Accession Number: 17010022. Available at: http://ieeexplore.ieee.org/document/7965755/?reload=true.

[34] Sławomir, J. (2016). "*Gattacking bluetooth smart devices*," white paper, 13 p., 2016. Available at: http://gattack.io/whitepaper.pdf, www.securing.pl.

[35] US-CERT, (2005). "*Spyware*," 8 p., 2016. Available at: https://www.us-cert.gov/sites/default/files/publications/spywarehome_0905.pdf.

[36] Threat hunting 2017 report presented by cybereason (2017). Available at: http://technodocbox.com/amp/67698931-Network_Security/Threat-hunting-2017-report-presented-by.html.

[37] Cisco. "*Cisco 5915 Embedded Service Router*." Available at: https://www.cisco.com/c/en/us/support/routers/5915-embedded-service-router/model.html.

[38] Kolisnyk, M., Kharchenko, V., Piskachova, I., and Bardis, N. (2017). "A Markov model of IoT system availability considering DDoS attacks and energy modes of server and router," in *ICTERI Conference*, Kiev, 14. Available at: http://ceur-ws.org/Vol-1844/10000699.pdf.

9

An Internet of Drone-based Multi-version Post-severe Accident Monitoring System: Structures and Reliability

Herman Fesenko[1,2,*], Vyacheslav Kharchenko[1,3,*], Anatoliy Sachenko[4,5], Robert Hiromoto[6,7] and Volodymyr Kochan[4]

[1]Department of Computer Systems, Networks and Cybersecurity, National Aerospace University "KhAI", Kharkiv, Ukraine
[2]O. M. Beketov National University of Urban Economy in Kharkiv, Kharkiv, Ukraine
[3]Centre for Safety Infrastructure Oriented Research and Analysis, Kharkiv, Ukraine
[4]Ternopil National Economic University, Ternopil, Ukraine
[5]Kazimierz Pulaski University of Technology and Humanities in Radom, Radom, Poland
[6]University of Idaho, Moscow, ID, United States
[7]Center for Advanced Energy Studies, Idaho Falls, ID, United States
*Corresponding Authors: h.fesenko@khai.edu; v.kharchenko@csn.khai.edu

A general structure and underlying principles for creating an Internet-of-Drone-based multi-version post-severe NPP accident monitoring system is described in this paper. The proposed design consists of an Internet-of-Things subsystem, a single-wired subsystem, and three drone-based wireless subsystems. Reliability block diagrams for the system and its subsystems are built based on considerations of different variants of sensor, communication, and decision-making systems. On the basis of reliability block diagrams, reliability models of the system and their subsystems are developed. The probability of failure-free operation that depends upon various system configurations and on the use of multiple redundant Wi-Fi communicating drones is obtained and analyzed.

9.1 Introduction

9.1.1 Motivation

Nuclear power plants (NPPs) play a significant role in ensuring the energy independence for many countries, as it is one of the most powerful and clean energy sources available. On the other side, NPPs are made up of very complex cyber physical systems that if compromised can result in potentially hazardous environmental consequences. It is, therefore, important to ensure the stability and reliability of NPPs by maintaining continual control and monitoring capabilities of equipment components that run the power units.

In case of an accident, a special highly resilient and survivable monitoring system is needed to measure critical parameters of the reactor's safety sub-components and provide mechanisms to physically assess criticality levels. In addition, the monitoring system must also maintain the decision-making policies for NPP recovery and minimal start-up sequences. In general, these issues can be mitigated by means of specialized equipment integrated into a post-accident monitoring system (PAMS). Such a system would typically consist of multiply redundant sensors and wired (cabled) networks connected to the crisis center (CrS). However, the Fukushima NPP accident confirmed that such types of equipment are vulnerable to both natural and man-made disasters. For example, at the Fukushima Daiichi NPP, a massive earthquake triggered a large tsunami that washed over its six nuclear power reactors. After the earthquake, the reactors were automatically shut down as planned. However, it was the tsunami that caused the main damage to the monitoring equipment, including the data channels. As a consequence, the design of a PAMS must tolerate equipment failures and maintain the ability to redirect critical information flows for proper decision making in real time.

9.1.2 State of the Art

After the Fukushima accident, the implementation of a reliable and survivable PAMS has become a priority requirement within the national and international regulatory bodies. A PAMS can also benefit other critical infrastructures (e.g., chemical enterprises, oil–gas transport systems, and so on). According to [1], such a system assures the functionality of accident and post-accident monitoring for any anticipated design-specific events. More importantly, the PAMS design must provide resilience under unforeseen and hazardous conditions such as earthquakes, severe fuel damage, and full de-energization of an NPP unit.

The diversity (multiversity) principle [2] incorporates the use of two or more options to perform the same functional operation. A PAMS, based on this principle of version-functional redundancies, is called a multi-version system. Kima et al. [3] proposed the use of PAMS based on this principle. They argued that in coping with severe accidents, such as Fukushima, a fully independent monitoring system is required, which is separated (isolated) from the conventional instrumentation and control system. Recent development in data management as well as modern solutions in aviation technology makes possible the exploitation by unmanned aerial vehicles (UAVs), often referred to as drones, for critical infrastructures' monitoring. The work [4] is devoted to a modular multi-mission airborne sensor system capable of performing operations from reconnaissance to radiological and nuclear surveillance. Towler et al. [5] present a remote sensing system for radiation detection and aerial imaging. Sanada and Torii [6] discuss a PAMS, which includes UAVs, unmanned observation boats, radioactivity radio-sonde, manned helicopters, and other devices needed for efficient monitoring an accident. Schneider et al. [7] note that UAVs can be deployed in a short time frame to map large areas (on the order of square kilometers) with regard to dose rate, surface activity, or radionuclide identification; they can collect vital data to be used by decision makers. The paper [8] proposes a drone-based solution to help in the search and rescue activities during disaster scenarios. The architecture proposed in this paper is composed of specialized drones that are designed to accomplish specific tasks and fitted within internal modules that are organized to accomplish their objectives.

The use of UAVs provides several advantages to the critical monitoring of accidents. In [9], the authors highlight the gain in terms of time and human resources, as they can free rescue teams from time-consuming data collection tasks, and assist research operations with more insightful and precise situation-aware guidance, thanks to today's advanced sensing capabilities. Tina et al. [10] present a novel approach of using UAVs to establish a communication infrastructure in case of disasters. The authors explain the details of the system in three aspects: end-to-end communication, localization and navigation, and coordination. An approach based on the extent of interaction between the UAV and terrestrially deployed wireless sensors is presented in [11]. The paper [12] considers the possibility of achieving the most efficient usage of a transmission channel capacity and the receiving of a generalized image (in a particular servicing object state) that relies on the entropy estimation of its sensors activities.

The coordination and the wireless network connectivity offered by UAVs can be regarded as an Internet of Drones (IoD). An IoD can be defined as a layered network control architecture designed for coordinating the access of drones in controlled airspaces and providing navigation services [13, 14]. The network consists of the following layers: airspace, node-to-node, end-to-end, services, and applications. In the layered IoD architecture, every layer can utilize services provided by all layers below it. At the present time, an IoD tends to be actively used in both civilian and military settings.

According to [15], although there are apparent benefits of such an architecture (e.g., avoiding airborne collisions, and guaranteeing safety and security through greater control over where drones can and cannot be), there are challenges that remain unsolved. Examples include effective routing and congestion control, and in particular security and privacy of data collected from the drones and outsourced to the cloud. Lin et al. [15] introduce a typical IoD architecture and describe several security and privacy challenges, such as privacy leakage, and the need for secure and efficient data sharing. They describe potential solutions suited to the nature of an IoD architecture. Other challenges related to the use of an IoD are energy provisioning, connectivity, and operations. Long et al. [16] conceptualize the energy neutral IoD (*en*IoD) to enable enhanced connectivity between drones by overcoming energy limitations for autonomous and continuous operation. The energy used in the system is originally gathered from renewable resources to minimize the gap between harvested and consumed energy in principle. Possible communication and networking solutions are presented to realize multi-dimensional objectives of *en*IoD. Among the issues relating to IoD applications, drones have limited processing and storage capabilities, and cannot cope with massive computational requirements for certain missions. To overcome these issues, Qureshi et al. [17] propose an integrated cloud-based architecture for an IoD, which integrates drones with the cloud to (1) virtualize access to drones and (2) offload heavy computations from the drones to the cloud.

Consequently, an IoD-based accident monitoring system for NPPs and other critical infrastructures offers a flexible alternative to more traditional approaches that lack real-time mobility and organizational diversity. However, the basic concepts of an IoD-based accident monitoring systems still require further study, analysis, and development before a fully operational system can be deployed.

9.1.3 The Goals and Structure

The goals of this chapter are to develop the principles, the design of scalable IoD-based communications infrastructures, and the corresponding reliability models that can support a PAMS for severe accident surveillance of NPPs. The chapter is structured as follows. Section 9.2 describes the principles of creating and structuring of the IoD-based monitoring systems. Section 9.3 develops the reliability models of PAMS and its subsystems. Section 9.4 explores the different options of IoD application for PAMS. Section 9.5 discusses the results and future steps of research.

9.2 Principles of Creating an Internet-of-drones-based Multi-version Post-severe Accident Monitoring System

9.2.1 Structure

Existing NPP PAMSs are based on wired networks that connect sensor areas directly with the crisis center. Reliability and survivability of such systems are assured by redundancy of equipment, cable communications, and other components. However, in the case of severe accidents, wired network-based PAMS can experience damaged sensors or broken cable connections. Under such conditions, the NPP PAMS is partially or totally rendered useless. To avoid such a problem, a multi-version approach is envisioned in which the wired network components and interfaces are expanded to include wireless communication components, which is more resilient to bridge physical failures.

To assure stability of a wireless network-based PAMS subsystem after an accident or a powerful jamming attack, a reliable transmission of data to support the possible failures of a wired network is deemed essential. Consequently, to improve the survivability of PAMS, the authors have introduced the use of a drone-fleet (Hiromoto et al. [18] and Kharchenko et al. [19]). An approach to the drone fleet subsystem survivability assessment that is based on a stochastic continuous-time algorithm is proposed in [20].

In the normal operational mode, data and command exchanges run through the wired network. If this process is damaged during an accident, a fleet of communication drones acting as an auxiliary wireless network is created to support these activities. Drones are launched in the event of a wired network failure or in the detection of a possible severe accident. The drones are designed to autonomously form a stable flight formation, which is configured in a master/slaves' arrangement. In order to conserve battery power or in the case of node dropout, the master node (which is given command responsibilities) can be reassigned to other slave nodes when deemed appropriate. From this vantage point, the formation of drones will cooperate to maintain the following functions: 1) to monitor and collect all data from sensor modules that are equipped with wireless connections; 2) to form a reliable mesh network for optimal data streaming between point-to-point transmissions; 3) to provide surveillance imaging for damage control, and search and rescue; 4) to summarize areas of contamination; and 5) to provide an unmanned observation platform for exploratory surveillance.

Finally, in a severe accident, it will be very important to minimize the power consumption of both the measurement and control modules, since both the wired and wireless microcontrollers can be activated. As a consequence, the power demands required by the wireless interfaces that interact with the drones must be analyzed from a signal strength perspective.

The structure of the proposed multi-version, post-sever accident monitoring system (MPSAMS) is shown in Figure 9.1. This system consists of the following components: 1) the NPP; 2) sensors composed of drones, Wi-Fi sensors (Wi-FiS), *light fidelity* sensors (Li-FiS) based on a bidirectional wireless technology similar to Wi-Fi, and Internet-of-Things sensors (IoTS); 3) several drone fleets (DF1 and DF2); and 4) a communication interface to a crisis center decision-making system (DMS), an autonomous decision-making support system (DMSS; groups of experts) to assure crisis center functionality, and an Internet cloud portal. Finally, the Internet cloud portal is made up of one IoT subsystem (IoT S) and three drones-of-things subsystems labeled DoT S1, DoT S2, and DoT S3.

9.2.2 Principles

The underlying principles for creating the MPSAMS (see Figure 9.1):

1. Diversity of sensor types: Wired and wireless sensors provide for greater flexibility under various failure modes. Three types of sensor modes are described here: wired sensors (WireS), sensors derived from the Internet of Things (IoTS), *light fidelity* sensors (Li-FiS), and drone-based wireless sensors that include drones employing Wi-Fi (Wi-FiS).
2. Diversity of transmission modes: Wired, wireless drone-based point-to-point and mesh routing.
3. Diversity of data types: Digital, light, and acoustic.
4. Mobility (drones) management.
5. Dynamical reconfiguration and redundancy (drone fleet).

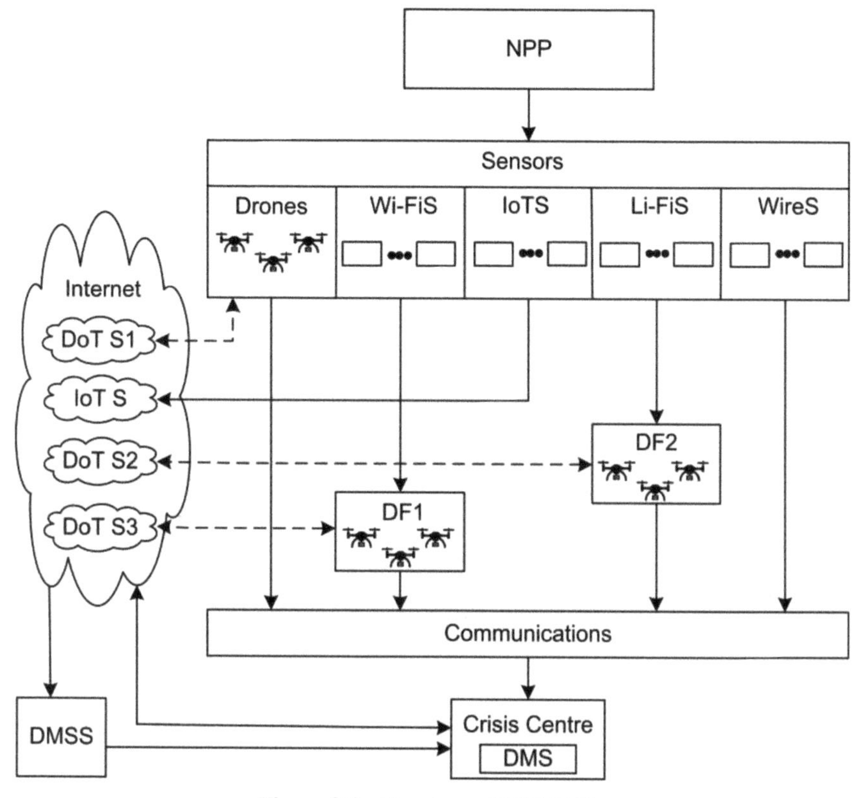

Figure 9.1 Structure of MPSAMS.

The primary task of sensors is to monitor physical parameters such as reactor temperature, reactor pressure, coolant flow, and radioactivity. Under a severe system failure, drone sensors replace the functionality of damaged wired sensors, which may monitor gamma dose rates within a 30-km radius of an NPP or provide views of inaccessible interior areas of a damaged NPP. The Internet subsystems IoT S, DoT S1, DoT S2, and DoT S3 collect real-time data, ensure their long-term storage, and provide current and retrospective data about accident related parameters. More importantly, DoT S1, DoT S2, and DoT S3 manage and coordinate the deployment and cooperation between drone-sensor fleets (e.g., DF1 and DF2) and ensure an additional channel for DMS or DMSS (groups of experts). DMS/DMSS shall perform the following main functions: 1) receiving notifications and initiating the response to an accident; 2) coordination and direction of on-site response actions; 3) providing technical and operational support to those personnel performing tasks at a facility and those personnel responding off the site; 3) direction of off-site response actions and coordination with on-site response actions; 4) coordination of monitoring, sampling, and analysis; and 5) coordination of communication with the public.

The drone fleet is located permanently at a considerable distance from the NPP and housed within the NPP as well. The communication network (wireless network and drone fleet systems) is deployed after the accident and drones fly to the designated accident zone. The drone fleet is divided into subsets: 1) repeaters (slave) that work together on a principle of "one leader." If the "leading drone-repeater" (master) is damaged, then another drone-repeater takes over the previous master's functions; 2) observers (equipped with a TV camera) when enabled runs the continuous visual monitoring of the accident location; and 3) additional sensors that can be carried by drones or placed in certain locations. Drones should be able to change their role by upgrading equipment at their base station.

9.3 Reliability Models for the Internet-of-drones-based Multi-version Post-severe Accident Monitoring System

9.3.1 Simplified Structure

Figure 9.2 illustrates a simplified structure (architecture) of a multi-version PAMS (MPSAMS) as a modified wired network that combines an IoT subsystem (SubIoT) with three wireless subsystems. The wired network is regarded as a subsystem denoted by SubG. In principle, the wireless components of the MPSAMS structure are divided into sensor components (left block) and their corresponding communication components (right block). The Wi-Fi subsystem (SubW), Li-Fi subsystem (SubL), and Wi-Fi drone-based subsystem (SubD) make up the wireless sensor components. Their corresponding data transmissions (communications) are indicated by arrows pointing to the right block. It should be noted that the drone fleet is integrated within the MPSAMS as both sensor components and communication components. The IoT subsystem provides additional sensor/drone capabilities not specified in the illustration. The MPSAMS architecture has the flexibility to easily incorporate *multi-version redundancy* of sensors and communication components.

It is important to note that the subsystems SubIoT, SubW, SubL, SubD, and the SubG are all diverse in terms of capabilities among one another.

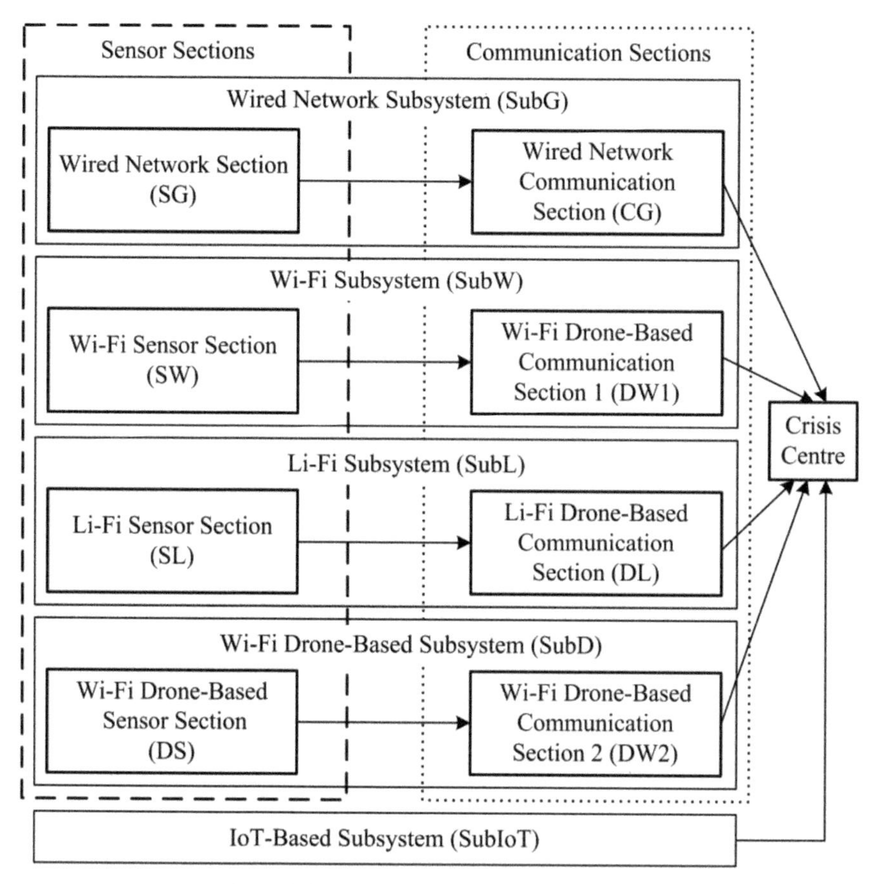

Figure 9.2 Simplified structure of MPSAMS.

This feature is critical in designing fail–safe systems. Also note that the communication sections for the Wi-Fi subsystem, the Li-Fi subsystem, and the Wi-Fi drone-based subsystem are all drone-based. Although not shown, portions of the IoT-based channels are also drone-based. This drone-based, communication capability is designed to support reliable transmission of data if the wired subsystem section happens to fail. Finally, to increase the MPSAMS' survivability, both sensor and communication sections of wireless network subsystems should be equipped with backup batteries and multiple blocks of wireless communication modules as well as self-testing and self-diagnostic systems. Thus, such a multi-version subsystem can be considered as an IoD-based MPSAMS.

9.3.2 Subsystems' Reliability Models

Figures 9.3–9.6 illustrate a reliability block diagram (RBD) for each MPSAMS subsystem described above (note that the IoT subsystem RBD will be taken up in Section 9.3.3). From these RBDs, an equation for the probability of failure-free operation (PFFO) can be obtained. The construction of the RBD is based on the following assumptions for each subsystem:

- Subsystems are unrecoverable.
- Each subsystem (drones/sensors) has two states: operational and non-operational.
- Drone/sensor failures are independent.
- A traditional PAMS is a wired network subsystem with a parallel–series chain structure.
- The sensor and communication sections of the Wi-Fi subsystem, the Li-Fi subsystem, and the Wi-Fi drone-based subsystem have a structure of type "k-out-of-n."
- All drones/sensors within their own group are identical.
- The sensor and communication sections are made up of main (primary) sensors and redundant sensors for backup.
- The switching process is "ideal," i.e., fault free.

Elements in Figure 9.3 are designated in the following way:

- n is the number of the sensors of the main/redundant chain of the sensor section.
- $\mathrm{m_{SG}}i$ is a sensor in sensor group i of the main chain of the sensor section $(i = 1, \ldots, n)$.
- $\mathrm{r}_{SG}i$ is a sensor in sensor group i of the redundant chain of the sensor section $(i = 1, \ldots, n)$.

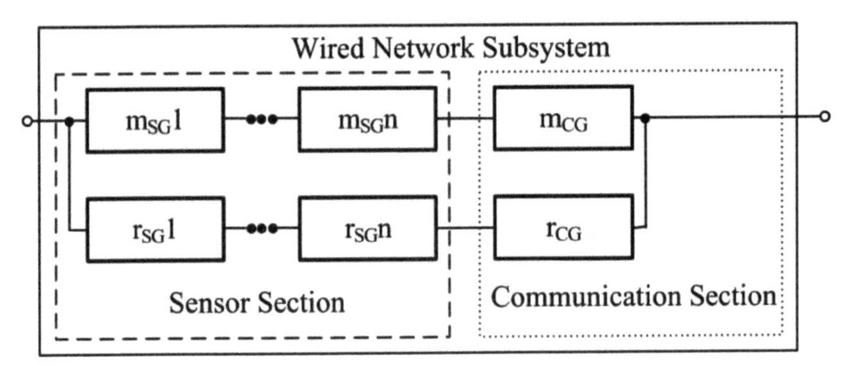

Figure 9.3 RBD for the wired network subsystem.

- m_{CG} is the main sensor of the communication section.
- r_{CG} is the redundant sensor of the communication section.

Elements in Figure 9.4 are designated in the following way:

- k is the number of groups of the Wi-Fi sensor section.
- $m_{SG}i\eta$ is a sensor η of the main chain of a group i of the Wi-Fi sensor section ($i = 1, \ldots, k$; $\eta = 1, \ldots, u_{mi}$, where u_{mi} is the number of such sensors).
- $r_{SG}ij$ is a sensor j of the redundant chain of a group i of the Wi-Fi sensor section ($i = 1, \ldots, k$; $j = 1, \ldots, u_{ri}$, where u_{ri} is the number of such sensors).
- $CU_{SW}i$ is the switch of group i of the Wi-Fi sensor section ($i = 1, \ldots, k$).
- $m_{DW1}i$ is a drone i of the main chain of the Wi-Fi drone-based communication Section 9.1 ($i = 1, \ldots, k$).
- $r_{DW1}b$ is a drone b of the redundant chain of the Wi-Fi drone-based communication Section 9.1 ($b = 1, \ldots, \beta$, where β is the number of such drones).
- CU_{DW1} is the switch of the Wi-Fi drone-based communication Section 9.1.

Elements in Figure 9.5 are designated in the following way:

- p is the number of groups of the Li-Fi sensor section.
- $m_{DS}i\eta$ is a sensor η of the main chain of a group i of the Li-Fi sensor section ($i = 1, \ldots, p$; $\eta = 1, \ldots, v_{mi}$, where v_{mi} is the number of such sensors).
- $r_{SL}ij$ is a sensor j of the redundant chain of a group i of the Li-Fi sensor section ($i = 1, \ldots, p$; $j = 1, \ldots, v_{ri}$, where v_{ri} is the number of such sensors).

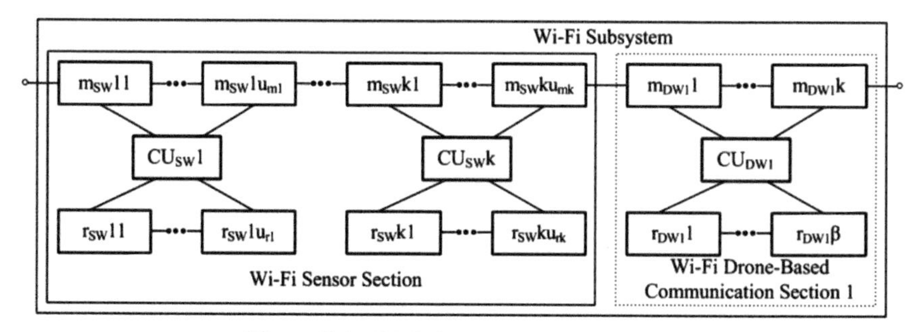

Figure 9.4 RBD for the Wi-Fi subsystem.

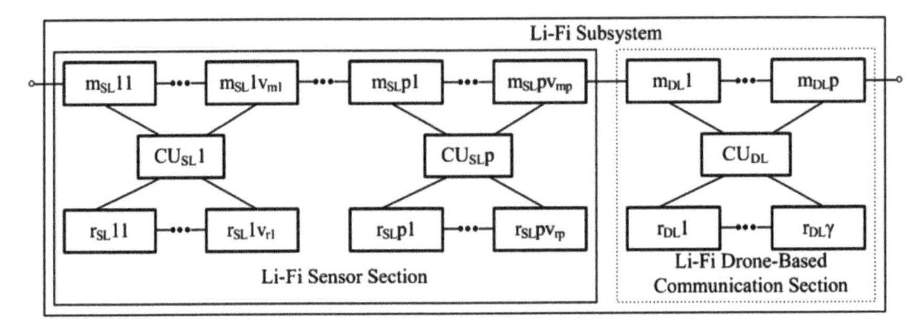

Figure 9.5 RBD for the Li-Fi subsystem.

- $\mathrm{CU_{SL}}i$ is the switch of a group i of the Li-Fi sensor section ($i = 1, \ldots, p$).
- $\mathrm{m_{DL}}i$ is a drone i of the main chain of the Li-Fi drone-based communication section ($i = 1, \ldots, p$).
- $\mathrm{r_{DL}}b$ is a drone b of the redundant chain of the Li-Fi drone-based communication section ($b = 1, \ldots, \gamma$, where γ is the number of such drones).
- $\mathrm{CU_{DL}}$ is the switch of the Li-Fi drone-based communication section.

Elements in Figure 9.6 are designated in the following way:

- $\mathrm{m_{DS}}i$ is a drone i of the main chain of the Wi-Fi drone-based sensor section ($i = 1, \ldots, f$, where f is the number of such drones).
- $\mathrm{r_{DS}}j$ is a drone j of the redundant chain of the Wi-Fi drone-based sensor section ($i = 1, \ldots, \omega$, where ω is the number of such drones).

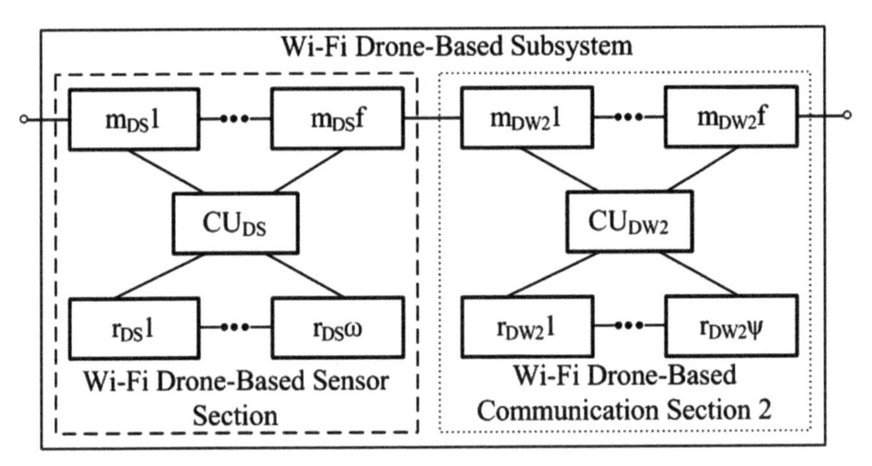

Figure 9.6 RBD for the Wi-Fi drone-based subsystem.

- $\mathrm{CU_{DS}}$ is the switch of the Wi-Fi drone-based sensor section.
- $\mathrm{m_{DW2}}i$ is a drone i of the main chain of the Wi-Fi drone-based communication Section 9.2 ($i = 1, \ldots, f$).
- $\mathrm{r}_{DW2}b$ is a drone b of the redundant chain of the Wi-Fi drone-based communication Section 9.2 ($b = 1, \ldots, \psi$, where ψ is the number of such drones).
- $\mathrm{CU_{DW2}}$ is the switch of the Wi-Fi drone-based communication Section 9.2.

Based on the above assumptions and the constructed RBD (see Figures 9.3–9.6), the PFFO can be calculated for: 1) the wired network subsystem using Equation (9.1) derived in accordance with the RBD shown in Figure 9.3; 2) the Wi-Fi subsystem using Equation (9.2) derived in accordance with the RBD shown in Figure 9.4; 3) the Li-Fi subsystem using Equation (9.3) derived in accordance with the RBD shown in Figure 9.5; and 4) the Wi-Fi drone-based subsystem using Equation (9.4) derived in accordance with the RBD shown in Figure 9.6.

$$P_{SubG} = 1 - \left(1 - p_{m_{CG}} \prod_{i=1}^{n} p_{m_{SG}i}\right)\left(1 - p_{r_{CG}} \prod_{i=1}^{n} p_{r_{SG}i}\right) \tag{9.1}$$

where P_{SubG} is the PFFO for the wired network subsystem, $p_{m_{CG}}$ is the PFFO for the main sensor of the wired network communication section, $p_{r_{CG}}$ is the PFFO for the redundant sensor of the wired network communication section, $p_{m_{SG}i}$ is the PFFO for a sensor *i* of the main chain of the wired network section, and $p_{r_{SG}i}$ is the PFFO for a sensor *i* of the redundant chain of the wired network section.

$$\begin{aligned} P_{SubW} = & \prod_{i=1}^{k} \sum_{j=0}^{u_{ri}} C^{j}_{u_{ri}+u_{mi}} (1 - p_{SWi})^{j} {p_{SWi}}^{u_{ri}+u_{mi}-j} \\ & \times \sum_{b=0}^{\beta} C^{b}_{\beta+k} (1 - p_{DW1})^{b} {p_{DW1}}^{\beta+k-b} \end{aligned} \tag{9.2}$$

where P_{SubW} is the PFFO for the Wi-Fi subsystem, P_{SWi} is the PFFO for each sensor of both the main and the redundant chain within its own group *i*, and p_{DW1} is the PFFO for each drone of both the main and the redundant chain within its own section.

$$P_{SubL} = \prod_{i=1}^{p} \sum_{j=0}^{v_{ri}} C_{v_{ri}+v_{mi}}^{j} (1 - p_{SLi})^{j} {p_{SLi}}^{v_{ri}+v_{mi}-j}$$

$$\sum_{b=0}^{\gamma} C_{\gamma+p}^{b} (1 - p_{DL})^{b} {p_{DL}}^{\gamma+p-b} \tag{9.3}$$

where P_{SubL} is the PFFO for the Li-Fi subsystem, p_{SLi} is the PFFO for each sensor of both the main and the redundant chain within its own group i, and p_{DL} is the PFFO for each drone of both the main and the redundant chain within its own section.

$$P_{SubD} = \sum_{j=1}^{\omega} C_{\omega+f}^{j} (1 - p_{DS})^{j} {p_{DS}}^{\omega+f-j}$$

$$\sum_{b=0}^{\psi} C_{\psi+f}^{b} (1 - p_{DW2})^{b} {p_{DW2}}^{\psi+f-b} \tag{9.4}$$

where P_{SubD} is the PFFO for the Wi-Fi drone-based subsystem, P_{DS} is the PFFO for each sensor of both the main and the redundant chain within its own section, and P_{DW2} is the PFFO for each drone of both the main and the redundant chain within its own section.

9.3.3 System Models

The proposed MPSAMS (see Figures 9.1 and 9.2) can be modified under NPPs features. Thus, MPSAMS can be based on 2 (Figures 9.7(a–c)), 3 (Figures 9.8(a–c)), and 4 subsystems (Figures 9.9(a–c)) without SubIoT, or 5 subsystems with SubIoT (Figure 9.10).

Based on the above assumptions and the constructed RBD (see Figures 9.7–9.10), the PFFO for MPSAMS can be calculated according to one of the equations presented in Table 9.1.

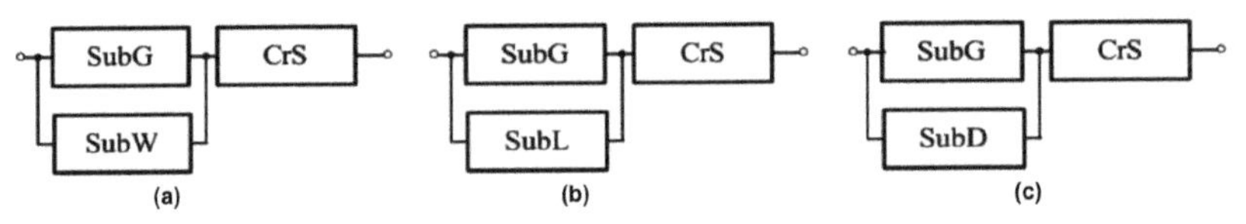

Figure 9.7 RBD for MPSAMS based on: (a) the wired network subsystem (SubG) and the Wi-Fi subsystem (SubW), (b) the wired network subsystem (SubG) and the Li-Fi subsystem (SubL), and (c) the wired network subsystem (SubG) and the Wi-Fi drone-based subsystem (SubD).

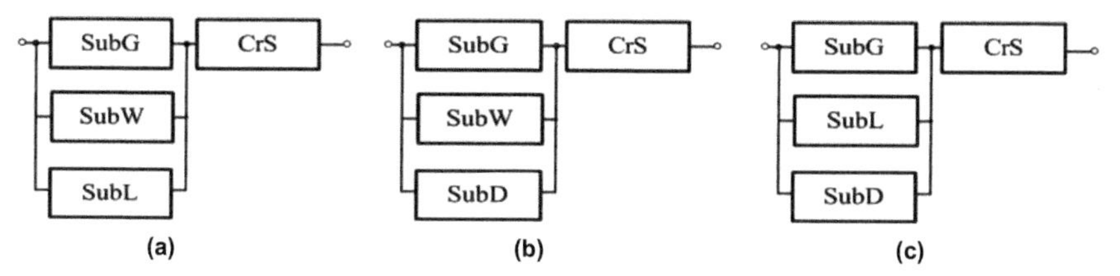

Figure 9.8 RBD for MPSAMS based on: (a) the wired network subsystem (SubG), the Wi-Fi subsystem (SubW), and the Li-Fi subsystem (SubL), (b) the wired network subsystem (SubG), the Wi-Fi subsystem (SubW), and the Wi-Fi drone-based subsystem (SubD), and (c) the wired network subsystem (SubG), the Li-Fi subsystem (SubL), and the Wi-Fi drone-based subsystem (SubD).

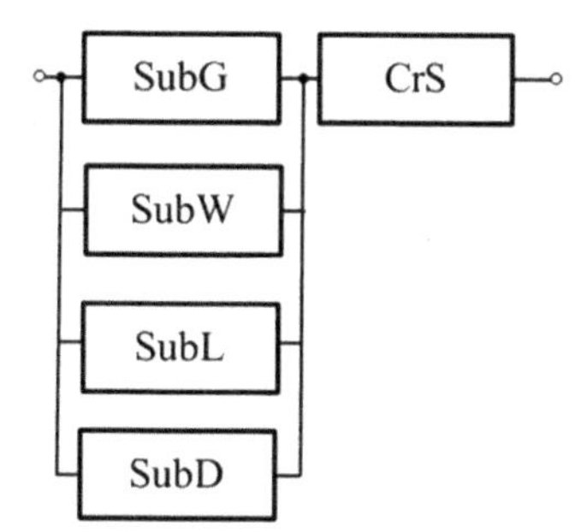

Figure 9.9 RBD for MPSAMS based on the wired network subsystem (SubG), the Wi-Fi subsystem (SubW), the Li-Fi subsystem (SubL) and the Wi-Fi drone-based subsystem (SubD).

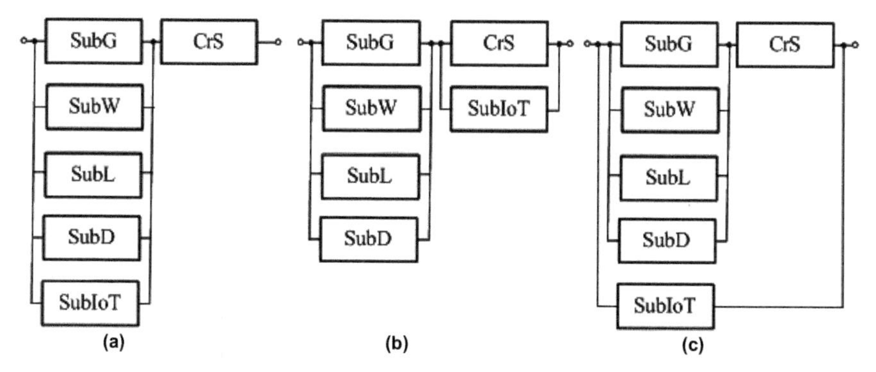

Figure 9.10 Various variants of RBD for MPSAMS based on the wired network subsystem (SubG), the Wi-Fi subsystem (SubW), the Li-Fi subsystem (SubL), the Wi-Fi drone-based subsystem (SubD), and the IoT-based subsystem (SubIoT).

Table 9.1 Equations for calculation of PFFO for various variants of MPSAMS in accordance with RBDs presented in Figures 9.7–9.10

Equation for Calculation of PFFO	Figure with RBD Based on which an Equation is Derived
$P_{S_{GW}} = [1-(1-P_{SubG})(1-P_{SubW})]\,P_{CrS}$	Figure 9.7(a)
$P_{S_{GW}} = [1-(1-P_{SubG})(1-P_{SubW})]\,P_{CrS}$	Figure 9.7(b)
$P_{S_{GD}} = [1-(1-P_{SubG})(1-P_{SubD})]\,P_{CrS}$	Figure 9.7(c)
$P_{S_{GWL}} = [1-(1-P_{SubG})(1-P_{SubW})(1-P_{SubL})]\,P_{CrS}$	Figure 9.8(a)
$P_{S_{GWD}} = [1-(1-P_{SubG})(1-P_{SubW})(1-P_{SubD})]\,P_{CrS}$	Figure 9.8(b)
$P_{S_{GLD}} = [1-(1-P_{SubG})(1-P_{SubL})(1-P_{SubD})]\,P_{CrS}$	Figure 9.8(c)
$P_{S_{GWLD}} = [1-(1-P_{SubG})(1-P_{SubW})(1-P_{SubL})(1-P_{SubD})]\,P_{CrS}$	Figure 9.9
$P_{S1_{GWLD\&IoT}} = 1-(1-P_{SubG})(1-P_{SubW})(1-P_{SubL}) \times (1-P_{SubD})(1-P_{SubIoT})\,P_{CrS}$	Figure 9.10(a)
$P_{S2_{GWLD\&IoT}} = [1-(1-P_{SubG})(1-P_{SubW})(1-P_{SubL}) \times (1-P_{SubD})] \times [1-(1-P_{SubIoT})(1-P_{CrS})]$	Figure 9.10(b)
$P_{S3_{GWLD\&IoT}} = 1-\{1-[1-(1-P_{SubG})(1-P_{SubW})(1-P_{SubL}) \times (1-P_{SubD})]P_{CrS}\} \times (1-P_{SubIoT})$	Figure 9.10(c)

9.4 Simulation

Figures 9.11–9.13 are designed by simulation using Equations (9.1)–(9.4) and the equations from Table 9.1, and they illustrate the ways of increasing the MPSAMS reliability. The first way is to increase the number of redundant elements (sensors or drones). For example, all plots (see Figures 9.11–9.13) show that the PFFO for the MPSAMS will increase if the number of the redundant drones grows from 0 to 3.

Note that for the considered cases:

- The best PFFO for the MPSAMS is achieved when the number of the drones under redundancy is 9 and the number of redundant drones is 3 (see Figures 9.11 and 9.12).
- The worst PFFO for the MPSAMS is achieved when the number of the drones under redundancy is 11 and the number of the redundant drones is 0 (see Figures 9.11 and 9.12).

The second way is to use more reliable elements for subsystems of MPSAMS. The depicted plots (see Figure 9.12) show that the PFFO for the MPSAMS

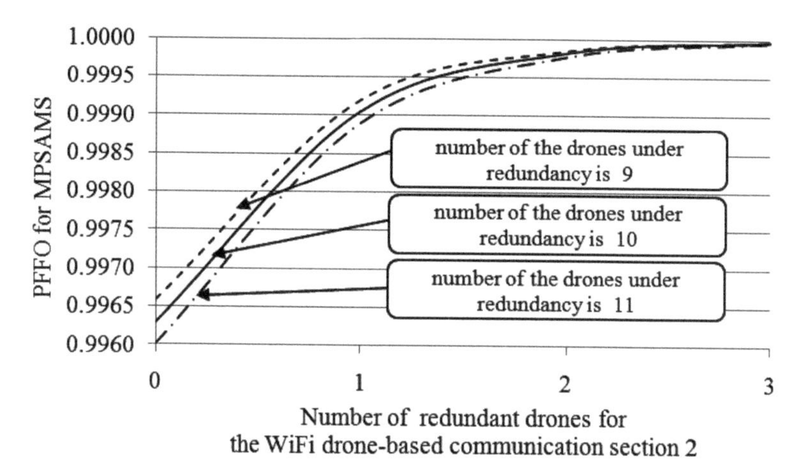

Figure 9.11 Dependence on PFFO for MPSAMS based on the wired network subsystem and the Wi-Fi subsystem on the number of the redundant drones for the Wi-Fi drone-based communication Section 9.2.

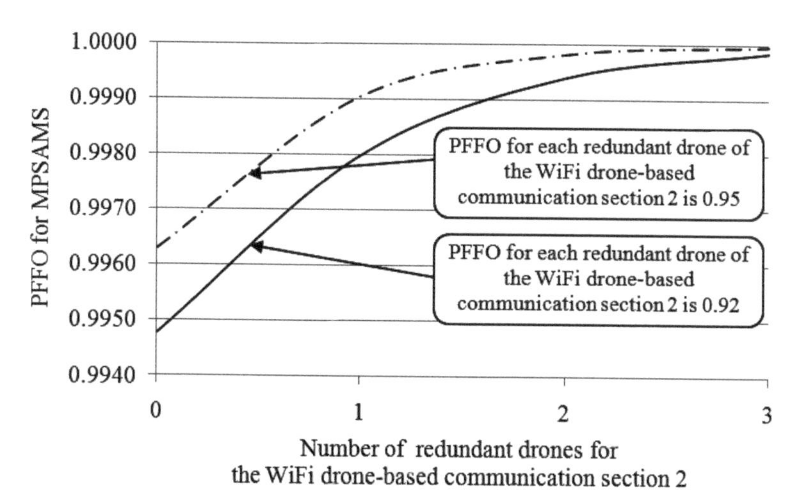

Figure 9.12 Dependence on PFFO for MPSAMS based on the wired network subsystem and the Wi-Fi subsystem on the number of redundant drones for the Wi-Fi drone-based communication Section 9.2 at different values of PFFO per each redundant drone.

will increase if the PFFO for each of the redundant elements grows from 0.92 to 0.95.

The third way is to increase the number of subsystems for MPSAMS. The MPSAMS based on SubG, SubW, SubL, and SubD (see Figure 9.13) has the best PFFO among other structure modes.

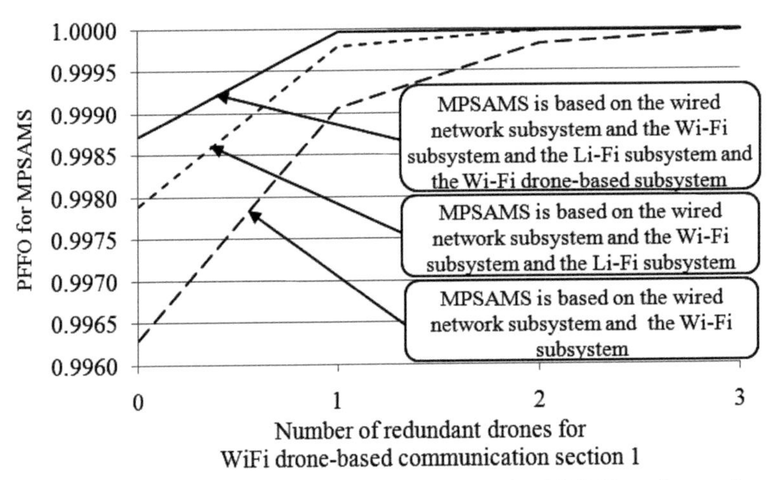

Figure 9.13 Dependence on PFFO for various variants of MPSAMS on the number of the redundant drones for the Wi-Fi drone-based communication Section 9.1.

9.5 Conclusion

1. The use of diverse data transmitted from the measurement and control modules as well as additional modules (sensors), which can be placed in areas not accessible by human operators, allows increasing trustworthiness of information about the reactor and the station area as whole.
2. The use of wireless connections with specified modules (described in point 1) is a fail–safe mechanism to maintain critical operational monitoring for a severely damaged NPP wired network. It is understood that the deployment of wireless communication within an NPP is prohibited by current regulatory standard. However, in the event of a major nuclear accident, all strategic technologies must be made available at that critical time.
3. Ensure an acceptable lifetime of accumulators for both operating measurement and control modules by relaying their signals. Thus, repeaters will be placed on the drones that reach NPP at the required time. The authors propose the use of MPSAMS to mitigate the potential hazards arising from severe NPP accidents. The MPSAMS includes one wired network subsystem (traditional PAMS) and three wireless network subsystems that are more resilient to cope with wired communication failures. In particular:
   - The number of MPSAMS' wireless network subsystems can be reduced according to the NPPs' characteristics.

- When all ground-based wired sensor sections are damaged, drone-sensors of the Wi-Fi drone-based subsystem type can be deployed to bridge and replace lost services.
- When the wired network subsystem fails, then each of the wireless network subsystems is deployed by drone-based communication sections to support the reliable transmission of data.

To increase the MPSAMS' survivability, both sensor and communication sections of wireless network subsystems are equipped with backup batteries and multiple blocks of wireless communication modules as well as self-testing and self-diagnostic systems. To provide the increment of MPSAMS' reliability, the number of redundant elements (drones or sensors) of subsystems should be increased and more reliable subsystem components should be used.

Finally, the MPSAMS does not belong to widely used class of systems. It is a system with the following properties:

- It is multi-functional because it performs many different tasks that are not homogeneous.
- It is extensible (scalable) because the number of components, including drones and measurement modules, can be changed dynamically.
- It is universal because it is invariant with respect to types of drones and modules employed, where these components can be defined by a set of minimum requirements only.
- It is a complete system that can be deployed and operated in many different environments.

References

[1] Accident and post-accident monitoring system (PAMS). (n.d). Available at: http://imp.lg.ua/index.php/en/pams-2.

[2] Kharchenko, V. (2016). "Diversity for safety and security of embedded and cyber physical systems: fundamentals review and industrial cases," in *Proceedings of the 15th Biennial Baltic Electronics Conference*, Tallinn, Estonia.

[3] Kima, C., Hurb, S., Sonc, K., and Jang, T. (2014). "Development of post-accident monitoring system for severe accidents," in *Proceedings of the 12th International Probabilistic Safety Assessment and Management Conference*, Honolulu.

[4] Pöllänen, R., Toivonena, H., Peräjärvia, K., Karhunen, T., Ilander, T., Lehtinen, J., et al. (2009). Radiation surveillance using an unmanned aerial vehicle. *Applied Radiation and Isotopes*. 2, 340–344. doi: 10.1016/j.apradiso.2008.10.008.

[5] Towler, J., Krawiec, B., and Kochersberger, K. (2015). Radiation mapping in post-disaster environments using an autonomous helicopter. *Remote Sensing*, 4(7), 1995–2015. doi: 10.3390/rs4071995.

[6] Sanada, Y., and Torii, T. (n.d.). *Radiation measurement by unmanned aircraft after Fukushima Daiichi nuclear power plant accident*. Available at: http://www.icao.int/Meetings/RPAS/RPASSymposiumPresentation/Day%201%20Session%202%20Massaki%20Nakadate.pdf.

[7] Schneider, F., Gaspers, B., Peräjärvi, K., and Gårdestig, M. (n.d). *Possible scenarios for radiation measurements and sampling using unmanned systems*. Available at: http://publications.jrc.ec.europa.eu/repository/handle/JRC95791.

[8] Camara, D. (2014). "Cavalry to the rescue: Drones fleet to help rescuers operations over disasters scenarios," in *Proceedings of the IEEE International Conference on Antenna Measurements and Applications*, Antibes Juan-les-Pins.

[9] Tanzi, T., Chandra, M., Isnard, J., Camara, D., Sebastien, O., and Harivelo, F. (2016). "Towards drone-borne disaster management: Future application scenarios," in *ISPRS Annals of the Photogrammetry, Remote Sensing and Spatial Information Sciences, XXIII ISPRS Congress*, Prague.

[10] Tuna, G., Veli, T., and Gulez, K. (2012). "Design strategies of unmanned aerial vehicle-aided communication for disaster recovery," in *Proceedings of the 9th International Conference on High Capacity Optical Networks and Enabling Technologies*, Istanbul.

[11] Erdelj, M., Natalizio, E., Chowdhury, K., and Akyildiz, I. (2017). Help from the sky: Leveraging UAVs for disaster management. *IEEE Pervasive Computing*, 16(1), 24–32. doi: 10.1109/MPRV.2017.11.

[12] Ivakhiv, O. (2016). Information state of system estimation. *International Journal of Computing*, 15(1), 31–39.

[13] Hall, R. J. (2016). An Internet of Drones. *IEEE Internet Computing*, 20(3), 68–73.

[14] Gharibi, M., Boutaba, R., and Waslander, S. L. (2016). Internet of Drones. *IEEE Access*, 4, 1148–1162. doi: 10.1109/ACCESS.2016.2537208.

[15] Long, T., Ozger, M., Cetinkaya, O., and Akan, O. (2018). Energy Neutral Internet of Drones. *IEEE Communications Magazine*, 56(1), 22–28. doi: 10.1109/MCOM.2017.1700454.

[16] Lin, Ch., He, D., Kumar, N., Choo, K-K. R., Vinel, A., and Huang, X. (2018). Security and privacy for the Internet of Drones: Challenges and solutions. *IEEE Communications Magazine*, 56(1), 64–69. doi: 10.1109/MCOM.2017.1700390.

[17] Qureshi, B., Koubâa, A., Sriti, M.-F., Javed, Y., and Alajlan, M. (2016). “Poster: Dronemap - A cloud-based architecture for the Internet-of-Drones,” in *Proceedings of The International Conference on Embedded Wireless Systems and Networks*, Graz.

[18] Hiromoto, R., Sachenko, A., Kochan, V., Koval, V., Turchenko, V., Roshchupkin, O., Yatskiv, V., and Kovalok, K. (2014). “Mobile Ad Hoc wireless network for pre- and post-emergency situations in nuclear power plant,” in *Proceedings of the 2nd IEEE International Symposium on Wireless Systems within the Conferences on Intelligent Data Acquisition and Advanced Computing Systems*, Offenburg.

[19] Kharchenko, V., Sachenko, A., Kochan, V., and Fesenko, H. (2016). “Reliability and survivability models of integrated drone-based systems for post emergency monitoring of NPPs,” in *Proceedings of The International Conference on Information and Digital Technologies*, Rzeszow.

[20] Kharchenko, V., Fesenko, H., and Doukas, N. (2017). A stochastic continues-time model of the drone fleet: research of survivability and choice of parameters. *International Journal of Instrumentation and Measurement*, 2, 25–30.

PART III

Architecting and Development

10

Virtualization of Embedded Nodes for Network System Characterization in IoT Applications

Manuel Schappacher, Artem Yushev, Mahbuba Moni and Axel Sikora

Institute of Reliable Embedded Systems
and Communication Electronics (ivESK), Offenburg University
of Applied Sciences, Offenburg, Germany
E-mail: manuel.schappacher@hs-offenburg.de;
artem.yushev@hs-offenburg.de; mahbuba.moni@hs-offenburg.de;
axel.sikora@hs-offenburg.de

The term Internet of Things (IoT) is used to portray an accumulation of innovations empowering embedded computing like connected objects or networking nodes accessible through the existing internet infrastructure. The validation, verification, and testing of the distributed network nodes in complex embedded networks for the IoT are an ongoing challenge. Therefore, a novel approach of a test environment for embedded networking nodes has been conceptualized and implemented. Its basis is the use of virtual nodes in a PC environment, where each node executes the original embedded code. Different nodes run in parallel, connected via so-called virtual channels (VCs). The environment allows modification of the behavior of the VCs as well as the overall topology during runtime to virtualize real-life networking scenarios. The presented approach is very efficient and allows a simple description of test cases without the need of a network simulator. Furthermore, it speeds up the process of developing new features as well as supporting the identification of bugs in wireless communication stacks. In combination with powerful test execution systems, it is possible to create a continuous development and integration flow.

10.1 Introduction

Nowadays, several so-called megatrends like Industry 4.0 and the (Industrial) Internet of Things (IoT or IIoT) are rapidly emerging and are leading to an increasing amount of new protocols and solutions to cover the manifold requirements of the different applications. Wireless technologies based on embedded systems play a major role for the local communication and are called short-range wireless networks [1] or narrow-band long-range networks [2].

With the advent of these newly defined or implemented protocols, an additional need for an efficient validation, verification, and testing of the distributed network nodes emerges. Although there are already many solutions available for these tasks, during our developments, we found out that the differences between the use of legacy network simulator tools and the implementation in real hardware are still too big to support a seamless design flow. The virtualized testbed for embedded networking nodes (VTENN) presented in this paper is especially useful, when hardware and firmware are already available, but when iterations between the virtualized world and the real embedded hardware are still needed, which is the typical case in the development of real solutions including conformance testing.

The structure of this paper is as follows. Sections 10.2 and 10.3 provide an overview of available solutions for simulation and emulation of embedded networks, discussing their advantages and disadvantages and explaining the need for the new VTENN approach. Section 10.4 gives an insight into our development objectives and our background and shows how VTENN fits into this picture. The virtual testbed and its implementation itself will be described in Sections 10.5 and 10.6, including the integration of the testing and controlling of the testbed. Section 10.7 describes how VTENN can be used in IoT applications. Finally, Section 10.8 gives a short summary and an outlook on future work.

10.2 Related Work

For the design, development, and implementation of embedded networks, there are already many approaches to estimate the performance and to validate and verify the implementation in early phases following the spiral process model [3]. These approaches are generally directed toward the use of prototypes and models, be it of the networking nodes or the communication channel. This paper is using the material from [4].

10.2.1 System Level Simulation

System level simulation contributes for the abstract evaluation of the network performance, e.g., regarding energy consumption or throughput. Typically, the design of these models takes place in the very early phase of the development and helps to identify the strategic direction of the development. Often, the models are very abstract and do not consider any hardware- or timing-related behavior. They might be implemented in general software environments, like using scripting languages or MATLAB. One example from our developments is used in [5].

10.2.2 Network Level Simulation

At the next level, the models become more detailed. If timing behavior of the network becomes interesting and if the communication properties of the network shall be evaluated, event-driven network simulators can be used.

Following the approach proposed in [6], two types of simulators can be observed: *specialized* and *generic*. However, we propose a third distinct type of simulators, and we call *nominal*.

1. *Specialized simulators* can emulate the behavior of specific nodes and their communication network. Numerous simulators of that type exist, whereas most mentionable in the world of IoT are OpenVisualizer with OpenSim [7] from OpenWSN, COOJA [8] from Contiki OS, and TOSSIM [9] for TinyOS. OpenWSN developers focused on one solution to run, monitor, configure, and control both real and virtual nodes (VNs) using a unified toolset (OpenVisualizer) together with OpenSim. However, it does provide the possibility neither to define specific topologies for the modeled network nor to change it dynamically.
   On the contrary, COOJA and TOSSIM simulators do provide a sophisticated toolset to control the network and execute automatic tests. However, in the latter cases, developers need to install complete and complex software simulators and need to know how to control and parameterize it. Many experiments and scientific papers follow this approach, although it might imply significant engineering and computational overhead.

In addition, the given environments are limited to the specific protocols and solutions only. Using them for any other solution or protocol beyond their original scope typically is not possible.

2. *Generic Simulators* – Simulators such as ns-2[1], ns-3[2], OMNET++[3], or OPNET, now known as Riverbed Modeler[4], are mainly designed for the analysis of large-scale networks and application-specific properties for a certain application model. They have no relation to specific hardware or operating systems what makes it difficult to use the same code on the simulator and in the real hardware. Mostly, a parallel code branch needs to be created to replace the hardware abstraction layer (HAL) by a simulator abstraction layer. Furthermore, deep changes within the basic paradigms of the code are required to fit into a simulator's programming model. Apart from our own activities in this field, we are only aware of Tazaki et al. [10], who proposed a way to execute code directly for ns-3. However, this approach is still relatively complicated and labor-intense.
3. *Nominal simulators* possess three major features: it assumes that the system under test (SUT) is a black box with no real output to media; it controls input and output of SUT's as well as a network topology; and optionally, it controls some system parameters like ID, state (turn on/off), etc. Comparing to other two types of simulators, the latter provides very basic inter-nodes' communication features, and used mainly as automatic/manual testing tool. Our approach also belongs to that category by providing a virtual, simplified abstraction of the radio communication while keeping the rest of the system as it is.

10.2.3 Network Level Emulation

Abstract software modeling is the basis for the first two approaches and is not sufficient for the validation of hardware-based implementations that come with new challenges. To verify such systems and to validate the performance of the network and the stability of the nodes, hardware testbed solutions with a controllable environment are extremely helpful. Therefore, we have developed an automated physical testbed (APTB), where we emulate the spatially distributed physical channels through dedicated and controllable

---

[1]http://www.isi.edu/nsnam/ns/
[2]https://www.nsnam.org/
[3]https://omnetpp.org/
[4]http://www.opnet.com

circuits. In [11], we present APTB and compare it against other testbed environments.

10.3 Requirements

During our research and development activities, we specify several requirements for our solution:

- It shall be possible to use a common code basis since it is necessary to iterate between virtual implementations in the simulator and the real implementation, e.g., for the integration of bug fixes or extensions.
- The solution must be flexible regarding the supported protocols and shall allow running any networking protocol.
- Debugging and monitoring functionality are mandatory. Many of the tools described above do not allow flexible monitoring of the processes in the network. Also debugging of individual nodes might be cumbersome. Therefore, the proposed solution shall provide a flexible and convenient interface to debug and test a protocol under development.
- Especially, if the development also includes industrial partners (or students), many simulation environments are by far too complex, because there is not enough time to become acquainted with the tools. Therefore, the proposed solution shall be easy to use.
- The definition of static and dynamic use cases and the control of the spatially or logically separated networks use scripting languages quite often. A significant effort goes into the definition of the notation. Therefore, the proposed shall use a given testing and test control notation, like TTCN.

For all of these reasons, we develop a novel approach that we call VTENN. It supports fast deployment and integration and provides flexible and convenient interfaces to debug and test a protocol under development. It provides an easy abstraction of the lower layer communication protocols so that developers and testers of the higher layers can fully concentrate on their parts.

This virtual testbed runs on a Linux host and allows the creation and execution of several virtual network nodes in parallel, executing the original embedded core, but communicating using so-called virtual channels (VCs). Since all the nodes can run within a virtual network environment on a single machine, the environment provides an easy access to internal data (e.g., states, buffer contents, etc.) of the nodes. An easy distribution is also possible since the environment is completely software based. Furthermore, the testbed

Table 10.1 Comparison between typical specialized, generic, and nominal simulators

Feature	COOJA	NS2/3	VTENN
Complexity	Average	High	Low
SUT independence	No	Limited	Full
Media simulation	Limited	Full	Limited
Testing support	Limited	No	Full
Connection to external systems (e.g., backend)	Limited	No	Full
Execution	Simulation time	Simulation time	Real time

allows the control and monitoring of the VNs as well as of the VCs during execution using a TTCN-3-based testing framework [12]. So, it becomes possible to easily control the use-case scenarios and topologies. Automatic regression tests can be executed. Table 10.1 shows some key characteristics of VTENN compared to other systems.

10.4 Background

The VTENN framework developed and presented herein is using, but not being limited to the 6LoWPAN communication stack emb::6.

10.4.1 The emb::6 Networking Stack

emb::6[5] is an open source wireless network stack supporting the 6LoWPAN communication protocol. The initial development of the emb::6 network stack started as a fork of Contiki OS[6] including μIPv6. In order to meet the requirements of a portable industry-grade network stack, core parts of Contiki have been removed or fundamentally reworked. It is maintained in our institute [13].

The basic architecture of the emb::6 wireless network stack consists of its networking core, a HAL, radio drivers (RF), and a separate so-called utility module that implements all common functionalities such as timer and event handling used by all other layers and modules. The networking core handles the network-related tasks, mainly the communication part, where different tasks are distributed over several layers. Beginning on top at the application layer (APL), usually serving as interface to the device application, the stack forwards requests layer by layer down to the radio driver, which is responsible

[5]https://github.com/hso-esk/emb6

[6]http://www.contiki-os.org/

for the implementation of the RF module drivers. In general, emb::6 covers the application layer, transport layer, network layer, data link layer, physical layer, and the radio driver.

To support different hardware, all hardware-dependent parts of emb::6 are encapsulated in a separate so-called board support package. This allows easy porting of emb::6 to different hardware platforms.

10.4.2 TTCN-3

Testing and Test Control Notation Version 3 (TTCN-3[7]) is a scripting language used in conformance and interoperability testing. This language being developed and maintained by ETSI does not define a specific communication interface to the SUT, but provides an abstraction of the communication port toward the SUT, which in turn should be implemented in another programming language. Furthermore, TTCN 3 has native support for ASN.1, IDL, and XML type definitions. The TTCN notation has been a major testing language for telecommunication protocols such as UMTS, 2G, 3G, and SIP, and now serves our testing tools as an interface for defining and running test cases (TCs).

10.5 VTENN Basics

10.5.1 General Architecture

The goal of VTENN is to provide an environment for development, tests, and verification of wireless communication systems that is easy to use and easy to distribute. To achieve this, the complete environment runs on a single Linux host including the virtual network, development tools, and testing tools. Figure 10.1 shows the overall VTENN architecture. Although the figure shows the whole setup within one host, it is also possible to distribute individual parts across different hosts, as long as they interconnect in a local network.

The key component is the virtual network (VNET) that includes the virtual communication nodes (N). Depending on the virtualized technology, a virtual edge router or gateway (ER/G) establishes the connection to the host network (HNET) which facilitates the connection of the virtual network with a real on, e.g., to access physical backend server. This allows, e.g., to test a whole system with a huge number of nodes. The number of such VNs

[7]http://www.ttcn-3.org/

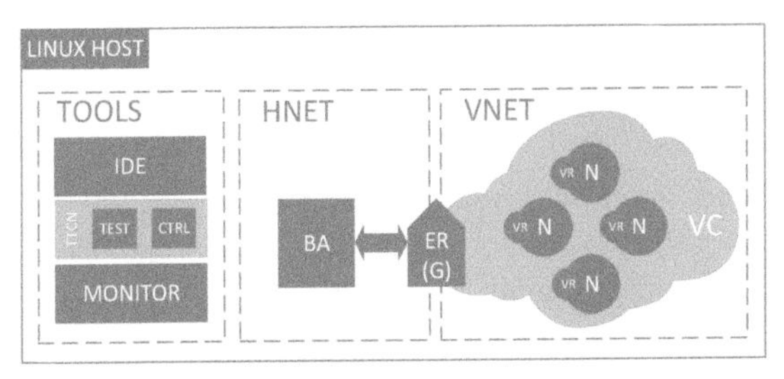

Figure 10.1 Overview of the virtual testbed architecture.

is flexible and nearly unlimited since each node is represented as a single host process that can be started and stopped on demand. To have a common behavior between VNs and real nodes, both run a firmware built from the same core source code. Additional monitoring and control software modules besides the actual application allow controlling, observing, and manipulating the nodes.

Communication between nodes takes place via so-called virtual radios (VRs) and VCs. Both can be configured, e.g., to represent a specific network topology or to emulate the characteristics of the channel such as its transmission quality. An additional module is used to control the VCs, e.g., to change the network topology on the fly.

Besides the nodes and its network, several tools contribute to the usability of the virtual testbed. IDEs help to develop features and to create bug fixes whereas changes in the code are applied immediately just by restarting the specific network nodes. Time-consuming reprogramming, as it is required for physical devices, is not necessary in this case, saving a lot of time especially when considering an update for a huge network with dozens of nodes. Finally, yet important, a set of test, control, and monitoring applications enable running tests on the system based on the TTCN framework. Additional software modules – like the ones mentioned above – provide according interfaces for the test framework to control the nodes and the network and allow collecting data from the network nodes.

10.5.2 Node Virtualization

One of the main benefits of the virtual network is to have the same code basis for both the VNs and the real physical devices. This avoids having an additional software branch that was special for the simulator. This saves maintenance and integrational work. Furthermore, nearly the same behavior for virtual and real devices can be expected – of course without the effects of the real communication channels. To have an environment with VNs, the firmware of the wireless nodes, including the core communication stack, needs to be adapted to support the host platform. This requires exchanging the hardware-dependent parts of the firmware.

A well-designed software architecture and a clear separation of the hardware-dependent and -independent parts of the code can simplify the process of porting the firmware. This usually can be achieved by using a HAL, as it is done in emb::6. A HAL provides a well-defined interface with a specific set of functions that are required by the firmware. When porting the firmware to a new platform, it is necessary to provide a corresponding HAL module that implements the according function set for the underlying hardware or operating system.

As described before, the virtual environment runs on a Linux host machine. Therefore, e.g., to run emb::6 nodes within this platform, an implementation of the according HAL functionality is required. In case of the Linux host, this affects the implementation of timer and delays, interface implementations such as UARTs that are mainly used for debugging or as a communication interface, and the implementation of the debugging outputs. Further adaptions are not required due to the nature of the Linux host. Usually, an SPI interface implementation is also required to access the radio transceivers. However, this is not necessary for the virtual environment since the nodes use a VR transceiver as explained in Section 10.5.3 and therefore provide a simplified interface.

10.5.3 Virtual Radio and Channel

In addition to the core firmware and the implementation of the HAL, a specific implementation of the transceiver driver is required for the virtual network environment.

Furthermore, VR channels need to be created that eventually allow a communication among the VNs via their VR drivers. Since the virtualization runs on a Linux host, a VN and a VC both are represented by single executables. Communication among several executables and therefore among the nodes is implemented using inter-process communication (IPC). Linux-based operating systems provide several common mechanisms to create IPC between applications such as shared-memory, pipes, semaphores, or sockets. However, those only represent basic mechanisms to transfer data without any higher logic whereas for the virtual network setup, it is required to access and model the communication paths more easily.

This includes, e.g.,:

- To define which and how nodes can communicate. That is an important requirement since this allows building complex topologies such as mesh or tree networks including routing scenarios.
- To manipulate the VCs. This becomes important, e.g., for routing algorithms.
- The possibility to monitor the VCs, e.g., using virtual sniffers and protocol analyzers.

The so-called Lightweight Communications and Marshalling (LCM[8]) module [14] is a set of libraries and tools for message passing and data marshaling, targeted at real-time systems where high bandwidth and low latency are critical. It provides a publish/subscribe message passing model and automatic marshaling/unmarshalling code generation with bindings for applications in a variety of programming languages.

The communications' aspect of LCM can be summarized as the publish–subscribe messaging system that uses UDP multicast as its underlying transport layer. Under the publish–subscribe model, each message is transmitted on a named channel, and modules subscribe to the channels required to complete their designated tasks. It is typically the case (though not enforced by LCM) that all the messages on a particular channel are of a single prespecified type [14]. Therefore, the radio driver implementations of the nodes that want to participate within a virtualized network have to subscribe and to publish to a specific set of channel names.

10.5.4 Virtual Topologies

One of the most valuable features of the virtual testbed is the possibility to control the underlying network including the channel characteristics and the topology. The VCs used for the communication among the VR drivers and therefore among the nodes allow creating any kind of topology. Even more important is the possibility to recreate such topologies, e.g., to reproduce TCs.

The mechanism to define an underlying topology either for TCs or during development follows a simple structure that defines which node is able to receive data from other nodes meaning that they share a common receive channel. Each VN specifies a single channel to which it publishes its data. Furthermore, the latter subscribes to several channels published by the nodes from which it shall receive data.

[8]https://github.com/lcm-proj/lcm

10.5.5 Monitoring and Control

Virtual platforms and simulators possess the ability to change the topology and connectivity and to control nodes behavior dynamically. Nevertheless, by focusing on medium properties of a connection, commonly used simulators overlook the deeper problem of a typical initial phase of a development where a programmer does not want to set up a complex simulation platform and tune it only to debug and test applications. To address this issue, the VTENN setup uses a lightweight model that suits not only to manual control but also allows to automatic functional or conformance tests.

By using a TTCN-3 client application, the functionality of VTENN is even increased. We can use it, e.g., to set up a topology for a particular TC and to control the nodes. It can create the nodes, execute all required TCs, and then shut down all the nodes automatically. Figure 10.2 depicts a possible TTCN-3 integration into the virtual network, on which one can find two hosts with TTCN-3 client and the VTENN environment. Before the actual test execution, the TTCN-3 client sets up a proper network of VNs by sending the according requests to the environment. Afterward, TTCN-3 TCs running on a client and associated with a current virtual network configuration can be executed via the so-called NECK protocol [12].

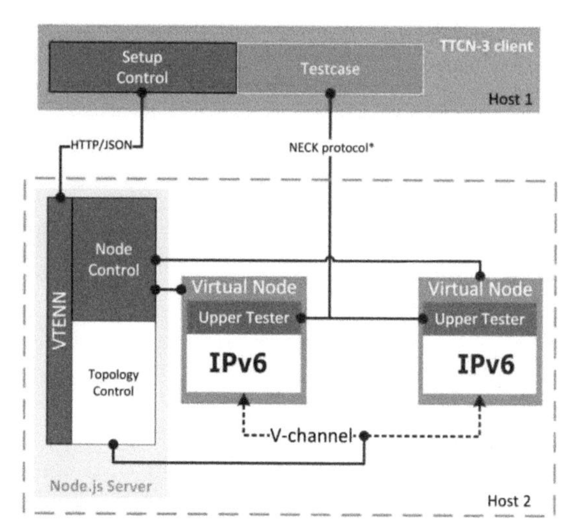

Figure 10.2 Integration of the TTCN-3 framework into VTENN.

10.6 Design and Implementation

The goal of the VTENN implementation is to provide an environment that is flexible enough to allow a controlled communication among the nodes whereas the management of the test setup shall remain as easy as possible. Therefore, the implementation of the VTENN follows a specific architecture as shown in Figure 10.3. It is split into several parts, namely, the test executor (TE), network manager (NM), VNs, and VC, each of them represented by a separate module running as an executable within the host system. The TE represents the highest instance within VTENN serving as a kind of the main controller for the whole environment and the tests. NM represents to manage the created VNs and channels. In general, the numbers of VNs and VCs are nearly unlimited depending on the resources of the host system.

As described, all of the communication within VTENN takes place using IPC with LCM as transport to provide a further level of abstraction. A communication path between two elements is referred to as a port. Such a port defines a bidirectional communication between two elements within VTENN, e.g., between a VC and its associated VN. Furthermore, ports can be separated into so-called data ports delivering actual data between nodes and therefore representing the virtualized wireless channel and control ports that are used to exchange management and control information.

With LCM as transport mechanism, VTENN uses the Standard Commands for Programmable Instruments (SCPI) as the communication protocol for management and control information. It defines a standard syntax and commands used in controllable programmable test and measurement devices. Therefore, for standardization and validation of messages and command passing systems between the VTENN elements, SCPI has been used as a communication protocol which covers standard message-handling protocols including error handling, unambiguous program, response-message syntactic structures [15], etc.

10.6.1 Test Executor

TE connects to the NM as well as to the VNs and VCs to send control information and to obtain status responses and statistical information via the control ports. It is responsible for the setup and maintenance of the network topology as well as for controlling the elements within VTENN. All this is realized by exchanging SCPI-based frames via the according control ports.

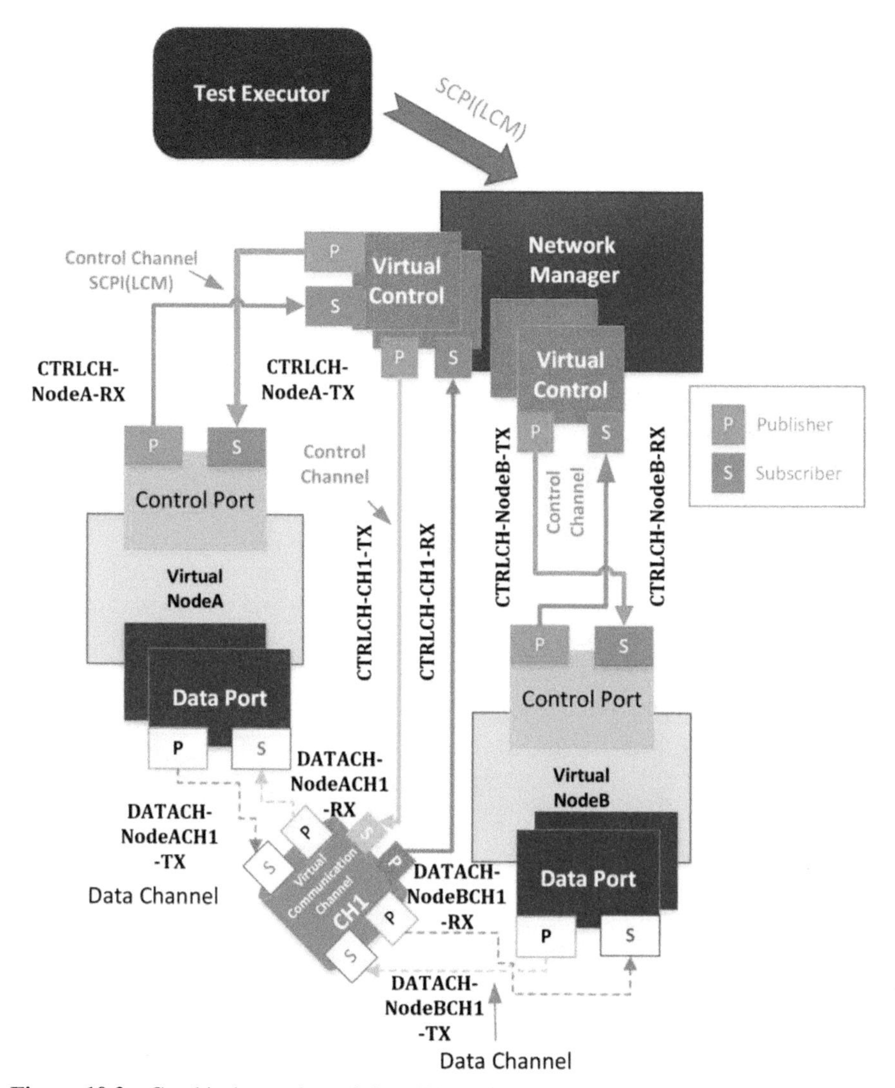

Figure 10.3 Graphical overview of the VTENN implementation including TE, NM, VNs, and VCs.

The TE is a flexible element within VTENN that can be exchanged with regard to the requirements and the complexity of the tests. Therefore, it can be represented, e.g., by a simple script or by a complex test framework such as TTCN as described in Section 10.4.2 as long as the required command set from the VTENN elements is supported.

10.6.2 Network Manager

VNs and VCs are controlled and managed by the NM. Therefore, the NM receives commands from the TE through a single control port. Upon the reception of command frames, the NM can, e.g., start or stop VNs, create or delete channels, and attach nodes to channels that finally allows communication among them. Furthermore, channel manipulations can also be controlled by the NM, e.g., to set channels' attenuation or delay.

Besides the control port to the TE, the NM also has further control ports to any network element within VTENN such as nodes and channels that are mainly used to exchange management commands and information. The different control ports are used to control the VNs as well as the VCs using a common naming convention. The NM publishes the commands to a VN using a control channel following the pattern of *CTRLCH-<Node-Name>-TX*. Furthermore, the VN subscribes to this channel to receive the command. Vice versa, the node publishes its messages or responses through the control channel named *CTRLCH-<Node-Name>-RX* that is subscribed by the NM to receive the message. The same mechanism applies for the communication between the NM and the VCs using the channel name as the unique identifier.

Figure 10.3 shows the connection from the NM to the VNs and channels, namely, *CTRLCH-NodeA-TX/RX*, *CTRLCH-NodeB-TX/RX*, and *CTRLCH-CH1-TX/RX* through the single control ports. The control ports will be created and connected after a VN or a VC has been started by the NM successfully. This is done by the NM by starting the according node or channel executables with their parameters such as the unique name upon the reception of the according command from the TE.

10.6.3 Virtual Nodes and Virtual Channels

VNs interconnect with each other through VCs, forming the virtual network topology. As described, the NM can start and stop the VNs on demand and attach them to VCs to establish a communication path. Data communication takes place using the LCM-based so-called data ports. Each channel has two data ports exactly to create a single communication path between two nodes. A node has multiple data ports and therefore can be attached to several channels making it possible to allow a node to communicate with several others.

Besides the data ports, a node also provides a single control port used for communicating with the NM. A VN can receive commands from the control port and publishes the according SCPI responses to the NM. After the creation of VNs and channels on demand, the NM can send specific *attach* and *detach* commands through the control ports to the VNs and channels for establishing or removing data connections. Just as control ports, data ports also use a common naming convention. VN publishes its data through the data channel named *DATACH-<Node-Name><Channel-Name>-TX* and receives messages from *DATACH-<Node-Name><Channel-Name>-RX*, respectively, using the prefix *DATACH*. By changing the names of the data channels to publish/subscribe to/from during runtime, changing a topology can easily be done using the according attach and detach commands sent by the NM.

The data connection between two nodes through a data channel can be seen from Figure 10.3. After the nodes and the channel have been created, the NM can establish a connection between the nodes by sending the according attach commands to the elements with the according counterpart to connect to as a parameter. After reception, the nodes and channels can publish and subscribe to the resulting ports and create the connection. Therefore, the VCs operate as some kind of tunnel delivering data received on the one data port to the other data port applying data manipulations according to its settings.

During the port connection between a VN and a VC, both check the availability of ports according to their internal list and start connecting if free ports are available. As mentioned before, a node can have many data ports, so the user can define to which port to connect a channel. Since channels provide only two ports to forward data from one to another, the connection will be done automatically using the next free port. Both procedures return the port number used for the connection. This number can be used later to remove the connection and to disconnect an element.

10.6.4 Sample Test Cases

According to the given approach and its implementation, it is possible to create nearly unlimited topologies of VNs and channels. A sample procedure executed by a python script as TE was created for establishing a simple test environment as it can be seen from Figure 10.4.

Here we have created three VNs (*NodeA, NodeB, and NodeC*) and three VCs (*CH1, CH2, and CH3*) at the beginning. Once the nodes and channels are up and running, the TE establishes connections between the VNs. This is done using the VCs according to the topology from Figure 10.4. The according

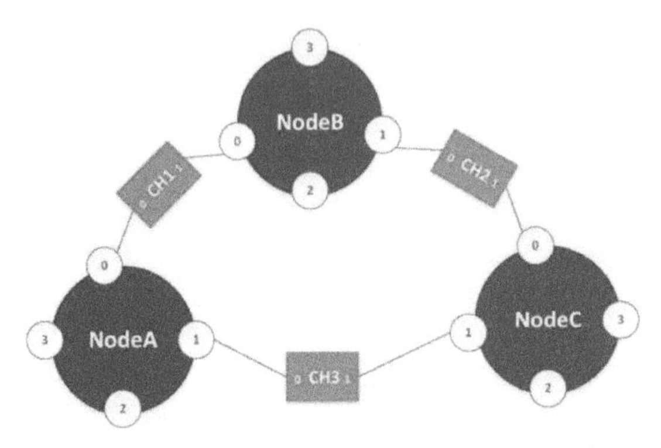

Figure 10.4 Test topology created by a sample TC using three nodes and three channels, whereas the path between NodeA and NodeB will be removed during the TC.

commands to establish the connection between *NodeA* and *NodeB* can be seen from the debug output in Figure 10.5. Once connected, the nodes can exchange data as shown in Figure 10.6. After a specific time, the connection between *NodeA* and *NodeB* will be closed by issuing the according detach command from the TE as shown in Figure 10.7. Afterward, the data communication between the two nodes stops.

```
[TestExe-C]-TEST-RX(20): NOde:CReate 'NodeA'
[TestExe-C]-TEST-RX(20): NOde:CReate 'NodeB'
[TestExe-C]-TEST-RX(20): NOde:CReate 'NodeC'
[TestExe-C]-TEST-RX(21): CHannel:CReate 'CH1'
[TestExe-C]-TEST-RX(21): CHannel:CReate 'CH2'
[TestExe-C]-TEST-RX(21): CHannel:CReate 'CH3'
[TestExe-C]-TEST-RX(29): ATTach 'NodeA','NodeB','CH1'
[NodeA  -C]-NODE-RX(17): POrts:AVailable?
[NodeA  -C]-NETM-RX(18): POrts:AVailable 0
[NodeB  -C]-NODE-RX(17): POrts:AVailable?
[NodeB  -C]-NETM-RX(18): POrts:AVailable 0
[CH1    -C]-CHAN-RX(17): POrts:AVailable?
[CH1    -C]-NETM-RX(18): POrts:AVailable 0
[NodeA  -C]-NODE-RX(21): POrt:COnnect 0,"CH1"
[NodeA  -C]-NETM-RX(22): POrt:COnnect:RES "OK"
[CH1    -C]-CHAN-RX(23): POrt:COnnect 0,"NodeA"
[CH1    -C]-NETM-RX(22): POrt:COnnect:RES "OK"
[NodeB  -C]-NODE-RX(21): POrt:COnnect 0,"CH1"
[NodeB  -C]-NETM-RX(22): POrt:COnnect:RES "OK"
[CH1    -C]-CHAN-RX(23): POrt:COnnect 1,"NodeB"
[CH1    -C]-NETM-RX(22): POrt:COnnect:RES "OK"
```

Figure 10.5 Attachment process of the example TC. At first, nodes and channels are created. After checking for available ports at the nodes and channels (*AVailable* request), the connections will be established using the *COnnect* commands.

```
[CH1    -D]-CHAN-RX(38): Data From Node [NodeA] Port [0]
[CH1    -D]-CHAN-RX(38): Data From Node [NodeB] Port [0]
[NodeA  -D]-NODE-RX(38): Data From Node [NodeB] Port [0]
[CH2    -D]-CHAN-RX(38): Data From Node [NodeB] Port [1]
[CH2    -D]-CHAN-RX(38): Data From Node [NodeC] Port [0]
[NodeB  -D]-NODE-RX(38): Data From Node [NodeA] Port [0]
[NodeB  -D]-NODE-RX(38): Data From Node [NodeC] Port [0]
[NodeC  -D]-NODE-RX(38): Data From Node [NodeB] Port [1]
```

Figure 10.6 Debug output after the successful establishment of the communication channels. Both nodes and the channels print debug messages for received data. The received data at a channel port will be forwarded to the respective connected node.

```
[TestExe-C]-TEST-RX(29): Detach 'NodeA','NodeB','CH1'
[NodeA  -C]-NODE-RX(18): POrt:DISconnect 0
[NodeA  -C]-NETM-RX(25): POrt:DISconnect:RES "OK"
[NodeB  -C]-NODE-RX(18): POrt:DISconnect 0
[NodeB  -C]-NETM-RX(25): POrt:DISconnect:RES "OK"
[CH1    -C]-CHAN-RX(18): POrt:DISconnect 0
[CH1    -C]-NETM-RX(25): POrt:DISconnect:RES "OK"
[CH1    -C]-CHAN-RX(18): POrt:DISconnect 1
[CH1    -C]-NETM-RX(25): POrt:DISconnect:RES "OK"
```

Figure 10.7 Debug output of the detachment process of the NodeA, NodeB, and CH1. After the detachment, no communication will take place anymore.

10.7 VTENN in IoT Applications

There are numerous utilize cases which will be influenced by IoT, covering all expectations and motivations in day-to-day life. The typical field of application of IoT is home and building automation, connected home, Industry 4.0, and so forth. In order to present the idea of VTENN in IoT environment, we have analyzed emb::6 – a scalable C-based 6LoWPAN stack which is a great option for IoT applications defined in Section 10.4.1. The emb::6-stack is optimized for the use in constrained devices without an operation system. By creating VNs by forming various topologies through VTENN and interfacing with emb::6 is a mechanism to compile and run the stack on a Linux machine, instead of an embedded device. This empowers developers and researchers to experiment with emb::6 without the need of necessary embedded hardware.

One of the basic usages of the components of emb::6 stack is the user datagram protocol (UDP) demo applications which show the use of standard UDP sockets. The client and server demo applications work together to exchange message back-and-forth. To create a specific topology, the nodes and the channel it subscribes to have to be created through VTENN to interface with emb::6. There are a couple of setups at the beginning stage that runs the UDP server and client demos in emb::6 using VTENN. We have created a simple topology consisting of two VNs in VTENN called NodeA which acts as an echo server and NodeB as an echo client connected through CH1. The server NodeA acts as a DAG-Root within the network so that no border router is required. UDP server demo always waits for the data to receive. This demo works together with a UDP client demo application. The client node continuously sends data packets to the server within a well-defined time interval. When the UDP server receives a random sequence number, it increments the received value by one and sends it back to the client. By running the demo applications in emb::6 and several VN topologies created from VTENN, the logging messages from the radio interface are used for further analyzing the stack.

There are some other use cases where the advantage of VTENN can be seen with the combination of automated TC script generated from TTCN. For this, we have configured the network to operate in a chain and mesh topology with four nodes and observed the difference in delay and how long it takes. In case of chain topology, if we terminate an intermediate node to cut the connection to the last node, the network connectivity will fail between the nodes until we restart the intermediate node, and for mesh topology, the network connectivity will fail between the nodes until a fallback route was found. All these steps can be done quite easily in VTENN with the combination of TTCN.

10.8 Conclusion and Future Work

This paper presents the architecture implementation and the actual state of the virtual test bed VTENN that allows easy and rapid development and tests of higher layer communication protocols by abstracting and providing the underlying lower layer communication. The actual status of the implementation allows creating and managing nodes, channels, and topologies dynamically as well as it provides a flexible usage of different TEs, such as simple scripts or even complex TTCN-based frameworks. Therefore, the VTENN environment provides an easy to use concept and framework to test and verify wireless network applications and provides several benefits.

The next steps will be the full usage of VTENN as one of the steps within a seamless continuous build and integration flow to support the development and verification of wireless communication stacks.

References

[1] Colombus, L. (2016). "*Gartner identifies the top 10 internet of things technologies for 2017 and 2018.*" Available at: http://www.gartner.com/newsroom/id/3221818

[2] DeLisle, J.-J. (2014). "*For Long-Range Communications, Is Narrowband "It"*? Microwaves and RF. Available at: http://mwrf.com/active-components/long-range-communications-narrowband-it.

[3] Sikora, A., Lill, D., Schappacher, M., Gutjahr, S., and Gerber, E. (2013)."Development of car2x communication and localization PHY and MAC protocol following iterative spiral model using simulation and emulation," in *5th Int'l Workshop on Communication Technologies for Vehicles*, Lille.

[4] Schappacher, M., Yushev, A., Moni, M., and Sikora, A. (2017). "VTENN – A virtualized testbed for embedded networking nodes," in *9th IEEE International Conference on Intelligent Data Acquisition and Advanced Computing Systems: Technology and Applications (IDAACS2017)*, 20–23, Bucharest, 797–803.

[5] Nguyen, P., Schappacher, M., Sikora, A., and Groza, V.F. (2016). "Extensions of the IEEE802.15.4 protocol for Ultra-Low energy real-time communication," in *Proc. I^2MTC, 2016 IEEE International Instrumentation and Measurement Technology Conference*, 23–26, Taipei.

[6] Brumbulli, M., and Gaudin, E. (2016). "Towards model-driven simulation of the internet of things," in *Complex Systems Design & Management Asia.* Springer International Publishing, 17–29.

[7] Watteyne, T., et al. (2012) OpenWSN: a standards-based low-power wireless development environment. *Transactions on Emerging Telecommunications Technologies*, 23(5), 480–493.

[8] Österlind, F., Dunkels, A., Eriksson, J., Finne, N. and Voigt, T. (2006). "Cross-level sensor network simulation with COOJA," in *Proceedings of 31st IEEE Conference on Local Computer Networks (LCN'06)*, Tampa, FL, 641–648.

[9] Levis, P., et al. (2003). "TOSSIM: Accurate and scalable simulation of entire tinyos applications," in: Akyildiz, I. F., Estrin, D., Culler, D. E., and Srivastava, M. B. (eds), *Proc. of SenSys* (New York, NY: ACM), 126–137.

[10] Tazaki, H., et al. (2013). "Direct code execution: revisiting library OS architecture for reproducible network experiments," in: Almeroth, K. C., et.al. (eds), *Proc. of 9th ACM Conf. on Emerging Networking Experiments and Technologies*, 217–228, CoNEXT'13.

[11] Sikora, A., Sebastian, E. J., Yushev, A., Schmitt, E., and Schappacher, M. (2016). "Automated physical testbeds for emulation of wireless networks," in *MATEC Web of Conferences, ICMIE 2016*, 75, 06006, 1–5. Available at: http://dx.doi.org/10.1051/matecconf/201675 06006.

[12] Yushev, A., Schappacher, M., and Sikora, A. (2016). "Titan TTCN-3 based test framework for resource constrained systems," in *MATEC Web of Conferences, ICMIE 2016*, 65, 06005, 1–5. Available at: http://dx.doi.org/10.1051/matecconf/20167506005.

[13] Schappacher, M., Schmitt, E., Sikora, A., Weber, P., and Yushev, A. (2015) "A flexible, modular, open-source implementation of 6LoWPAN," in *8th IEEE International Conference on Intelligent Data Acquisition and Advanced Computing Systems: Technology and Applications (IDAACS2015)*, Warsaw, 838–844.

[14] Huang, A. S., Olson, E., and Moore, D. C. (2010). "LCM: Lightweight communications and marshalling," in *2010 IEEE/RSJ International Conference on Intelligent Robots and Systems (IROS)*.

[15] IEEE Standards Association. (1988). Available at: http://standards.ieee.org/findstds/standard/488.2-1992.html.

11

IoT Meets Opportunities and Challenges: Edge Computing in Deep Urban Environment

Marta Chinnici[1,*] and Saverio De Vito[2]

[1]ENEA, R.C. Casaccia, Energy Technologies Department, ICT Division (DTE-ICT), Rome, Italy
[2]ENEA, R.C. Portici, Energy Technologies Department, Photovoltaic and Smart Network Division (DTE-FSN), Portici, Italy
*Corresponding Author: marta.chinnici@enea.it

Internet of Things (IoT) is a novel proposed paradigm as a roadmap to address the critical trends concerning the smart environment in the foreseeable future. The vision of IoT environment based on smart "things" represents an essential strategy for the progress on challenges related to the urban context: system architecture, design and development, and human involvement. On the other hand, with the introduction of IoT paradigm in the urban environment, it is witnessing an explosive proliferation of endpoints, and consequently, a continuous flow of the heterogeneous data produced by various applications. In this view, the cloud computing paradigm presents, *de facto*, several limitations (e.g., high transmission latency), and therefore, a new computing paradigm called edge computing (a.k.a. fog computing) in the era of IoT could be the candidate for a new architecture model in the urban environment. This chapter explores advantages of edge computing architecture in the urban environment and proposes a hierarchical distributed structure that extends from the edge of the network to the core. Moreover, a novel proposal of smart city concept, called "deep urban environment" is introduced. In this context, the authors imagine a city based on the new dimension achieved through IoT, which adds to Big Data and data analytics through a massively distributed number of sources at the edge. Eventually, pervasive urban air quality monitoring is analyzed as an application scenario that can significantly benefit from the implementation of this concept. In particular, on-board (edge implementation) calibration algorithms are proposed.

11.1 Introduction

Thanks to the increasing implementation of the Internet within society, new sources of vast volumes of data have shown up in the last few years. The term Big Data was introduced to describe the data explosion in the digital universe, and their role is widely recognized as being one of the most powerful drivers to promote energy efficiency, productivity, and sustainability, and to support innovation within the cities. According to the report by Cisco [1] and Ericsson [2], 50 billion of heterogeneous devices will be connected to the Internet by 2020. These devices are getting increasingly smarter, connected to the Internet, and among themselves, giving value to the Internet of Things (IoT) paradigm that is/will generating/e an unprecedented volume and variety of data. In [1], the authors have estimated that the data generated by physical objects would reach 507.5 zettabytes (ZB) per year (42.3 ZB per month) by 2019 – 269 times higher than the amount of data transmitted to data centers (DCs) from end-user devices and 49 times higher than total DCs' traffic. The IoT paradigm due to its nature represents an ecosystem made by cyber-physical systems (CPS) where a set of physical objects (things) – such as cameras, sensors, and actuators – are interconnected with virtual resources available in a cyber world (without the human intervention), to an ubiquitous computing environment [3]. The dissemination of the IoT paradigm has a vast potential that is visible through several possible applications and in the novel value-added services to make life easier and healthier, to increase productivity and to build more intelligent and sustainable environments. Thanks to the use of IoT paradigm, a plethora of applications that make use of the enormous amount and variety of data generated by such "things" to provide new services are available. Therefore, the application of the IoT paradigm in the urban environment is of particular relevance as it responds to important societal needs and trends [4]. Indeed, modeling a complex ecosystem as an urban context is of particular interest as it responds to the ongoing push of many national governments in adopting ICT solutions in the management of public affairs, thus realizing the so-called smart city concept [5]. In concrete, *what does IoT mean in urban environment?* Even if the application of the IoT paradigm in the urban environment is of particular relevance, however, many challenges, related to IoT devices issues, have still to be tackled. Indeed, these devices often lack of enough resources to perform compute-intensive tasks such as big sensor data analytics for detecting essential events (e.g., traffic congestion and anomalous state of power grid). A *naive* approach consists in uploading for each IoT device, all sensor data to the cloud to perform compute-intensive tasks and data analytics processes. However, this method is not applicable to IoT applications with real-time constraints, such as smart transportation, electric grid management, or surveillance for public safety. Even if an IoT application does not have stringent timing constraints, uploading all big sensor data to cloud imposes a significant burden on the Internet with the limited upload bandwidth. A promising emerging technology that can address this challenge is edge computing (a.k.a. fog computing), where computing resources are pushed to and made available at the edge of the Internet. For instance, a cloudlet (a DC in a box) at the network edge may perform extensive data analytics and uploads only the added-value information derived from sensor data to the cloud. By supporting compute-intensive services (e.g., big sensor data analytics, computer vision/image processing, or data security/privacy measures) near sensors, edge computing can significantly alleviate the latency and bandwidth concerns, while enhancing the quality and availability of services. To deeply envision the potential applications of IoT paradigm in the urban context, it is necessary to go beyond the mere concept of cloud and provide more efforts for introducing a novel framework which is composed of cloud servers and IoT devices. The edge computing could be a possible candidate to get a step forward in this direction. This chapter goes through the recent research advances in IoT paradigm (e.g., devices, architecture, limitation of communication protocols, lack of standards, etc.) and its application to urban environment and discusses the benefit of the edge computing. The recent technologies in response to IoT challenges (such as energy efficiency perspective, standards, and metrics) are analyzed and discussed. Moreover, the authors introduce an advanced smart city concept, called "deep urban environment". It is conceived to address the challenges linked to the cyber-physical vision of the city and to achieve the desired functionalities through the cooperation of different systems thanks to the mutual relation among IoT – big data and data analytics. An explanation of these concepts and a use case of edge computing are provided, consisting of a system developed by ENEA which provides a relevant example for cooperative air quality monitoring applications in the urban environment. In details, on-board or edge implementation of calibration algorithm is proposed as a way to efficiently solve latency and computational load issues emerging in the urban context.

11.2 The Role of Big Data in IoT Era

11.2.1 Big Data Generation

Data-intensive computing is more and more to be considered as the basis for a new, fourth paradigm for science [6]. Two factors are encouraging this trend: (1) the vast amount of data are becoming available in more and more application areas and (2) the infrastructures allowing storing these data for sharing and processing are becoming a reality. The technical and scientific issues related to this topic have been designated as Big Data challenges. Big Data refer to not only the realization of information explosion but also technologies, which ensure value generated from the analysis of these massive data [7]. Big data have been defined as a broad term for the collection of massive datasets with a great diversity of types so that it becomes difficult to process by using state-of-the-art data processing approaches or traditional data processing platforms (such as Processing Big Trajectory Data). The main characteristics of big data have been presented and discussed since 2001 when D. Laney from META Group in [8] proposed the first characterization based on volume, velocity, and variety. In 2012, Gartner retrieved and gave a more detailed definition of big data based on high-volume, high-velocity, and high-variety information due to assessment that requires new forms of processing to enable enhanced decision making, insight discovery, and process optimization. Recently, even more, eight Vs have been introduced and elaborated such as volume, velocity, variety, veracity, variability, viscosity, and virality. However, not all types of big data have all the features for each of the characteristics aforementioned. The markets for big data involve three layers [7]: (1) the *infrastructure layer* (hardware components), (2) *the data organization, analytics,* and *management layer* (software components), and (3) the *services layer* (big data applications). The infrastructure layer, which mainly includes external storage systems, servers, DC networking infrastructure, and cloud infrastructure, is the foundation of the big data technologies. The data organization, analytics, and management layer, typically implemented by means of software, are in charge of storing, processing, and analyzing various structured and unstructured data, which could be in offline, real time, or both. The services layer stands for the big data relevant external interfaces and applications, such as business consulting, project services, integration services, data storage services security supports, and technical training. To each layer mentioned above, respectively correspond three specific issues related to: generation, acquisition, and communications of big data. Data generation is the first step of big data. According to [9], principal sources of big data are enterprise data, IoT data (presented in Section 11.2.1), and biomedical data. Not all raw data generated are useful for extracting values, and excessive data generation could cause massive burden in energy and resource consumptions, and thus it is necessary to achieve adequate big data generations toward energy efficiency objectives.

11.2.2 IoT Data and Big Data Analytics

Nowadays, the combination of the Internet and emerging technologies such as embedded sensors, near-field communications, and real-time localization transforms everyday simple objects into smart objects capable of understanding and reacting to their environment. These objects enable new computing applications and represent the base of the vision of a global infrastructure composed of networked physical objects (things) known today as the IoT. *Why the term things?* The origin of the definition of "things" is associated with an object, being, or entity. Serbanati et al. [10] proposed the association of this term to Internet context and introduced the term "smart object" while Magerkurth [11] as objects that are connected directly or indirectly with their environment, and can describe their possible interactions. The increasing volume, variety, and velocity of data produced by the IoT will continue to fuel the explosion of data base for the near future: the data are the new oil. With estimates ranging from 16 to 50 billion Internet-connected devices by 2020, the hardest challenge for large-scale context-aware applications and the smart environment is to tap into diverse and ever-growing data streams originating from everyday devices and to extract hidden but relevant and meaningful information and hard to detect behavioral out of it [6]. The vision of the IoT built from smart objects raises several important research questions as presented in this chapter, in particular regarding the system architecture, design and development, and human involvement. In this section, the attention is focused on big data generated by IoT technology. Data generation is the first step of big data, and IoT has become a crucial source of big data and high business value. IoT and big data could be imagined as two sides of the same coin: big data hold extensive information generated by the IoT technology with the use of IT, which serves a wide range of applications in several and different domains. Considering that the vital component of IoT is the sensor, most IoT big data gathered by radio frequency identification (RFID) and sensor network (SN) technologies exist in many industries, such as agriculture, traffic, transportation, medical care, public departments, families, and so on. In SNs, IoT devices may be embedded with sensors. Meanwhile, the physical object as smart devices and wireless sensing sensor networks (WSNs) are unit expected to be connected shortly. In particular, WSNs are available in various applications and services and most organizations, including public and private, especially in the medical field and healthcare. With the recent advances in communication technology, more and more data are generated and collected. Therefore, the big data will grow exponentially, and consequently, increase the challenges of extracting and retrieving the valuable hidden data. The mutual relation between IoT and big data analytics can be expressed considering the following aspects [12]:

- *Data Collection and Process*: Big data analytics considers which data are collected, how data are processed, and which dataset is used to obtain the information.
- *Data-set Target*: Help to find the target of a data collection at the sampling frequency, and at all times, limit the use of the data to the defined destination.
- *Data Destination*: Collect and store only the amount of data necessary for the intended lawful purpose.
- *Data Access*: Allow individuals' access to data maintained about them, information on the source of the data, critical inputs into their profile, and adopted some algorithms to develop their profile.
- *Information Control*: Allow all individuals to correct and control their information.
- *Privacy*:
    - Conduct a privacy impact assessment for each user: consider data anonymization at the top to the bottom level to make synchronization between them.
    - Limit and carefully control access to personal data.
- *Ethical Issue*: Verify if results from profiling are "responsible, fair, and ethical and compatible with and proportionate to the purpose for which the profiles are being used."

11.2.3 IoT System Architecture

As stated in Section 11.2.1, IoT infrastructure architecture can be considered as a three-layer system that consists of IoT enabling technologies, IoT software, and IoT applications and services. This infrastructure is event driven and real time, supporting the context sensing, processing, and exchange with other things and the environment. The IoT infrastructure, for its nature, is complex due to the vast number (50–100 trillion) of heterogeneous (possibly) things that dynamically join and leave the mobile IoT, generate, and consume billions of parallel and simultaneous events geographically distributed all over the world [6]. Therefore, it is difficult to represent, interpret, process, and predict the diversity of possible contexts. However, the IoT infrastructure must fulfill important characteristics such as reliability, safety and security, survivability, fault tolerance, and user-friendly interconnection of different layers to provide the QoS for a vast amount of users. Also, the IoT infrastructure must manage the communication, storage, and compute resources. Indeed, the primary function of the IoT infrastructure is to support communication among things (and/or other entities, e.g., people, applications, etc.); this function must be flexible and adapted to the vast variety of things (from simple sensors to sophisticated smart objects). In detail, things need a communication infrastructure that is low data-rate, low power, and low complexity. In this section, a brief discussion of some essential characteristics in the IoT system architecture is addressed.

- *Reliability*: The reliability of IoT system means that the system should be able to deliver IoT service under different circumstances. This feature is more critical in the field of emergency response applications (e.g., medical). Indeed, a reliable IoT system should utilize a failure-resilient communication network, which distributes the information successfully throughout all the IoT layers. An unreliable network leads to long delays and data loss, which finally ends to wrong decisions.
- *Security and privacy*: In heterogeneous networks as used in an IoT system, it is difficult to guarantee a high level of privacy and security for users. One of the reasons for this issue is the lack of the universal standard for IoT security. Within the Internet infrastructure, the objects distribute information including the keys and passwords that should be protected and encrypted with a high level of security. The most crucial security requirements include authentication and data tracking, mutual trust, data and information integrity, and digital forgetting. Moreover, the privacy is the most important feature, which lets the users trust the system. The pervasive use of smart objects with private information about a user necessitates using appropriate content privacy techniques to protect clients' data. Despite many technologies introduced to achieve customers' data privacy, still, IoT systems' privacy and security need to be frequently upgraded to enhance the information privacy techniques.

- *Interoperability:* Even if the individual three layers of the IoT infrastructure can be designed independently, however, their interconnections cause some issues and challenges such as interoperability of heterogeneous systems. Indeed, the vast number of the heterogeneous smart objects with different platforms both at the software and hardware layers should be integrated into an IoT infrastructure. However, due to the lack of adoption of universal standards among the different platforms, the interoperability becomes a critical feature in design and builds of IoT services to meet the user requirements. However, steps forward have been made at different interoperability level, for example, at data and service levels, the main integration technologies currently available are based on the service-oriented architecture (SOA) and REST architectural styles, with web services and REST on HTTP as the most used implementations and text-based data (e.g., XML and JSON).

11.3 Deep Urban Environment

11.3.1 Urban Paradigm

New technological developments continue to penetrate states of all regions of the world, as more and more people and objects are getting connected to the Internet. Most countries are reaching a critical mass regarding ICT access driven by the spread of the mobile Internet [13]. The acceleration of ICT diffusion enables the development of smart environments. As the network availability and speed are improving at a steady rate and computers become smaller, more energy-efficient and lower priced Internet-connected devices with sensors and actuators are deployed on ever larger scales in our environment, creating an IoT [6]. Despite the significant interest in IoT, there is yet to be an agreed definition of such a concept because its nature is difficult to capture and shape. Indeed, there are several definitions for the IoT concept as reviewed in [14]. However, to underline the nature of IoT paradigm as CPS (within smart environment) and to emphasize both the integration with the physical environment and the communication aspects, the term IoT can be defined as follows: "... *describes the pervasive presence of a variety of devices – such as sensors, actuators, and mobile phones – which, through unique addressing schemes, are able to interact and cooperate with each other to reach common goals*" [15]. This definition is not comprehensive. The term IoT and more general, in IoT' vision, physical objects (things):

- are equipped with *sensors* (such that sensed information is available on the Internet) and/or *actuators* (such that things can be controlled from the Internet) capturing environmental variables and reacting to various external stimuli.
- Can be addressable controlled and monitored over the Internet.
- Can communicate with other physical and/or virtual resources.
- Provide variable data for applications users or as input for data analytics stream processing systems.

The vision affects a large amount of businesses [16]. Indeed, the exploitation of IoT technology for production has been called *Industry 4.0* [17], which targets a more flexible output for which the IoT supports the entire lifecycle from the design phase onward. A representative application category of IoT paradigm is the smart environment such as smart cities which cover a vast range of needs related to public safety, water and energy management, smart buildings, government and agency administration, transportation, health, education, etc. [6]. Indeed, the variety of applications reflects the very nature of IoT paradigm: *a complex ecosystem composed not only of a variety of elements and of several possibilities of interaction models that arise in a smart environment*. During the last years, this application class of IoT paradigm is the subject of many R&D activities. Moreover, the mutual relation between the IoT paradigm and Big Data management has become the priority for government strategic plans. This reciprocal connection is responsible for the production of an ever increasing amount of data created by Internet-connected devices (or things) and consists of device states (properties) and data collected by its embedded sensors and by humans using applications running on these devices, e.g., smartphones. Consequently, the generated big data require software computational intelligence techniques for data analysis and keeping, retrieving, storing, and sending the information using a specific type of technology such as the computer, computer networks, mobile phones, etc. Thus, the data hold extensive information generated by IoT technology with the use of IT, which serves a wide range of applications in several domains. The purpose of big data analytics has grown tremendously in the past few years directing to the next generation of intelligence for big data analytics and smart systems. At the same time, the IoT has entered in the public consciousness, sparking people's imaginations on what a fully connected world can offer [12]. Hence, both IoT and big data trends give a plenty of reasons for the improvement regarding urban' applications; the combination multiplies the benefits. An urban environment that is running on big data will shortly become fully immersed in the IoT where the role of data

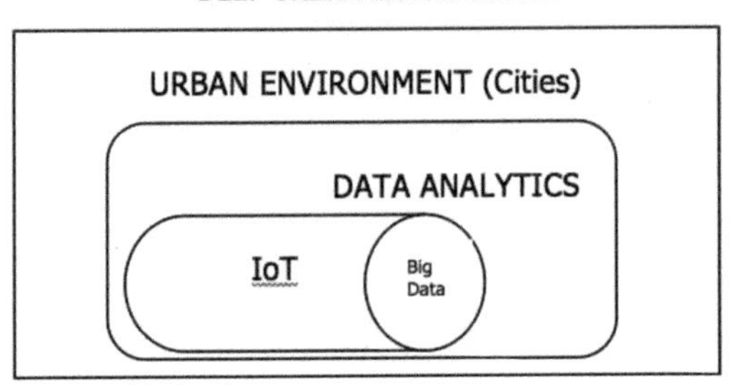

Figure 11.1 Sketch of deep urban environment.

analytics allows interpreting the knowledge to understand the urban context, which is becoming a deep urban environment. The concept of "deep urban environment" introduced in this chapter by the authors represents an advance of the smart city. The deep urban environment is based on the IoT paradigm and is built on a unified architecture where heterogeneous things coexist and where a wide variety of big data are produced. A promising technology that could support the umbrella of IoT applications in a deep city is the so-called edge computing. In this scenario, the role of data analytics is fundamental to analyze big data produced by IoT devices. A **deep urban environment** is defined as an advanced smart city concept regarding the combination of three fundamental pillars: *IoT – Big Data – Data Analytics.* The deep urban environment is a holistic concept based on collaborative, cross-sectorial, and interdisciplinary research: and this thinking is needed to design it and to develop it further.

The urban environment will demand a deep and broad transformation of its context and IoT paradigm applying in-depth analytics through software, will turn extensive data into powerful new insight and intelligence, and this will create a new environment that the author called deep urban environment (Figure 11.1). The deep urban environment concept will not only result in the continuous evolution and formation of the smartness of the city but also in the emergence and development of data information dimension provided through data analytics that can make more powerful in solving problems for human beings.

11.3.2 Urban IoT Applications

A plethora of IoT applications (automation, telemedicine and healthcare, energy consumption, military/aerospace control, industrial system, and so on) is presented in the urban scenario even if many of them are in potential and not

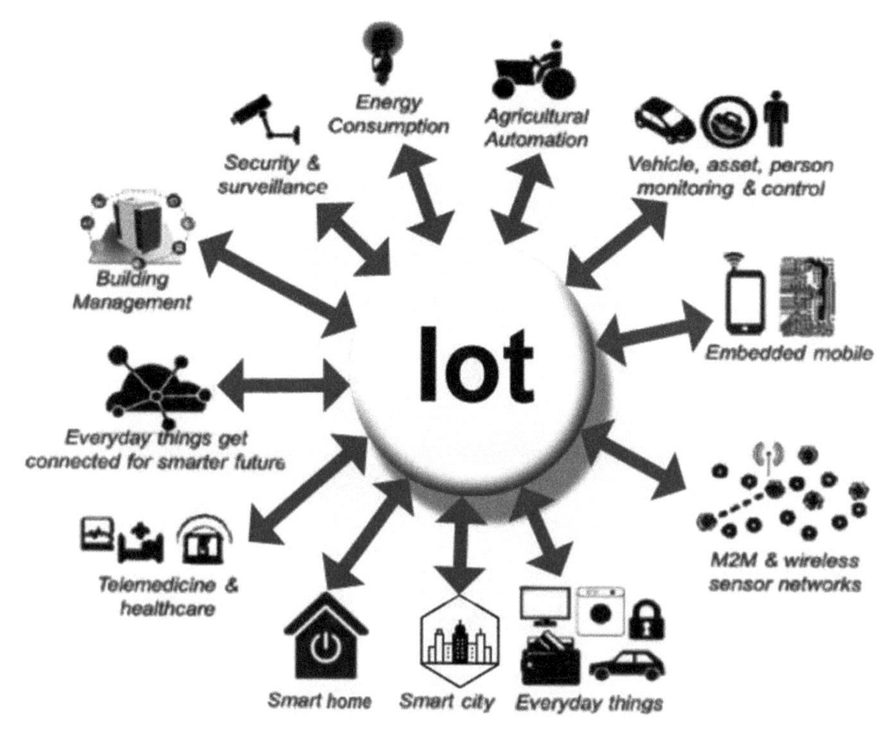

Figure 11.2 Some IoT applications in urban context.

in act deployed. In Figure 11.2, some of these applications are shown. Smart cities represent an important scenario to use the IoT paradigm not only to improve the management of urban flows but also to allow real-time response to several issues and challenges in the meantime.

In particular, a smart city can be considered as a general application category in which other domains such as smart home [18], smart grid, smart automotive and traffic management, etc., are included. The Internet-connected devices can be applied everywhere and can be used and shared in many use cases in an urban environment. In Figure 11.3, is reported an overview of 57 IoT use cases and 20 categories, namely, Libelium by [19], e.g., cities, security and emergencies, logistics, retail, industrial control, agriculture, home automation, eHealth, etc. Instead, research by Beecham [20] has provided nine categories of uses cases in an urban environment.

The services provided by IoT regarding, for example, the *Data capturing* (using microsystems, sensors, and actuators); *QoS&Latency* (applying wireless and Bluetooth); network (using IPv6, lower latencies); and *Security&*

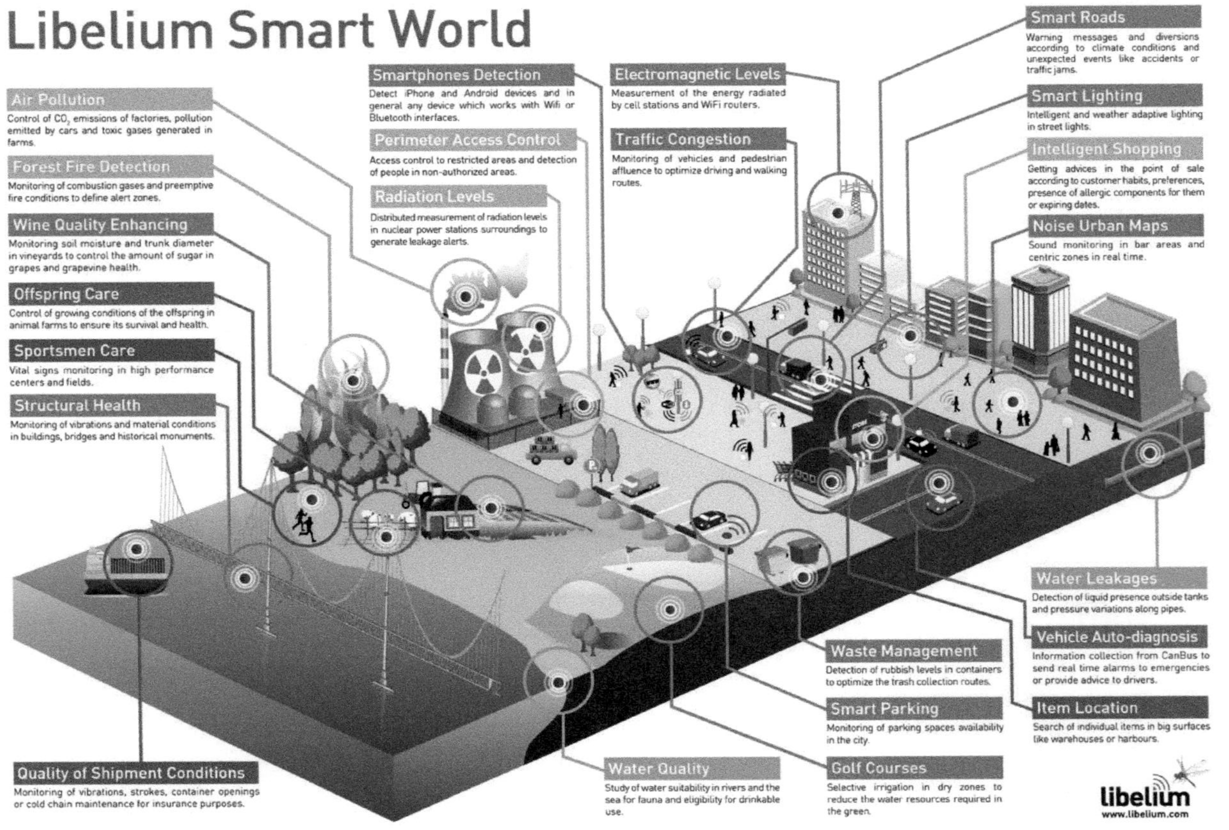

Figure 11.3 Libelium[1] smart world: IoT use case [19].

---

[1]"©Libelium Communicaciones Distribuidas S. L. – http://www.libelium.com"

Privacy (adopting cloud technologies). The principal benefits of IoT system applications in an urban environment concern mainly with the data collection and data processing. However, the use of IoT paradigm can also be extended as a support to production decisions in terms of allows quick activation, the proactive monitoring system, troubleshooting, and automatic alerts [12]. Therefore, with the ever-increasing trend toward development of IoT paradigm – and hence of the smart objects, the pioneering technique of big data analysis – the modern urban environment is evolving with the concept of the broader term "deep" as the author call "deep urban environment." A significant, though not exhaustive, list of IoT applications to urban environment is presented in [12]. In the following, some of the industrial companies which represented the drivers of IoT applications revolution in the urban contest are reported: Cisco (can benefit to any IoT system by developing the real-time data analysis); Siemens and other similar company (to convert the work into M2M); General Electric, (sensors to be operated automatically and provide the data at specific places); and Harley Davidson (to build real-time performance management systems); Bosh. In general, many manufacturers are working on smart systems and devices to make their product to work with the IoT paradigm.

11.4 The Emergence of Edge Computing in Urban Context Beyond the Cloud

The progress in science and technology has enabled the development of smaller, cheaper, and faster computing devices capable of sensing the environment, communicating, and actuating remotely which resulted in an increased interest of applying the IoT [21] to several aspects of life. The introduction of IoT paradigm to urban context involves a vast number of connected devices, which require a scalable architecture to host them without any degradation of the quality of service demanded by applications. Moreover, many devices that make up the IoT are resource constrained [22]; indeed, for example, there is the lack of computing power, bandwidth, storage, and energy resources. The constraints due to scarcity of these resources notable limit the development of IoT applications' domain. For this reason – to provide several services – the cloud computing has become *de facto* the computing platform for application processing in the era of IoT [23]. However, there are the limitations of the cloud model such as the high transmission latency and high costs. Indeed, from one side, the computing company mostly

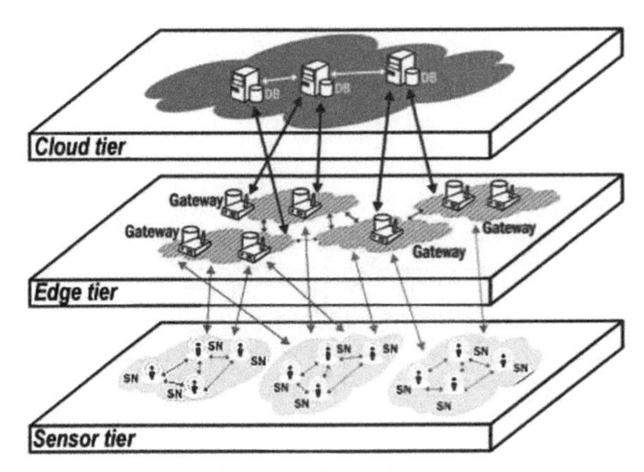

Figure 11.4 Edge tier: architecture' scale.

relies on hauling all the data to the cloud to leverage the cost benefits and efficiency of high-capacity storage and compute platform in the DCs [24]. On the other side, several applications mostly rely on local computation for their decision making because of stringent low latency requirements. Indeed, in an urban context, the deployment of IoT technologies poses several inherent conceptual and technical challenges that are not resolved by cloud architectures. To overcome the constraints, the development of IoT architecture evolves and expands; some of these IoT challenges and efforts in each area are summarized in [25]. Currently, as a remedy to cloud computing limitations, a promising computing paradigm which extends the functions of cloud computing closer to the IoT devices, called edge computing (also known as fog computing [26]) has recently been advocated. The main idea behind edge computing is to extend the cloud layer with the introduction of an intermediate layer (tier) with the data processing capability in between the two layers of IoT such as the cloud and sensors (Figure 11.4).

In Figure 11.4 are shown through a high-level architecture vision both the further independency among three-tier – sensor, edge, and cloud – and the correspondent local connectivity on each layer – devices, gateway, and DCs. The concept of edge computing has recently emerged to push the frontier of applications, data, and services away from the centralized cloud nodes (the core) to the periphery of the network (the edge). Since the edge tier is situated close to the things tier, the data processing at such tier can significantly reduce the communication delay, shorten the response time of IoT applications, and generate less traffic over the Internet [23]. In general, the idea of introducing bridging layers in IoT paradigm is referred using several different terms and accents: fog computing, micro-cloud, and cloudlet. Specifically, fog computing, as the representative paradigm of edge computing, keeps the data processing close to end-users to reduce the communication delay over the Internet and minimize the bandwidth burden by not entirely offloading the generalized data to the cloud. The similar vector of challenges motivates the choice of terms. Since the potential set of services, which can be integrated within the new paradigm, is vast, in this section, the benefits consequent to the application of this approach in the urban context are motivated through the discussion of the features, organization, and functions of this tier.

11.4.1 Edge Vision

In an urban IoT scenario, centralized real-time processing of a large and heterogeneous set of data streams is not feasible. The urgency of a new architecture emerges from the need of handling the vast amount of generated data. Edge computing enables a new breed of applications and services, and that there is a fruitful interplay between the cloud and the sensors, particularly when it comes to data management and analytics. As aforementioned, the edge computing extends the cloud-computing paradigm to the edge of the network. While edge and cloud use the same resources (networking, servers, and storage), and share many of the same mechanisms and attributes (such as virtualization and multitenancy), the extension is non-trivial in that there exist some fundamental differences that stem from the edge *raison d'etre* [6]. In contrast to the cloud layer, the edge is closer to the perception layer, and this provides a number of advantages. The immediate benefit is its location-awareness: thanks to the large-scale geographical distribution of the devices that make up to the edge tier. This location awareness can be utilized to address multiple requirements of IoT urban applications (e.g., mobility and security). Summarizing, the edge paradigm vision was conceived to address applications and services that do not fit well the cloud-computing paradigm. In detail:

- Applications that require very low and predictable latency – the cloud frees the user from many implementation details, including the precise knowledge of where the computation or storage takes place. This freedom from choice, welcome in many circumstances becomes a liability when latency is at premium (gaming and video conferencing).
- Geo-distributed applications (pipeline monitoring and SNs to monitor the environment).
- Fast mobile applications (smart connected vehicle and connected rail).

- Large-scale distributed control systems (smart grid, connected rail, and smart traffic light systems).

Edge computing is an architecture that uses end-user clients and one or more near-user edge devices collaboratively to store substantial amount of data, process compute-intensive tasks, communicate jointly to reduce interference, and carry out management tasks cooperatively to improve the application performance. In such edge computing architectures, any device with computing, storage, and networking capabilities can serve as a near-user edge device. The end-user clients and various edge devices can exist in a hierarchy alongside the existing cloud-based architecture to improve the overall system performance. This notion of edge computing is also referred as "edge analytics" [27]. Edge computing solves four critical challenges by executing tasks near the edge of local access networks.

1. It pools underutilized resources of edge devices regarding storage or computer capabilities and minimizes the network overhead of hauling data to the cloud.
2. It provides context awareness, as application-level details are available near the client at the network edge.
3. It enables real-time response with latency in the order of tens of milliseconds by processing near the network edge instead of relying on the cloud, where the multihop architecture of the network core may result in undesirable delays.
4. The software stacks on edge devices can be upgraded in an agile manner without modifying the software stacks in the cloud or core network.

To accommodate the computing, networking, and storage services and to address the requirements of IoT systems, the edge layer offers the characteristics discussed in the following. As suggested by Cisco [28], the main requirements for new computing paradigm are: *minimize latency, conserve network bandwidth, address security concern, operate reliably, collect and secure data across a wide geographic area with different environmental conditions, and move data to the best place for processing*. All requirements mentioned above are not all met by the traditional cloud computing architectures.

11.4.2 Application in Urban Environment: Pollution Monitoring

A deep urban environment will implement hundreds of different application scenarios producing data that should be transformed in semantic content for a variety of end users. As we have seen, "things" concepts will pervade and deeply transform appliances, infrastructure, and mobile personal systems. Interaction among things will occur at different semantic layers requiring the capability to extract meaningful information along the path connecting the sensor to the remote cloud infrastructure and back to the user. The same application will be sustained by different systems whose data production should be made accessible to cooperation systems. Cooperation among smart CPS, actually, will represent another driver for the distribution of semantic extraction components along the chain [29]. Vertical application scenarios will moreover be interconnected with partial insights reached along the network and being shared among them. A specific example of the involved challenges may be reported using smart air quality monitoring systems. Nowadays, air quality is a paramount concern for citizens and governmental agencies throughout the globe. Mainly driven by non-sustainable mobility habits and patterns, air pollution in cities is strongly affecting citizens' health determining significant losses in terms of lives and large economic impacts. Air quality information is inherently distributed due to the underlying phenomenon nature, and the distributed interaction that air quality has with humans (e.g., personal pollutant exposure) and their infrastructures [30]. Air pollution is generated in several spatially distributed sources like cars on roads, gas-based heating systems, in cities electric power plants, and so on. Moreover, pollutant concentrations show large spatial variability due to complex fluid dynamic effects emerging in cities, like the well-known urban canyon effect. The current sparse regulatory monitoring network is clearly not adequate to cope with this variability and with the emerging needs of the population that include the assessment of personal exposure. In this scenario, experts forecast the deployment and interconnection of multiple systems ranging from low accuracy level wearable or mobile systems to fixed smart multisensor systems and conventional reference analyzers [31]. All these systems will have different vendors and different accuracies and will try to cope with chemical sensors issues in different ways (Figure 11.5).

Moreover, each of them should be networked and equipped with local processing components for local feedback and mutual calibration capabilities [29]. Chemical sensors are actually affected by multiple issues that undermine the field accuracy of out of the shelf-calibrated systems. Chemical sensors are actually affected by multiple issues that undermine the field accuracy of out of the shelf-calibrated systems [32]. In fact, they suffer from cross sensitivities that hamper their capability to estimate precisely the concentration of their target gas when an interferent gas is simultaneously present in the gas mixture under analysis. A typical cross sensitivity may be observed in EC cells

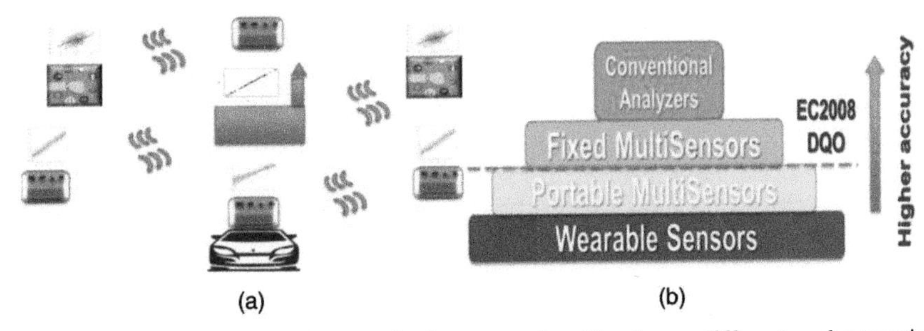

Figure 11.5 Future air quality monitoring network will rely on different and pervasive sources of information (a) ranging from wearable system to high-accuracy conventional analyzers and fixed multisensors meeting regulatory data quality objectives (b).

targeted to NO2 in the presence of ozone[2]. Environmental variables interference may cause similar issues modifying the zero-target gas response and the sensitivity of the sensor to the target gas invalidating simple calibration schemes [33]. Several recent results have highlighted the capabilities of data-driven (black box) multivariate model to solve these issues when tuned with field-recorded data [34]. These methods, usually implemented with machine learning tools, allow the smart air quality system to provide precise and accurate concentration estimation. The implementation of these methodologies far from where the data have been sensed, i.e., in a cloud-based setting, could generate an impressive computational burden if the number of deployed sensors will increase as it is expected to provide an effective service for air quality mapping. Cloud servicing hardware could face the need to implement concentration estimation algorithms in a real-time fashion, for thousands or hundreds of thousands of devices providing multisensory data at fast pace (1 s to 1 min sampling rate). On-board or edge implementation will allow a much more even distribution of the computational load putting the computation close to where concentration data are needed. Actually, the benefit of this method, if on-board implemented, is much more interesting for the end user. Such an intelligent air quality-monitoring unit may be capable of locally outputting a concentration estimation resulting in immediate and real-time feedback to the user. Backend harmonization is another significant advantage for on-board or local gateway implementation of calibration functions. The amount of additional information regarding each single node to be stored in the cloud will be extremely reduced if we do not need to store calibration data

[2]Alphasense NO2-A4 Sensor Datasheet – Visited on Dec, 2017 – Available at: http://www.alphasense.com/WEB1213/wp-content/uploads/2016/04/NO2-A43F.pdf

anymore. Moreover, the networking load will be positively affected avoiding the transmission of necessary but redundant and ancillary data especially for selected scenarios. EC sensors again provide a typical example. The cloud execution of calibration function for a single target gas may require the transmission of both working and auxiliary electrodes' signals for the relevant sensor as well as system internal temperature signal, all of them are required to correct for part of sensor drift due to aging and environmental interference. Local or edge let-based concentration estimation will allow for transmission of concentration information only, immediately reducing by a factor of two the amount of information sent. The next section will focus on the latter positive outcomes providing a specific application example with dramatic impacts.

11.4.3 Network Load Improvements

Local execution of a multivariate calibration function for gas multisensors will strongly limit the network load with significant benefits for the user (costs) and network operators that are easily graspable. Such a system should, however, be able to locally implement a machine-learning tool such a neural network, managing network transmission, and application level handshaking. The benefit of on-board computing on network load reduction may become impressive in certain specific scenarios. Actually, safety and security scenarios are characterized by a few significant events representing alarms for surpassing certain thresholds set on dangerous gas concentration. The capability to recognize this event on the edge of the network by on-board or local processing will put to near zero the expected WAN traffic load. This scenario has been in fact analyzed by several authors. Among them, De Vito et al. [35] have proposed the on-board execution of a neural network component for their pervasive electronic nose, called TinyNose, based on a polymeric chemical sensor array and a TelosB microcontroller node. In their paper, the authors choose a relatively small neural network whose weights were offline computed and then offloaded to the TelosB node by firmware update. The network analyzed the response of the chemical sensor array in search of a specific pollutant (Ethanol) while subjected to the presence of a relevant intereferent (acetic acid). The false positive rate of the network when considering a 10 ppm threshold for ethanol was found to be 0.05. That means that we can expect a strong reduction of unwanted network transmission at the node level. Local processing unit may take the burden of detecting uncorrelated events further reducing the WAN load to values of few data packets

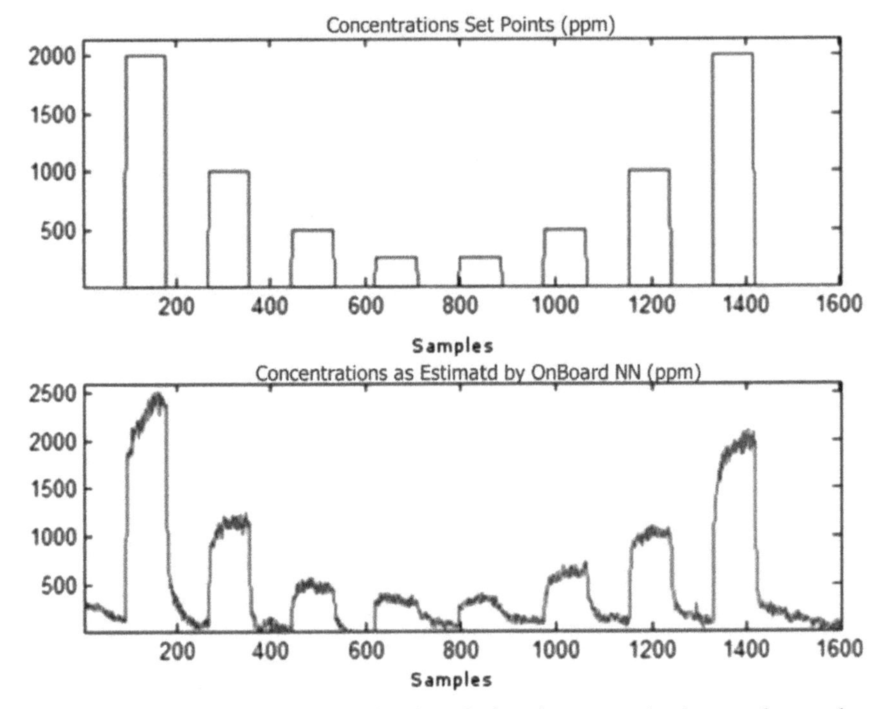

Figure 11.6 TinyNose on-board estimation of ethanol concentration by neural networks.

a day without compromising responsiveness and accuracy of detection. With an increase of 35% of memory requirements, they showed that the system would become more energy efficient starting from a probability of relevant event equaling 0.97, while the expected one is close to zero in safety and security scenarios. The estimated lifespan of the system in this condition considering the false positive rate was found to be in excess of 100 days (Figure 11.6).

11.4.4 Network Local Estimation of Concentration for Immediate Exposure Feedback

As above mentioned, information on personal exposure to pollutants is becoming more and more relevant. The MONICA system developed by ENEA gives a relevant example for cooperative air quality monitoring applications [36]. The systems rely on a smartphone or an ancillary SBC-based

Figure 11.7 (a) The Monica cloud interface where information coming from multiple deployed multisensors is collected and fused together. (b) The smartphone real-time mapping utility allows for immediate and locally compute feedback on personal pollutant exposure.

system for long-range transmission, geo-location, and local concentration estimation capabilities. Based on an array of electrochemical sensors, it needs to sample WE and AE signals for each of the sensor plus temperature and relative humidity signals. Based on these signals, a neural network component located and run by an Android smartphone application is capable of locally estimating the concentration of ozone, NO_2, and carbon monoxide in the air. Being a wearable/portable device, the system can feed the user with a feedback on personal exposure to pollutant while moving in the city. By exploiting the GPS signal, the APP is capable to associate averaged measurements to location coloring the route followed by a user during a mobility session with colors reflecting the actual pollutant concentration. The MONICA App (Figure 11.7) is designed so as to store the calibration function hyperparameters for all the devices locally. At the connection time, the appropriate hyperparameter set is selected in order to implement the proper calibration function for each of the connected device. In an edge computing fashion, the MONICA app acts as a second layer local information processing system, supporting the execution of several different calibration schemes as required by the connecting device.

Table 11.1 A tradeoff comparison among accuracy and complexity of machine learning techniques for smart air quality calibration scenarios

ML Technique	Input No.	Main Storage Parameter Value	Storage Complexity	Computational Complexity	Accuracy
NN	63	15	960	960	1.1 ppb
SVR	63	13095	>80K	13095	1.1 ppb
GP	63	30240	>900M	30240	1.06 ppb

It is now clear that the on-board calibration scheme is at the center of a pervasive air quality monitoring solution. In an attempt to establish the most suitable on-board calibration scheme, one must aim to achieve high accuracy simultaneously trying to reduce the computational and storage impacts. In a recent work [37], the authors have provided a comprehensive test framework exploiting two air quality datasets trying to quantify both accuracy and computational/storage load of machine learning techniques in this specific scenario for on-board implementation, in particular, results based on a dataset featuring several months of measurements taken by an electrochemical sensor-based, fast sampling (20s), smart air quality monitoring system. A training set of 30,240 samples has been set apart for training purposes while the remaining samples have been used for validation (10,085 samples) and testing (11,485 samples). The validation set has been used for optimizing each technique's hyper parameters values. Table 11.1 shows an original view of the obtained results. Here, actually, we compare SVR, Gaussian processes, and NN techniques in their dynamic version, i.e., when coupled with a 3-sample-long tapped delay line.

It can be seen that NN provides for the best tradeoff among accuracy and computational/storage complexity. NN prediction complexity actually depends on input dimensionality and hidden layer neurons number. For a typical smart air quality scenario, this accounts for less than 1000 weights and related multiplications. For SVR, it is well known that hyperparameters' optimization usually lead to a significantly high number of support vectors. In our case, more than one-third of the training samples have been taken as a support vector. This lead to a storage complexity exceeding 80K double numbers while the prediction computational complexity is bound to the number of support vectors. Gaussian processes being the best performing technique, it also accounts for the most significant complexity both in terms of storage and computing being them bound, respectively, to the square of the number of training samples and to the number of training samples. In the overall, despite being suboptimal, FFNN appears as among the best candidates for on-board or pervasive implementation of smart air quality multisensors' multivariate calibration components.

11.4.5 Dependability: Reliability, Security, and Maintenance

Edge computing components will undoubtedly impact on CPS dependability. Self-assessment and graceful performance degradation components can be hosted on edge to provide for enhanced reliability of single devices. Cryptographic services and fraudulent injection detection systems can positively affect security. At the maintenance level, we'd like to stress the positive impacts of edge calibration introduced in the MONICA system. Sensor and concept drifts in chemical sensing systems challenge the durability of their calibration schemes. In particular, slow changes in sensors sensitivity and baseline response may lead to the obsolescence of calibration parameters whether seasonal changes in temperatures may lead the calibration to work outside of its safe envelope, i.e., the range of target gases and interferents for which its parameters have been optimized. As such calibration parameters should be periodically updated and mutual calibration or semi-supervised learning schemes, introduced respectively by Hasenfratz [38] and De Vito [39], may help to implement the update using online gathered data. When two mobile air pollution analyzers come close to each other, they can exchange information on their concentration estimations, and this may help to correct for drift effects. The correction may be computed locally or more appropriately when sufficient amounts of data have been gathered, on cloud facilities. Corrected parameters should then be sent to the calibration component executor (e.g., device, smartphone, or cloudlet).

As previously mentioned, the MONICA architecture allows for edge execution of multivariate concentration estimation components relieving cloud-based backend from this task (Figure 11.8). Whether executed as a part of the smartphone App or on Cloudlets, the components' parameters are downloaded from the cloud at each device start-up time. In this way, each device can have its parameters updated as soon as a new, enhanced, calibration parameters' set becomes available. Calibration maintenance can, in this way, be made possible transparently to the user without the need to specifically initiate this action or sending the device back to the factory for firmware updates. This result can, of course, be achieved with a full cloud implementation of the concentration estimation component, but again, its edge execution allows for reduced latency for local pollution levels and decreased data bandwidth needs. Practically edge computing provides for

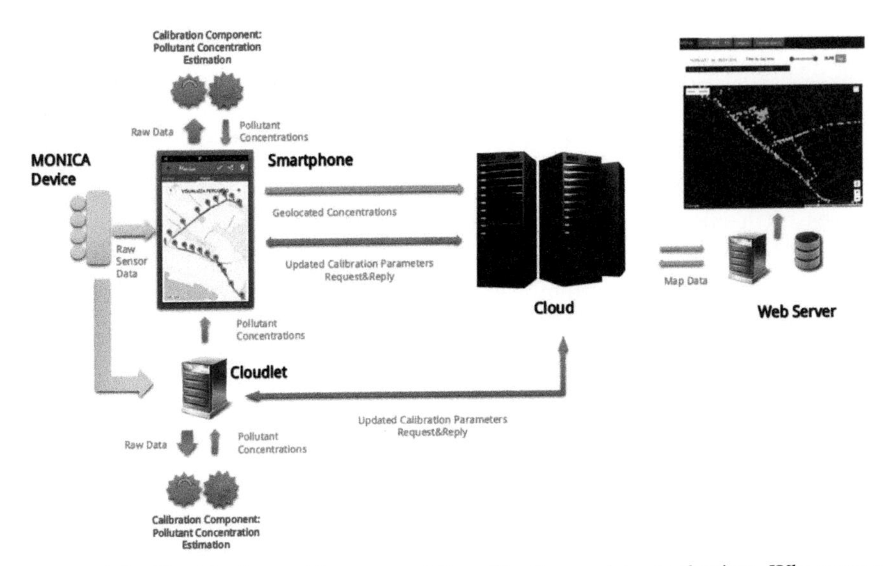

Figure 11.8 The Monica architecture and its calibration update mechanism. Whenever a Monica device becomes connected to a smartphone/cloudlet/gateway, the latter asks the cloud-based backend for updated calibration parameters to maintain the calibration components. In this way, raw data are always processed with the most accurate calibration available.

retaining maintenance flexibility offered by cloud computing calibration components' execution without the requirements to sacrifice performance.

11.5 Challenges

Recent years have witnessed an explosion of the IoT paradigm in the urban context. The vision of IoT built from smart objects raises several essential research questions regarding system architecture, design and development, and human involvement. Some of the research proposals target specific application domains in an urban environment (e.g., integrated service framework focusing on a smart house environment). However, several research works aimed at providing a holistic solution for general IoT architecture. Indeed, many studies are addressed to build the right balance for the distribution of functionality between smart things and the supporting infrastructure. Related to this issue as the cloud computing could become true infrastructure enablers for IoT architecture. Indeed, during the past few years, IT has seen different significant paradigm shifts in technology. These changes have interested primarily the areas of cloud computing, big data, and IoT. The convergence of these areas already creates a new platform for the market that could allow enterprises to develop new business and mission capabilities. These capabilities can enable a more integrated view of its architecture as we have explored in this chapter. Despite the efforts, the development of a unified architecture remains a significant gap due to a lack of standards on the structure of IoT. In this sense, the status quo is that there is an intranet of things with many islands of field networks existing in silos, each supporting different IoT applications. This phenomenon is also due to the heterogeneity of devices and addressing their interoperability remains a great challenge. Hence, it becomes mandatory to provide a unified architecture to seamlessly integrate these decentralized environments with different kinds of things for Internet scale applications in the urban environment. In summary, the significant critical challenges (that occur when dealing with a vast amount of information continuously flowing from things) for a unified IoT architecture in the urban environment are as follows:

- **Standardization:** Even if the industry leaders are trying to develop specified standards, at present, there is a lack of standards of IoT and the hurdles facing standardization can be divided into three interlinked categories: *platform, connectivity, and applications:*
    a. *Platform:* regarding form and design of the products (UI/UX), analytics tools used to deal with the massive data securely streaming from all products, and scalability.
    b. *Connectivity:* includes all parts of the consumer behavior (e.g., smart cars, smart homes, and smart cities). From the business perspective, the connectivity using IIoT (industrial IoT) and in this category the M2M communications dominating this field.
    c. *Applications:* this category includes three functions: control "things," collect "data," and analyze "data." IoT needs killer applications to drive the business model using a unified platform.
- **More Connectivity and More Devices.** At present, maintenance and real-time processing of so many data streams are not feasible for many application developers with a limited budget. It is necessary to improve the connectivity concerning the rapid proliferation of IoT devices. In the meantime, the heterogeneity of things requires software developers to have a vast variety of expertise to deal with all kinds of data (e.g., physics, medical images, etc.), which is overly demanding in practice. Moreover, networks' bandwidth and computation are wanted

if data from things are continuously transferred over the network to the applications for computation.

- **Edge computing:** To overcome the IoT paradigm challenges and meet the requirement of the application domain, the demand for an intermediate computing layer as the edge becomes evident. The edge tier represents, in the meantime, a synergic integration of both the IoT and cloud (cloud things: **CoT**) and of the cloud and sensors (cloud of sensors: **CoS**). Deploying edge computing can help to quickly implement platform and applications by offering SOAs in the way that each customer needs. To execute all applications at edge –nodes open new issue as, for example, the resources' allocation (using, e.g., stochastic model and Lyapunov optimization) and management.
- **Security and Privacy:** Security is a major concern and vulnerabilities need to be addressed. The introduction of security and privacy policies and centralized control issue is necessary at different levels of the architecture, such as edge-tier and data, in particular, to analyze sensitive data locally eliminating transmitting to the cloud for better privacy.
- **New business models: IoT-as-a-Service (IoT-a-a-S):** Novel services provisioning models will be developed in many IoT verticals supported by Big Data and artificial intelligence tools. These models and the correlated data are addressed to end customers. However, the potential for IoT business model transformation extends beyond this, to encompass an increasing variety of more complex, as-a-service business models such as smart cities.
- **Skills in IoT's Big Data Analytics:** Dynamic data sharing is pivotal of IoT, and hence, the Big Data Analytics will be instrumental in building responsive applications. Consequently, the need for Big Data skills will rise for most IoT service providers.

An additional challenge that arises in drawing a unified IoT architecture for the urban environment is related to the energy efficiency theme. In this chapter, the authors decline the theme of the energy efficiency from two perspectives: *physical* and *methodological.* The first side includes the category of IoT baseline (actuators, sensors, and devices) and the high-architectures levels (on edge tier, on cloud of things, and cloud of sensors). Regarding, for example, devices, the emerging edge devices (smartphones) are potentially utilized in data generation since these devices can reduce the deployment of corresponding sensors, which is considered environmentally friendly. For example, several authors introduced mobile crowd sensing, where individuals with mobile devices collectively share data and extract information to benefit IoT applications. Utilizing mobile devices for sensing may significantly reduce the manufacture and deployment costs. RAMSES [40] considered using existing portable devices for IoT data generation in the hardware design of devices, and through implementing an RF energy-harvesting circuit to deliver sensor data to standard RFID readers wirelessly, the system consumes less energy in the process of data generation. The edge computing techniques have been considered efficient from the energy side since it reduces the load generated by the central communication and computation infrastructure. However, the solutions explored so far solely focus on data processing, without an in-depth analysis of information acquisition, representation, and transportation solutions needed to increase the efficiency and achieve a sustainable technology. At edge node in particular, energy consumption of sensor nodes and latency of gathered data presented at end-users' terminals must be carefully considered. The second side concerns the physical nodes. The main challenges regarding them are related to the energy efficiency theme; indeed, currently many research works are addressed in this direction [21]. The *methodological* issue of energy efficiency theme, in particular, is related to the indicators, metrics. Energy efficiency metrics are essential tools for the monitoring of energy use and environmental conditions at different architecture scales. These parameters make it possible to quantify the performance of the integrated systems and compare various technologies and strategies. Taking into account the high level of the architecture scale, the existing metrics that have been developed to measure the energy efficiency and thermal performance of DCs are not always capable of tracking improvements at different levels of a DC. However, in a recent work [41], the authors provided both a holistic framework that helps the reader to simultaneously take into account the effects of different metrics at different DC scales and the current efforts to achieve common policies, standards, and regulatory strategies for energy efficiency in DCs. Furthermore, a step forward has been made on DCs' energy efficiency measurement in recent years, regarding metrics related to the "useful work" or "work done" [42–45] that cover the application level in a DC. In the DC scenario, the cloud computing has become a *de facto* approach for service provisioning over the Internet. However, none of the existing metrics is precise enough to distinguish and analyze the performance of DC communication systems from IT equipment and then to assess energy efficiency strategies of cloud computing systems. As for the cloud computing also with the introduction of edge computing paradigm (a.k.a. fog computing), the need to introduce and to adopt new metric to measuring energy efficiency is becoming a priority. The challenges aforementioned converge in a new perspective of the city as discussed in this chapter by the authors: a "deep" urban environment, an evolution of urban environment featured by the new dimension provided by the mutual relation among IoT, Big Data, and Data Analytics. The authors imagine this trend since these pillars are fundamental for the ongoing challenges in the city. The use of deep is also used by the architecture level, for example, to the cloud paradigm, "deep cloud" in order to name an intelligent platform by the community and for community. The concept of deep urban environment presented in this chapter could be linked to the emerging trend: Cybermatics [46]. It is proposed as such a holistic field for the systematic study of cyber entities in cyberspace and cyber world, and their properties, functions, and conjugations with entities in conventional spaces/worlds.

11.6 Conclusion

In this chapter, the authors set out to explain the necessity, the rationale and significance of, the proposed trend of the deep urban environment, what it is and what it encompasses, and how it is related to other fields and areas. In this work, the authors introduced and defined the deep urban environment concept as the next evolutionary step toward the urban context of the future based on a converging nexus among IoT, big data, and data analytics (e.g., machine learning) through the introduction of edge computing paradigm and smart cities technologies. In this concept, pervasive sensors and actuators systems, managed by computational intelligence components, interact with the city physical and chemical environment producing a vast amount of data. We argue that these processes cannot be efficiently managed without re-distributing the computing and networking load by the use of the emerging edge computing paradigm. The resulting architecture is capable of hosting and running computational intelligence components devoted to extracting semantic content and selecting actuating strategies at several layers including nodes, local gateways, and a cloud-based backend. In analogy with deep learning architectures, the concept is based on a pyramid of consecutive layers of interacting components that gradually enrich the semantic information content in an efficient way. Network latency and load minimization, local feedback capabilities, and computing scalability are among the primary benefits that can be experienced using the concept. Some of the most relevant applications in smart cities today are based on the air quality monitoring infrastructure. Future air quality monitoring networks primarily characterized by extreme pervasive offer a significant example for the use of the deep urban environment paradigm. Pervasivity will be reached by deploying hundreds of intelligent wearable/mobile/fixed solid-state sensor-based air quality monitoring each one featuring different levels of accuracy. Efficient semantic extraction including concentration estimation, low latency feedback to the user for personal exposure evaluation, geographical sensor fusion, and air quality mapping can significantly benefit from a design that exploits the load distribution among all the network edge components up until the cloud backend. There, the most valued information for a community, i.e., a high-resolution air quality map, could be eventually extracted based on local concentration estimation performed close, if not on board, of the smart sensing device. We have shown two different applicative examples, in this scenario, in which the benefit of the defined concept could be particularly focusing on. Finally, we have identified the most relevant challenges that still hamper the development of complete applications relying on the deep urban environment concept.

Acknowledgments

Marta Chinnici would like to express her deep and sincere gratitude to Prof. Alfonso Capozzoli, from the Politecnico of Torino, for his valuable help in the critical discussion, his constructive commentaries, and insightful suggestions.

References

[1] Dave, E. (2011). "*The internet of things how the next evolution of the internet Is changing everythings*," White Paper, Cisco.

[2] Hans, V. (2010). "*CEO to shareholders: 50 billion connections 2020*," White Paper, Ericsson.

[3] Miorandi, D., et al. (2012). Internet of things: vision, applications and research challenges. *Ad Hoc Netw.*, 10, 1497–1516.

[4] Zanella, A., et al. (2014). Internet of things for smart cities. *IEEE Internet Things Journal*, 1, 22–32.

[5] Schaffers, H., et al. (2011). Smart cities and the future internet: towards cooperation frameworks for open innovation. *Future Internet, LNCS*, 6656, 431–446.

[6] Bessis, N., and Dobre, C. (2014). "*Big data and internet of things: a roadmap for smart environments*," Studies in Computational Intelligence, Springer.
[7] Wu, J., et al. (2016). Big data meet green challenges: greening big data. *IEEE Systems Journal*, 10, 873–887.
[8] Laney, D. (2001) "*3D Data management: controlling data volume, velocity, and variety*," Application Delivery Strategies, META Group.
[9] Chen, M., et al. (2014). "*Big data: related technologies challenges and future prospects*," Springer.
[10] Serbanati, A., Medaglia, C.M., and Ceipidor, U. B. (2011). "*Building bllocks of the internet of things: state of the art and beyond*," Deploying RFID, Chapter 20, IntechOpen.
[11] Magerkurth, C. (2012). "*Internet of things-architecture*," Deliverable D1.4, European Project IoT-A (257521).
[12] Dey N., et al. (2018). "*Internet of things and big data analytics*," Springer.
[13] ITU. (2012). "*Measuring the Information Society*," Report, Geneva.
[14] Minerva, R., Biru, A., and Rotondi, D. (2015). "*Towards a definition of the internet of things*," Report, IEEE.
[15] Giusto, D., et al. (2010). "The internet of things', in *20th Tyrrhenian Workshop on Digital communications*, Springer, Berlin.
[16] Marwedel, P. (2018). "*Embedded system design – embedded systems, foundations of cyber-physical systems, and the Internet of Things*," Springer.
[17] Brettel, M., et al., (2014). "How virtualization, decentralization and network building change the manufacturing landscape: an industry 4.0 perspective," in *International Scholarly and Scientific Research and Innovation*, 8, 37–44,
[18] Capozzoli, A., Cerquitelli, T., and Piscitelli, M. S. (2016),"Enhancing energy efficiency in buildings through innovative data analytics technologies," In *Pervasive Computing Next Generation Platforms for Intelligent Data Collection*, Academic Press, 353–389.
[19] Libelium. (2013). "*50 Sensor applications for a smarter world*," Report.
[20] Beecham. (2013) "*Sector map showing segmentation of the M2M market*,".
[21] Rahmani, A. M., et al. (2018). "*Fog computing in the internet of things, intelligence at the edge*," Springer.

[22] Aazam, M., and Huh, E. N. (2014). "Fog computing and smart gateway based communication for cloud of things," in *International Conference on Future Internet of Things and Cloud (FiCloud)*, 464–470, IEEE.
[23] Nan, Y., et al. (2017). "Adaptive energy-aware computation offloading for cloud of things systems," in IEEE Access, 5, 23947–23957.
[24] Mitton, N., et al. (2012). Combining cloud and sensors in a smart city environment. *EURASIP J. Wirel. Commun. Netw.*, 2012, 1–10.
[25] Al-Fuqaha, A., et al. (2015). "Internet of things: a survey on enabling technologies, protocols, and applications," in *IEEE Communications Surveys and Tutorials*, 2347–2376.
[26] Bonomi, F., et al. (2012). "Fog computing and its role in the internet of things," in *Proc. In Mobile Cloud Computin. Workshop MCC*, 13–16.
[27] Satyanarayanan, M. (2017). The emergence of edge computing. *Computer*, 5, 30–39.
[28] Cisco, (2015). "*Fog computing and the internet of things: extend the cloud to where the things are*," White Paper.
[29] Arfire, A., Marjovi, A., and Martinoli, A. (2015). "Model-based rendezvous calibration of mobile sensor networks for monitoring air quality," in *IEEE Sensors*, 1–4.
[30] Xinghe X., et al. (2017). "A review of urban air pollution monitoring and exposure assessment methods," in *Proc. Int. J. Geo-Inf. Intern (ISPRS)*, 6(12), 389.
[31] De Vito S., et al. (2016). "Machine learning for future intelligent air quality networks," in *Proceedings of 6th Scientific Meeting EuNetAir*, 38–41.
[32] Castell, N., et al. (2017). "Can commercial low-cost sensor platforms contribute to air quality monitoring and exposure estimates?" *Environment International*, 99, 293–302.
[33] Esposito, E., et al. (2017). "Is on field calibration strategy robust to relocation?" in *ISOCS/IEEE International Symposium on Olfaction and Electronic Nose*, Montreal, QC.
[34] Borrego, C. et al. (2016). Assessment of air quality micro sensors versus reference methods: the EuNetAir joint exercise. *Atmospheric Environment*, 147, 246–263.
[35] De Vito, S., et al. (2011). "Wireless sensor networks for distributed chemical sensing: addressing power consumption limits with on-board intelligence," in *IEEE Sensors*, 11, 947–955.
[36] Capezzuto, L., et al. (2014). "A maker friendly mobile and social sensing approach to urban air quality monitoring," *IEEE Sensors*, 12–16.

[37] De Vito, S., et al. (2018). Calibrating chemical multisensory devices for real world applications: an in-depth comparison of quantitative machine learning approaches. *Sensors and Actuators B: Chemical*, 255, 1191–1210.

[38] Hasenfratz, D., Saukh, O., and Thiele, L., (2012). "On-the-fly calibration of low-cost gas sensors," *Wireless Sensor Networks, LNCS*, 7158, Springer.

[39] De Vito, S., et al. (2012). "Semi-supervised learning techniques in artificial olfaction: a novel approach to classification problems and drift counteraction," *IEEE Sensors*, 12, 3215–3224.

[40] De Donno, D., Catarinucci, L., and Tarricone, L. (2014). "RAMSES: RFID augmented module for smart environmental sensing," in *IEEE Transactions on Instrumentation and Measurement*, 63, 1701–1708.

[41] Chinnici, M., Capozzoli, A., and Serale, G. (2016). "Measuring energy efficiency in data centers," in *Pervasive Computing Next Generation Platforms for Intelligent Data Collection*, Academic Press, 299–351.

[42] Quintiliani, A., Chinnici, M., and De Chiara, D. (2016). "Understanding 'workload related' metrics for energy efficiency in data center," in *Proceedings of 20th Int. Conf. on System Theory Control and Computing*, 830–837.

[43] Chinnici, M., and Quintiliani, A. (2013). "An example of methodology to assess energy efficiency in data centers," in *Proceedings of Third International Conference on Cloud Green Computing*, Karlsruhe, IEEE, 459–463.

[44] Chinnici, M., De Chiara, D., and Quintiliani, A. (2018). "Data center, a cyber-physical system: improving energy efficiency through the power management," in *Proceedings of DASC/PiCom/DataCom/CyberSciTech, Orlando'2017*, IEEE.

[45] Chinnici, M., and De Chiara, D. (2019). "An hpc-data center case study on the power consumption of workload," in Lecture Notes in Electrical Engineering Springer Nature.

[46] Ma, J., et al. (2015). "Cybermatics: a holistic field for systematic study of cyber-enabled new worlds," *IEEE Access*, 3, 2270–2280.

12

Hybrid Control System of Mobile Objects for IoT

Anzhelika Parkhomenko*, Dmytro Kravchenko, Oleksii Kravchenko and Olga Gladkova

Software Tools Department, Zaporizhzhia National Technical University, Zaporizhzhia, Ukraine
*Corresponding Author: parhom@zntu.edu.ua

More often, the concept of Internet of Things is inseparably connected with something smart: smart transport, smart industry, smart city, smart business, and so on. Smart solutions are interesting and useful for society, as they allow to make our life more comfortable and to provide resource saving. The components of such systems are both – stationary and mobile objects. Nowadays, a lot of types of mobile objects are successfully used in different areas of human activity: mobile intelligent robots and robotics systems, intelligent transportation systems, remotely controlled technological systems, etc. However, the issues of ensuring of mobile objects control systems' dependability, effective control, and security require further research and development. This chapter is focused on the features of realizations of mobile objects' hybrid control systems based on two different modes – the remote and the autonomous. Thus, all the advantages of both approaches can be used to ensure the system efficiency in various operation conditions.

12.1 Introduction

Nowadays, there are many types of mobile objects and a lot of variants of their control systems development [1–4]. All types of mobile objects can be classified depending on the environment of exploitation (ground, water, air, and space) and, according to the [5], on the mechanisms of movement realization. There are five levels of mobile objects' automation with a detailed description of their characteristics proposed in [6].

The most perspective principles of mobile objects' control are:

- Remote, guided by human through various interaction interfaces (communication channels);
- Autonomous, based on various sensors and "brain" – artificial intelligence of these objects.

Remotely (distance) controlled systems differ by the type of this channel:

- Mechanical channels are used where the objects are removed from each other at a relatively small distance or required to provide undistorted instantaneous reaction.
- Electrical channels: wired channel (is used where there is no possibility to apply wireless channels); radio channel (is used primarily for mobile objects controlling); ultrasonic channel (is used rarely, to control mobile and stationary objects at a relatively short distance); and infrared channel (is usually used for consumer electronics).

The radio channel is one of the main possibilities for remote control of remote mobile object [7]. It can be used for the FM, AM, or SSB modulation and, the frequency diapason can also be different. Depending on the channel and frequency, the mobile object's work distance (radius) is different. Together with the development of the Internet of Things technologies, the way of remote control object through Wi-Fi is appeared.

Remote control of mobile objects is a very promising field, which allows improving the convenience and safety of human society life. According to the fact that the operator can be at a large distance from the controlled object, such objects could be used in dangerous or inaccessible places for people, for example, in areas of environmental disaster or military actions [7–9].

Although remote control provides a lot of benefits, in some cases, it is more acceptable to use autonomous control to exclude user errors [10–12]. Therefore, providing hybrid (autonomous and remote) control of the mobile objects is an actual task. However, safety, reliability, usability, and many other groups of requirements should be taken into account when creating such a special variant of control system [13].

Therefore, the issues of hybrid control systems' effective design are still relevant. The goal of this work is the realization of mobile objects' hybrid control system based on two different modes – the remote and the autonomous. The main tasks are: to develop the architecture of the system and to offer the methods of implementation that will provide the required range of the system operation, low power consumption, good video signal, efficiency of the operation, as well as will reduce the cost of the system.

Further, this chapter is organized as follows. Section 12.2 discusses the related works in the field of mobile object control with the analysis of the disadvantages of the similar systems. Section 12.3 discusses the features of the proposed system architecture. Section 12.4 discusses the methods of implementation and the results of system development and evaluation. Section 12.5 discusses main useful findings with outlook on future work. Section 12.6 concludes this chapter.

12.2 Related Work

The authors of [8] proposed an automobile race control system that consists of a radio-controlled car model and a special software for a mobile phone. Radio-controlled car, in turn, contains an accelerometer and a Bluetooth receiver. The necessary software is installed into a mobile phone, which must also have a Bluetooth transmitter. The operator, with the help of the software interface installed on the phone, manages a radio-controlled car. In addition, the system transmits information about the speed of the car to the mobile phone in real time.

The main disadvantage of this system is a small range of radio transmitter (up to 10 m), which does not allow using of a system for mobile objects' control over long distances. In addition, the system does not allow the setting of mobile objects, as well as remote control of several objects.

The system of automated control by means of radio control [9] consists of the working body of the transport network and the transport route with loading and unloading posts. While the working body of the transport network has installed memory block with the corresponding commands of the block of communication, this block is connected to the control panel of the remote control with the integrated microprocessor. The memory block is connected through the communication unit to the control panel of the system and the information output block. The specified system works at the frequency of 27 MHz radio waves.

The disadvantages of this system are the lack of receiving/transmitting of a video image to control the motion process as well as the usage of the Citizen's Band (27 MHz), which has a wide range of operations, but does not guarantee a stable and reliable connection, since it is sensitive to the influence of domestic and industrial obstacles. In addition, it has a weak penetrating ability, and therefore, is not intended for use in urban development. Also, there is practically no compact equipment for operating at this frequency, which does not allow changing the dislocation of the system fast.

Thus, the existing approaches to the implementation of the automated system of remote control of mobile objects do not provide opportunities for working with a group of objects and the required range of remote control (50 m and more), as well as mobility, noise immunity, and reliability of system operation.

Consequently, the task of creating an automated system of remote control of mobile objects that enables the control of the movement of several objects at the same time and the setting of their parameters is relevant. The system must allow creating the effect of the operator presence inside the study area to control the trajectory and traffic conditions, as well as to provide relevant information about the current state of mobile objects to the dispatcher.

As known, for the realization of the effective interaction of a mobile object with the environment, various types of sensors and actuators are used. In the works [14, 15], different sensors are proposed for usage in autonomous and remotely controlled systems and provided immediate information about the environment is considered. CCD/CMOS cameras, due to their ability to recognize the color, contrast, and optical symbols, give a new set of features completely different from all other devices. However, the result of recognition essentially depends on the characteristics of the hardware, software, and on the implemented algorithms.

Therefore, the second task is to provide the possibility of mobile objects' autonomous control based on the usage of effective algorithms of obstacle recognition and motion planning. Thus, the system of hybrid remote and autonomous control of the group of mobile objects is realized.

12.3 Methodology

The generalized architecture of FPV-Auto project (FPV – First Person View) [16, 17] is presented in Figure 12.1. This mobile object control system consists of a mobile object with a camera (FPV platform), a control station (CS), video glasses, and a joystick. An operator using the CS and joystick carries out the direction of the mobile object movement, speed, and other functions. The data exchange occurs through radio communication. The effect of user's presence inside the car has been created due to video glasses. That is why the quality of video stream is one of the main metrics of the project assessment.

FPV Auto CS is the key module of the system, which allows controlling three mobile objects simultaneously. CS general and internal functions: object

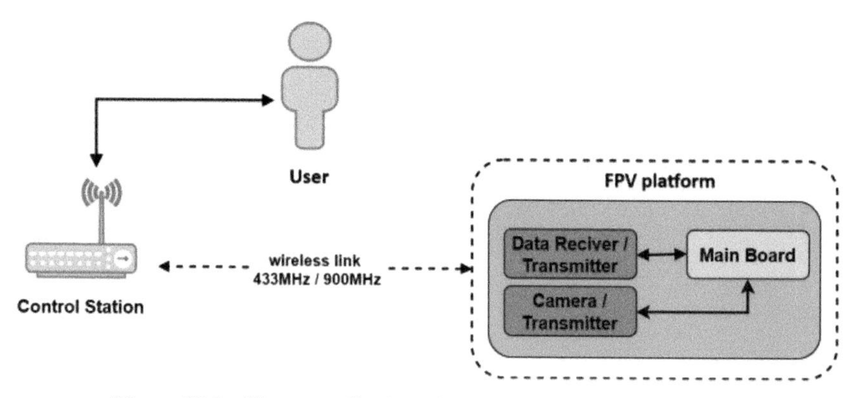

Figure 12.1 The generalized architecture of the FPV Auto project.

binding to radio channel; speed mode selection; control channel activation (turn ON/OFF radio transmitting); front wheel position adjusting; starting/stopping race; UART initialization and transmitting; ADC initialization and reading; and display of control information on the LCD screen.

Previous CS realization was based on Atmega 8 microcontrollers and originally designed PCB. However, this implementation had a lot of drawbacks: low reliability of wired connections was caused by using of perfboards, point-to-point construction, and non-latching connectors; strong electromagnetic interference; high power consumption; short range of radio control (10 times smaller compared with the calculated one); and interference in video stream. Therefore, the task of the investigation of CS parameters and characteristics improvement methods was relevant.

For the realization of mobile object control systems' autonomous mode, it was decided to create an additional subsystem with two modules: obstacle recognition module and motion planning module. The work of this subsystem is aimed at finding and constructing a path for the movement of the mobile object with the recognition of obstacles and their avoidance.

Based on the peculiarities of the FPV Auto system existing prototype implementation, the requirements for the subsystem have been created. The obstacle recognition module must perform multiple recognitions of the objects (traffic lights, pedestrians, and other mobile objects) on images in various formats and resolution no more than 1280×768 pixels. The program should visualize the results of work by overlaying colored frames with the value of probability of classes for the found objects on top of the original image. The motion planning module should identify possible collisions with obstacles of different types, sizes, and locations. The program has to build a

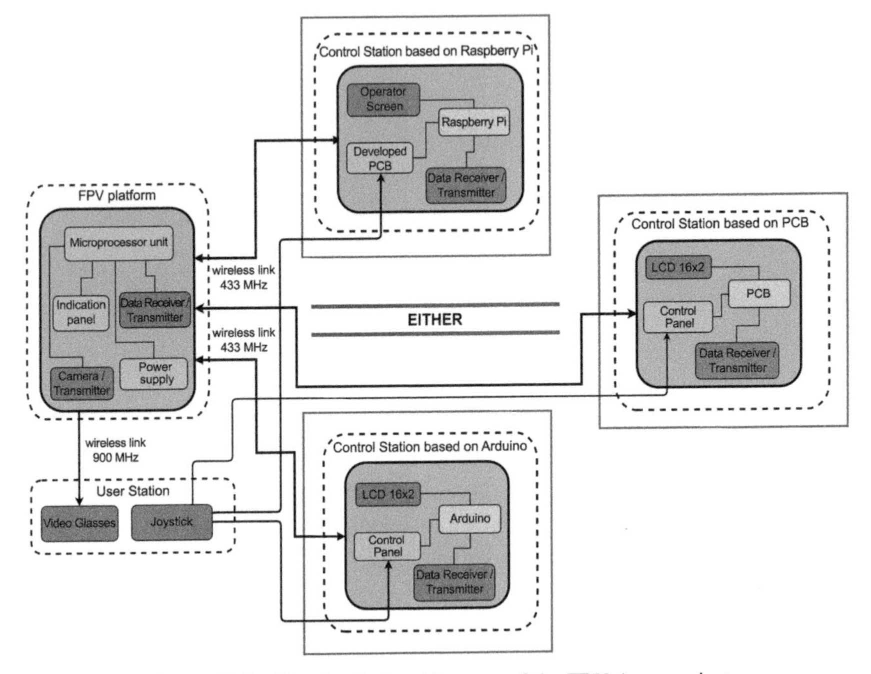

Figure 12.2 The detailed architecture of the FPV Auto project.

roadmap based on the minimum number of nodes and to provide high-speed performance.

The decision about the hardware/software implementation of mobile objects' control systems is a complex task. That is why the design technologies of such systems are often based on reusable solutions, which allow to reduce time and the cost of the project [18].

FPV Auto CS was modernized with using three approaches – based on two different modern hardware/software platforms (Raspberry Pi and Arduino) as well as originally designed PCB (Figure 12.2).

12.4 Implementation and Evaluation of the Hybrid Control System

12.4.1 Subsystem of Remote Control

The first approach was based on Raspberry Pi platform and allowed to implement user interaction with the system only through the software-defined solution.

The Raspberry Pi B2 gives a lot of functions and possibilities, but there is no analog to digital converter necessary for this project. Therefore, extra PCB has been created to work with analog devices. The PCB consists of an eight-channel analog-to-digital converter MCP3008 and a logic level converter, built on two MOSFET transistors. This board allows handling analog signals from the joystick and converting the logic levels of 5- and 3.3-V signals obtained from the radio transmitter. Altium Designer software has been chosen to solve the design task of electronic schematic diagram, wire routing, and three-dimensional model of PCB creating.

The software has been written in C^{++} language on Raspbian operating system using Qt Creator environment. Created program provides such functions as channel binding, joystick calibration, speed mode selection, and simultaneously operating three mobile objects.

The program has the user-friendly interface, which helps the user to determine the channel status by the color indicated by each of them (Figure 12.2). To ensure the reliability of operation, the program provides exception handling when the operator attempts to start the race without users, without connected joysticks, without channel binding to the FPV platform, or when invalid values are entered.

To ensure simultaneous control of three mobile objects at the start of the race, a separate thread is created for each active channel, which is possible due to the full operating system. Since the range of values in joysticks controllers' potentiometers may differ, the extra window for more precise control calibration has been created (Figure 12.3).

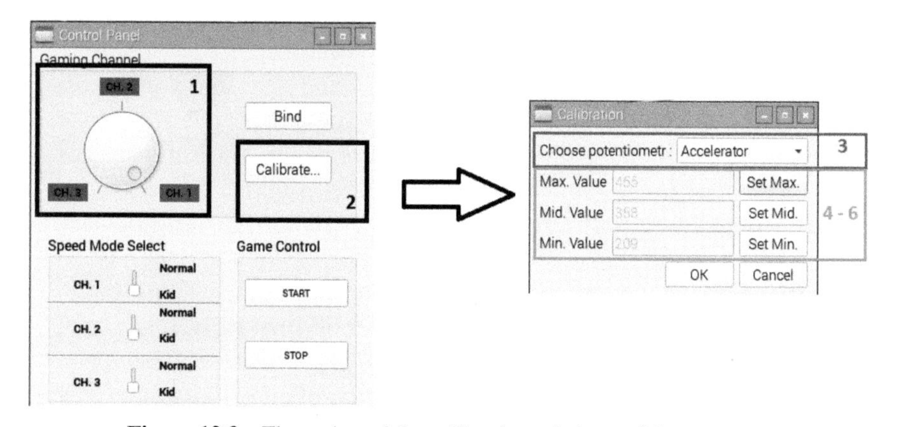

Figure 12.3 The main and the calibration windows of the program.

The second approach was based on the Arduino platform and allowed to create a more compact CS. The ability to run on battery power is also important. It was decided to use the Proteus ECAD system, which provides the ability to perform simulation of microcontrollers by uploading file with the extension "hex" to their models. It allows detecting programming errors or incorrectly connected components and check system behavior in general, before real prototype creation. The CS model and the simplified model of the mobile object were created in Proteus to test their interaction.

Software has been developed using Arduino IDE. It handles keystrokes, changes the position of the channel selector, displays corresponding messages on the LCD, reads analog values from the joystick, normalizes them, and transmits the information to the mobile object by the radio transmitter, using UART protocol. The program also allows choosing speed mode, joystick calibrating, and simultaneously controlling of three mobile objects.

The CS based on Arduino requires the use of breadboard in order to increase the number of ground and power terminals and mock connections to interact with the Arduino platform. It reduces the reliability and compactness of the system. To eliminate these shortcomings, we decided to design our original PCB.

After completing the analysis of the existing microcontrollers, the microcontroller Atmega16 was chosen as a basis and the following design decisions have been made: in order to increase the stability of MCU and to reduce the impact of voltage fluctuations in the power supply, the circuit uses the linear voltage NCP1117 with low voltage drop (similar to those which are used in the Arduino platform). The shift register 74HC165 is used to increase the number of terminals. In addition, all pull-up resistors are located on the motherboard, thus avoiding unnecessary usage of point-to-point construction.

The electronic schematic was tested in Proteus, and after fixing mistakes, the PCB for CS was created in Altium Designer. The next step was to make the PCB and its connection to the operator control panel. The software has been developed after successful simulation in Atmel Studio 7.0. It provides the same functionality as software developed for Arduino. But additionally this version provides the ability of wheels' calibration.

The software for mobile object is universal and can work with all implementations of CS on Raspberry Pi, Arduino, or our original PCB. It has been developed in Atmel Studio 7 in C language and it uses two interrupt handlers. The first handler is necessary for the implementation of signal loss timer that stops the mobile object if the control commands stop coming. The main part of the program is concentrated in the second interrupt, which is triggered when the data arrive at UART. It recognizes the data received from the radio receiver and executes corresponding actions, such as binding the car to the channel, setting the wheel speed and rotation angle, and starting or stopping the race. The developed programs were loaded into the corresponding models, and then system's work was simulated in Proteus.

The real prototypes of CS have been created after mistakes fixing. Functional testing has shown that three implemented CS prototypes perform all the above-mentioned general and internal functions correctly. The results of the comparative analysis of the previous CS implementation and new ones are given below (Table 12.1).

Testing was conducted indoors according two scenarios and it consisted of estimation of systems' working distances – Scenario 1: measuring the distances on which the car completely loses the control signal (in this case the car stops and special signal LED lights up) and Scenario 2: measuring the distance at which the quality of the transmitted signal on the video glasses

Table 12.1 The comparison of CS implementations

	Variants of CS Realization			
Specification	Previous PCB	New PCB	Raspberry Pi 2 Model B	Arduino Mega 2560
SoC/Micro-controller	ATmega 8 PU 8 bit	ATmega 16A 8 bit	Broadcom BCM2836 32 bit	ATmega 2560 8 bit
Max operating frequency	16 MHz	16 MHz	900 MHz	16 MHz
SDRAM	1 KB	1 KB	1 GB	8 KB
Low-level peripherals	SPI, I2C, UART, 3 PWM pins	SPI, I2C, UART, JTAG, 4 PWM pins	SPI, I2C, UART, 2 PWM pins	4 UART, I2C, SPI 15 PWM pins
GPIO pins	23 pins / 6 ADC channels	32 pins / 8 ADC channels	40 digital pins	54 digital pins / 16 ADE channels
Power consumption	15 mA	13 mA	800 mA	38 mA
Persistent memory (FLASH)	8 KB	16 KB	Up to 32 GB (microSD slot)	256 KB of which 8 KB used by bootloader
Other peripherals	No	No	4 USB, HDMI, Ethernet, CSI, DSI	USB Type B

becomes insufficient for controlling the machine (appearance of continuous strong interference on the transmitted image or complete absence of a signal). All the three developed versions of the CS showed almost identical results approximately 50 m, which is higher than the values obtained for CS previous implementation.

For video signal quality improvement estimation, MSU Video Quality Measurement Tool 9.1 was used [19]. As the video sources available for comparison were filmed with the same camera but in different conditions, the suitable list of applicable quality metrics is quite limited. Two most informative metrics were chosen – MSU brightness flicking (Figure 12.4) and MSU noise estimation (Figure 12.5), which allowed us to compare completely different footage. Brightness flicking metric allows us to measure flicking quantity between neighboring frames of the sequence. Noise estimation metric allows us to estimate the noise level for each frame of the video sequence. Resulting plots demonstrate an obvious lowering of noise level and flickering amount for new CS realization based on own designed PCB compared to the version on the previous designed PCB.

The taxonomy, proposed by Carlson and Murphy [20], was used for FPV Auto project reliability testing. The results of failure analysis of the new CS implementation are given in Tables 12.2 and 12.3. Table 12.3 shows that communications are the most common source of physical failures, followed

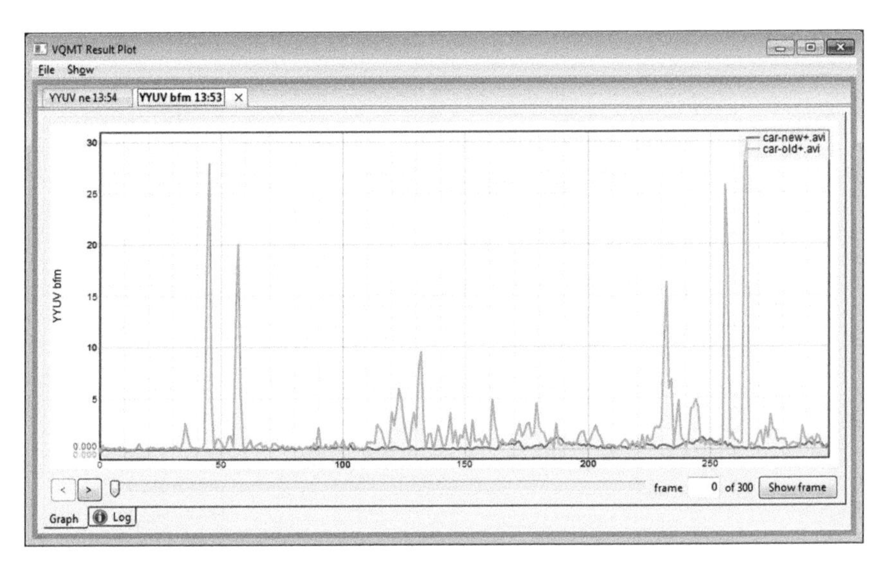

Figure 12.4 MSU brightness flicking.

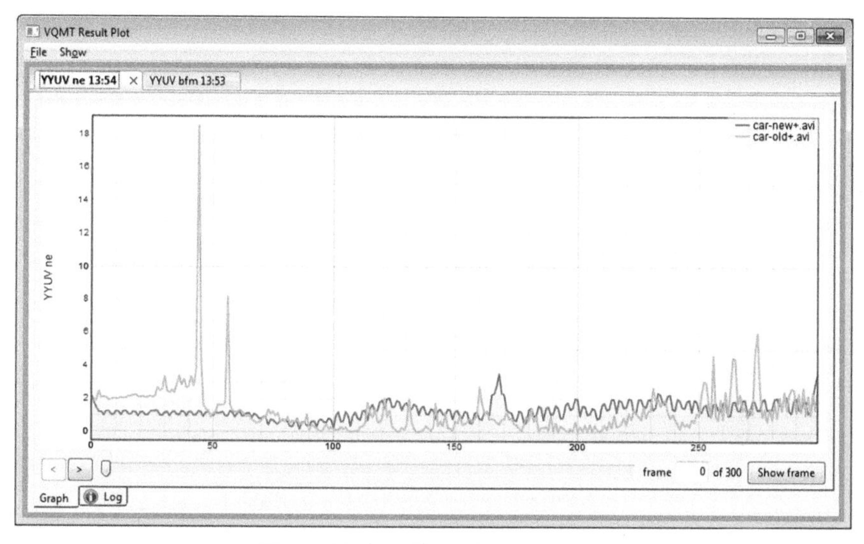

Figure 12.5 MSU noise estimation.

Table 12.2 The summary of the reliability testing

Name	% of Usage	Mean Time between Failures (h)	Availability	Average Downtime (h)
FPV Auto	76%	2.04	73%	4

Table 12.3 The probability of physical class failures

Manufacturer	Effector	Control System	Power	Communications
FPV Auto	0.27	0.13	0.06	0.54

by effectors, control system, and power. So, it should be improved in future implementation.

12.4.2 Subsystem of Autonomous Control

As known, Raspberry Pi 2 B with a dual-core processor and 1 GB of RAM under the control of the Raspbian OS has acceptable computation power and memory capacity. That is why the realization of FPV Auto system based on Raspberry Pi 2 B minicomputer was chosen as a basis for solving the task of providing system autonomous operation.

The review of existing systems and methods for object detection [21–28] showed that they could be roughly divided into two categories: systems based on the so-called traditional models (without the usage of neural networks) and systems that are using neural networks.

The studies [21–23] show that methods based on recognition without the usage of neural networks do not provide the desired accuracy of work and they require a preliminary analysis and selection of features. The creating of recurrent networks is a complex process, and their training requires many resources. Therefore, it was decided to use a convolutional neural network [24].

According to the technical limitations of the FPV Auto project, the convolutional neural network should have as few weight parameters as possible in order to ensure acceptable speed. The analysis of existing solutions [24–28] proved that the most suitable architecture is MobileNet, which is proposed in [24]. For object localization, it was decided to use the single-shot multibox detector (SSD) [28] method because it has a fairly high speed.

In practice, it is hard to find cases when the entire convolutional neural network is trained from scratch, with setting of weight parameters' values randomly, due to the insufficient size of the dataset. Instead of this, first, it is common to pretrain the network on a very large set of data, and then to use the resulting network as an initialization or as a features' extractor from the image for the target task. After reviewing the existing transfer learning methods for training such networks [29, 30], it was decided to use the fine-tuning of the network. As a base for the fine-tuning, it used the MobileNet+SSD network, pretrained on the Microsoft Common Objects in Context dataset [31]. As the initial dataset is similar to the new one, it makes no sense to change the original network architecture. For the network training, the Google Tensorflow framework was used [32].

For the realization of obstacle recognition subsystem, the algorithm for the neural network fine-tuning in TensorFlow was proposed that includes the following steps:

- To collect the set of images with targeted objects;
- To highlight and put class labels on each object in the images;
- To split data into training and test dataset;
- To generate TFrecord;
- To create a file for mapping the class indexes till their string representation (it must have *.pbtxt extension);
- To set up network's configuration file;
- To train the network;
- To export the static inference graph from the new model.

The software was implemented on Python3 language in the PyCharm environment, and the OpenCV 3.3 library was used to work with video and images.

The testing of the created subsystem was held on a dataset with 50 images with different resolutions, which contained people, cars on the road, traffic lights, and other objects (Figures 12.6 and 12.7). To evaluate the impact of lighting characteristics on the ability to recognize objects, images were added with low lighting (at night, during rain) to the test dataset. We also used four videos to test the accuracy.

The testing showed that the recognition accuracy by metric Top-1 is 68.3% and the average recognition time is 683.6 ms, which is suitable for usage in FPV Auto project.

Figure 12.6 The result of object recognition on the image with good lighting.

Figure 12.7 The result of object recognition on the image taken during rain.

The collision detection is a key element in path planning, which greatly influences the performance of the entire algorithm.

Taking into account the fact that a mobile object can usually move with an arbitrary trajectory, as well as system requirements and a small computing power of the used Raspberry Pi 2 B minicomputer, it was decided to use a bounding sphere for the performance of a mobile object [33]. It was also decided to represent obstacles with two kinds of bounding volumes, namely: bounding sphere and bounding box. This solution will allow us to perform the collision detection of a mobile object with obstacles quickly.

The sweep and prune approach was used to increase the collision detection speed, which requires the ability to sort objects by their position in the environment.

Among the motion planning approaches, the potential field method [34] was considered, in which the mobile object is represented as a point and is modeled as a particle that moves under the influence of an artificial potential field. However, the main problem of the algorithm is the situation when the mobile object gets stuck in the local minimums.

The visibility graphs and approximate cell decomposition [33] use a combinatorial approach to path planning in which the connections between collision-free states is fixed into a graph, and then the solutions are found using graph search.

The visibility graphs method calculates the possible paths by checking the visibility of every vertex in the map from each other. If other vertices are visible, this is a possible path. This path is then saved into a graph.

This will be done until every possible path is calculated. The disadvantages of this algorithm include the fact that the algorithm tries to build the path as close to the obstacles as possible.

The approximate decomposition algorithms split collision-free space into a finite set of regions called cells. In the next step, the cells are classified as full, partially full, and free of obstacles. Then, using the algorithm to search the path on the graphs, an attempt to find the path through free cells is made, or in case of impossibility, it is made through partially filled. The main disadvantage of this algorithm is the necessity to store a matrix that describes the entire configuration space.

With the growing size of the configuration space, the usage of such techniques becomes impractical because they require an explicit representation of collision-free space geometry. Therefore, it was decided that these algorithms are not suitable for solving the task because their usage requires a large amount of memory.

It was also taken into account that the algorithm will be used in a dynamic environment where obstacles can change their positions, and for algorithms that use the combinatorial approach, this requires complete reorganization of the path graph.

The basic idea of sampling-based motion planning algorithms [33] is to avoid the explicit construction of the obstacle region and instead to conduct a search that probes the configuration space with a sampling scheme. This probing is provided by a collision detection module, in which the motion planning algorithm considers as a "black box." This enables the development of planning algorithms that are independent of the particular geometric models.

The probabilistic roadmap method (PRM) builds a graph as follows: generates vertices that correspond to the collision-free configurations of the mobile object randomly and connects each of them with neighboring vertices, while trying to create a single connected component that connects start and target vertices. The disadvantage of this algorithm is the necessity of a large number of attempts to create nodes' connections to construct a linked graph.

For implementation, the rapidly exploring random tree (RRT) algorithm was chosen, which is based on the incrementally construction of a search tree by movement in a random direction. The algorithm can support the connected structure with the minimal number of edges and it has a better performance than PRM [35] because it does not require a large number of attempts to create connections between nodes to construct a linked graph.

In order to improve the speed of the path search in trivial cases, a ratio was added to the RRT algorithm, which determines the percentage of cases in which the moving object moves toward the target.

A simple local planner was chosen. It moves directly along the straight line segment between two points in C-space, performing the check of collision detection at uniform intervals on the line segment.

In order to make the subsystem independent from the operating system and the hardware, software implementation was performed in Java using the Swing library for the visualization of the program's work.

The developed program has been tested on various environments. Some of them test the ability of algorithm to find the path in the "narrow passages" (Figure 12.8). Others check the algorithm's ability to find a path in situations where a mobile object needs to change the direction of movement sharply (Figure 12.9).

According to the testing results, the average time of the algorithm is 3.7 ms, and the average number of nodes in the tree is 50 that is acceptable for usage in FPV Auto project.

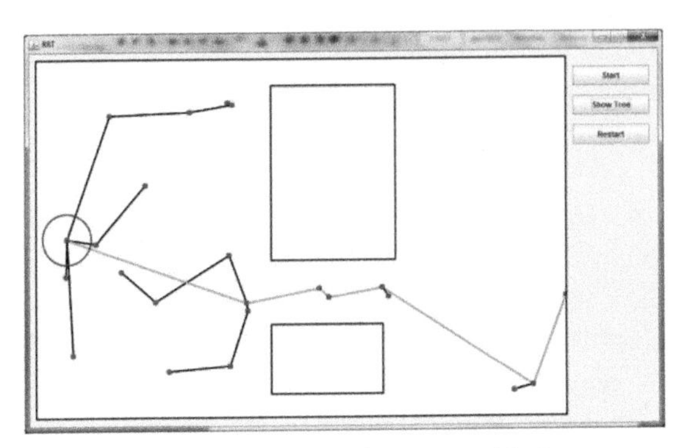

Figure 12.8 The example of the environment with the narrow passage.

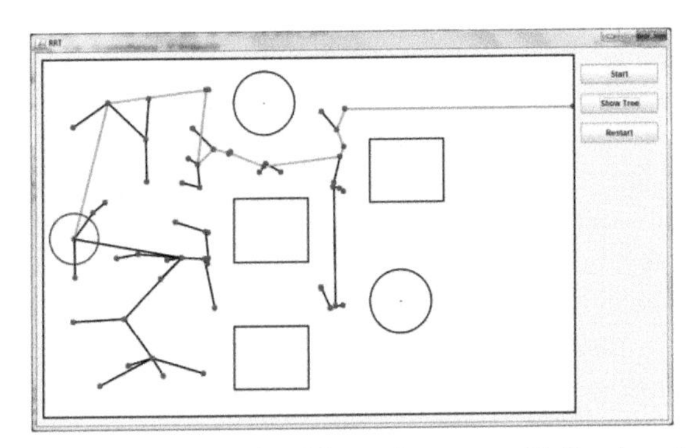

Figure 12.9 The example of the environment with the sharp turns.

12.5 Results and Further Work

Therefore, we can summarize that each variant of implementation of mobile objects control system has its own disadvantages and advantages and each platform has its features.

The Arduino Mega platform has more internal periphery and I/O ports, but does not have such processing power and memory size as the Raspberry Pi. Despite the fact that Arduino is well adapted to work with other hardware devices, it was necessary to use breadboard to increase the amount of power and ground pins for a large amount of connected devices, but it is not convenient. Raspberry Pi usage opens up a lot of possibilities for system's further modernization, but the lack of built-in analog-to-digital converter as well as of logical level converter requires creating of additional board or purchasing of extra shields. The usage of the new designed PCB allows to get rid of unreliable, mock connections and it makes CS realization cheaper (the cost of the new PCB is approximately 8 euros, which is three to four times less than the cost of Raspberry Pi 2 Model B or Arduino Mega 2560 boards). Despite the fact that our new designed PCB is not factory made, it provides good parameters, for example, low power consumption. To ensure system portability, we recommend that the usage of CS based on the new designed PCB as it is reliable has flexible power requirements, and does not require a keyboard, mouse, and monitor, like the system based on Raspberry Pi. However, for realization of hybrid control with the implementation of autonomous and remote modes, the realization on Raspberry Pi is more appropriative.

Future work will be aimed at improving the algorithms of obstacle recognition and constructing an optimal path of mobile objects' movement in the autonomous mode.

12.6 Conclusion

The integrated FPV Auto project based on two approaches (remote and autonomous) for mobile objects' control system was implemented. The architecture of the system was developed, and the implementation of hardware and software parts that meet the stated requirements was proposed. The proposed variants of realization provide the required range of system operation, low power consumption, good video signal, efficiency of the operation, and low cost. The application of reusable solutions based on ready software-hardware platforms, frameworks, and libraries as a rule allows us to accelerate the design process, but it is not a guarantee of implementation of all requirements to the project. Therefore, the development of new original hardware and software components and modernization and improvement of existing algorithms are still relevant. That is why methods and tools for mobile objects control systems design are constantly updated to ensure safety, efficiency, reliability, and usability of solutions.

Such practically oriented projects are the excellent way of testing and comparing different design strategies as well as approbation of research results. They give the necessary experience in the field of development of modern mobile object control systems for IoT.

Acknowledgments

This work was supported by the Erasmus+project ALIOT – "Internet of Things: Emerging Curriculum for Industry and Human Applications" (573818-EPP-1-2016-1-UK-EPPKA2-CBHE-JP).

References

[1] Huawei. Connecting to the smart future. Smart Transportation, 2016. Available at: http://www.huawei.com/minisite/hwmbbf16/insights/smart_transportation_16Nov_PRINT_spread.pdf.

[2] Bangar, P. Y., Pacharne, S. B., Kabade, S. S., and Rajarapollu, P. R. (2016). "Design and implementation of next generation Smart car," in *Proceedings of Int. Conf. on Automatic Control and Dynamic Optimization Techniques.*

[3] White, J., Clarke, S., Groba, C., Dougherty, B., Thompson, C., and Schmidt, D. C. "*R&D challenges and solutions for mobile cyber-physical applications and supporting internet services.*" Springer Journal of Internet Services and Applications, Volume 1, Number 1, 2010, pp. 45–56.

[4] Shupik, I. "*How to choose an automodel.*" 2003. Available at: http://www.rcdesign.ru/articles/auto/cars_intro (In Russian).

[5] Zhou, Y., Xu, G., Qin, F., Xu, K., Wang, G., Ou, Y., et al. (2013). "The prospect of smart cars: intelligent structure and human-machine interaction," in *Proc. Int. Conf. on Robotics and Biomimetics.*

[6] *Concerning Automated Vehicles.* 2018. Availiable at: https://www.nhtsa.gov/staticfiles/rulemaking/pdf/Automated_Vehicles_Policy.pdf

[7] *Vehicle control unit based on radio communication.* 2017. Available at: https://worldwide.espacenet.com/publicationDetails/biblio?adjacent=true&locale=en_EP&FT=D&date=20170329&CC=CN&NR=206049553U&KC=U%23#.

[8] *Radio controlled car APP platform race system.* 2017. Available at: https://worldwide.espacenet.com/publicationDetails/biblio?DB=EPODOC&II=1&ND=3&adjacent=true&locale=en_EP&FT=D&date=20170426&CC=CN&NR=106600433A&KC=A#.

[9] *Automated system of transport network working body radio control.* 2008. Available at: http://library.uipv.org/document?fund=2&id=126978&to_fund=2.

[10] Vaidya, A., Kolte, M. T. (2015). Design and Implementation of Intelligent vehicle system based on ARM cortex. *Int. Journal of Advanced Research in Computer and Communication Engineering*, 4(6), 223–225.

[11] Suwannakom, A., Wiengmoon, B., Tathawee, T. (2016). "The development of motion control for unmanned ground vehicle navigation," in *Proc. Int. MultiConf. of Engineers and Computer Scientists*, II, 630–633.

[12] Zhou, Y., Xu, G., Qin, F., Xu, K., Wang, G., Ou, Y. et al., (2013). "The prospect of smart cars: intelligent structure and human-machine interaction," in *Proc. Int. Conf. on Robotics and Biomimetics.*

[13] Phuoc-Nguyen, Nguyen-Huu, Titus, J. (2009). "*Reliability failure in unmanned ground vehicle (UGV),*" Ground Robotics Research Center. Available at: http://arc.engin.umich.edu/grrc/techreports/200901_ReliabilityUGV.pdf.

[14] Barnard, M. (2016). *Tesla and Google Disagree About LIDAR — Which Is Right?* Available at: https://cleantechnica.com/2016/07/29/tesla-google-disagree-lidar-right/.

[15] Guizzo, E. (2011). "How google's self-driving car," in *IEEE Spectrum.* Available at: https://spectrum.ieee.org/automaton/robotics/artificial-intelligence/how-google-self-driving-car-works

[16] Parkhomenko, A. V., and Gladkova, O. N. (2014). "Virtual tools and collaborative working environment in embedded system design," in *Proc. Int. Conf. on Remote Engineering and Virtual Instrumentation*, 91–93.

[17] Parkhomenko, A., Kravchenko, O., Kravchenko, D., and Gladkova, O. (2016). "Modernization of mobile object control system based on raspberry Pi and arduino platforms," in *Proc. Int. Symp. on Embedded Systems and Trends in Teaching Engineering*, 249–253.

[18] Parkhomenko, A., Gladkova, O., Sokolyanskii, A., Shepelenko, V., and Zalyubovskiy, Y. (2016). "Reusable solutions for embedded systems'design," in *Proc. Int. Conf. on Remote Engineering and Virtual Instrumentation*, 313–317.

[19] MSU Graphics and Media Lab (Video Group). *MSU Video Quality Measurement Tool.* 2018. Available at: http://www.compression.ru/video/quality_measure/video_measurement_tool.html.

[20] Carlson, J., and Murphy, R. (2009). "*Reliability and failure in unmanned ground vehicle (UGV),*" University of Michigan. Availiable at: http://arc.engin.umich.edu/grrc/techreports/200901_ReliabilityUGV.pdf.

[21] Sánchez, J., and Perronnin, F., (2011). "High-dimensional signature compression for large-scale image classification," *Int. Conf. on Computer Vision and Pattern Recognition*, 1665–1672.

[22] Lu, Z., Wang, L., and Wen, J.-R. (2016). Image classification by visual bag-of-words refinement and reduction. *Neurocomputing*, 173, 373–384.

[23] Gklezakos, D. C., and Rao, P. N. (2017). Transformational sparse coding. *arXiv:1712.03257.*

[24] Howard, A. G., Zhu, M., Chen, B., Kalenichenko, D., Wang, W., Weyand, T., et al. (2017). Efficient convolutional neural networks for mobile vision applications. *arXiv preprint arXiv:1704.04861.*

[25] Krizhevsky, A., Sutskever, I., and Hinton, G. E. (2012). Imagenet classification with deep convolutional neural networks. *Advances in Neural Information Processing Systems*, 1097–1105.

[26] Iandola, F. N., Moskewicz, M. W., Ashraf, K., Han, S., Dally, W. J., and Keutzer, K. (2016). Squeezenet: alexnet-level accuracy with 50x fewer parameters and <1mb model size. *CoRR,abs/1602.07360.*

[27] Simonyan, K., and Zisserman, A. (2014). Very deep convolutional networks for large-scale image recognition. *preprint arXiv:1409.1556.*

[28] Liu, W., Anguelov, D., Erhan, D. Szegedy, C., Reed, S., Fu, C.-Y., et al. (2016). "SSD: Single shot multibox detector," in *European Conf. on Computer Vision*, 21–37.

[29] Bengio, Y., (2012). "Deep learning of representations for unsupervised and transfer learning," in *JMLR W&CP: Proc. Unsupervised and Transfer Learning*, 27, 17–37.

[30] Yosinski, J., Clune, J., Bengio, Y., and Lipson, H. (2014). "How transferable are features in deep neural networks?," in *Proc. Advances in Neural Information Systems*, 27, 3320–3328.

[31] Microsoft. *Common Objects in Context Dataset.* 2018. Availiable at: http://cocodataset.org/.

[32] Google. *Tensorflow framework.* 2018. Available at: https://www.tensorflow.org/.

[33] LaValle, S. M. (2006). *Motion Planning.* Cambridge University Press, 2006.

[34] Choset, H. (2005). *Principles of Robot Motion: Theory, Algorithms, and Implementations.* The MIT Press, 626.

13

Software Architecture for Smart Cities and Technical Solutions with Emerging Technologies' Internet of Things

Dinesh Kumar Saini*, Kashif Zia* and Arshad Muhammad

Faculty of Computing and Information Technology, Sohar University, Sohar, Oman
*Corresponding Authors: dinesh@soharuni.edu.om; kzia@soharuni.edu.om

Smart cities are much needed in today's world and it is based on software systems and software architecture. Software architecture is the backbone of smart cities. All the services and systems running in smart cities are basically running on core software architecture and components. Information and communication technologies' tools and techniques are running based on a particular software architecture like client–server or Blackboard. Smart cities help in improving the living standard of the people and make them smart citizens which enable them economic, mobility, and smart services. Internet of Things (IoT) is going to play a very important role in building smart cities. There are certain issues like dependability and reliability of IoT devices in the smart cities which need attention. In this chapter, we addressed the software architecture for smart city which can be dependable and reliable. This is an emerging research issue and effort is made in this chapter to address the issues for building dependable IoT infrastructure for smart cities. The goal of this chapter is to provide an overview of software architecture with dependability issues in designing smart cities.

13.1 Introduction

A smart city is an environmentally conscious city that uses information technology (IT) to utilize energy and other resources efficiently and provide seamless services to the citizens and help city administrators to control the whole city form central locations [1–3].

The characteristics of a smart city are as follows:

- It provides well-established infrastructure and smart services in all areas of life.
- It reduces the usage of environmental capital and support smart growth.
- It promotes the use of information and communication technologies.
- It provides high quality of life.

A smart city uses digital technologies or ICT to enhance the performance and well-being, to reduce costs and resources' consumption, and to engage more effectively and actively with citizens. To enhance the quality and performance of urban services, sectors have been developing smart city technology including government services, transport, and traffic management, energy, healthcare, water, and waste [5–7].

13.1.1 Challenges in a Smart City

The foundation elements in a smart city would include:

- Well-organized resources and management of resources
- Power generation and distribution
- Sanitation and waste management,
- Public transport services like trains, tram, tubes, buses, and taxi system
- Smart homes and houses
- Availability of bandwidth and Internet everywhere
- E-governance and E-business
- Security – cyber and physical
- Health system
- Smart solutions for environment.

This is not, however, an overall list, and cities are free to add more applications, which are shown in Figure 13.1 that describes the software architecture of the city.

13.1.2 Software Architecture for a Smart City

Software architecture is a mix of layers with blackboard in smart cities. The goal of this chapter is to provide an overview of software architecture with dependability issues in designing a smart city. Figure 13.1 shows the software components where dependability is a major concern.

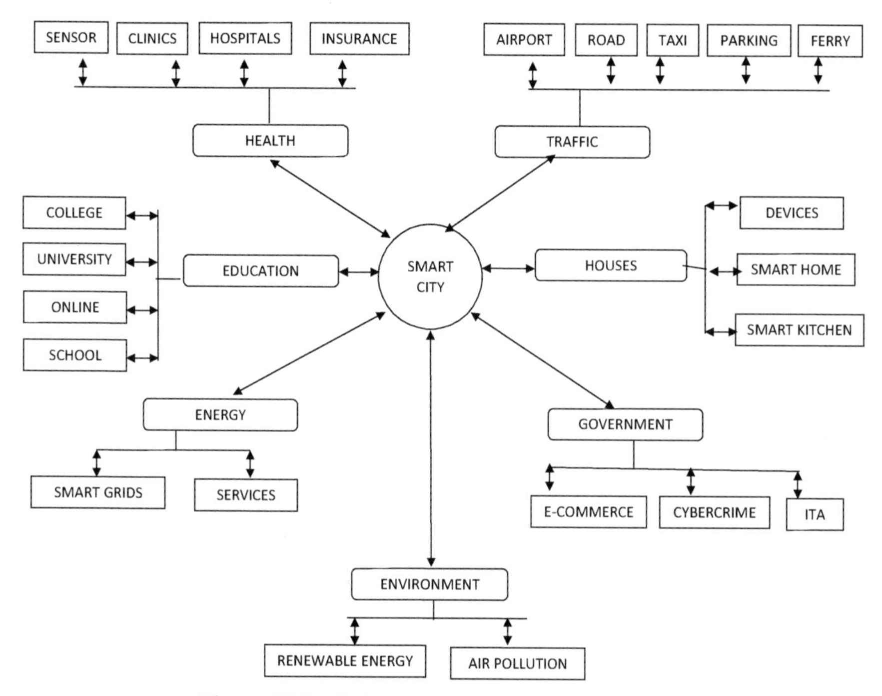

Figure 13.1 Software architecture for a smart city.

13.1.3 Smart City Governance: Example of Oman

All the regions of the country are divided into various regions and then various elements of the city are shown as connected. Software architecture is the backbone of the smart city; all the applications, systems, and services running are based on the software architecture which is shown in the Figure 13.2 [8, 9].

13.1.4 Examples of Services Like Intelligent Transport System or Smart Transportation

One feature of the smart city solution is the intelligent transport system or smart transportation. It provides a transportation network that ensures easy and fast mobility by the following ways:

- The city will also have its own transport network which may include air, trains, metro, trams, ferry, boats, buses, and taxi networks. It may have zoning.

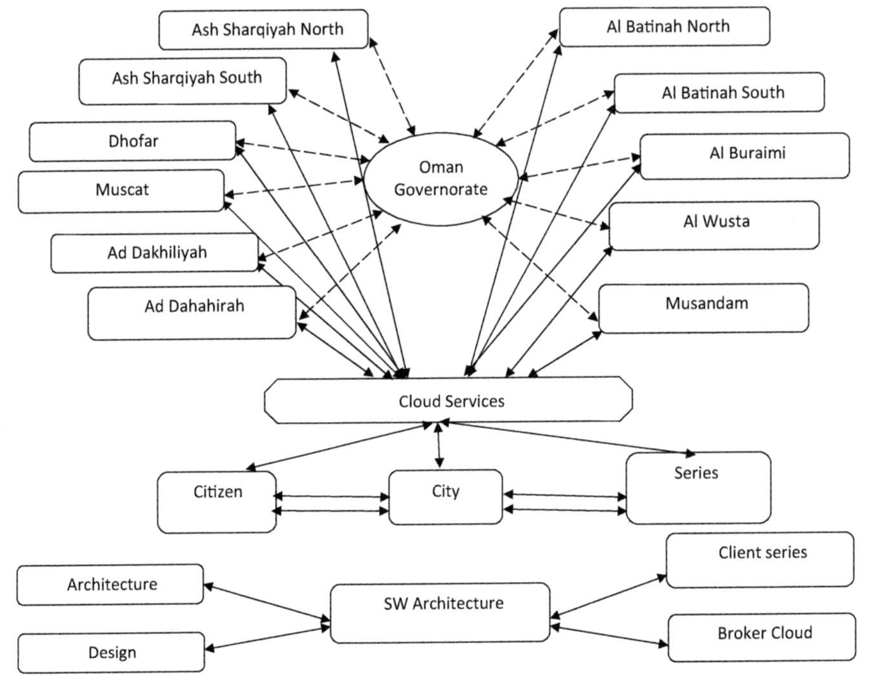

Figure 13.2 Software architecture as the backbone of the smart city.

- It uses a multimodal transport system:
    - Intercity transport
    - Intra-city transport.
- The railway line also passes the city.
- By Elevated walkway, every building will be directly connected to the metro station.

The term smart transportation system refers to information and communication technology, which makes the systems smart and offers better services to the citizens.Applied to transport infrastructure and vehicle, that improve transport outcomes such as safety, productivity, reliability, and environmental performance. Metro trains are made smarter by offering better e-services in smart cities and made automated and more powerful in safety and security systems.

A smart vehicle or modern vehicle is the vehicle which equipped with a driver assistance system can "feel" (by sensors), "see" (by cameras), and "speak" (by communication system) which mainly runs on the clean energy

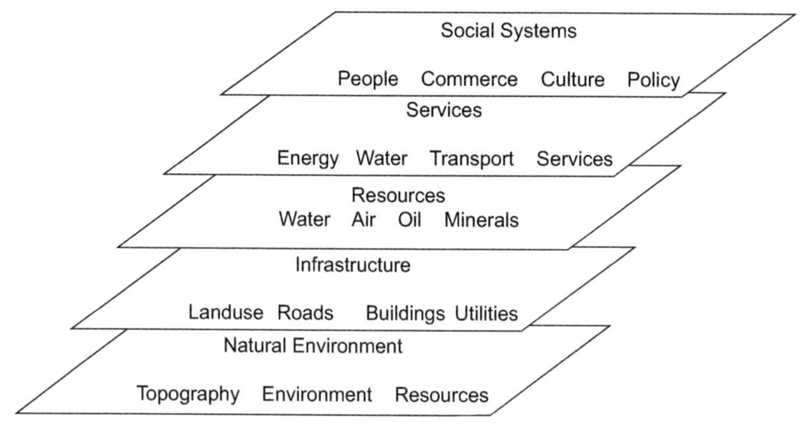

Figure 13.3 Smart urban modeling.

source. The presence of the electric powertrain is intended to achieve either better fuel economy than a conventional vehicle or better performance. A hybrid electric vehicle is a type of electric vehicle that combines a conventional engine (ICE) propulsion system with an electric propulsion system (hybrid vehicle drivetrain).

13.1.5 Smart Urban Modeling

The layered architecture is used to design urban mobility; natural environment is the basic core layer; on top of this, infrastructures like houses, buildings, and utilities are built. The next layer is of energy resources. The next layer will be services which include energy, water, transportation, etc. The topmost layer is about social systems which connect society, people, business, and culture. The whole systems are well intact in a modern smart city which is based on the layered architecture. Figure 13.3 shows the layered architecture of the smart city.

13.2 Security in a Smart City

A smart city is based on smart services which are offered by various systems which are based on software architectures. Most of the systems and applications which run services need security in place. Security is a major issue in today's systems and smart cities need it very much. In this chapter, the security architecture is proposed which is shown in Figure 13.4.

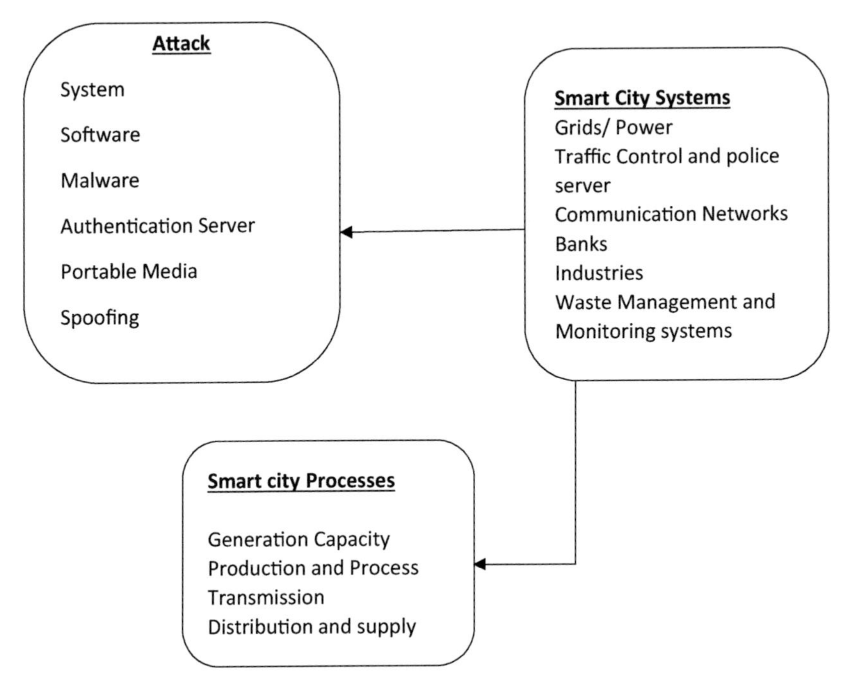

Figure 13.4 Major destination and possible attack point in a smart city.

Major destinations and possible venerable places in the city are marked in Figure 13.5 for security.

Regarding security issues making sure the security, dependability, resilience, and balance of Internet applications and services are crucial and attacks need to be studied in detail. making sure the security, dependability, resilience, and balance of Internet applications and services are crucial and attacks need to be studied in detail. The Internet of Things (IoT) is no different in this value, and security in IoT is linked to the ability of users to trust their environment fundamentally [10]. If people do not believe their connected devices and their information is reasonably secure from misuse or harm, the resulting destruction of trust causes a disinclination for Internet use. This has total effects on electronic commerce, technical innovation, free speech, and every other aspect of online activities practically. Indeed, guaranteeing security in IoT products should be considered a top priority for the sector.

The device also needs a firewall or deep packet inspection capability to control traffic that is destined to terminate at the device. Why is a host-based firewall or IPS required if network-based appliances are in place? Deeply

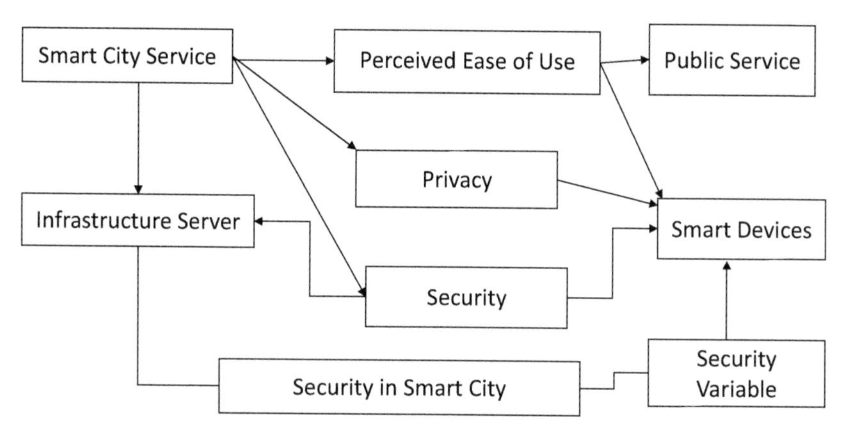

Figure 13.5 Security architecture for a smart city.

embedded devices have unique protocols, distinct from enterprise IT protocols. For instance, the smart energy grid has its own set of protocols governing how devices talk to each other. That is why industry-specific protocol filtering and deep packet inspection capabilities are needed to identify malicious payloads hiding in non-IT protocols. The device need not concern itself with filtering higher level, common Internet traffic the network appliances should take care of that, but it does need to filter the specific data destined to terminate on that device in a way that makes optimal use of the limited computational resources available.

13.2.1 Attack Analysis

There is a need of attack analysis and monitoring in smart cities (see Figure 13.6). Various servers are always attacked and security administrators have to monitor attacks on a continuous basis; risk categorization is needed on a continuous basis. Attack exposure framework must be in place to monitor information sensitivity and trust boundaries are maintained in smart city systems.

13.2.2 Cyber-Physical Systems in Smart Cities

Today's smart cities become a dream for each country, and there are many types of research spoke and magnified this topic and infrastructure that supporting it, from this point our chapter research in one of the essential supports services of smart city which is Smart grid. "The smart grid is a

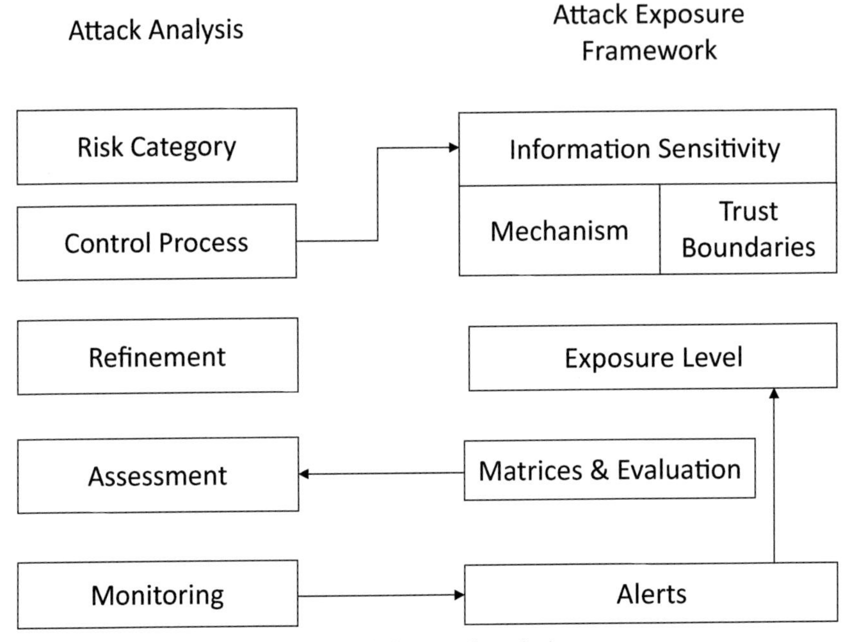

Figure 13.6 Attack analysis.

technological innovation that improves efficiency, reliability, economics, and sustainability of electricity services. It plays a crucial role in smart city energy infrastructure."

The main challenges of smart grids, however, are how to manage different types of front-end intelligent devices such as power assets and smart meters efficiently and how to process a huge amount of data received from these devices "big data."

The benefits of smart grid in smart cities and how it affects the people:

- the smart grid and how it is required for smart cities;
- the increasing consumption of electricity in smart cities, the consumption will be increasing rapidly because all things will depend on electricity even transportation.
- how the smart metering affects smart grid and customer's satisfaction, is the reading by human need, the reading could be in minus, and how we can deal with this kind of situation;
- what is the possible technologies to support two-way communication in smart grid, as we know the electricity flow and data working in one

direction nowadays, is it possible to come in two ways, and there is an existing experience on this;
- the software's and clouding computing presentation and efficiency, how the customer can be contacted with the electricity company, and how the electricity company can manage their assets, electricity feeding, and outage management will be in the smart grid.

Smart grid financing has empowered numerous noteworthy utility managers to execute new power establishment, programming and hardware stages, and operational procedures and models. The smart grids are recognizing vital redesigns in operational profitability, cost, and customer advantage. The smart grid will drive the electric business' change from a concentrated, creator controlled framework to the one that isn't so much united yet rather more client shrewd [11–13].

Depending on the National Institute of Standards and Technology statistics, smart grids will provide:

- Enhance power scalability and reliability.
- Increase utilization and manage backup to handle peak load efficiently.
- Increase electric power networks' capacity and availability.
- Improve strength to disturbance.
- Predictive support and self-mending reactions to framework unsettling influences.
- Depend on new sources of energy like renewable sources.
- Automate processes like maintenance and operations.
- Environmental benefits where find new power sources to support electric vehicles.
- Increase security level on grids.
- Increase customer choices and satisfaction.

Transmission and distribution management systems are utilized to gage the trustworthiness and proficiency of the transmission and dispersion electrical systems (focal office to substation).Expanded mechanization and two-way correspondences are required changes that happen in this piece of the smart grid. With support from NERC and Federal subsidizing, very progressed Syncro phasors create a consistent information stream that best in class examination motors use to arrange, plan, and work the transmission organize. These advances require successful capacity procedures for dealing with their little however ceaseless information stream.

- Demand/response management systems make an interpretation of meter peruses into load necessities, bringing about advanced vitality

circulation among substation and end-client. Supervisory control and data acquisition is developing to bolster more proactive, powerful, remote system administration. These applications empower self-mending from power unsettling influences, yet they likewise present new worries for digital security, as they develop to IP-and web-based structures.
- Outage management systems give crisis controls and actuate extraordinary assets and procedures in case of a blackout.
- Alternative energy management systems are new frameworks that will enable utility system administrators to use circulated era from sustainable sources (the wind, sun based, geothermal, and hydro).
- Geographic information systems coordinate occasions to topographical areas, which is helpful in scene upkeep errands like tree development administration, and also blackout recognizable proof and get out improvement.
- Utility enterprise applications enable all workforce to deal with the developing area of collective client connections. These frameworks incorporate charging, client mind, client utilization entrances, and customary IT stages.
- Data analytics applications exploit new sorts of information from the smart grid and additionally recorded information.

With the improvement of modem data innovation, the information that should be prepared in the current power framework turns out to be increasingly substantial and complex. The information sources, the information arrangements, and the speed of information social occasion are expanding significantly. Keeping in mind the two-way communication channel between the customer and the network should be online and reliable.

13.3 IoT Solutions for a Smart City

IoT is a very promising solution for smart cities. It makes smart city monitoring easier, helps in controlling the cyber infrastructure and managing the city infrastructure, and makes the real-time data processing and actionable information available for controlling [14–16].

Smart cities need proper integration of comprehensive information systems with proper security in place. Various sensors and actuators are installed in smart cities like RFIDs, IR, GPS, etc. Most of the devices will be connected through the Internet and special protocols [17].

Smart cities must be instrumented, interconnected, and intelligent, and IoT enables all these features of smartness in the modern cities [11, 18].

IoT involves embedded systems, cloud, and real-time processing for smart cities; various legacy systems are moving toward new systems, and various communication protocols are emerging and allowing various devices and machines to communicate and interconnect.

Various sectors like health, environment, power, social, and personnel are growing like anything and new devices and machines coming into existence. A new paradigm is emerging for people and it is called Internet of People. Most of the vehicles are connected and monitored through the Internet which is called Internet of Vehicles.

In smart cities, IoT enables wearable devices for head, eye, wrist, and fingertips which will help to monitor people in the city. These wearable sensor devices can also help to monitor patients in the smart cities. Healthcare industry is heavily depending on IoT in smart cities.

Security of data and data protection is a very challenging task in smart cities' IoT devices. If security breach happens in smart cities, it will create a serious problem in monitoring and managing.

13.4 Conclusion

In this chapter, the authors provided a software architecture for smart cities, while proposing a smart city for Oman. Smart cities have a lot of services to offer to the citizens and will improve the lives of people in the cities. There are certain challenges in the smart cities which are discussed in the chapter. Security of data and other services used in a smart city is a major challenge in the smart city. Attack analysis and security solution are proposed in the chapter. IoT is discussed in detail and possible solutions it is offering to smart cities are discussed in detail. IoT is offering a lot of services in smart cities and it will help to control and monitor all services needed in smart cities.

Acknowledgments

The authors would like to thank Sohar University, University Research Center, and Faculty of Computing and Information Technology colleagues for valuable suggestions and useful feedbacks.

References

[1] Bakící, T., Almirall, E., and Wareham, J. (2013). A smart city initiative: the case of Barcelona. *Journal of the Knowledge Economy*, 4(2), 135–148.

[2] Alawadhi, S., Aldama-Nalda, A., Chourabi, H., Gil-García, J., Leung, S., Mellouli, S., et al. (2012). Building understanding of smart city initiatives. *Electronic Government,* 2012, 40–53.

[3] Lee, J. H., Phaal, R., and Lee, S.-H. (2013). *Technological Forecasting and Social Change*, 80(2), 286–306.

[4] Saini, S. L., Saini, D. K., Yousif, J. H., and Khandage, S. V. (2011). "*Cloud computing and enterprise resource planning systems*," in *Proceedings of the World Congress on Engineering*, 1, 6–8.

[5] Cocchia, A. (2014). Smart and digital city: A systematic literature review. *Smart City*, 13–43. Springer International Publishing.

[6] Hollands, R. G. (2008). Will the real smart city please stand up? Intelligent, progressive or entrepreneurial? *City*, 12(3), 303–320.

[7] Nam, T., and Pardo, T. A. (2011). "Conceptualizing smart city with dimensions of technology, people, and institutions," in *Proceedings of the 12th annual international digital government research conference: digital government innovation in challenging times*, 282–291.

[8] Hall, R. E., Bowerman, B., Braverman, J., Taylor, J., Todosow H., and Von Wimmersperg, U. (2000). The vision of a smart city. *Paper Presented at the 2nd International Life Extension Technology Workshop*, Upton, NY.

[9] Washburn, D., Sindhu U, Balaouras, S., Dines, R. A., Hayes, N., and Nelson, L. E. (2009). Helping CIOs understand "smart city" initiatives. *Growth,* 17(2), 1–17.

[10] Saini, D. K. (2012). Security concerns of object-oriented software architectures. *International Journal of Computer Applications*, 40(11), 41–48.

[11] Bou-Harb, E., Fachkha, C., Pourzandi, M., Debbabi, M., and Assi, C. (2013). Communication security for smart grid distribution networks. *IEEE Communications Magazine*, 51(1), 42–49.

[12] Rolfes, C., Poschmann, A., Leander, G., and Paar, C. (2008). "Ultralightweight implementations for smart devices–security for 1000 gate equivalents," in *Proceedings of CARDIS*, 5189, 89–103.

[13] Saini, H., and Saini, D. (2007). Proactive cyber defense and reconfigurable framework for cybersecurity. *Int. Rev. Comput. Softw.*, 2, 89–98.

[14] Kim, T.-h., Ramos, C., and Mohammed, S. (2017). Smart city and IoT. *Future Generation Computer Systems*, 2017, 159–162.

[15] Sanchez, L., Muñoz, L., Galache, J. A., Sotres, P., Santana, J. R., Gutierrez, V., et al. (2014). SmartSantander: IoT experimentation over a smart city testbed. *Computer Networks*, 61, 217–238.

[16] Jin, J., Gubbi, J., Marusic, S., and Palaniswami, M. (2014). An information framework for creating a smart city through the internet of things. *IEEE Internet of Things Journal,* 1(2), 112–121.

[17] Gubbi, J., Buyya, R, Marusic, S, and Palaniswami, M. (2013). Internet of Things (IoT): A vision, architectural elements, and future directions. *Future Generation Computer Systems,* 29(7), 1645–1660.

[18] Robles, R. J., Kim, T.-h., Cook, D., and Das, S. (2010). A review on security in smart home development. *International Journal of Advanced Science and Technology,* 15, 13–22.

[19] Mishra, B. K., and Saini, D. K. (2007). SEIRS epidemic model with a delay for transmission of malicious objects in a computer network. *Applied Mathematics and Computation*, 188(2), 1476–1482.

[20] Miorandi, D., Sicari, S., De Pellegrini, F., and Chlamtac, I. (2012). Internet of things: Vision, applications and research challenges. *Ad Hoc Networks,* 10(7), 1497–1516.

14

Approaches and Techniques to Improve IoT Dependability

Nikolaos G. Bardis[1,*], Nikolaos Doukas[1,*], Vyacheslav Kharchenko[2], Vladimir Sklyar[2] and Svitlana Yaremchuk[3]

[1]Department of Computer Engineering, Hellenic Military Academy, Vari, Greece
[2]Department of Computer Systems, Networks and Cybersecurity, National Aerospace University "KhAI", Kharkiv, Ukraine
[3]Department of General Scientific Disciplines, Danube Institute, National University "Odessa Maritime Academy", Odessa, Ukraine
[*]Corresponding Authors: bardis@ieee.org; nd@ieee.org

The ever-increasing availability, cost effectiveness, and miniaturization of powerful processing equipment have given rise to a class of systems, collectively known as the Internet of Things (IoT). Such systems span a variety of application areas from household appliances to safety critical devices and advanced military hardware, used in highly sensitive contexts. Successful operation presumes qualities such as robustness, reliability, availability, and security, attributes that may be classified under the title of dependability. Dependability covers the ability of critical systems working for sensitive applications to provide uninterrupted operation under severe disturbances, gracefully degrade when limiting conditions are reached, and maintain the ability to resume normal service once the disturbances have been removed. Survivability as an attribute of dependability is a virtue of systems that contributes to their overall reliability. Reliability could be considered as a broader term, covering the numbers of faults appearing, the existence of design flaws, and implementation errors that create favorable conditions for the appearance of operational faults or malicious intervention and could be hence considered as an indication of the lack of proneness of the system to problematic behavior and failure. This chapter highlights technologies presented by the authors that promote dependability and other attributes in IoT systems. First, a method is presented that enhances the cyber security capabilities of equipment by enabling the execution of cryptographic calculations in cloud resources. Next, a scheme for the effective design of IoT information processing is presented that is based on assurance case considerations for safety and other dependability attributes including security. Finally, a metric for software requirements' correctness to assess and assure software reliability is analyzed.

14.1 Introduction

14.1.1 Motivation

The ongoing information revolution is based on advancements in software, microelectronics, storage, quantum, cloud, and soft computing that drive the field of artificial intelligence and will soon be powered by nanotechnology. The technological evolution gave rise to the concept of Internet of Things (IoT) where increasing and impressive numbers of diverse devices are networked and given capabilities of collaborating with each other. This in turn rendered existing design approaches and processing concepts that were targeted to the development of single apparatus or closed systems, incapable of satisfying the current requirements of designing systems consisting of independent devices possessing different information processing capabilities with parts of the system operating at remote locations, being mobile and being exposed to a variety of possible interventions with malicious intent. Crucial factors for the success of an IoT system are its level of reliability, security, and survivability as attributes of dependability. It is hence becoming apparent that such characteristics of the final systems need to be pursued very early in their development, starting with the algorithmic design of aspects such as cyber security. Additionally, they may be enhanced at the systemic design phase by employing suitable design techniques that predict problematic combinations of operational conditions. Finally, the employment of quantitative methods for software (SW) and general design requirements correctness allows to assess and assure SW and system reliability.

Survivability as an attribute of dependability is a non-functional yet critical lifecycle property of engineering systems which must be robust to disturbances and thus survivable [1]. The survivability properties cover the ability of both critical and non-critical systems working for applications that are anyway important to their users to provide uninterrupted operation under severe disturbances, gracefully degrade when limiting conditions are reached, and maintain the ability to resume normal service once the disturbances have been removed. IoT systems are a case where survivability acquires additional importance, given the inherently distributed nature of systems that gives rise to increased probability of failure or unavailability of random subcomponents. Survivability is a virtue of systems that contributes to their overall reliability. Reliability may be described as an essentially broader property, illustrating the numbers of faults appearing and the existence of design flaws and implementation errors that create favorable conditions for the appearance of operational faults or malicious intervention, even if such problems never materialize. Reliability may hence be considered as a property indicating the absence of proneness of the system to problematic behavior and failure.

The issue of survivability has already been extensively studied in the past in the context critical networks systems [2–4]. When talking about survivability of mission critical IoT systems, reference is always made to cases where the uninterrupted operation of the system under consideration is mandatory, regardless of the type of fault or attack. The need for dependability in systems was perceived in the early stages of the evolution of information technologies and the identification of threats to dependability has been within the focus of researchers; a taxonomy of such threats was compiled by A. Avižienis et al [5]. The problem of defining dependability and survivability has been extensively studied [6–12] and efforts have been devoted to deriving quantitative survivability parameters leading to specific metrics [4, 6, 13–18]. Such metrics need to incorporate a variety of parameters. The first such parameter necessary to be incorporated is time. The time element will have to demonstrate the capability of the system to complete its mission within the set time constraints, irrespective of attacks or failures. The concept of degraded operation may hence be introduced, whereby the completion of the mission occurs at a time later than normal, but still within the given deadlines. A quantitative such measure is hence proposed in [1] as the difference between the time it takes any critical mission parameter to attain a given goal in the absence of disturbance, compared to the same time in the presence of disturbance. Consequently, survivability is about maintaining essential services at a quantitatively measurable level of performance in the presence of disturbance, such that the mission is accomplished in a timely fashion.

14.1.2 Objectives and Structure

The objective is to propose methods for increasing of IoT systems' dependability by the implementation of additional assurance and assessment measures on requirements and design stages. In this chapter, recently presented technologies are reviewed, which are able to promote the goal of achieving increased dependability. As it will be illustrated in the subsequent sections, dependability, along with the associated attributes of reliability and robustness, may be pursued by the employment of suitable techniques in three different stages of their development. More specifically:

- in the design of algorithmic implementations of security algorithms using distributed and cloud resources, so as to achieve the necessary security levels across the entire system, irrespective of the particular hardware of each subsystem and its possible limitations [19];
- at the initial implementation stage by means of the assurance case driven design (ACDD) [20];
- in the acceptance stage, where software requirements' correctness verification will be presented as a means for promoting the required attributes in designs [21].

Section 14.2 hence presents a secure implementation of modular arithmetic operations for IoT and cloud computing applications [22]. Modular arithmetic operations formulate the core of the calculations necessary for most fundamental cryptographic algorithms. Modular arithmetic is in general highly computationally intensive. The algorithm presented enables information processing devices that are for any reason unable to perform such operations locally, to delegate them to cloud or base computing resources, without compromising security.

Section 14.3 presents the use of ACDD for computer systems [20, 21]. The case methodology enables the recognition of security and safety gaps in computer systems in early stages of their implementation, thereby driving the final system implementation to circumvent such problems. The motivation for using ACDD in the context of enhancing survivability and reliability is twofold: (i) it considers security and safety features of critical control and communication systems and networks as an integral property and (ii) it is highly suited for the case of embedded components, such as field-programmable gates arrays (FPGAs) and microprocessor units (MCUs), which are applicable for IoT solutions.

Section 14.4 presents a metric-based method for software requirements' correctness improvement [23]. As the complexity of user requirements rises, the complexity of software systems increases. Hence, the number of implementation defects increases, software system reliability and dependability as a whole decrease, and the design costs increase. The technique measures the dependence of requirements on multiplicities of other requirements, on internal subsystems, and sensitivity to business rules thus extracting a complexity metric. The metric is highly relevant for the case of IoT dependability, where such interdependencies may be varying in a highly dynamic fashion.

The analyzed technological advances are particularly relevant to the design of dependable IoT systems, principally in the context of critical missions. Conclusions as to the influence of the presented technologies for IoT system survivability are hence highlighted.

14.2 Secure Implementation of Modular Arithmetic Operations for IoT and Cloud Applications

14.2.1 Modular Arithmetic Operation for IoT and Cloud Security

Cloud technologies provide a solution to the perpetual problem of limited computing resources being available to the user: the computing power of thousands of processors and virtually unlimited amount of memory can be provided to anyone that requires it. On the application side, this evolution is mirrored in the concept of IoT; diverse devices communicate with each other in order to provide an integrated user service experience. However, the emergence and increasing use of cloud and IoT technologies significantly affects the security of computer information processing in all areas of human activity. The possibilities of using the powerful computing resources of modern supercomputers provided by cloud technologies allow potential attackers to multiply the effectiveness of overcoming existing information security systems. Most of the existing methods for achieving a security breach, irrespective of the elaborateness of the security system, use brute force attack. This type of attack can be inherently parallelized and thus effectively implemented on multiprocessor computer systems, access to which is provided by cloud technologies.

The computational implementation of a large class of these security algorithms is based on the operations of modular exponentiation or involution and modular multiplication [24]. Contemporary technologies of this sort include public key algorithms for encryption (RSA and ECC), the Diffie–Hellman key exchange algorithm, digital signature algorithms, and the digital signature Standard. In order to provide an adequate security level against possible attacks, the above algorithms use numbers with word lengths that may in certain cases exceed 1000 bits [25]. More specifically, current security level requirements dictate that that the word lengths to be used for the algorithms based on elliptical curves (ECC) should range from 128 to 256 bits, while the word lengths used for algorithms that are based on exponential transformation should range from 1024 to 2048 bits [26].

Software implementations of modular multiplication for numbers of such long word lengths on general-purpose processors or micro-controllers with much smaller word length architectures (from legacy 8-bit microcontrollers to state-of-the-art 64-bit processors) inevitably involve a large volume of overhead processing and are hence considered extremely computationally complex. Consequently, the software implementations of public key-based information security algorithms on general purpose processors become several orders of magnitude slower in comparison with symmetric algorithms (such as DES or AES) if they are required to provide protection against security attacks of similar intensity. Naturally, the speed of computation of modular arithmetic operations on large numbers becomes an especially acute problem in the case of low word length microcontrollers [27].

A second fundamental operation of the modular arithmetic used in algorithms of this class is the modular involution, i.e., the calculation A^E mod M. In the majority of implementations, modular involution computations are carried out by the method of "squaring and multiplications" [27], which uses a number of multiplications close to the theoretical minimum. Based on this fact, in order to increase the performance of the algorithms implemented in software, it is important to decrease the time needed for performing modular multiplication. Consequently, research for algorithms to increase the performance of modular multiplications, for use in software implementations of encryption schemes on general purpose processors and micro-controllers, is vitally important.

The proposed algorithms are shown to be capable of answering two important concerns, namely, security and cost. Since security becomes cheaper and faster to implement, efficient security algorithms may be incorporated in IoT applications and applications that exploit cloud computing resources for additional processing power. Apart from its other characteristics that satisfy the requirements set so far, the new algorithm is shown to be robust and capable of being used in very demanding applications such as those concerned with defense and public safety systems.

In this section, a method is also presented for the distribution of the operations necessary for the direct calculation of the modular exponentiation in distributed processing resources, with applications to high-performance computing resources available as cloud-based services [19]. This innovative method produces extremely significant acceleration of the overall calculation process, which may be estimated to a factor of 128 times in typical situations. Hence, when cloud resources are available, devices with limited computing resources are capable of realistically implementing strong security algorithms.

14.2.2 Shortfalls of Methods for Secure Remote Implementation of Modular Exponentiation

Researchers have not paid adequate attention to the fact that among the principle criteria for estimating the efficiency of the use of remote processing resources, the attainment of high computational rates plays an important part. For this purpose, it is necessary to propose algorithms for the secure remote calculation of the exponent that will enable the exploitation of the main advantage of cloud computing systems, which is the large numbers of processors that may be operating in parallel [19, 22, 27].

For the secure distributed implementation of the proposed method for the modular exponentiation, the random split of the code of the exponent E in fragments is necessary. Following that, the code A is transformed and the data are sent to the distributed processing nodes, for calculating the exponentiation. The fragments of the original code of the exponent are used as codes for this calculation. The method proposes a special procedure for the combination of the received results so as to obtain the correct overall result. The advantage of this method is that it enables the implementation based on the principle strength of the concept of cloud systems, the ability for parallel processing. The principle disadvantage of the method is the relatively small speed improvement – approximately by a factor of three.

In the remainder of this section, subsections A and B present the analysis of the methodology presented in [19] that provide the foundation for further development of the algorithms and with the final purpose of reducing the volume of the calculation needed to be executed on user equipment, by several orders of magnitude.

14.2.3 Secure Parallel Modular Exponentiation

For attaining the target, an m-bit exponent has been proposed [19], $E = \{e_0, e_1, \ldots, e_{m-2}, e_{m-1}\}, \forall i \in \{0, 1, \ldots, m-1\} : e_i \in \{0, 1\}$, such that:

$$E = \sum_{i=0}^{m-1} e_i \cdot 2^i \tag{14.1}$$

This is randomly divided into h groups $\delta_0, \delta_1, \ldots, \delta_{h-1}$ of rearranged bits that contain $n_0, n_1, \ldots, n_{h-1}$ bits, respectively. For this $n_0 + n_1 + \cdots + n_{h-1} = m$ and the group δ_0 contains the first n_0 bits of the exponent: $\delta_0 = \{e_0, e_1, \ldots, e_{n_0-1}\}$, the second group δ_1 contains the next n_1 bits: $\delta_1 = \{e_{n_0}, e_{n_0+1}, \ldots, e_{n_1-1}\}$ and similarly for the remaining groups.

Using this notation, the bits of the group δ_0 correspond to the number g_0, the bits of group δ_1 correspond to the number g_1, and similarly until the bits of group δ_{h-1} that correspond to the number g_{h-1}:

$$\forall l \in \{0, 1, \ldots, h-1\} : g_l = \sum_{j=0}^{n_l-1} e_{n_0+n_1+\cdots+n_{l-1}+j} \cdot 2^j \tag{14.2}$$

Hence, the code of the exponent *E* may be represented as the sum:

$$E = g_0 + g_1 \cdot 2^{n_0} + g_2 \cdot 2^{n_0+n_1} + \cdots + g_{h-1} \cdot 2^{m-n_{h-1}} = \sum_{l=0}^{h-1} g_l \cdot 2^{\sum_{j=0}^{l-1} n_j} \tag{14.3}$$

By introducing the representation

$$w_0 = 1, w_1 = 2^{n_0}, w_2 = 2^{n_0+n_1}, \ldots, w_{h-1} = 2^{n_0+n_1+n_2+\cdots+n_{h-2}},$$

then the exponent *E* may be rearranged in the form:

$$E = \sum_{l=0}^{h-1} g_l \cdot w_l \tag{14.4}$$

In this case, A^E mod *M* may be represented in the form:

$$\begin{aligned} A^E \text{mod} M &= ((A^{g_0} \text{mod} M)^{w_0} \cdot (A^{g_1} \text{mod} M)^{w_1} \cdots \\ &\quad (A^{g_{h-1}} \text{mod} M)^{w_{h-1}}) \text{mod} M \\ &= (\prod_{l=0}^{h-1} (A^{g_l} \text{mod} M)^{w_l} \text{mod} M) \text{mod} M. \end{aligned} \tag{14.5}$$

14.2.4 Secure Modular Exponentiation in Cloud Infrastructure

In order to solve the problem of securely calculating exponentiation in remote resources, a method is proposed [19] and [22] that is based on replacing the exponent *A* by a multiplication of the form $A \cdot G$ mod M. The number

$G < M$ is picked at random by the user and is stored as a secret number. For simple modulo *M,* according to Fermat's little theorem, this is true for any *G*: $G^{M-1}\text{mod}M = 1$. For most cryptographic algorithms and more specifically for the RSA, the value of *M* is the product of two prime numbers, *p* and $q : M = \mathrm{p} \cdot \mathrm{q}$. In this case, according to Euler's generalization of Fermat's little theorem, it is true that: $G^{(\mathrm{p}-1)\cdot(\mathrm{q}-1)}$ mod $M = 1$. In any case, it is possible to simply define a value $\xi < M$, such that $G^{\xi}\text{mod M} = 1$. The user may then calculate the constant $C = G^{\xi-E}\text{mod } M$ for $\xi > E$. By defining *S* as the product $S = A \cdot G \text{ mod } M$, then:

$$E = g_0 + g_1 \cdot 2^{n_0} + g_2 \cdot 2^{n_0+n_1} + \cdots + g_{h-1} \cdot 2^{m-n_{h-1}} = \sum_{l=0}^{h-1} g_l \cdot 2^{\sum_{j=0}^{l-1} n_j} \tag{14.6}$$

The above calculation is the foundation for the secure calculation of A^E mod *M* in cloud infrastructure. The procedure consists of the sequence of steps described below.

Before the procedure:

1. The user selects a secret number *G* at random, and based on its characteristic modulo *M*, a value ξ is defined such that $G^{\xi}\text{mod } M = 1$.
2. For the selected *G,* the user calculates the constant $C = G^{\xi-E}\text{mod } M$ that is also kept secret and stored.

For the calculation of the exponent $A^E\text{mod } M$*:*

1. The modular multiplication $S = G \cdot A\text{mod } M$ is calculated.
2. A random $u < E$ is selected, the magnitude of which is significantly smaller than the magnitude of the code of the exponent *E*. The quantity $d = E - u$, the magnitude of which coincides with the magnitude of *E*.
3. The values of *S* and *d* are transmitted to the cloud resources where the exponentiation $B = S^d$ mod *M* is calculated. The value of *B* is returned to the user.
4. The user calculates the value $Y = S^{\mathrm{u}} \text{ mod } M$.
5. The user calculates the value of the result in the form $R = C \cdot B \cdot Y\text{mod M}$.

This form is calculated according to the following expression:

$$\begin{aligned} R &= C \cdot B \cdot Y \text{ mod } M = (G^{\xi-E}\text{mod } M \cdot G^d \cdot A^d \cdot G^{E-d} \cdot A^{E-d})\text{mod } M \\ &= G^{\xi-E+d+E-d} \cdot A^{d+E-d}\text{mod } M = G^{\xi}\text{mod } M \cdot A^E\text{mod } M \\ &= 1 \cdot A^E\text{mod } M = A^E\text{mod } M. \end{aligned} \tag{14.7}$$

It is shown that the overall total number of cases that are necessary to be tested for a brute force attack is $2^{m-1+\eta}$. In practical terms, for $m = 4096$, this number becomes $2^{4095+32} = 2^{4127}$. In comparison, cracking the RSA via exhaustive testing requires the analysis of 2^{4096} cases, while the proposed method requires 2^{4127} cases. Consequently, the proposed method for modular exponentiation in distributed computing systems provides an adequately high level of security from attacks that attempt to retrieve or restore the secret codes *A* and *E*.

14.3 Security and Safety Case Driven Design for IoT Systems

14.3.1 Concept of Assurance Case Driven Design

A goal of security and safety analysis for dependable IoT systems is not only proving a conformance with requirements but mostly discovering gaps in such a conformance assessment approach. Security and safety case methodology contains a potential for improvement assurance analysis techniques and tools. We name a set of assurance case-based techniques and tools as ACDD. A practical use of ACDD lays in improvement of certification and licensing processes for critical systems including IoT-based ones [20, 21].

From this prospective, assurance case may be implemented for the earliest stages of life cycle activities to drive safety implementation from the scratch.

The main motivation of ACDD is the following:

- To develop a technique to assess safety and security features as soon as possible during the development of a system concept (specification and design);
- To develop a technique to develop a system concept (specification and design) in a safe and secure manner;
- AC DC also supports the following important topics:
- Research of integral security and safety features of modern critical control and communication systems and networks as an integral property; security importance increasing requests' implementation of security requirements as a part of licensing issues; such an approach is named security informed safety case [20, 21]; such an approach is targeted to analyze safety and security in a structured way and creating security informed safety case that provides justification of safety taking into particular consideration the impact of security;

- Research of different type of embedded components, such as FPGAs and MCUs, which are applicable for IoT solutions;
- Research applications for specific market, for example, cloud computing and big data analytics with high-level requirements to safety, security, and quality of service.

At the present assurance case methodology progress lays in multidisciplinary dissemination of theory and experience [20, 21]. Experts from different areas may develop a general and cross-platform security and safety assurance approaches. At the same time, there are some potential areas for assurance case improvement, such as:

- Assurance case should be faster to find gaps in compliance with requirements than to demonstrate such compliance.
- It is reasonable to implement assurance case from the earliest stage of life cycle; one more reason to do it is a prospective idea to combine of assurance case with an argument-based design approach, what is a basis for elimination a board between design and modeling.
- Assurance case should provide as many details as it is needed for comprehensive analysis.
- Assurance case should support re-using of system safety and security files during system operation and maintenance.
- Assurance case should support cost effectiveness of system life cycle.
- It is reasonable to improve formalism of assurance case against empirics in descriptions.

Assurance case has two sides of description and implementation:

1. A static part which describes an approach to combine arguments for assurance support;
2. A dynamic part to support a static part movement between stages of the analyzed system life cycle.

There are the following notations for the assurance case static part:

- Basic Toulmin notation, developed by the author as an extension for a classic implication operation;
- Claim-argument-evidence (CAE) notation and goal structuring notation (GSN) based on Toulmin notation.

CAE and GSN formalisms are based on classical set theory, graph theory, and relation algebra. Such relations' tracing allows us to propose extensions for existing notations. In this article, we discuss an approach to develop claim-argument-evidence-criteria notation as an extension of CAE notation.

The second side of assurance case implementation is dynamic application via life cycle stages. IDEF0 notation is considered as a fundamental for a formal description. Application of set theory and graph theory has been considered as a basis for IDEF0 notation. It allows proposing development-verification and validation-assurance case notation for description of dynamic assurance case application.

14.3.2 Approach to Implement ACDD

The new methodology implementation requires not only technical measures but also organizational efforts to improve involved parts' collaboration.

A chart in Figure 14.1 demonstrates such collaboration of the following three parts during ACDD implementation:

1. Design team responsible for a product development;
2. Quality assurance and/or safety and security management team responsible for following all quality, safety, and security procedures during development, verification and validation (V&V), configuration management, audits, and other relevant activities;
3. Assessment and certification team as a third part responsible for independent safety or security assessment of a product usually with issuing of a formal conformance document.

After establishment of organization and collaboration aspects, let's analyze a general ACDD framework (see Figure 14.2).

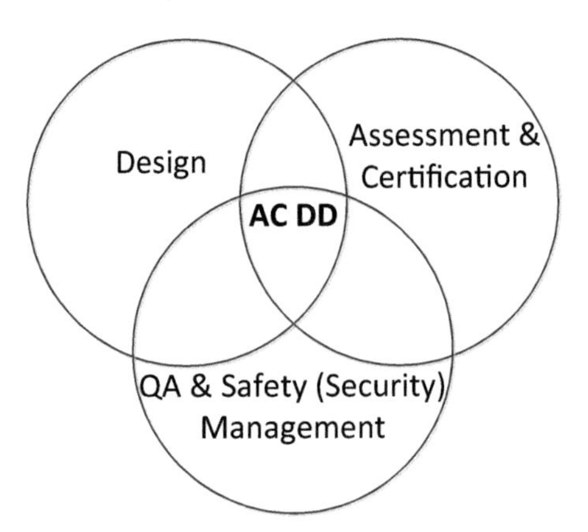

Figure 14.1 ACDD collaboration chart.

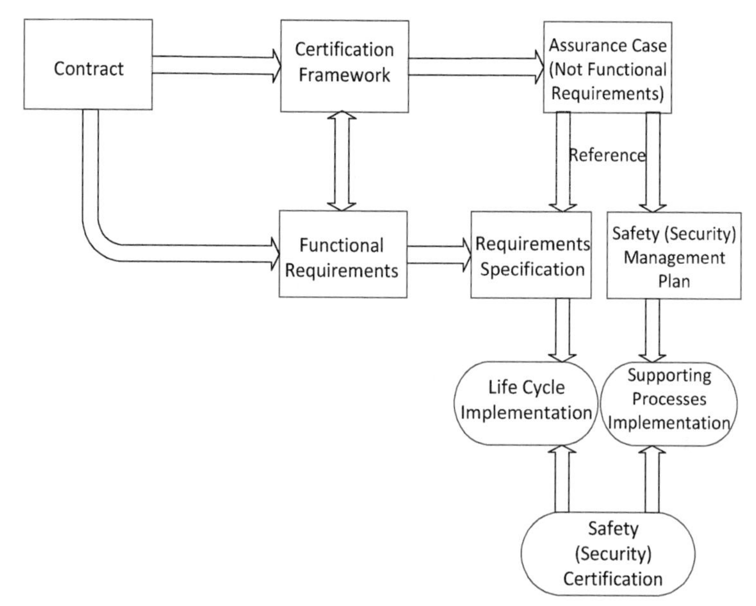

Figure 14.2 General framework for ACDD.

Usually, the first step in any system development is signing a contract. This contract is an input for system functional requirement as well as certification or licensing framework for safety and security critical applications. The requirement specification has to be developed on the base of contractual functional requirements.

Safety and security critical systems shall have an important addition to the requirement specification describing not functional requirements targeted to implement system integrity. The ACDD approach proposes to present such requirements in a view of a preliminary assurance case. Such preliminary assurance case is not a result of assessment but a target which has to be achieved after the system implementation. Not functional requirements of assurance case are an input for safety or security management plan which has cover life cycle description with all development support processes. Some parts of not functional requirements (for example, self-diagnostic requirements) may affect the requirement specification. After that staged life cycle with V&V and other supporting processes, activities (project management, configuration management, and others) have to be implemented in accordance with safety (security) management plan. After the contract and the requirement specification stages, life cycle usually includes design, implementation, integration, validation, installation, and commissioning stages. Assurance case activities have to be implemented after each of the stages. Safety or security certification has to finalize the system life cycle before transfer it in operation at the customer site. Also, during the operation, a periodical assessment or certification has to be done with associated update of assurance case.

Assurance case structure depends from a type of the application. For example, a typical structure of assurance case for industrial functional safety-related application includes: security activities coordinated with safety, process implementation and assessment, and product implementation and assessment (Figure 14.2). Assessment can be done in a view of deterministic analysis, probabilistic analysis, or demonstration.

Apart from guiding the system design process, the assurance case driven approach provides the foundation and framework for the application of purpose defined correctness metrics for the assessment of implementations. The use of such metrics provides increased assurances that dependability achievements due to algorithmic advances or design techniques are not jeopardized by implementation defects. A technique for software requirements correctness is presented in the following section.

14.4 Software Requirements Correctness Improvement for IoT Reliability

14.4.1 Challenges for Software Systems' Reliability

The use of software tools and software systems (SWS) for different IoT applications is ever increasing and incorporates ever more complex and diverse requirements, having considerable influence on reliability and safety. In its turn, the complexity of SWS increases. Consequently, the number of faults imposed by developers increases, the SWS reliability decreases, and the design costs increase. It's one of the key challenges in modern IoT engineering encouraging increased software reliability [21].

Research has yielded a number of models, methods, and software application to evaluate and manage the SWS reliability. However, these methods don't take into account variety and specific features of SWS development business processes and require great expenditures for adaptation and implementation. The above factors prevent to implement widely available model methods and reliability aids in software engineering routine practice, making such an implementation impossible in a number of cases. Therefore, the first principle of the (SRM) provides adaptability, simplicity, and low costs contributing to easy implementation of the assessment technique into business processes of software corporations.

Each SWS development starts from collection and analysis of requirements. This process includes clarifying and documenting what should be done by an application and what features it should possess. This process involves all the concerned parties – investors, purchasers, users, from the one side, and managers, analysts, designers, codifiers, testers, i.e., SWS developers, at the other side. In the course of extended joint work, errors may occur with any party category at both sides. As a result, incorrect requirements contain defects of various types, such as erroneous data, unclearness, ambiguities, incompleteness, doubling, and contradictions. Imperfect requirements may cost very much. The requirements defects are 20–50 times more expensive to correct once the defects got into the development process. Inspection of one full-scale software project showed that problems with requirements led to US$ 600.000.000 development budget exceeding, 8-year time delay and reduced functionality. Another research [23] showed that it takes about 30 min to correct error detected at the stage of requirement processing, whereas it takes 5–17 h to correct an error detected in the course of system testing. As identified in [28], eliminating requirements' defects is 100 times expensive than those detected in the course of requirements development, if indicated by the customers. Evidently, methods directed to identify as much defects, as possible, in requirements save a plenty of time and costs. Therefore, the method of improvement of SWS requirements' correction is suggested. In the following, standards and terms applicable at the stage of requirements construction are described.

14.4.2 Methods of SWS Requirements' Correctness Improvement

Modern SWS include a plenty of multitype requirements. They include functional requirements, requirements to user and software interfaces, hardware interfaces, communications interfaces, requirements to capacity, data safety storage, system safety, etc. Therefore, the SWS specification is a bulky document of hundreds of pages. Theoretical approach supposes total review of all these requirements. In practice, "even in medium-size specification, experts review thoroughly only the first part, some of the most stubborn may look through half, but it is unlikely that anyone reaches the end" [29]. Entire SWS requirement review is impossible due to lack of time, financial, and human resources. In this view, the offered method or requirements' correctness improvement consists in extracting the most priority and complicated requirements from their entire set to be thoroughly reviewed to detect the highest possible quantity of faults in prevailing restricted resources situation. Priorities are assigned to requirements after their collection and documenting jointly by the manager, analyst, and customers' representatives. Requirements' complexity is determined by means of metrics.

Publications [29–31] and standards [32, 33] have been studied for the analysis of existing metrics. These standards contain numerical evaluations for about 50 metrics. They include metrics for completeness; precise detailed requirements; non-elementary requirements; requirements not coordinated with other requirements; undetectable and untestable requirements; missed or skipped requirements; non-obvious requirements; defective requirements detected for 1 h of checking; detailed requirement value; and metrics for deletion, addition, or modification of the requirement. Their analysis shows that such metrics are designated for quality control, requirements' review, and analysis efficiency. However, there is no metric evaluating requirement complexity in these standards.

The authors of [34] apply requirement metric as a number of faults per documentation page and provide its interpretation and permissible value as 0.2 defects per page of the requirements' testing description, i.e., one defect per five pages. It is an unclear definition since the number of requirements per page is unknown.

The authors of [19] stress upon correctness and traceability as the most important feature of requirements without outlining evaluation formulae. The authors define correctness as a precise description of desirable functionality. Traceability is defined as a possibility of analysis both in backward direction to initial sources of requirements and forward to design elements, initial code, tests, scope of application, and other SWS elements. The elements include requirements of different types, business rules, architecture and design components, initial code modules, test versions, and reference files; traceability is of great importance for requirements' management. It enables to determine all the workable artifacts which should be modified once the requirement changes. As per complexity, the authors mark the complexity of the requirements' detection process, complexity of their statement in foreign language, complexity of understanding, and technical implementation of requirements with complicated Boolean logics (combination of operators AND, OR, and NOT) having, however, refrained from providing numerical evaluation aids for these features.

The authors of [20] and [22] propose more advance set of metrics. In [20], they are numerical metrics of unique, understandable, misinterpreted, modified requirements and requirements under testing.

The authors of [22] added metrics of ambiguity, logical linkage, and capability to check the requirements starting from the initial source. Yet, the work does not contain complexity metrics.

Thus, the review of publications and standards showed the lack of metrics enabling to evaluate requirement complexity. The referred enlisted metrics describe construction process and various requirements' features, saving the complexity. With such metric missing, the most complicated requirements cannot be identified and selected for inspection and review to detect higher defects quantity with prevailing restricted resources situation. Reference [19] notes that the requirement complexity is evaluated by experts in a subjective manner. The analysis results and author's personal experience in SWS development enable to state that requirements' complexity is not numerically evaluated in the software engineering. Evaluations are not applied to improve requirements' correctness and the SWS reliability. Therefore, it is necessary to develop a metric for precise numerical evaluation of requirements' complexity.

14.4.3 Proposed Metric for Requirement Complexity Evaluation

The new metric should take into account internal and external complexity inherent with the requirement. Internal complexity is determined by the quantity of components. External complexity is represented by links connecting a particular requirement with other requirements, business rules, etc. These links are illustrated in Figure14.3 representing an entity relationship (ER) diagram of logical contacts of the requirement with other elements.

The ER diagram shows that the requirement may include one element, or more; may affect, or does not affect several dependent requirements; may depend, or not, on some affecting requirements; may depend (or not) on some affecting business rules; and may depend, or not, on some affecting entities, strictly specified for each SWS being developed. The higher the quantity of elements in requirement and links of the requirement with other requirements, business rules, and other entities is, the higher the requirement and its complexity rate by metric are.

Mathematical representation of the requirement complexity evaluation by metric (requirement complexity metric, RCM) looks, as below:

$$\mathrm{RCM} = \mathrm{EQ} + \mathrm{IBRQ} + \mathrm{IDRQ} + \mathrm{IDEQ} \tag{14.8}$$

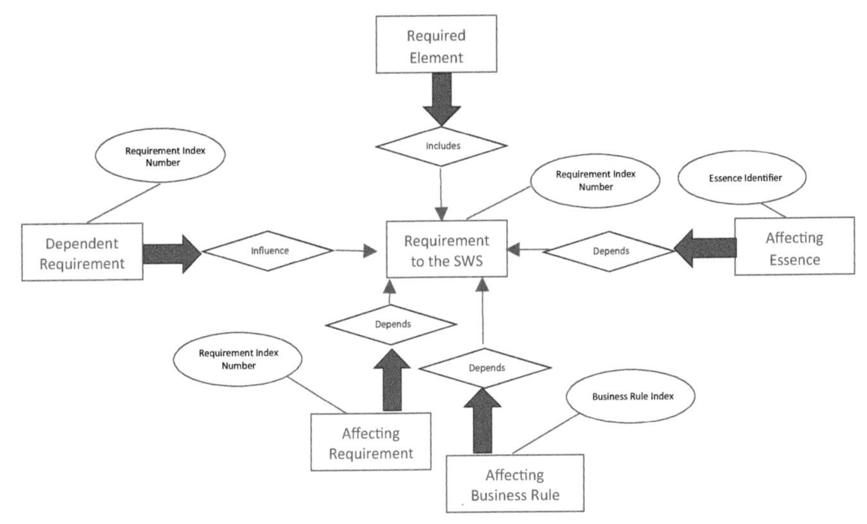

Figure 14.3 ER diagram of logical interconnections of the requirement with other components.

EQ – the elements' quantity (column 2), IBRQ – the influencing business rules' quantity (column 3), IDRQ – the quantity of the affecting and dependent requirements (column 4), and IDEQ – the quantity of the influencing and dependent entities.

Thus, the method of improving requirements correctness SWS for different IoT applications has been offered. The method applies an admission that more complex requirements contain more actual and potential defects. The method applies the proposed RCM metric to evaluate requirements' complexity. The method applies a double sorting procedure based on requirements' priority and complexity evaluation. The method is advantageous comparatively to the total verification due to possibility to select the most complicated requirements with top priority by means of generating them through ranked script. The method enables to improve requirements' correctness and defectiveness.

14.5 Conclusion

In this chapter, the non-functional lifecycle requirement of dependability was introduced as interpreted in the context of IoT equipment and software. Dependability attributes such as reliability, safety, survivability, and security were seen to contribute to the common goal of designing IoT systems that possess high tolerance rates in terms of their ability to maintain an acceptable level of service during disturbances, such as cyberattacks gracefully degrade in the case when the disturbances become extremely severe, while still preventing data loss, and rapidly return to normal operation when the disturbance is removed.

Survivability was examined as a virtue of systems that contributes to their overall reliability. Reliability was defined as a measure of the lack of proneness of the system to problematic behavior and failure. Technologies suitable for promoting increased survivability were reviewed that have been proposed by the authors. First, a method was presented for enhancing the information security capabilities of equipment by enabling the execution of cryptographic calculations in powerful, distributed, and ubiquitous cloud resources.

Additionally, a scheme for the effective design of IoT information processing was presented that was based on assurance case considerations and was hence able to support a robust design process. As a final contribution to system reliability, a metric for software requirements correctness was examined, which is able to partially quantify this commonly intuitive part of IoT implementation.

References

[1] Stavroulakis, P. (ed.). (2003). *Reliability, survivability and quality of large scale telecommunication systems: case study: olympic games*. Hoboken, NJ: John Wiley & Sons.

[2] Ellison, R. J., et al. (1997). *Survivable network systems: An emerging discipline*. CARNEGIE-mellon University, Pittsburgh, PA: Software Engineering Inst.

[3] Dou, B., Wang, X., and Zhang, S. (2009). "Research on survivability of networked information system," in *International Conference on Signal Processing Systems*, IEEE, 56–60.

[4] Yun, L. and Trivedi, K. S. (2006). Survivability quantification: the analytical modeling approach. *International Journal of Performability Engineering*, 2(1), 29–44.

[5] Avizienis, A., et al. (2004). "Basic concepts and taxonomy of dependable and secure computing. dependable and secure computing," in *IEEE Transactions on Dependable and Secure Computing*, 1(1), 11–33.

[6] Krings, A. W., and Zhanshan, M. (2006). "Fault-models in wireless communication: towards survivable ad hoc networks," in *Military Communications Conference, (MILCOM 2006)*, IEEE, 1–7.

[7] Neumann, P. G. (2003). "Achieving principled assuredly trustworthy composable systems and networks," in *DISCEX*, 2, 182–187.

[8] Avizienis, A., et al. (2001). *Fundamental concepts of dependability*. University of Newcastle upon Tyne, Computing Science.

[9] Neumann, P. G., et al. (2016). *Modular research-based composably trustworthy mission-oriented resilient clouds (MRC2)*. SRI INTERNATIONAL, Menlo Park, CA.

[10] Westmark, V. R. (2004). "A definition for information system survivability," in *Proceedings of the 37th Annual Hawaii International Conference on System Sciences*, IEEE, 10.

[11] Neumann, P. G. (2006). "System and network trustworthiness in perspective," in *Proceedings of the 13th ACM Conference on Computer and Communications security*, ACM, 1–5.

[12] Neumann, P. G. (1978). "Computer system security evaluation," in *National Computer Conference*.

[13] Heegaard, P. E., and Trivedi, K. S. (2008). "Survivability quantification of communication services," in *IEEE International Conference on Dependable Systems and Networks with FTCS and DCC*, 462–471.

[14] Knight, J. C., Strunk, E. A., and Sullivan, K. J. (2003). "Towards a rigorous definition of information system survivability," in *Proceedings of DARPA Information Survivability Conference and Exposition*, 78–89.

[15] Ma, Z., and Krings, A. W. (2008). "Competing risks analysis of reliability, survivability, and prognostics and health management (PHM)," in *Proceedings of IEEE Aerospace Conference*, 1–21.

[16] Knight, J. C., and Sullivan, K. J. (2000). *On the definition of survivability*. University of Virginia, Department of Computer Science, Technical Report CS-TR-33-00.

[17] Knight, J. C., Strunk, E. A., and Sullivan, K. J. (2003). "Towards a rigorous definition of information system survivability," in *Proceedings DARPA Information Survivability Conference and Exposition*, IEEE, 78–89.

[18] Kyamakya, K., Jobman, K., and Michael, M. (2000). "Security and survivability of distributed systems: an overview," in *Proceedings of MILCOM 21st Century Military Communications Conf*erence, IEEE, 449–454.

[19] Bardis, N. (2017). "Secure, green implementation of modular arithmetic operations for IoT and cloud applications," in *Proceedings of Green IT Engineering: Components, Networks and Systems Implementation*, Springer, Cham, 43–64.

[20] Sklyar, V., and Vyacheslav, K. (2016). "Assurance case driven design for computer systems: graphical notations versus mathematical methods," in Third *International Conference on Mathematics and Computers in Sciences and in Industry (MCSI)*, IEEE, 308–312.

[21] Yaremchuk, S., Nikolaos, B., and Kharchenko, V. (2017). "Metric-based method of software requirements correctness improvement," in *ITM Web of Conferences*, EDP Sciences, 3–9.

[22] Markovskyi, O. P., et al. (2015). "Secure modular exponentiation in cloud systems," in *Proceedings of the Congress on Information Technology, Computational and Experimental Physics (CITCEP 2015)*, 18–20.

[23] Kelly, J. C., Sherif, J. S., Jonathan, H. (1992). An analysis of defect densities found during software inspections. *Journal of Systems and Software*, 17(2), 111–117.

[24] Schneider, B. (1996). *Applied cryptography: protocols, algorithms, and source code in C*. Hoboken, NJ: John Wiley & Sons.

[25] Wang, H., et al. (2008). "Comparing symmetric-key and public-key based security schemes in sensor networks: A case study of user access control," in *Proceedings of the 28th International Conference on Distributed Computing Systems, ICDCS'08*, IEEE, 11–18.

[26] Bardis, N. G., et al. (2009). Optimised information system reliability techniques for knowledge society acceptance. *International Journal of Knowledge and Learning*, 5, 207–221.

[27] Bardis, N., et al. (2010). "Accelerated modular multiplication algorithm of large word length numbers with a fixed module," in *World Summit on Knowledge Society* (Berlin: Springer), 497–505.

[28] Hofmann, H. F., and Franz, L. (2001). Requirements engineering as a success factor in software projects. *IEEE Software*, 18(4), 58.

[29] Wiegers, K., and Joy. B. (2013). *Software requirements*. New York, NY: Pearson Education.

[30] Iqbal, S., and Naeem Ahmed Khan, M. (2012). Yet another Set of Requirement Metrics for Software Projects. *International Journal of Software Engineering and Its Applications*, 6(1), 19–28.

[31] Haleem, M., et al. (2013). Overview of Impact of Requirement Metrics in Software Development Environment. *International Journal of Advanced Research in Computer Engineering & Technology (IJARCET)*, 2(5), 1811–1815.

[32] IEEE/ISO/IEC 29148-2011. (2011). "*Systems and software engineering – Life cycle processes – Requirements engineering.*"

[33] IEEE Std.1061. (1992). *Std. 1061 for a Software Quality Metrics Methodology*. New York, NY: Institute of Electrical and Electronics Engineers.

[34] Braude, E. (2001). *Software Engineering: an Object-Oriented Perspective*. Hoboken, NJ: Wiley.

PART IV

Implementation and Industry Cases

15

Holistic Systems Engineering Methodology for Intelligent Energy Systems – with a Case Study from "ruhrvalley"

Carsten Wolff[1,*], Torben Lippmann[2] and Uwe Jahn[1]

[1]Fachhochschule Dortmund, Dortmund, Germany
[2]Westfälische Hochschule, Gelsenkirchen, Germany
[*]Corresponding Author: carsten.wolff@fh-dortmund.de

Cities and metropolitan region are becoming the main pattern of organizations for civilizations. Therefore, they form socio-economic systems of growing importance. These systems show a high complexity and lack a centralized hierarchical structure. Therefore, distributed approaches for the management and patterns of self-organization are required. On a technical level, smart system approaches like the smart city or smart grid are applied using the advances in IT, e.g., the Internet of Things. Relevant fields of application are metropolitan mobility and renewable energy systems. Specifically, in the energy sector – apart from smart electricity grids – the heating grids are an important topic for "smartification." This paper describes a holistic systems engineering approach for the future energy (and mobility) systems of metropolitan regions. The industry-university cluster "ruhrvalley" formed by more than 50 partners in one of the largest European metropolitan regions – the German Ruhr Valley – has set out the target to develop such a systems engineering methodology. It is based on the combination of innovation management approaches with a structured process for the development of intelligent technical systems. The resulting "smartification process" is applied as a case study to the design of a flexible and controllable organic rankine cycle turbine as a component for such systems.

15.1 Introduction

Cities and metropolitan regions are complex and heterogeneous systems. They are complex to manage since they consist of a lot of subsystems and components which are not necessarily compatible and usually not centrally planned or aligned. This is the case on the socio-economic level, but also in most of the technical subsystems. The management with distributed and to some extent self-organizing approaches allows a dynamic development and growth and the inclusion of very diverse people and systems. This makes metropolitan regions a very successful organizational pattern with a fast growing importance [1].

For the technical implementation of the basic supply systems of a city, the management and control is done by so-called smart systems (e.g., smart grid, smart building, smart city) [2] which equip the technical components of such systems with embedded systems, connect them to middleware and cloud systems via the *Internet of Things (IoT)* [3, 4], and operate them with several service functions. The network of such intelligent technical systems and the information driven operation is forming an *information supply chain* [5].

Interconnected physical systems with an inherent intelligence (implemented in software and embedded systems) form *cyber-physical systems (CPSs)* [6, 7]. These CPSs have become the technical solution in a variety of applications. Designing and implementing CPSs for the energy and mobility supply on a metropolitan scale is a complex problem, since additional targets (see Figure 15.2) come into play. The industry-university cluster ruhrvalley[1] [8] addresses these issues in one of the largest metropolitan regions of Europe.

The Ruhr Valley used to be the industrial heartland of Germany. Today, it forms a cluster of larger and smaller cities with approximately 6–7 Mio. citizens [9, 10]. Since steel mills and coal mining are gone, the patterns of mobility and energy consumption and supply are changing. Special challenges are the conversion of traffic to eMobility and the provision of energy by renewable sources. In Ruhr Valley, heat is an important topic. It used to come from the large power plants and was distributed via a central heat transmission system. Today, these heat grids need to be cut into smaller systems (loosely coupled) and fed by renewable heat sources. Developing the future systems for mobility and energy in Ruhr Valley is the aim of the *industry-university cluster ruhrvalley*. This requires a common language delivered by a joint systems engineering approach.

[1] www.ruhrvalley.de

This contribution outlines a holistic systems engineering approach for complex energy systems based on comprehensive digitalization. Within this approach, the IoT forms the main backbone of the information supply chain. Since energy systems are critical infrastructure with challenging requirements in terms of safety, security, and dependability, the systems engineering approach has to support the design of the IoT-based information supply chain in a respective way, using an appropriate system architecture and design flow. The following sections introduce the relevant state of the art in terms of system architecture and systems engineering methodology, the case study of an organic rankine cycle (ORC) turbine, the IoT-based system, and software architecture for this case study, and based on that as the main result the outline of the applied system engineering process.

This contribution is based on a previous paper published at IEEE IDAACS 2018 [11]. Furthermore, it includes descriptions of IoT-based SW architectures in the same context [8]. It is intended to describe the generic architecture model and development process. Furthermore, the methodology is demonstrated in the case study of the ORC turbine.

15.2 Systems Engineering for Intelligent Energy Systems – Literature Review and State of the Art

Systems engineering is a term used for a combination of project management and the engineering of technical things. It can be defined as a set of processes, methods, and tools (see the handbooks of INOCSE [12] and the respective body of knowledge [13]). The systems engineering approach of ruhrvalley is reaching a bit further. For the systems of a metropolis, the socio-economic systems play a vital role. The way decisions are taken, the wishes and preferences of the citizens, the regulatory framework, economic considerations, and environmental responsibility have to be taken into account to come to sustainable goals. Therefore, target setting processes are very important. The connection of all kinds of systems (e.g., technical or information systems) has to utilize public infrastructure leading to open standards and open protocols. All such considerations have to be part of the systems engineering approach. In ruhrvalley, the basic architecture of the systems is derived from the very general view of Ropohl [14].

Ropohl groups so-called activity systems into a three-layer structure (see Figure 15.1). The lowest layer (execution system) can be a physical or technical system (like the heat-to-power conversion system of the ORC

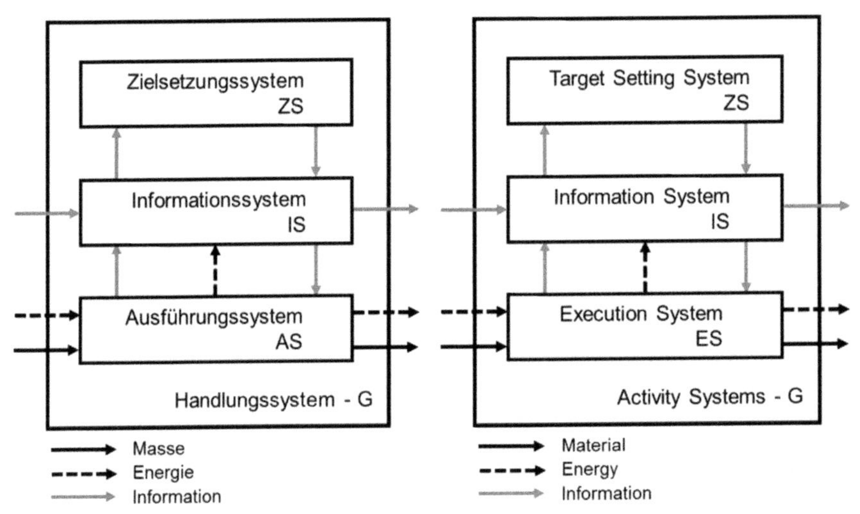

Figure 15.1 Original German version (left) of the technology stack according to Ropohl [14] and own English translation (right).

turbine, see Section 15.3). With the middle layer (information system), this would form a CPS. In our case study, the information system is based on IoT technology. This has several reasons and implications:

- The information layer at the same time separates and connects the physical, technical execution system from and with the (more or less) intelligent target setting system. Therefore, clear and possibly *standardized (IoT) interfaces* are important.
- In a metropolitan region, energy systems are distributed. The information system uses the public network infrastructure (the Internet) which raises requirements in terms of *security, safety, and reliability*. Some can be solved with IoT technology, while some have to be considered in the design of an execution system which is safe also without IoT connection. Real-time requirements are another challenge to be addressed.
- The target setting system should be able to control and optimize the *operation of a high number of execution systems*. This leads to the need for device management, authentication, access rights' management, eventually payment management, etc., which is available and partly standardized in IoT systems and technology.

The target setting system on the third layer forms the planning, optimization, or learning components. In the terminology of IoT systems and with respect

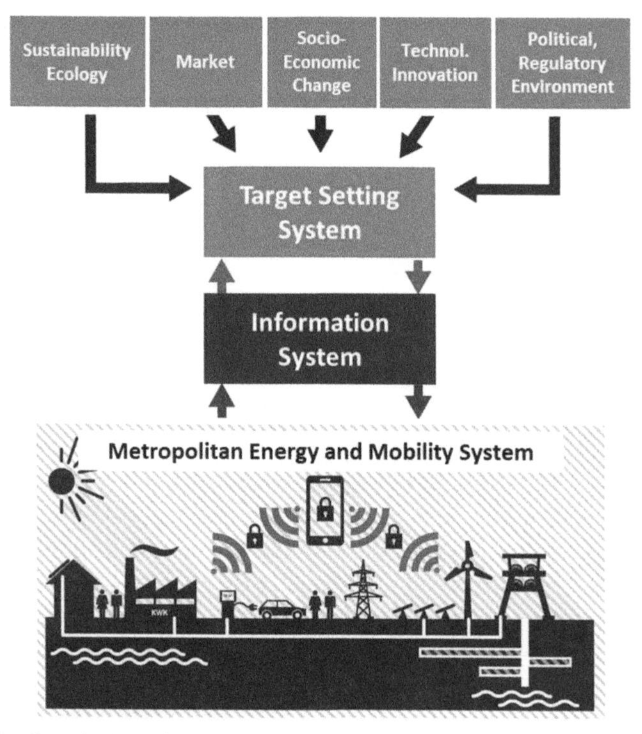

Figure 15.2 Generic three-layer architecture model for metropolitan energy and mobility systems based on the architecture model of Ropohl.

to the case study, this can be seen as cloud-based services for energy management, system maintenance, optimized operation (e.g., based on energy prices), and user interaction.

In ruhrvalley, the target setting system is considered to be a much more comprehensive and holistic area. It intends to connect the goals and targets of the socio-economic system of the metropolitan region to the development and operation of the energy systems. Therefore, approaches from innovation management [15] are incorporated on that layer to drive on a large scale the change of the old metropolitan region into a modern and sustainable mobility and energy system. The idea is to derive the targets for a technical system from the target- and goal-setting process of the environment, including the ecological and economic factors, the market analysis, the socio-economic environment, the innovation trends, and the political and regulatory framework. There are established ways and processes to analyze the trends, demands, and targets from these domains. Nevertheless, it is still new to utilize this knowledge about high-level targets to really steer technological development and innovation. Even more, it is new to incorporate these targets to steer complex technical systems: Breaking the targets down and operationalize them for steering technical systems is a challenge.

Ropohl's three-layer architecture is not only a pattern for socio-economic systems or metropolitan systems at large. It is also usable on the scale of the technical systems and sub-systems. With its holistic and still flexible features, it is an ideal basis of a comprehensive systems engineering approach reaching from large scale down to the actual technical systems (e.g., a heat pump or an eCar). Therefore, it can serve as a basis for the common language of the ruhrvalley while it is still close enough to the engineering world to be usable as a pattern for innovative energy and mobility systems.

The German research cluster "Self-optimizing systems in mechanical engineering (SFB614)" [18] developed the operator controller module (OCM) [16, 17] as a common architecture for mechatronic systems [19]. The respective three-layer model is an obvious variant of Ropohl's general view on systems. The OCM technology stack (see Figure 15.3) is based on the three-layer approach and supports the separation of controllers with hard real-time requirements from more complex and strategy oriented planning and controlling tasks. On the lowest level, the *motor loop* is controlled by "classical" controllers. To make the approach flexible, these controllers can be configured with different parameters. The parameters of a controller and the selection of the appropriate controller type are typical work in the domain of control system engineering. Controllers are developed and parameterized according to the characteristics of the controlled system and to the intended behavior for the operation of the controlled system. Usually, one or several (efficient and effective) operation points are intended for the controlled system. The parameters of the controller are set according to the operation mode (given by the selected operation point) and the parameter (or even controller) change may be done during operation (re-configuration). Since all tasks in the motor loop have direct impact on the controlled system, hard real-time requirements and safety and dependability requirements have to be fulfilled. The development and the technology stack for re-configurable motor controllers is state of the art [18]. The re-configuration is controlled and set by the next layer, the *reflective operator*. This is a rule-based system with state machines and service functions, e.g., for emergency notification. The reflective controller is not allowed to interact directly with the technical system but only via the controllers of the motor loop. Therefore, the motor controllers can be designed to continue operation or fail safe even if the

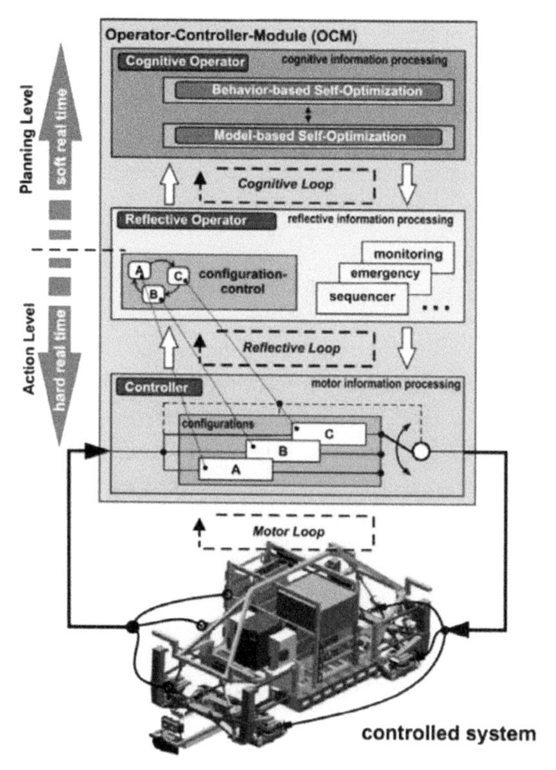

Figure 15.3 Technology stack of the operator-controller-module (OCM) [16, 17], generalized structure of intelligent technical systems.

reflective operator is not working anymore. With the reflective operator, the overall system can be operated stand-alone and it can be connected to other systems in a network or to a smart grid. This layer can be equivalent to an IoT or M2M middleware system with typical functionality [3]. As mentioned, the IoT layer should form the interface between the single technical systems and the overall energy system management. It needs to fulfil real-time requirements and requirements for safety, security, and dependability to a certain degree, but still de-couples the overall system from the single technical system (with much harder requirements). The partitioning of functionality to the three layers and the balance between autonomy of the lower layer systems versus connectivity and coupling for an overall optimum operation are relevant design goals for the systems engineering. For intelligent technical systems, a more sophisticated layer for planning, reasoning, and learning is added, the *cognitive operator*. This layer can include strategies for self-optimization and machine learning [20]. The user interaction can be added to this layer or to the reflective operator. For the case study, this layer is a cloud-based energy management and system maintenance solution. In general, the implementation of this technology stack and the combination with a (M2M or IoT based) networking module turns autonomous mechatronics systems (AMS) into networked mechatronic systems (NMS) [16, 17].

15.3 Case Study: ORC Turbine

As a demonstration of an implementation of an energy system based on the three-layer OCM stack, the design of an ORC turbine was implemented [21]. The ORC is a thermodynamic cycle for the conversion of low temperature heat into mechanical work (and via a generator into electricity). It is used to convert exhaust (waste) heat into renewable electricity which supports the green energy trend [22]. In renewable energy systems, the ORC turbine needs to be adaptable to different heat profiles, it needs to be fast in ramping down and up (due to fluctuations in heat and electricity demands) and it needs to be controllable within larger energy systems. Our ORC demonstrator supports all requirements. It is a two-stage process [23] with the option to use different thermo fluids [24, 25] and a very high efficiency in different operation points (compared to other ORC technology [26]). The two-stage approach means that the system is using two different turbines (HT TuGen – high temperature turbo generator, LT TuGen – low temperature turbo generator, see Figure 15.4.) These two turbines operate with different thermo fluids to convert heat from two different temperature ranges into electricity. This gives more flexibility in addressing different temperature profiles of exhaust heat and it gives a better efficiency in utilizing the energy contained in the exhaust heat. For a typical 500 kW (electrical power) block heat and power plant (with approximately 600 kW thermal power), the electrical power can be increased by 10–15% with this approach. In addition, the temperature profile of the remaining heat can be tuned to a 90°C output for the hot water from motor and liquefier which makes it fully usable for heating purposes. This allows maximum efficiency for the utilization of the heat.

The mechanical setup of the system is implemented in four modules (liquefier, two containers with HT TuGen and LT TuGen, direct vaporizer, see Figure 15.5). This allows easy configuration and assembly of different version of the ORC turbine system. The intention of the modular setup is that different heat sources (with different heat profiles, e.g., temperatures) require tailored ORC turbine solutions to be efficient.

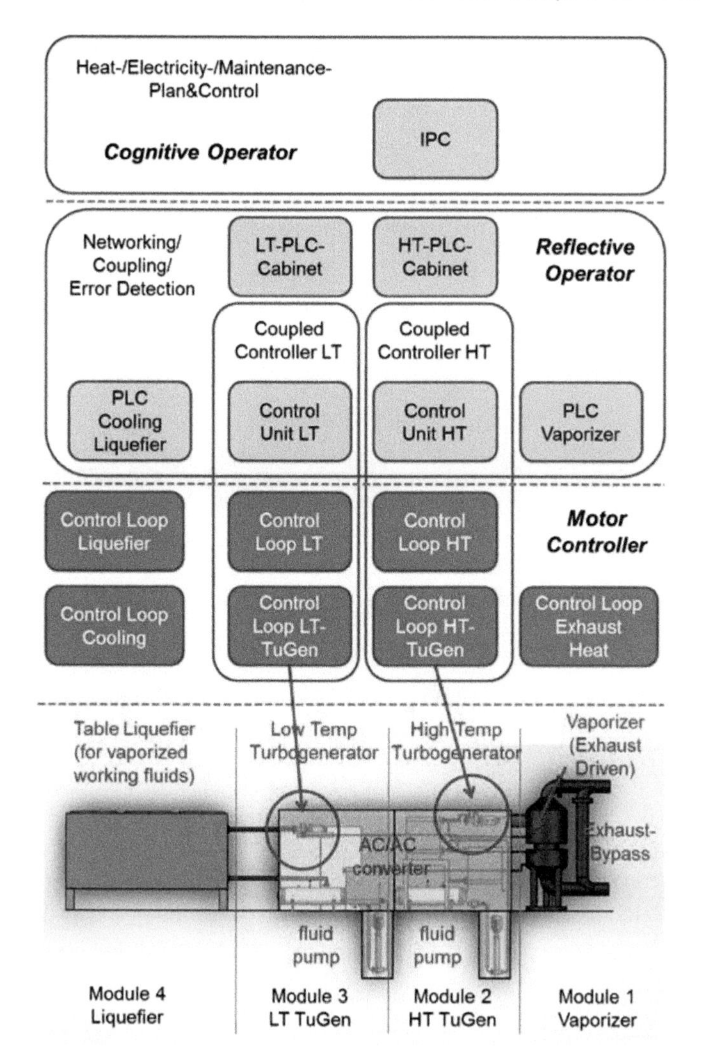

Figure 15.4 Technology stack of the ORC turbine control system, matched to the three layers of the OCM.

The flexibility to configure different ORC turbine setups according to different heat sources is a very important feature for placing the system in the market. Existing ORC solutions suffer from the small range of usable heat sources, or – on the other side – high cost for customizing the solution. To support the tailoring, the control systems and control software have to follow a modular, configurable design, too.

Figure 15.5 Demonstrator setup of the ORC turbine (left, background: stacked two containers with HT TuGen and LT TuGen, middle: direct vaporizer, right: liquefier module).

15.4 Software Architecture for an IoT System based on OCM

The system and software architecture for the ORC turbine [8] is based on the three-layer model of the OCM technology stack. The thermodynamic system (see Figure 15.6) with its vaporizer, the two turbines (NT and HT), the pumps and tubes, and the condenser plus the electrical system with the generators and the electrical converters are considered to be the *controlled system.* The *motor loop* with its controllers is based on a real-time control via a real-time programmable logic controller (PLC) and operates the system autonomously around the desired operation points. The *reflective operator* implements the state machines for driving the machine from one operation point to another operation point (and from power on to operation and again back to power off). Furthermore, it implements the coupling and synchronization to the electric grid and the service, monitoring, and alarm function via the IoT and up to the planning layer. The *cognitive operator* plans the operating sequence of the ORC turbine system based on the demand for heat and electricity. The controller knows the capabilities of the ORC system and may plan and optimize, e.g., daily sequences of operation modes. It can take maintenance intervals into account and it can use self-optimizing and learning strategies. The cognitive operator is connected to smart power grid and smart heating grid interfaces and to the overall optimization instance of the regional energy

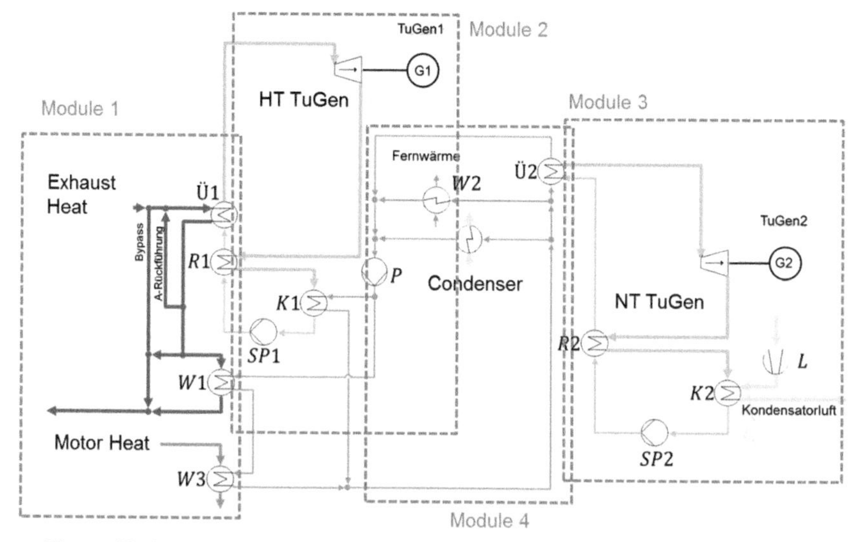

Figure 15.6 Thermodynamic circuits of the four modules of the ORC turbine [21].

system. On the different layers of the system, different architectural concepts, description of components, and languages are deployed. The implementation is done on a distributed PLC and Industry-PC system [8, 20] and connected via IoT technology to an IoT cloud [21].

The *motor controller* loops (lowest level of the OCM, see Figure 15.2) of the ORC turbine are part of the respective TuGen modules. The controllers are split into a high-speed part which is directly connected to the generator and the attached power converters (for feeding into the public grid). These controllers are based on MPP trackers and control the rounds per minute (rpp) of the turbine directly. The rpp are set by the controllers of the respective thermal circuits, in this case the high temperature (HT) circuit and the low temperature (LT) circuit.

These controllers for the thermodynamic system apart from the turbo generators (rpm) follow the model predictive control (MPC) approach (see Figure 15.7). The thermal circuits with all their elements are translated into Matlab/Simulink models which are simulated online to predict the correct settings for the thermal circuits based on the sensor inputs. Additional thermal circuits are the motor cooling circuit and the direct vaporizer circuit. The (simplified) mathematical model of the thermodynamic and electro-mechanical processes of the ORC turbine (controlled system) is simulated online at any time. With this model, the controller can calculate (predict)

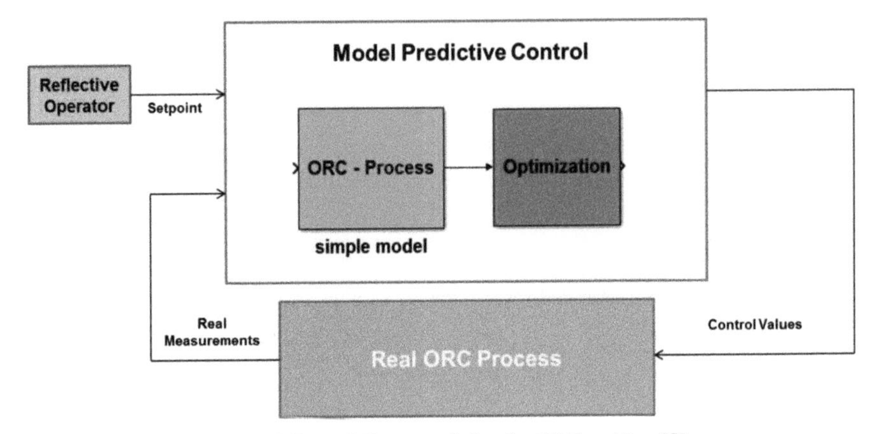

Figure 15.7 MPC approach for the ORC turbine [8].

what actuator values need to be set to achieve a certain system behavior based on the current state of the system (which is signaled by the sensors). Furthermore, based on the model, the optimum placement of the actuators and sensors can be derived and the dynamics of the controlled system can be simulated. The MPC approach is quite powerful but consumes a lot of processing power. Therefore, it is better suited for slower processes like thermodynamic processes. For the high-speed supersonic turbine, a faster independent control loop is implemented. The controllers are initiated and started by the state machines of the reflective controller. They read their parameters from a shared memory (SVI, see Figure 15.8) controlled by the reflective controller. All controllers and the system models are described and validated in Matlab/Simulink and deployed via automatic code generation into C-Functions of the PLC [8].

The *reflective operator* (middle level of the OCM) of the ORC turbine is implemented within an industry standard Bachmann PLC [20] together with the motor controllers (with the MPC approach). The reflective controller is designed based on state charts which trigger the transition between the operations points of the motor controllers, react to events and errors, and implement alarm and reporting functions. The state chart description of these control flows differs from the data flow driven design of the motor controllers. In addition, it is less critical in terms of real-time operation since the motor controllers "buffer" changes in the control flow from one state to the other. The state chart descriptions of the reflective controller (plus several data processing tasks) are implemented with CoDeSys on the Bachmann PLC in

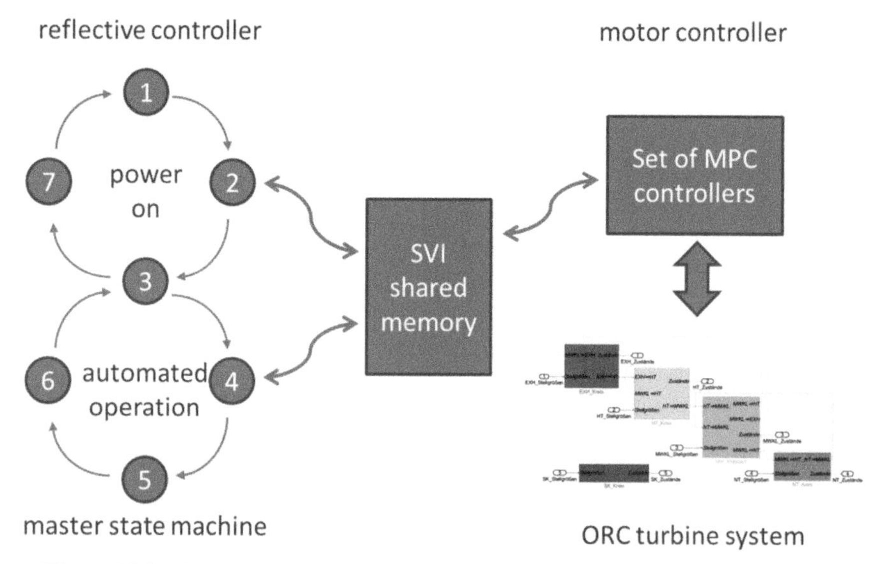

Figure 15.8 Coupling of state machines and controllers via SVI shared memory [8].

a subsystem which is separated from the motor controllers. Communication between the different control flows and the motor controller is done via a synchronized and protected shared memory (called SVI). The SVI contains all states of the system and all relevant values for sensors and actuators. Furthermore, it is used to set the operation points for the controllers. The data structure of the SVI is automatically generated from a master system scheme which contains the main thermodynamic circuits and defines the coupling points. The MPC controllers read their parameters from the SVI but can operate the system safely afterward independently. The SVI implements connectivity via OPC-UA [8], too.

The *cognitive operator* sits in the third layer on top of the technology stack of the OCM and controls the reflective operator via the cognitive loop (see Figure 15.3). It has no direct connection to the motor controllers or to any sensors and actuators. Therefore, it faces only very soft real time requirements. The cognitive operator plans the operation of the ORC turbine on a very abstract level. This also involves the user interaction via the HMI of the turbine and the connection to an IoT cloud for data storage and processing. Via the cognitive operator, the ORC turbine receives the demand projections for heat and electricity, pricing data (e.g., from smart grid) and other operational plans. Based on this, the strategy of the ORC turbine operation is planned.

The implementation of the ORC turbine based on the underlying structure of the three layer OCM stack helped to implement the complex control system. The motor controllers, the state charts for the reflective operator, the thermal and mechanical system, and the cloud connection were implemented by different teams using different technologies for modeling and programming. The encapsulation of the modules and different layers was implemented in protected subsystems with communication via a dedicated synchronized and protected shared memory. Therefore, the approach was very useful for the system design of highly dependable systems. In addition, it delivers a blue print for a more generalized design process for similar intelligent technical systems.

15.5 Smartification Process for Intelligent Technical Systems

The design process of the ORC turbine based on the three-layer OCM model allows distributed and concurrent development of complex intelligent technical systems by engineers and computer scientists with different domain knowledge and different methods, processes, and tools. Furthermore, it facilitates the transition from AMS to NMS [18] which are the basis of CPSs. A consistent and comprehensive engineering process is an important step toward the development of complex CPS [6]. Therefore, the outlined design process based on the OCM is intended to form a blue print for the future systems engineering methodology of complex CPS.

NMSs or CPSs are a very challenging technology due to the real-time, dependability and safety, and security requirements. As a ground rule, communication should be partitioned in such a way that the single communication tasks are always done at the least critical level. Networking in the OCM technology stack is possible on all three layers. The motor controller layer is most critical for networking since it involves real-time and safety critical tasks. Communication between systems is usually not required on this layer. The ground rule is to avoid breaking motor control loops up and split them over several systems or levels. If such motor control loops are broken up, dependable real-time communication technology (e.g., real-time Ethernet) has to be deployed.

The layer of the reflective operator is the domain of M2M or IoT networking. It involves mainly exchange of state changes between different state machines (or state charts) which is less critical since the controlled

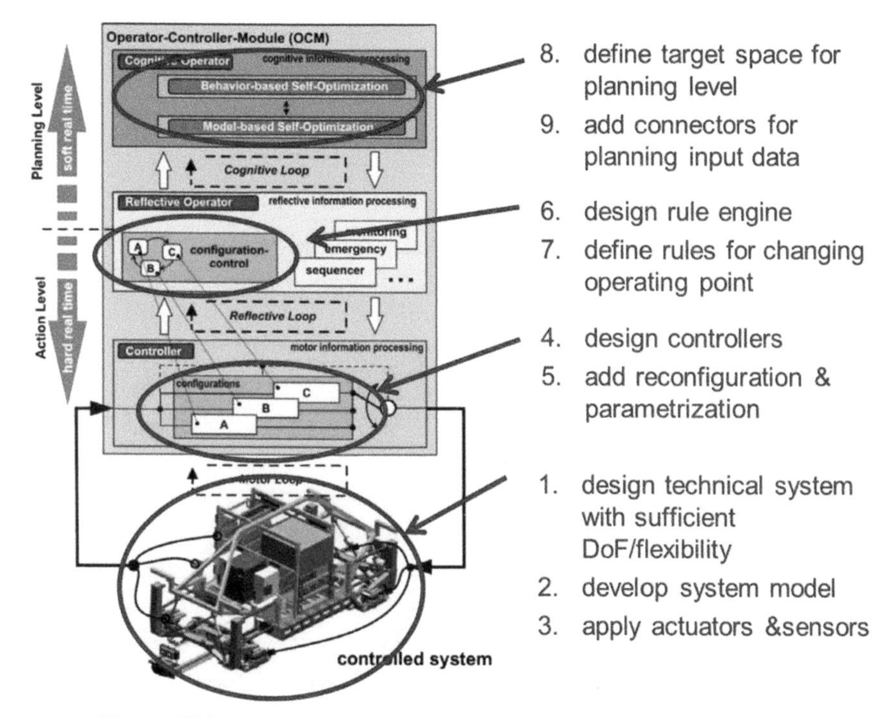

Figure 15.9 Nine-stage "smartification process" based on the OCM [21].

(technical) system is still under full control of the motor controllers. Both message passing and shared memory coupling can be deployed.

The cognitive controller can even be implemented in a cloud system or be part of planning tools (e.g., from energy management systems or production planning systems).

The design process of the ORC turbine is following a nine-stage sequence which is forming a "smartification process" for intelligent technical systems [21, 27]. This process allows the parallel work in teams from different engineering domains (see Figure 15.9):

- In stage 1, mechanical and thermodynamic engineers work together with proven and mature methods and tools from their domains.
- In stage 2, the thermodynamic model was developed, simulated, and verified. Based on that, the control engineers developed the mathematical model of the ORC turbine with Matlab/Simulink. Again, simulation and verification take place.

- In stage 3, electrical engineering, instrumentation, and measurement come into play, a classical engineering domain.
- Stages 4 and 5 are control engineering domains. Engineers can use their elaborated model-based tools and methods, e.g., based on the model derived on layer 2.
- Stages 6 and 7 use processes, methods, and tools from computer science. The underlying technical systems of the motor information processing are abstracted to interfaces for setting and checking operation points – without too much consideration about real time. Existing IoT and M2M software systems can be used (e.g., open source).
- In stages 8 and 9, planning and strategy are the main topics. Interdisciplinary teams can focus on developing the best strategy for the systems operation without diving too deep into the technical system.

Within the systems engineering approach for ruhrvalley, this gives the opportunity to develop the components, the network, and the overall system behavior of networked, distributed energy and mobility systems within a consistent and comprehensive framework which connects engineering processes, methods, and tools within one holistic approach. With the inclusion of the concept of the reflective operator, it opens the door to IoT and CPS. With the cognitive operator, real interdisciplinary development work is possible with the involvement of non-technical people, even citizens, and politics. With the separation of the planning layer from the technical implementation layers, the development of business models (or deriving business models from strategy options of the technical system) becomes quite simple and straightforward.

With respect to innovation and research, the framework helps to structure the possible fields of innovation and therefore to steer research into the intended directions. It helps to select innovations from different domains (e.g., in a morphological box) to form a novel system in a structured way:

- On stage 1, innovations like new materials, innovative mechanical or thermodynamic approaches, and construction and manufacturing aspects are relevant.
- On stage 3, innovations from the sensor and actuator technology allow new ways for controlling and steering the systems behavior.
- Modern control theory influences the innovations on stages 2, 4, and 5. Due to the availability of ever faster electronics, control systems can become more sophisticated, adaptable, and powerful.
- Stages 6 and 7 are driven by the technology leaps in IoT, M2M, and CPS. Together with innovations in the respective fields of security, this

allows large and still flexible systems of distributed energy and mobility components. Dependability and safety are further research topics in this field.

- On stages 8 and 9, the innovation fields of data analytics (including big data), self-optimizing and learning systems (e.g., deep learning), and user interaction technologies are relevant. The development of new business models finds its field here, too.

The industry-university cluster ruhrvalley is composed of companies, university institutes, and research partners on all levels of the smartification process. The full technology chain and the required expertise are available. Companies and startups in ruhrvalley form complete value chains based on the different levels of the process and the respective technology chain. Innovation management [15] will orchestrate the project portfolio of ruhrvalley with the required innovations in all fields. It will use the common language of systems engineering to decide and prioritize the innovation projects. Furthermore, it will support the development of a comprehensive information supply chain supporting the development processes, the operation of the technology chains, and the establishment of the respective value chains. Innovation in ruhrvalley intends to incorporate a broad variety of data sources for taking the required decisions, involving classical research on innovation potentials and user demand, but also by making users, citizens, and even politics co-producers of innovations.

15.6 Conclusion

This chapter presents and outlines of a novel systems engineering approach based on a common architecture model (OCM) which scales from large socio-economic systems (e.g., the metropolis Ruhr Valley) down to technical systems like a flexible ORC turbine. The "smartification process" has been demonstrated on the architecture of the ORC turbine. Within the industry-university cluster ruhrvalley, the systems engineering methodology is further developed. It is planned to develop methods and tools (e.g., a tool chain with software tools for the different steps). The cluster uses a sequence of research and innovation projects to develop missing elements and a portfolio of solutions and a technology tool box for the development of such systems. IoT middleware is operated based on own developments [3] and based on the Eclipse IoT platform [8]. An IoT gateway system is developed and a security concept is implemented. Research infrastructures, demonstrators, and field tests in real-world environments are operated and maintained. The first application is on a smart heating grid, while further applications will address the management of eMobility charging infrastructure. It is intended to generate new business models for existing partner companies and to develop start-up companies if required. Furthermore – within the Ruhr Master School – the staff of the future is trained and educated [28]. With this approach, a novel and comprehensive system of technology chains, value chains, and information supply chains is generated which supports the transition toward sustainable energy and mobility systems for complex and heterogeneous metropolitan regions – as the Ruhr Valley is one.

Acknowledgments

This research is funded by the German Federal Ministry of Education and Research (BMBF) within the program FH IMPULS (grant agreement number for FH Dortmund: FKZ: 03FH0M11IA).

References

[1] United Nations (Hrsg.). (2015). *World Urbanization Prospects.*

[2] EPosSS. (2013). "*Strategic research agenda of the european technology platform on smart systems integration,*" Version 2. Available at: www.smart-systems-integration.org.

[3] Rademacher, F., Peters, M., and Sachweh, S., (2015). "Model-driven architecture for the internet of things," in *Proceedings of the 41st Euromicro Conference on Software Engineering and Advanced Applications,* Madeira.

[4] Vermesan, O., Friess, P., Guillemin, P., Gusmeroli, S., et al. (eds). (2011). "Internet of things strategic research roadmap," *Cluster of European Research Projects on the Internet of Things, CERP-IoT.*

[5] Sun, S., and Yen, J., (2005). "Information supply chain: a unified framework for information-sharing," Kantor, P. et al. (eds): ISI 2005, LNCS 3495, 422–428.

[6] Acatech. (2012). *AgendaCPS: Integrierte Forschungsagenda Cyber Physical Systems.* German Acatech, Berlin: Springer-Verlag.

[7] Lee, E. A., (2008). "Cyber physical systems: design challenges," in *Proceedings of the International Symposium on Object/Component/Service-Oriented Real-Time Distributed Computing (ISORC)*, Orlando, FL.

[8] Wolff, C., Knirr, M., Pallwitz, T., Igci, H., Priebe, K.-P. Schulz, P., et al. (2017). "Software architecture for an ORC turbine – case study for an intelligent technical system in the era of the internet of things," in *Proceedings of the 23rd International Conference on Information and Software Technologies (ICIST 2017)*, Springer CCIS, Druskininkai.

[9] Kriegesmann, B., Bötcher, M., and Lippmann, T. (2015). "*Wissenschaftsregion ruhr – wirtschaftliche bedeutung, fachkrfräteeffekte und innovationsimpulse der hochschulen und außeruniversitären forschungseinrichtungen in der metropole ruhr*," Regionalverband Ruhr.

[10] Wirtschaftsförderung metropoleruhr. (2016). *Wirtschaftsbericht Ruhr 2015.*

[11] Wolff, C., Lippmann, T., and Lauschner, U. (2017). "Systems engineering for metropolitan energy systems – ruhrvalley," in *Proceedings of the 9th IEEE International Conference on Intelligent Data Acquisition and Advanced Computing Systems: Technology and Applications*, Bukarest.

[12] INOCSE. (2011). "*INCOSE Systems Engineering Handbook v. 3.2.2.*" INCOSE–TP–2003–002–03.2.2.

[13] INCOSE. (2013). "*Guide to the systems engineering body of knowledge – G2SEBoK.*" Available at: http://g2sebok.incose.org/app/mss/menu/index.cfm.

[14] Ropohl, G., (2009). "*Allgemeine technologie, eine systemtheorie der technik*," 3. überarbeitete Auflage, Universitätsverlag Karlsruhe.

[15] Kriegesmann, B., and Kerka, F., (Hrsg.). (2007). "*Innovationskulturen für den aufbruch zu neuem – missverständnisse – praktische erfahrungen – handlungsfelder des Innovationsmanagements*," Wiesbaden.

[16] Gausemeier, J., Steffen, D., Donoth, J., and Kahl, S., (2009). "Conceptual design of modularized advanced mechatronic systems," in *Proceedings of the 17th International Conference on Engineering Design (ICED'09)*, Stanford, CA.

[17] Gausemeier, J., and Kahl, S. (2010). "Architecture and design methodology of self-optimizing mechatronic systems," in *Mechatronic Systems Simulation Modeling and Control*, InTech, 255–286.

[18] Adelt, P., Donoth, J., Gausemeier, J., Geisler, J., Henkler, S., Kahl, S., et al. (2008). "*Selbstoptimierende systeme des maschinenbaus – definitionen, anwendungen, konzepte*," HNI-Verlagsschriftenreihe, Band 234, Paderborn.

[19] Lückel, J., Hestermeyer, T., and Liu-Henke, X. (2001). "Generalization of the cascade principle in view of a structured form of mechatronic

systems," in *Proceedings of the IEEE/ASME International Conference on Advanced Intelligent Mechatronics (AIM 2001)*, Como.

[20] Bachmann. (2018). "*Bachmann M1 Automation System.*" Available at: http://www.bachmann.info/branchen/erneuerbare-energien/

[21] Wolff, C., Knirr, M., Schulz, P., Priebe, K.-P. Strumberg, J., and Vrabec, J., (2016). "Flexible and controllable ORC turbine for smart energy systems," in *Proceedings of the IEEE EnergyCon 2016*, IEEE Xplore, Leuven.

[22] Nakicenovic, N., Grubler, A., Inaba, A., Messner, S., Nilsson, S., Nishimura, Y., et al. (1993). Long term strategies for mitigating global warming. *Energy International Journal*, 18(5), 401–609.

[23] Dubberke, F. H., Priebe, K.-P., Vrabec, J. Roedder, M., and Neef, M. (2015). "Thermodynamic simulation and experimental validation of a cascaded two-stage organic rankine cycle turbine," in *Proceedings of the 3rd International Seminar on ORC Power Systems*, Belgium.

[24] Chen, H., Goswami, D. Y., and Stefanakos, E. K. (2010). A review of thermodynamic cycles and working fluids for the conversion of low-grade heat. *Renewable and Sustainable Energy Reviews*, 14, 3059–3067.

[25] Roedder, M., Neef, M., Laux, C., and Priebe, K.-P. (2015). "Systematic fluid selection for organic rankine cycles (ORC) and performance analysis for a combined high and low temperature cycle," in *Proceedings of the ASME Turbo Expo: Turbine Technical Conference and Exposition*, Montreal.

[26] Quoilin, S., Van Den Broek, M., Declaye, S., Dewallef, P., and Lemort, V. (2013). Techno-economic survey of organic rankine cycle (ORC) systems. *Renewable and Sustain. Energy Reviews,* 22, 168–186.

[27] Wolff, C., Strumberg, J., Knirr, M., Foerst, I., Tekampe, J., and Hensen, C. (2015). "Smart biogas power plants," in *Proceedings of the Dortmund International Research Conference*, Dortmund.

[28] Wolff, C. (2016). "Ruhrvalley and ruhr master school: two pillars of a new cooperative university of applied sciences in engineering," in *International Symposium on Embedded Systems and Trends in Teaching Engineering*, Nitra.

16

Smart Waste Management System: A Fusion of IoT and Blockchain Technology

Manish Lamichhane[1,*], Oleg Sadov[2] and Arkady Zaslavsky[3]

[1]PERCCOM Masters Student (Previous), Software Developer (Current), Germany
[2]Leader at NauLinux & Linux Cyrillic Edition projects, Russia
[3]Senior Principal Research Scientist, Data61, CSIRO, Australia
*Corresponding Author: manishlamichhane@gmail.com

Blockchain technology and Internet of Things (IoT) are two of the most popular technologies today. IoT is an interconnection of devices that can communicate with each other. It can help create smart solutions that can enhance the quality of life of people. Likewise, blockchain is distributed database systems that promise high level of security and availability of data with least transaction overhead. In this paper, we attempt to bring together these two technologies to develop a smart waste management system. The scope of implementation of such smart solutions to real-life problems is limited by the lack of proper payment infrastructure that can support micro-payments in return of services. Current financial systems have problems dealing with micropayments due to large overhead cost of transactions. Blockchain technology could be a reasonable solution to overcome such a problem. It can help lower the transaction cost and time thus lowering cost of services in general, which can specially impact developing countries. The proposed smart waste management (SWM) system uses latest development in blockchain like smart contracts, decentralized autonomous organization, and its own cryptocurrency to handle the investment and service charges while it uses simulated smart garbage bin with QR-reader that communicates to the central server using MQTT, a popular IoT protocol. Furthermore, a Telegram Bot running in telegram messaging application helps user interact with the

SWM system. Measurement of transaction times for blockchain in two different networks, i.e., a private network and a public test network provides an outline of resource allocation and speed of transaction using blockchain.

16.1 Introduction

World Bank's data from 2012 suggest that by 2025, there are going to be 4.3 billion urban residents generating about 1.42 kg/capital/day of municipal solid waste. The waste management cost is projected to increase to about $375.5 billion in 2025 from an annual of about $205.4 billion in 2012 [1]. At present, approximately 50% of total population of world are residing in cities. These data display that the waste management is going to be a very expensive aspect of governance.

Developed countries have some sort of waste management system already in place. Many developing countries, however, are struggling with serious problems related to waste management [2]. Various IoT-based approaches have been proposed to establish a proper waste management system. However, they all rely on third-party services for the payment of service. Majority of the population in many developing world countries don't own a bank account, let alone online banking services [3]. This chapter presents a proof of concept of a system that brings together two powerful technologies: IoT and blockchain technology, to address real-world problems in (but not limited to) developing countries. The focus of this chapter will be on the implementation of decentralized autonomous organization (DAO), which will be dealt in detail in coming sections. The proposed solution is named "Thrift and Green" and will be called TAG, for convenience, hereafter. TAG is a proof of concept of a system that brings together two cutting edge technologies: IoT and blockchain to propose the solution to the problem of waste management.

The massive overhead cost of current financial systems makes micropayments difficult [4, 5] Blockchain technology [6] enables financial transactions beyond borders and completely displacing intermediaries in terms of security and transaction overhead cost. The implementation of smart contracts [7] on top of blockchain largely widened the scope of the impact of blockchain. DAO [8] is a concept of a democratic organization that can exist on Ethereum blockchain whose objective can be defined as a collection of smart contracts. It then runs on its own without human intervention. Any change made to the organization has to be passed through voting of its members or else it won't take place. A DAO can execute other smart contracts and also interact with other DAOs. This could give rise to totally new business models and can have a very wide implementation scope. One can even imagine government bodies and different companies running on top of blockchain using the concept of DAO.

16.2 IoT, Blockchain and Dependability

CSIRO describes Internet of Things (IoT) as, "The IoT is the latest Internet evolution that incorporates a diverse range of things such as sensors, actuators, and services deployed by different organizations and individuals to support a variety of applications." It is a network of devices that can sense, collect, process, and relay information across the network, and hence it is one of the strongest building blocks of future cities or smart cities. Gartner's Hype Cycle shows IoT as one of the most popular buzz words in 2017 [9]. But along with the possibility, a question of dependability arises. A research conducted by Mckinsey [10] shows that security is the major concern of consumers when it comes to IoT. Similarly, the maintenance overhead of IoT devices can snowball into a significant liability in part of the vendors. A good example can be rolling out the software updates to IoT devices. Thus, IoT has a high usability on the one hand while it also has low dependability. An appropriate solution to the dependability issues inherent in IoT can be blockchain. It is a network of decentralized, peer-to-peer nodes that provide security, transparency, and reliability. We can refer to the above example of deployment of updates by IoT vendors to understand how blockchain can play a significant role in enhancing the dependability of IoT. The authors have taken Christidis's research on IoT and blockchain [11] to establish the idea. Vendors run a private blockchain network where all the IoT devices/sold produced by them are connected or will be connected. A smart contract that can store the hash id of the location latest firmware update is deployed in the same private network. The IoT devices have this address of smart contract built into them during manufacture or they can discover this address by a discovery mechanism. The firmware updates can be stored in some distributed file system (e.g., IPFS) which provides a hash of the location of storage or they can be served by the vendors themselves. In their research, the authors have assumed that IoT devices possess the capacity to share the binary of the firmware. When enough nodes will have fetched the firmware, the vendors can stop serving it themselves as the devices can now share the firmware update among themselves. Certain mechanism can be written in the smart contract to enforce a duration or number of queries after which the smart contract will remove the data on the firmware update. This example depicts how a combination of IoT and blockchain technology enables the possibility of developing dependable systems with high usability.

16.3 Background and Motivation

The process of solid waste management can be divided into two parts: 1. collection and transport and 2. disposal and recycle. World Bank describes waste management as the collection of solid wastes from the point of production to the point of treatment or disposal. In developed countries, most of the existing waste management systems include multiple third-party waste collectors that transport the waste disposed in waste bins to a recycle plant or landfill sites. Most of these systems use fixed cost pricing strategy, i.e., whatever amount of waste the users produce, a fixed amount is charged to the users. There is no motivation whatsoever to produce lesser waste or even consider producing less waste. On top of this, the waste data are not reliable as well. A study conducted in the same domain in Sweden [12] reflects the data reliability problem; 30 waste management companies were keeping records of the waste data which amounted to 500,000 entries. Lack of quality control and appropriate pre-defined criteria to record the data resulted in usability of only a fraction of data from the dataset.

To address these problems, different solutions using IoT devices have been forwarded (some discussed in literature review). However, these solutions do not use IoT protocols designed for resource constrained environments. The use of proper IoT protocols designed for constrained environment like MQTT or CoAP could have enhanced the energy efficiency.

Also, if we consider the deployment of this kind of system in developing countries, the main problem that arises would be payments. As discussed in Section 16.1, the main hindrance would be the lack of proper payment infrastructure for the payment of waste management services. Department of Computer Science, Stanford University, outlines several technical, social, and economic challenges in current micropayment systems [13]. One of the main barriers of explosion of micropayments structure is the incapability of minimizing the transaction cost in relation to the actual incurred cost. In case of cryptocurrencies, there is no third-party involvement, and hence the transaction overhead is minimal. Micropayments can be executed in a simplified, transparent way.

Many governments in developing countries do not even have funds to invest in developing a proper waste management system. However, a different approach to solve this can be local community coming together and raising fund to invest into solutions to their problems. DAO bypasses the need of a third party or Government in general for lawful regulation and asset security. Because smart contracts always will behave the way they are designed to behave before execution. The codes cannot be tampered with or changed once it is deployed on the blockchain network. Like any normal organization, a DAO may need investment. The investments are all performed in cryptocurrencies. Cryptocurrencies can be purchased from different exchanges or can be generated using another DAO that can act as a central bank. TAG uses its own cryptocurrency regulated by its own central bank. The bank can inflate or deflate the currency as per the need and consensus of its regulatory members.

For a common user, knowing all these technical terms in detail is less comforting and impractical. Here, we present TAG from a common user's perspective. A user who wants to dispose waste using TAG needs two things: a Telegram App and an Ethereum account. The Ethereum account can simply be created by downloading an Ethereum Wallet [14]. An alternative to use Ether or to buy cryptocurrency would be minting, which is discussed in Section 16.5. TAG has an entire ecosystem of DAOs that runs using a common currency regulated by a central bank. Users then need to transfer some funds to this central bank account's address to use TAG's services. Once the user has account with some balance in it, they can generate a QR-code using the Telegram app by sending their account number to a Telegram Bot. The Telegram Bot also sends the location of the smart garbage bin (SGB) closest to the user's location with information about the level of waste, the current exchange rate, and the charging policies. The user can open the SGB by scanning the QR-code. After disposing the waste, the system will measure the weight of the waste user deposited and transfer PercCoins from the user's account in the central bank to the SGB owner's account in the central bank that equals the amount 5% of the weight of the waste that user disposed. So, if the user disposed 1000 grams of waste, the cost would be 50 (5% of 1000) cents converted in PercCoin. The pricing of PercCoin is beyond the scope of this research. We have used a constant rate of exchange for TAG to simplify things.

16.4 Related Work

As a solution to this problem of waste management, a lot of projects using technologies like IoT, RFID, WSN, etc., are already in place.

16.4.1 Sensing Waste Levels

Shahabdeen has patented a SGB [15] that uses sensors mounted in the cover (range finder) or bottom (weight sensor) of the SGB that can sense the amount of items deposited in the SGB. This system takes into consideration this very approach for calculating the amount of waste present in the SGB. For the purpose of experimentation, a simulation of SGB is used. The above-mentioned approach is suitable for deployment to real world. Folianto, Low, and Yeow, in their paper Smartbin: Smart waste management (SWM) system [16], have proposed a Smartbin that detects the level of waste in smart bins using ultrasound sensors. A mesh topology of wireless sensor network is used to collect the data from the sensors in the bins (nodes) and deliver it to the backend server from the gateway nodes using the Internet. Based on the 6 months deployment period, the authors claim that their system provides a data delivery rate of 99.25%. Also, an approach based on image processing that takes timely snapshot of level of waste in SGBs and processes them using microcontroller to identify the level of waste and send information to the central office using the GSM module has been in place [17]. From a sustainability perspective, this approach, however, is not very appropriate as the SGB in case of TAG will already have a camera that acts as a QR-reader. Another camera for image processing could be very energy intensive.

16.4.2 Sorting Waste at Source

Another approach where selective sorting of different types of waste is enabled using RFID technology [18] is already in place. The quality of waste generated can be significantly improved by the use of this approach. To perform model sorting at source technologically in our system is currently not possible. Including this capability can enhance the impact of TAG. This can be included in future prototypes of TAG.

16.4.3 WSN-based Architectures

Wireless sensor networks are used to check the filling of the bin by the use of sensors.

These sensors feed data to data transfer nodes which enable long-range transmission of data to a decision support server [19]. This idea of SGB interacting with the remote server is used in TAG. The most interesting of all is a WSN and RFID-based SGB which introduces the idea of HSGB (Head SGB) that monitors other SGB within its area using a mesh topology [20].

An edge router is responsible for assigning HSGB using the battery/power status of a particular SGB over the span of 7 days. This approach introduces an "adaptive user charge policy" that motivates users to produce less waste by charging users with a rate based on the 10% of the total waste production by the user on a monthly basis. Since the concept of volume-rate-based waste disposal system is implemented and has shown a 33% reduction in the amount of food waste produced, TAG uses this approach as well.

16.5 Architecture Overview

The blockchain component contains all the smart contract and DAOs, the details of which are explained in following subsections. Figure 16.1 shows the overview of different components of TAG.

The SWM server consists of a Telegram Bot that users use to interact with the SWM system using Telegram [21] mobile application. Since blockchain transactions are relatively slower compared to centralized databases, a Mon-goDB database is used to store SGB information and transaction information. All the financial information is segregated to blockchain.

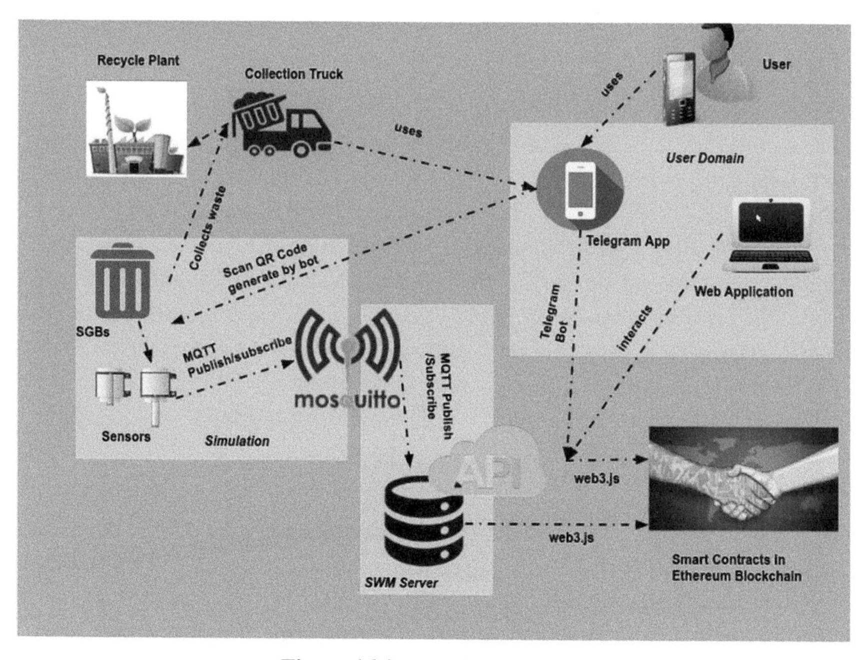

Figure 16.1 TAG architecture.

This enhances the efficiency of the system and keeps the financial data secure, powered by blockchain. User domain consists of the Telegram application and a web application. The Telegram application, as mentioned above, is the means to interact with the Telegram Bot. The web application provides information about the user and system transactions, the amount of waste users produced, and a god-view to see all the SGBs in a map with related information. In place of actual smart waste bin, a simulation of waste bin is used, which consists of a QR scanner and a MQTT publish and subscribe client to interact with the SWM server. The following is the detail description of some of the components:

16.5.1 Blockchain

TAG is powered by Ethereum blockchain [22]. The main motivation behind this is the implementation of smart contracts [7] and the concept of DAO [8] provided by Ethereum. Solidity [23] is a Java-like language, most developed and supported by the Ethereum community. Smart contracts are written in Solidity 0.4.8 for TAG. Truffle [24] v2.1.2 is used to deploy the smart contracts to Ethereum private and test networks. TAG interacts with three smart contracts. The detailed description of each is listed below.

16.5.1.1 The bank

DAO is the representation of any real-world organization on top of blockchain. One of the crucial aspects of it is that once it is deployed, the code can't be tampered with. Members can change the rules or objectives using the mechanism of voting. The Bank is one such smart contract which represents a Bank-like organization from real world in blockchain. The Bank has a registry which is an array of addresses and balances. The addresses can be the account number of users or other DAOs. This bank can be understood as a Central Bank regulating the economy of an eco-system of DAOs. All other DAOs use a common currency of this central bank. The currency used by this central bank is named "PercCoin." There can be two methods to create currency in the Bank.

1. *External Transfer*
   Exchange rates can be defined for different currencies to be used in the bank. The economics of the process is not within the scope of this research. For TAG, whenever a certain amount of external transfer of Ether is made to the Bank, it records the sender account in its registry

and assigns an equivalent amount of PercCoin with an exchange rate 1 Wei = 1 PercCoin to this account. If it is an existing account, it adds the equivalent amount of PercCoin to the existing balance of the account.

2. *Minting*
   Purchase of Ether or mining of Ether might not be a best idea in developing countries, especially in rural communities. It takes a lot of computing resources and the investment on devices can be substantial with the increase in time, as the difficulty of network also increases over time. Anything that is of value or even fiat currency can be used to buy PercCoin. The Bank can mint an equivalent amount of currency as per the provided asset of value and add it to the user's account.

All the codes for TAG are available in Github under a GPL license. The project components can be accessed at ***https://github.com/itmo-swm***

16.5.1.2 Community DAO

Community DAO can be understood as an organization running with the objective of investing and managing in waste management. Any member of this Community DAO can propose a proposal which is subjected to voting. The voting rules can be set during the first deployment and later a new proposal can be passed to change the voting rules, if necessary. Other different templates can be used to create a DAO which have different flavors of democracy. The main points to understand in this type of DAO, as mentioned in Ethereum website, are:

minimumQuorum
It is the minimum number of votes required to pass a proposal. If there are five members and this parameter is set to three, at least three people should vote to pass the proposal.

debatingPeriodInMinutes
It is the time window (measured in minutes) for the voting period. If set to 10, the members must cast their vote within these 10 min.

majorityMargin
A proposal passes if there are more than 50% votes plus the majority margin. If there are four members and this parameter is set to one, then 50% of total possible votes would be two and majority margin = 1. So if three votes (50% of 4 + majorityMargin) are received, the proposal passes. It should be left to 0 for a simple majority and total possible votes −1 for absolute majority.

16.5.1.3 SGBFactory

SGBFactory represents an account for SGB producers. It can be understood as a market place. The ComDAO can raise fund. The members can decide upon the investment amounts for SGB and pass a proposal to transfer fund to this SGBFactory. Upon reception of the fund, the SGB producers can deliver the SGB. This represents one of the fundamental capabilities of blockchain, i.e., movement of goods and services in real world based on the transaction in blockchain. Though a pretty straightforward concept is defined here, in actual implementation, an escrow fund can be created which can hold the transfer from ComDAO and only transfer the fund to SGBFactory account upon the reception of SGB by ComDAO members in real world.

16.5.2 SWM Server

Even though blockchain has massive potential, it should be accepted that this technology is in its dawn. Due to delayed processing time, blockchain technology is not efficient enough to work with applications that need swift read write. Due to this very drawback of blockchain, TAG is designed as a hybrid system, where a central server, named SWM server, handles components that need fast read write and blockchain technology to take care of the financial transactions. The SWM has the following components.

16.5.2.1 MongoDB

The SGB information like owner account, quantity of waste, location information, etc., is stored in the MongoDB. One of the advantages of using MongoDB is the easier processing of geoJSON data. When supplied with a location value and radius (in radian), MongoDB can return the location of all the SGB lying within the circular area of given radius. This information is requested by the Telegram Bot when the user selects to throw the waste, in Telegram Bot menu.

The transaction information is also stored in MongoDB in another collection. It contains information like user's account, who disposed the waste, the amount of waste disposed, the SGB in which the waste is disposed, and the amount of PercCoin transferred from the user's account to the SGB owner's account.

16.5.2.2 Telegram Bot

A Telegram Bot named "perccomanitmo" is deployed in Telegram application. It can be searched in Telegram in the same way users look up to other users. Typing "/hello" will start the conversation. The user can interact with it to find the balance or to generate QR-code that is needed to open the SGB. If the users have enough balance in their account, this bot generates a QR-code and asks for the users' location. Once the location is provided, it is considered as the center of the circle and the closest SGB within a circular area with radius 3 km is returned to the user. MongoDB's geoJSON query comes in handy at this point. Users can then go to the SGB and scan the QR-code generated earlier. If the QR-code is not re-used and has been generated by the Telegram bot, the SGB opens.

16.5.3 SGB Simulation

An SGB simulator with QR-code scanner and MQTT publishing and subscribing capability is developed for the convenience of experimentation.

This simulator scans the QR-code generated by the Telegram Bot in user's Telegram application and communicates with the server using the MQTT protocol. If the decoded QR sent by this SGB is verified by the server, the lid of SGB opens.

16.5.4 User Domain

The user domain consists of two components that users can use to interact with TAG. They are as follows.

16.5.4.1 Telegram

Users can use features of TAG via Telegram Application. A Telegram Bot named "perccomanitmo" works as an intermediary between the user's Telegram and the TAG. The QR-code generated by TAG is sent to the user in their Telegram account. The user can use this QR to open the smart waste bin.

16.5.4.2 Web application

Users can look up information about themselves (like the amount of waste they have produced and the amount of money they have spent) in a web application. Apart from this, the web application provides information about ways to reduce waste. It also shows different information to users that might motivate them to produce less waste in a responsible way.

16.6 MQTT Message Exchange Architecture

MQTT is one of the most popular IoT protocols. It is efficient for resource-constrained, high traffic environment. In comparison among different protocols, MQTT stands out in high traffic, resource-constrained environment [25]. With scalability in mind, MQTT is used as a protocol of choice to interact between the SGB simulation and the SWM server.

16.6.1 MQTT Topic Hierarchy

The first level of topic hierarchy is classified according to the devices using them. Hence, the first level consists of: SGB and server. The next level of classification of message is based on the action each device does. Each device uses MQTT for publishing and subscribing: (i) authentication message and (ii) waste message. Figure 16.2 is the description of message exchange. Table 16.1 shows the mapping of device with MQTT Topic and the respective action it performs.

16.6.2 Authentication

When SGB scans the QR-code, it publishes the decoded value to SG-B/authenticate topic. To receive this message, the server is subscribed to SGB/authenticate. Once the server receives the message, it checks whether the decoded value is the same one that was generated by the server when the user requested to throw waste using the Telegram app. If verified, the server

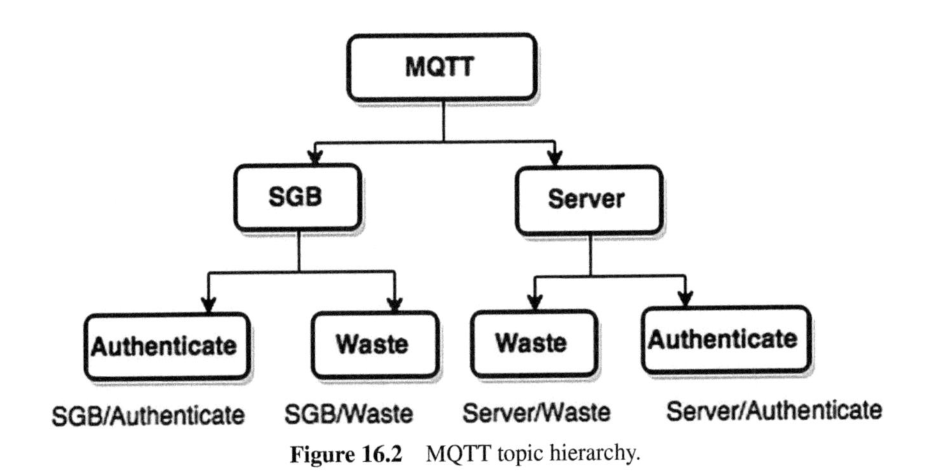

Figure 16.2 MQTT topic hierarchy.

Table 16.1 MQTT topic table

Device	Publish	Subscribe	Action
SGB	SGB/Authenticate	Server/Authenticate	Authenticate User
Server	Server/Authenticate	SGB/Authenticate	Authenticate User
SGB	SGB/Waste	Server/Waste	Send Waste Info
Server	Server/Waste	SGB/Waste	Sends Transaction success info

will publish a JSON data with appropriate response to the server/authenticate topic. The SGB is subscribed to the server/authenticate topic to receive the message from the server. If the user is verified, the SGB opens.

16.6.3 Waste

When the user puts the waste in SGB, the sensors calculate the total weight of waste in SGB (including the weight of waste generated by the user). Upon closing of the lid, the SGB publishes the waste information to the SGB/waste topic. The server will receive the message by being subscribed to the SGB/waste topic. The amount of waste generated by the user is calculated as the difference between the total waste amount sent by SGB and the total waste amount present in the server's database. PercCoin equals to 5% of the waste amount generated by the user is transferred from the user's account to the SGB owner's account and the transaction information is updated in the server's database. The status of this operation (successful or unsuccessful) is published by the server to the topic server/waste. For simulation environment, SGB can be subscribed to the server/waste topic, but it is not significant to real-world deployment of SGB.

16.7 Limitation and Future Work

To increase the quality of waste generated, sorting of waste at source is vital. Users could be rewarded by recycle companies based on the quality and type of waste they produce, which in turn would motivate users to sort waste and check the waste they generate, for example, rewarding users for using more recyclable products. For this feature to work, associating users with the waste they produced becomes important. This is a challenge that the proposed system cannot address as of now. We are looking for ways to address this scenario in TAG in such a way that the implementation will impose least technological complexity in part of users. Another interesting use case could be the development of Uber-like waste management system, where users are rewarded for delivering the waste from pickup locations to recycle facilities. This concept can be represented with a DAO using smart contracts that can interact with other Recycle Plant DAO and Community DAO.

16.8 Conclusion

In this chapter, we have proposed an IoT- and blockchain-based waste management system. To the best of our knowledge, there has not been another implementation of similar kind, till date. Volume rate-based system relies largely on stream of micropayments. The proposed smart waste management system is able to handle micropayments by the virtue of blockchain and smart contracts. Furthermore, this chapter outlines that the concept of DAOs has strong possibilities to create different business models which can lower the penetration cost and boost innovation. People can come together, raise fund, and invest into solutions that can address their problems and can also be sure that they are in control of their own investment.

Blockchain technology is a relatively new concept. Ethereum blockchain is gaining popularity but it is still a system under development and yet to be mature. One of the drawbacks it imposes is that creating a system whose entire database is running on top of blockchain is not very applicable at the moment because of the processing time the network needs to mine data into blocks. Bitcoin's network alone consumes as much electricity as the country of Ireland as a whole [26]. A change of algorithm from proof of work to proof of stake might possibly make the network less resource intensive and enhance the efficiency. Another drawback is the lack of knowledge of blockchain technologies among common population. With the increase in the amount of applications built on top of blockchain and the improvement of efficiency of the network, the appeal of blockchain might grow.

Acknowledgments

The research reported here was supported and funded by the PERCCOM Erasmus Mundus Program of the European Union [27]. Part of this work has been carried out in the scope of the project bIoTope [28] which is co-funded by the European Commission under Horizon-2020 program, contract number H2020-ICT-2015/ 688203 – bIoTope. The research has been carried out with the financial support of the Ministry of Education and Science of the Russian Federation under grant agreement RFMEFI58716X0031.

References

[1] Hoornweg, D., and Bhada-Tata, P. (2012). WHAT A WASTE-A Global Review of Solid Waste Management. *Urban development series; knowledge papers no. 15*. Washington, DC: World Bank.

[2] Idris, A., Inanc, B., and Hassan, M. N. (2004). Overview of waste disposal and landfills/dumps in Asian countries. *Springer-Verlag*, 322(6), 104–110.

[3] Kshetri, N. (2017). Will blockchain emerge as a tool to break the poverty chain in the global south? *Third World Quarterly*, 38(8), 1710–1732.

[4] Jovic, Z., Coric, G., and Pejovic, I. (2016). "Challenges of modern electronic banking," in *Sinteza International Scientific Conference on ICT and E-Business Related Research.*

[5] Song, W., Kou, W., and Tan, C. (2016). "An investigation on multiple e-payments and micro-payment," in *Proceedings 16th International Parallel and Distributed Processing Symposium*, IEEE, Ft. Lauderdale, FL.

[6] Nakamoto, S. (2008). *Bitcoin: A peer-to-peer electronic cash system.*

[7] Buterin, V., et al. (2014). "*A next-generation smart contract and decentralized application platform*," White paper.

[8] Jentzsch, C. (2016). Decentralized autonomous organization to automate governance. *Final Draft-Under Review.*

[9] Gartner. (2017). *Gartner Hype Cycle 2017.*

[10] McKinsey (2015). *Security in internet of things.*

[11] Christidis, K., and Devetsikiotis, M. (2016). "*Blockchains and smart contracts for the internet of things*," IEEE Access, 4, 2292–2303.

[12] Shahrokni, H. B., Van der H., Lazarevic, D., and Brandt, N. (2014). "Big data gis analytics towards efficient waste management in stockholm," in *Proceedings of the 2014 conference ICT for Sustainability,* Atlantis Press, 140–147.

[13] D. o. C. S. Stanford University. *Micropayments: A viable business model?* [accessed May 16, 2017].

[14] Foundation, E. (2017). *Ethereum wallet (0.8.9).*

[15] Shahabdeen, J. (2016). *Smart garbage bin*, US Patent App. 14/578, 184.

[16] Folianto. F., Low, Y. S., and Yeow, W. L. (2015). "Smartbin: smart waste management system," in *2015 IEEE Tenth International Conference on Intelligent Sensors, Sensor Networks and Information Processing (ISSNIP)*, 1–2.

[17] Prajakta, G., Kalyani, J., and Snehal, M. (2015). "Smart garbage collection system in residential area," in *IJRET: International Journal of Research in Engineering and Technology eISSN*, 2319–1163.

[18] Glouche, Y., and Couderc, P. (2013). "A smart waste management with self-describing objects," in *The Second International Conference on*

Smart Systems, Devices and Technologies (SMART'13), (eds) Leister, W., Jeung, H., and Koskelainen, P. (IARIA: Rome).

[19] Longhi, S., Marzioni, D., Alidori, E., Buo, G. D., Prist, M., Grisostomi, M., et al. (2012). "Solid waste management architecture using wireless sensor network technology," in *2012 5th International Conference on New Technologies, Mobility and Security (NTMS)*, 1–5.

[20] Hong, I., Park, S., Lee, B., Lee, J., Jeong, D., and Park, S. (2014). Iot-based smart garbage system for efficient food waste management. *The Scientific World Journal,* 2014:646953.

[21] Telegram Messenger, L. *Telegram,* 3, 2015.

[22] Wood, G. (2014). "*Ethereum: a secure decentralised generalised transaction ledger,*" Ethereum Project Yellow Paper, 151.

[23] Gavin Wood, A. B. L. H. Y. H., Reitwiessner, C. (2014). *Solidity.*

[24] Truffle. Available at: http://truffleframework.com/.

[25] Karagiannis, V., Chatzimisios, P., Vazquez-Gallego, F., and Alonso-Zarate, J. (2015). A survey on application layer protocols for the internet of things. *Transaction on IoT and Cloud Computing*, 3(1), 11–17.

[26] O'Dwyer, K. J., and Malone, D. (2014). Bitcoin mining and its energy footprint. *ISSC 2014/CIICT 2014*, Limerick.

[27] Klimova, A., Rondeau, E., Andersson, K., Porras, J., Rybin, V., and Zaslavsky, A. (2016). An international master's program in green ict as a contribution to sustainable development. *Journal of Cleaner Production*, 135, 223–223.

[28] Funded Horizon, E. (2020). *Program.* biotope [accessed May 18, 2017].

[29] Collina, M., Bartolucci, M., Vanelli-Coralli, A., and Corazza, G. E. (2014). "Internet of things application layer protocol analysis over error and delay prone links," in *2014 7th Advanced satellite multimedia systems conference and the 13th signal processing for space communications workshop (ASMS/SPSC)*, 398–404.

17

Automation of Control Processes in Specialized Pyrolysis Complexes Based on Industrial Internet of Things

Yuriy Kondratenko[1], Oleksiy Kozlov[2], Andriy Topalov[2], Oleksiy Korobko[2] and Oleksandr Gerasin[2]

[1]Petro Mohyla Black Sea National University, Mykolaiv, Ukraine
[2]Admiral Makarov National University of Shipbuilding, Mykolaiv, Ukraine
E-mail: yuriy.kondratenko@chmnu.edu.ua; oleksiy.kozlov@nuos.edu.ua; topalov_ua@ukr.net; oleksii.korobko@nuos.edu.ua; oleksandr.gerasin@nuos.edu.ua

This paper describes Industrial Internet of Things (IIoT) system, designed by the authors, for monitoring and automation of control processes in the specialized pyrolysis complexes (SPC) for municipal polymeric waste (MPW) utilization. The functional structure and main components of the generalized specialized pyrolysis complexes are given. Proposed by the authors, the approach is based on the usage of remote Programmable Logic Controller (PLC)-based control units that can perform local control operations along with the cloud-based server that is used for continuous monitoring and optimization of PLCs' parameters. The specific example of the proposed control system for pyrolysis complex based on the IIoT approach and industry 4.0 aspects is presented. Considerable attention is given to particular qualities of the functional structure, software, and hardware implementation as well as multilevel human–machine interface (HMI) for constant monitoring of working process of the developed control system.

17.1 Introduction

Deficit and high cost of natural organic fuel as well as environmental pollution by harmful waste are major contemporary energy and environmental problems in many countries around the world. The use of alternative fuels is the most effective approach for solving such problems. The implementation of municipal polymeric waste utilization technologies is one of the ways for obtaining alternative fuels. The most effective municipal polymeric waste utilization technology is the multiloop circulating pyrolysis (MCP) technology that allows the total utilization of the entire volume of the polymeric solid waste in environmentally friendly and energy saving modes [1, 2]. Additionally, the usage of this technology allows obtaining (at the output) the alternative fuel (liquid and gaseous fractions), which can be further used in such domains, as fuel and petrochemical industries, electric power production, etc. [1–3]. The main drawback of the existing technological equipment for MPW disposal in accordance with the MCP technology is the low level of automation, which leads to its inefficient use, inability to work in the optimal modes, and consequently, lower quality of output obtained products.

The specialized pyrolysis complexes, which are used for the implementation of the MCP utilization technology, are complex multicomponent control objects with randomly changing conditions of functioning. Measurement, monitoring, and control of their operating parameters cause several difficulties due to the peculiarities of the MCP technological process. Thus, for the effective operation with high energy/economic indicators, the given SPCs should use the advanced monitoring and control systems (typically with a hierarchical branched structure) based on data, received during the execution of the current technological processes in real-time mode [2, 3]. Automated posts of operator, process controllers, industrial computers, and facilities for industrial networks implementation are the main components of such systems. Additionally, compatibility of software with most digital controllers, presented on the market, is the important problem arisen at the designing of SPCs' monitoring and control systems. Thus, the user has a wide range for choosing of the appropriate hardware able to integrate with the operating system or multiple operating systems. Nowadays, Supervisory Control And Data Acquisition (SCADA) systems [4–7] are successfully used in this area. Focus on the industrial sector requires such systems to have advanced performance stability and reliability. Requirements for cybersecurity and risk assessment methodology for industrial information control systems are adapted from the requirements for IT systems. This category has published by a large number of National Institute of Standards and Technology (NIST) guidance documents.

Among them, NIST SP 800-82 describes the difference between IT systems and information control systems and provides guidance for protecting systems, including SCADA systems, distributed control systems, and other systems that perform control functions. Hot standby replication, constant data backup, and real-time control of equipment working process increase the reliability of SCADA systems. Modular architecture along with standardized hardware and software decreases the cost and terms of system maintenance.

Intensive developments of computer technologies and automatic control theory as well as the practical results of complicated technological processes' automation in industry produce SCADA conception [4, 5]. A high level of automation is achieved by SCADA technology application at solving the tasks of the monitoring and control systems designing, such as: data acquisition, transmission, processing, display, and storage. HMI is an important part of any SCADA system. Clarity and completeness of the appearing on the screen information, the presence of control means, and built-in decisions support features provide friendliness of HMI. The efficiency of a dispatcher interaction with the system increases and his possible critical control errors substantially reduce thanks to the above-mentioned SCADA systems' advantages.

However, today, the task of processing operational information in remote workplaces and controlling the technological process with the help of the Internet and wireless technologies requires substantial expansion of the functions of the existing SCADA systems. Almost all leading companies-developers of SCADA systems are moving to the concept of Industry 4.0 [8] and actively develop products in the field of Industrial Internet of Things (IIoT).

The aim of this work is to develop the remote, dependable control system for specialized pyrolysis complex with the use of industry 4.0 approach. Thus, the main objectives can be listed as follows: (a) analyze the possibilities and practicability of IIoT approach implementation into SPC control system; (b) design the software and choose appropriate hardware for system maintaining the focus on the dependability and scalability of the designed control system; and (c) conduct series of tests and analyze results of the system implementation.

17.2 Industrial Internet of Things Approach and Its Implementation

In recent years, innovations in hardware, connectivity, big data analytics, and machine learning have converged to generate huge opportunities for industries. Hardware innovations mean that sensors are cheaper, more powerful, and run longer on battery life. Connectivity innovations mean that it is cheaper and easier to send the data from these sensors to the cloud. Big data analytics and machine-learning innovations mean that once sensor data are collected, it is possible to gain incredible insight into manufacturing processes.

These insights, also known as industry 4.0 concept, can lead to massive increases in productivity and drastic reductions in cost. Whatever is being manufactured, it can be done faster, with fewer resources, and at lower cost.

The Internet of Things (IoT) enables significant transformation across multiple market segments, from consumer, enterprise, agriculture, healthcare, manufacturing, and utilities to government and cities. IIoT (a subset of the larger IoT) focuses on the specialized requirements of industrial applications, such as manufacturing, oil and gas, utilities, and others.

The definition of IIoT has different meanings, but IIoT can be considered as a global system for collecting and exchanging data where physical objects and devices are interconnected via cloud-based web servers. Moreover, IIoT plays a decisive role in the further development of the info-communication industry. This is confirmed both by the position of the International Telecommunication Union and the European Union in this matter, and by the inclusion of the IoT in the list of breakthrough technologies in the United States, China, and other countries [8, 9].

Due to IIoT technologies, manufacturing companies will be able to optimize everything – from warehouse work to performance directly to production tasks, if every industrial construction, vehicle, and even instrument will be equipped with sensors and will regularly send a report on their condition, location, and other characteristics. For example, a utility company can save large sums by continuing to maintain the reliability and integrity of the grid instead of purchasing new equipment, since real-time data and predata from various sources of information, such as output power of distributed generators, current load of electrical grids, predicted and statistical use of smart household appliances processed by the Internet-based definition of things. Planning specialists and engineers in the field of industrial transportation use the IIoT technology to process data from sensors to analyze the transport patterns of trucks and loaders by improving the flow of traffic in real time. In the system of water supply, IIoT technologies, together with pressure and flow sensors, are used to find out the general state of the water supply system and, if necessary, to carry out repairs on time.

Consequently, the introduction of IIoT technologies in SCADA enables the creation of a highly effective system for monitoring and managing technological processes that will work through the Internet. Today, global manufacturers of SCADA systems consider IIoT for the implementation of monitoring and management systems in a cloud server with the display of real-time imitators based on web technologies. In this case, SCADA monitoring and control are implemented through a usual browser acting as a thin client. Similar systems include IIoT server and client terminals: personal computers (PCs), personal digital assistants, or smartphones [8–11]. Customers can connect to the IIoT server by Internet/Intranet and interact with applied automation tasks through web or WAP pages [12].

In particular, SIEMENS developed the WebNavigator package for the WinCC SCADA system; another example is the Adastra company that developed the Trace Mode Data Center web server for the SCADA system Trace Mode 6. All of these solutions provide access to the project for remote clients. Thin client technology allows you to view and make adjustments to operational and archival information from any remote workplace through any browser.

It should be noted that the cloud-based IIoT design approach can lead to significant decrease of system cost along with implementation of advanced control algorithms. This can be achieved by replacement of expensive PLC's that are currently using on-site with cheaper and less powerful IIoT-based controllers. For example, ICP DAS is actively cooperating with world-class cloud technology leaders and launches IIoT controllers with support for cloud-based servers, Microsoft Azure and IBM Bluemix. These controllers will implement basic control algorithm and will receive any adjustments and corrections to their program from the main cloud-based server. Thus, the main server will implement advanced control algorithms and provide additional control and optimization layers, such as parametric optimization through the real-time data analysis and prediction techniques that cannot be implemented on PLC-based systems.

The main difficulty of using the IIoT approach in the industry lies in the safety of the operation of industrial equipment such as lathes, load-lifting cranes, conveyors, heating shades, etc. For reliable operation of this kind of equipment, the real-time high-quality data processing technologies should be applied [7–10]. Any technical malfunctions, inaccuracies in the questionnaire of sensors, or other mistakes may result in improper handling of technological processes or accidents that could harm human health, reduce the productivity of the enterprise, and cause economic losses to the whole.

Therefore, security issues in the IIoT play a significant role. The sphere of information security is actively developing with the use of ever more powerful encryption tools, highly reliable firewalls, and a variety of Virtual Private Network (VPN) technologies. However, the reliability of the data obtained is difficult to assume as 100%, due to the lack of visual control over the course of the technological process by the remote operator. In this regard, recently, in the IIoT, considerable attention is paid to video monitoring of technological processes and objects. This is due to the technological advances of visualization and increased bandwidth of communication channels. The introduction of such systems enables monitoring of objects (e.g., plant sites), remote diagnostics, which allows them to achieve higher quality products while reducing equipment downtime (which can be identified promptly with appropriate technical personnel qualifications). Existing industrial solutions consist in the use of video web servers (for example, SIVICON of SIEMENS company), or universal web-servers (for example, the web-studio INDUSTOFT).

Thus, the IIoT systems should include highly efficient software and hardware means for the implementation of specialized algorithms of monitoring and automatic control. In addition, such systems should have an increased level of reliability, performance, and information security. Thus, the development of the IIoT systems for the SPCs monitoring and control processes automation is rather an important task.

17.3 Generalized IIoT-Based Pyrolysis Complex Control System

The functional structure of the generalized SPC's IIoT system, proposed by the authors, is shown in Figure 17.1, where the following notations are accepted: UL – upper level; AL – average level; LL – lower level; PLC1–PLC5 – programmable logic controllers; GAOM1–GAOM5 – groups of analog output modules; GDAM1–GDAM5 – groups of data acquisition modules; GA1–GA5 – groups of actuating mechanisms; GS1–GS5 – groups of sensors; U_{A1}–U_{A5} – control signals of actuating mechanisms; U_{S1}–U_{S5} – sensors' output signals; Y_1–Y_5 – output values of actuating mechanisms; X_1–X_5 – SPC process parameters; WLU – waste loading unit; SGM – steam–gas mixture; SR – the solid residue; WULU – waste unloading unit; MCS – multiloop circulation system; CFSU – condensation and fractions separation unit; GF – gaseous fractions; and LF – liquid fractions.

The SPC IIoT system is designed for monitoring, automatic control, and visualization of the SPC main process parameters in real-time mode at the

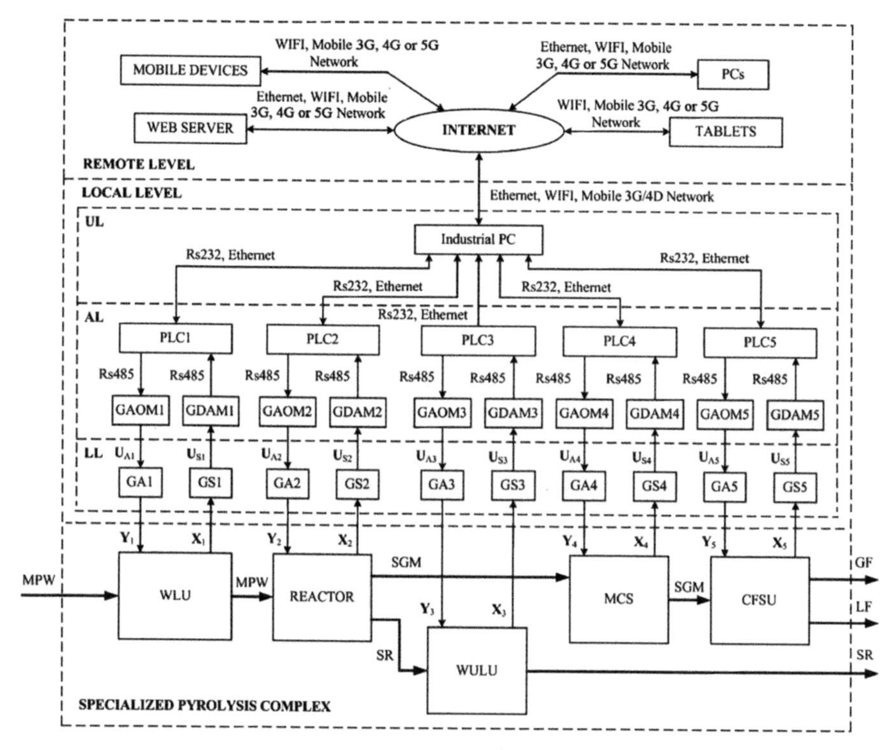

Figure 17.1 The functional structure of the generalized SPC's IIoT system.

local level and at the remote level via the Internet. It provides visual display of information about the state of the SPC main components with clear indication of current states and emergencies as well as the ability of integration into existing large-scaled systems of industrial plants, enterprises, or factories. Furthermore, the SPC IIoT system allows automatic data sending in case of emergency to selected users of the enterprise network. In addition, it has modern and highly integrated programming environment, which is flexible and easy to extend, flexible network designs and ability of connection of online expert system for more detailed data analysis.

The proposed generalized SPC IIoT system has two levels of automatic control and monitoring: local level and remote level. Local level, in turn, is divided into three main hierarchical levels:

1) lower level – the level of sensors and actuating means of the SPC, that is involved in the performance of an MPW utilization technological process;

2) average level – the level of the peripheral monitoring and control units (includes groups of data acquisition and analog output modules as well as PLCs);
3) upper level – the HMI level (includes industrial PC).

Lower level (Figure 17.1) includes the automatic control subsystems (ACSS) performing automatic regulation of the given values of SPC main controlled parameters in the presence of various operation disturbances [2, 7].

Industrial PC of the upper level runs as the operator station. It includes the specialized HMI in the form of a control panel of the MPW utilization process with the images of the SPC technical parameters measured by different sensors [2] and SPC main components. Additionally, the given values of the control parameters (in fact, they are inputs of the SPC's ACSS) are set on the industrial PC according to a control program. PLCs of the average level are processing data (received from the groups of sensors) and sending the information to the industrial PC. In addition, they get the set values of the control parameters of MPW utilization technological process and perform automatic control functions. Signals from sensors are received by groups of data acquisition modules and then transmitted to the PLCs in an appropriate form for subsequent processing [7]. Groups of analog output modules receive digital control signals from the PLCs and produce required analog control signals, which are directly sent to the SPC's actuating mechanisms.

The monitoring and automatic control operations in the remote level are implemented by means of powerful web servers, remote computers, tablets, and different types of mobile devices [2, 7]. In this case, the industrial PC with the help of wired (Ethernet) or wireless (WiFi, mobile network 3G, 4G, and others) connection technologies and family of protocols TCP/IP exchange data via the Internet with a specialized web server that is placed on a powerful computer. The specialized web server, in turn, receives data process parameters and provides web access to other users (remote computers, tablets, and mobile devices). Moreover, using the web server, the access to the MPW utilization technological process data can be given from any PC of the industrial plant, enterprise, or factory, that is running under any operating system (Windows, Linux, Mac OS, etc.), and if desired, from any PC in the world connected to the Internet.

In turn, the specialized HMI in the form of a control panel of the MPW utilization technological process is installed on the server and on the all computers of the industrial plant, enterprise, or factory, where it is necessary, with all available functions of monitoring and control. Additionally, the specialized server can implement the function of automatic data sending to all the necessary users in case of emergency and include the online expert system for data analysis and forming the further control goals.

The mathematical models of the SPC main components are developed and presented by the authors in [3] and [13]. The ability of the load level's measurement for pyrolysis reactor by the implementation of indirect method [2] is the important particularity of the developed SPC IIoT system.

The direct measurements of reactor load level L_{R} values by the level sensor application cause many problems associated with the conditions of the MPW utilization process, in particular, the level sensor operating conditions in the pyrolysis reactor. Thus, high temperature (more than 500°C), high viscosity, the aggregate state and density changings of wastes during the operating process, wastes boiling, and, as a result, the possibility of wastes' solidification and sticking to the sensing element [2] are the main limitations of the level sensor using in the reactor cavity.

The indirect level measuring method helps to solve this problem. It is based on the loading level L_{R} value calculation by measuring of another SPC parameters [2]. The proposed SPC IIoT system uses flow rate values of fuel gaseous Q_{GF} and liquid fractions Q_{LF}, temperature value T_{R} in the working area of the reactor, as well as the linear movement value Y_{HD} of the loading hydraulic actuator piston and calculates the current value of load level of the pyrolysis reactor by the following equations:

$$L_{\mathrm{R}} = \frac{M_{\mathrm{LW}} - (M_{\mathrm{GF}} + M_{\mathrm{LF}})}{\rho_{\mathrm{WR}}\,(T_{\mathrm{R}})\,S_{\mathrm{R}}}; \tag{17.1}$$

$$M_{\mathrm{LW}} = \rho_{\mathrm{LW}} S_{\mathrm{P}} Y_{\mathrm{HD}}; \tag{17.2}$$

$$M_{\mathrm{GF}} = \rho_{\mathrm{GF}} V_{\mathrm{GF}} = \rho_{\mathrm{GF}} \int Q_{\mathrm{GF}} dt; \tag{17.3}$$

$$M_{\mathrm{LF}} = \rho_{\mathrm{LF}} V_{\mathrm{LF}} = \rho_{\mathrm{LF}} \int Q_{\mathrm{LF}} dt, \tag{17.4}$$

where M_{LW} – the mass of waste, which is loading into the reactor; M_{GF} – the fuel gaseous fractions mass, which stands out during the MPW utilization process; M_{LF} – the fuel liquid fractions mass, which stands out during the SPC working process; ρ_{LW} and ρ_{WR} – the values of the MPW density which are loaded by the hydraulic piston actuator and MPW which are located inside the reactor at functioning, respectively; ρ_{GF} and ρ_{LF} – the fuel gaseous and liquid fractions density values, respectively; V_{GF} and V_{LF} – the values of volume of the fuel gaseous and liquid fractions, respectively; and S_P and S_R – the values of cross-sectional area of the hydraulic piston actuator and the reactor, respectively.

Values of density ($\rho_{LW}, \rho_{WR}, \rho_{GF}$, and ρ_{LF}) are determined experimentally depending on the MPW type (polystyrene, polypropylene, and polyethylene) [2].

Furthermore, the proposed SPC IIoT system can implement intelligent principles and technologies of automatic control based on the theory of artificial neural networks and fuzzy logic [14–16]. This allows achieving higher quality indicators and accuracy at automatic control of main MPW utilization process parameters. The monitoring and automatic control systems, developed based on the artificial neural networks and fuzzy logic, are currently successfully applied in such areas as: technological processes and transport control, financial management, medical and technical diagnostics, pattern recognition, stock forecast, etc. [13, 14, 17].

As a concrete example of the given above-generalized SPC IIoT system implementation, let us consider the IIoT system designed by the authors for the specialized pyrolysis complex MCP-5.

17.4 Implementation of the IIoT System for the SPC MCP-5

IIoT system for the specialized pyrolysis complex MCP-5 is implemented with the modular structure that includes monitoring and automatic control of main technological parameters of the MPW utilization process in the local and remote levels using powerful tools for data storage and analysis. This system provides execution of the following functional tasks: (a) monitoring, automatic control, and visualization of the pyrolysis complex main process parameters in real-time mode at the local level and at the remote level via the Internet; (b) visual display of information about the state of the SPC main components with clear indication of current states and emergencies; (c) the ability of integration into existing large-scaled systems of industrial plants, enterprises, or factories; and (d) automatic data sending in case of emergency to all users of the enterprise.

17.4.1 Functional Structure of the IIoT System for the SPC MCP-5

The structure of the proposed IIoT system by the authors for the specialized pyrolysis complex MCP-5 is shown in Figure 17.2. The developed Web

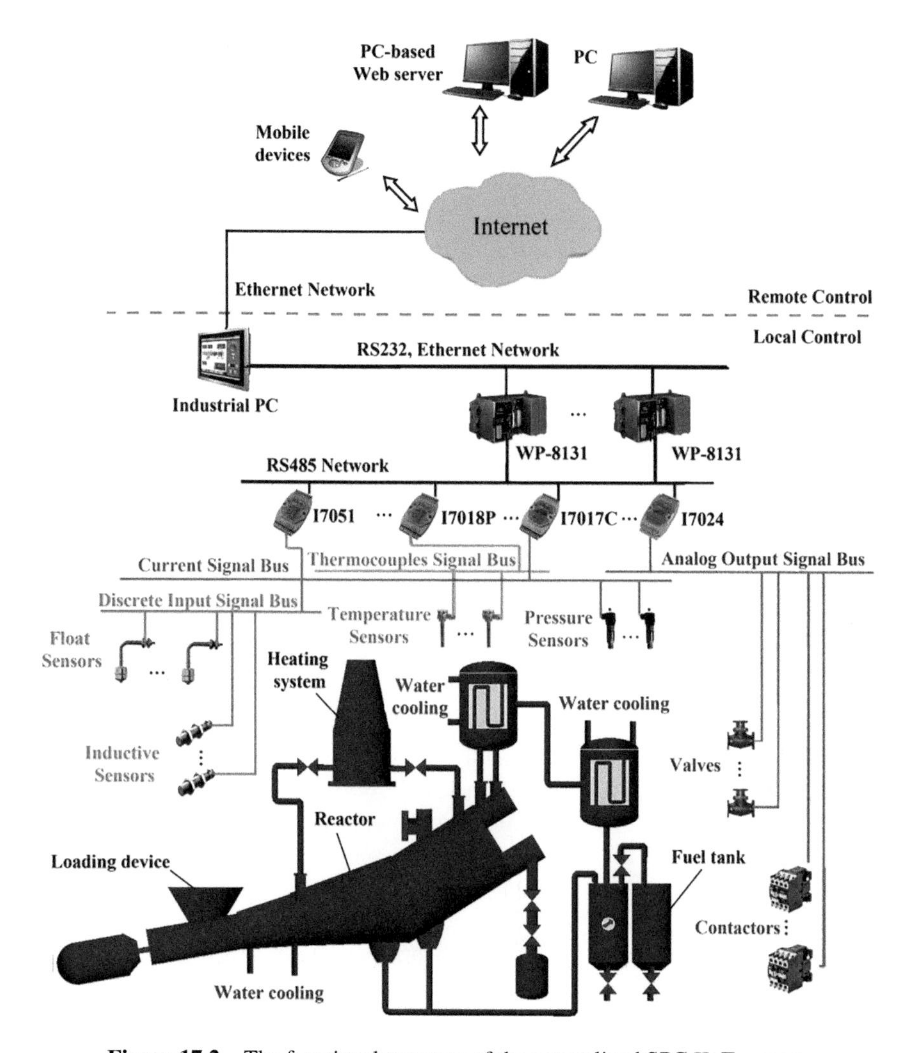

Figure 17.2 The functional structure of the generalized SPC IIoT system.

IIoT system structure (Figure 17.2) is built considering the necessity of the simultaneous measurement of various physical quantities, functioning in real-time mode and the need to build a modular system that has the ability of integration into existing large-scaled systems. In turn, the designed structure allows to measure and control the main technological parameters of the MPW utilization process, such as: pressure, temperature and load level values in the reactor, fuel liquid fractions' level in the fuel tank, and a set of temperature values in key points of SPC. Optionally, the given IIoT system could be additionally equipped with two float inductive sensors (upper and lower level values indicating) and one radar sensor (intermediate level values determining) for load level value measurement in the SPC reactor that can increase system dependability and fault tolerance.

According to Figure 17.2, the main actuating mechanisms of SPC include the water-cooling pumps, the hydraulic drive piston for MPW loading into the reactor, fuel pumps, normally closed Jaksa D224 valves, AC gear motors as a part of MPW unloading unit, fans for cooling, and flue gases blowing. The mentioned actuating mechanisms are powered from the AC mains using contactors PML 1160M type.

17.4.2 Description of the MCP-5 IIoT System Hardware

For hardware implementation of the above-mentioned system for monitoring and automatic control of SPC main process parameters, the authors used the next measuring means: pressure sensors PD100 type, thermo-couples of the K- and L-types, inductive sensors SN04-N and PIP-8-3 types, and discrete float level sensors PDU 1.1 type.

ICP DAS I-7018Z modules are used as the data acquisition modules for the thermocouples and pressure sensors readings. The ICP DAS I7041PD modules for 14 inputs are used as the data acquisition modules for the discrete input signals. In addition, the ICP DAS I-7061D modules with 12 relay power outputs are used for actuating mechanisms automatic control. ICP DAS I7561 modules provide USB to RS-232/422/485 converting for modules connection to PC at testing and debugging of all systems [18, 19].

The control unit of the IIoT system, developed for specialized pyrolysis complex MCP-5 using the above-mentioned devices is presented in Figure 17.3.

PLCs ICP DAS WP-8131 are used in the current IIoT system as the main execution units. The specifications of the PLC are: CPU PXA270 or Comparable (32 bits, 520 MHz), RW Memory SDRAM 128 Mb, Zeropower SRAM 512 Kb, Flash memory 128 Mb, and also supports storage map of 16 GB microSD [19]. PLC includes VGA, Ethernet, USB 1.1, RS232, RS485, and RS 482 interfaces.

Communication between data modules and PLC is made with the use of DCON protocol and data exchange between the data server and PLCs is made with the use of proprietary SCADA TRACE MODE data exchange mechanism.

Figure 17.3 Control unit of the specialized pyrolysis complex MCP-5 IIoT system.

17.4.3 Description of the MCP-5 IIoT System Software

For controlling and managing the processes in specialized pyrolysis complexes at the local level, a modern computerized SCADA-based system, which has a hierarchical structure with a vertical form of control of the elements included in them, was developed. In fact, such systems can be represented as pyramids, each level of which is subject to a higher level. The upper level defines the task (plan) for the lower level and, in turn, depending on the state of the lower level, the plan can be adjusted.

The use of SCADA systems allows to gather information about the SPC process, provide an interface with the operator, accumulate the database, and automatically process the data. SCADA systems' SPCs present the following basic requirements: work in real time, reliability of the system (technological and functional); safety management; processing accuracy and data representation; and simplicity of system expansion.

The SPC control algorithms in the PLC subsystem are realized using the tools, which are provided by the SCADA-system TRACE MODE 6, which is a versatile and compact SCADA system (all system editors are called from one application). It contains a set of programming languages, which enable the programming of corresponding software and hardware. These languages are compatible with the IEC 61131-3 guidelines.

A special web server TRACE MODE Data Center [20] is used for remote control of SPC processes. The main objective of this software product is to provide users with the ability to control the process through a web browser using Internet/Intranet networks or wirelessly (GPRS, Wi-Fi, Bluetooth, etc.). This new web server enables to access real-time data from any computer using different operating systems (Windows, Linux, Mac OS, etc.). With the help of TRACE MODE Data Center, the system operator can not only monitor the process, but also manage it online – changing the task control to the temperature and pressure regulators on the screen. It is also possible to generate reports on the technological process.

TRACE MODE Data Center has a security system integrated into the TRACE MODE 6 security system. Therefore, only authorized users can access the web server data. The rights of various users are flexibly configured and administered (in real time). To ensure more safety, TRACE MODE Data Center supports reliable encrypted VPN connection from external PCs connected to the Internet.

TRACE MODE Data Center licensing is based on concurrent connections of the client devices (PCs, pocket PCs, mobile phones, etc.). The program is available in versions for 4, 8, 16, 32, 64, and unlimited workplaces.

When using TRACE MODE, the data center does not require the installation of Web servers from other manufacturers (such as Microsoft IIS), which distinguishes this system from the solutions used by some competing SCADA developers. Moreover, TRACE MODE Data Center requires a separate PC.

In the given SPC IIoT system, the TRACE MODE Data Center serves the local SPC control system. The TRACE MODE Data Center receives real-time data from the local system on the base of TRACE MODE 6 over TCP/IP and provides access to data via the Internet for computers and mobile devices (smartphones). To control the current technological process, the operator logs on to the Web browser page that supports the virtual Java machine and opens the SPC project based on IIoT [21–23]. Moreover, the graphical capabilities of the real-time monitor on the web page almost coincide with the real-time monitor of the local system.

17.4.4 HMI of the MCP-5 IIoT System

In the IIoT system for remote monitoring and automatic control of SPC main process parameters [24, 25], the HMI is realized using the tools, which are provided by the basic version of the SCADA system TRACE MODE 6 [20]. The designed HMI has a multi-window interface [26, 27]. The main screen (Figure 17.4) provides the visualization of the main indicators of the control system process parameters on the operator control display as well as grants an ability to set needed ranges of the parameters.

The indication of SPC modes is provided by changing the color of the displayed value from red (means manual) to green (means automatic, placed in the upper left corner in Figure 17.4). The current state of the discrete sensors is displayed in the same color way: (a) the green color corresponds the open state of discrete sensors (there is no liquid fuel at this level or the hydraulic drive piston at this position) and (b) the red color corresponds to the closed state (the sensor has been submerged into the liquid or the hydraulic drive piston has achieved this position). Additionally, the designed software includes the graph screens, which indicate the dynamic of main indicators' changes and are called by clicking on the "Graphs" button. For example, the graph screen, presented in Figure 17.5, shows the dynamic of the temperature changes of the SPC reactor main control points. The graphs shown in the figure are obtained at the process of the reactor initial heating.

The required data processing algorithms for values' conversion, range checking, temperature, and actuating mechanisms' automatic control are executed directly in the PLC [28–30] using the implemented programs in FBD and ST languages. All control algorithms of developed system are implemented using the SCADA TraceMode 6 software. It should be noted that the developed SCADA Trace Mode 6 HMI of the MCP-5 IIoT system can be easily reconfigured directly in the PLCs using the program possibilities of Micro Trace Mode 6.

As a result, the IIoT system developed by the authors for the specialized pyrolysis complex MCP-5 allows increasing the accuracy and reliability of measurements due to self-diagnosis and elimination of operator errors at the system setting up and calibration. In addition, it gives the opportunity to reach a zero value of the relative deviations of the real values from the set values for main controlled parameters of the complex (temperature, pressure, and load level values in the reactor, level of the fuel liquid fractions in the fuel tank, and

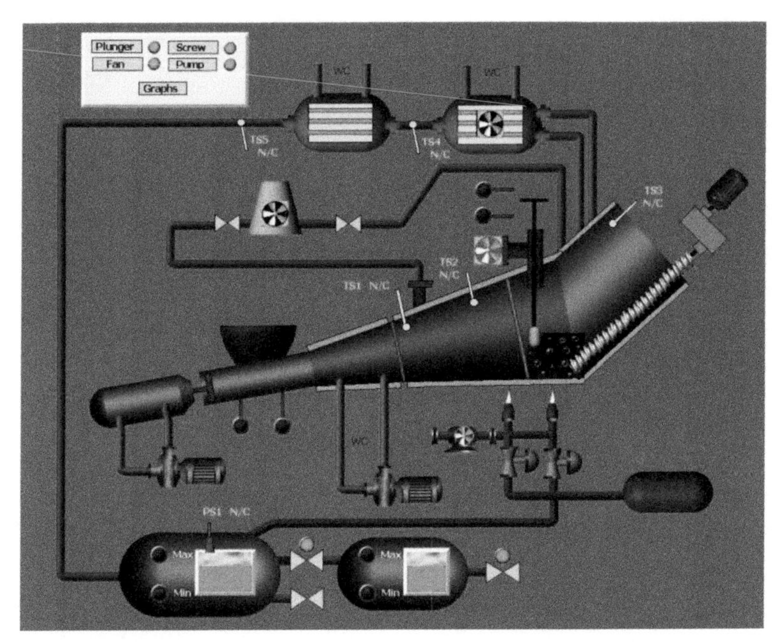

Figure 17.4 Main screen of the HMI of the remote SCADA system for the specialized pyrolysis complex MCP-5.

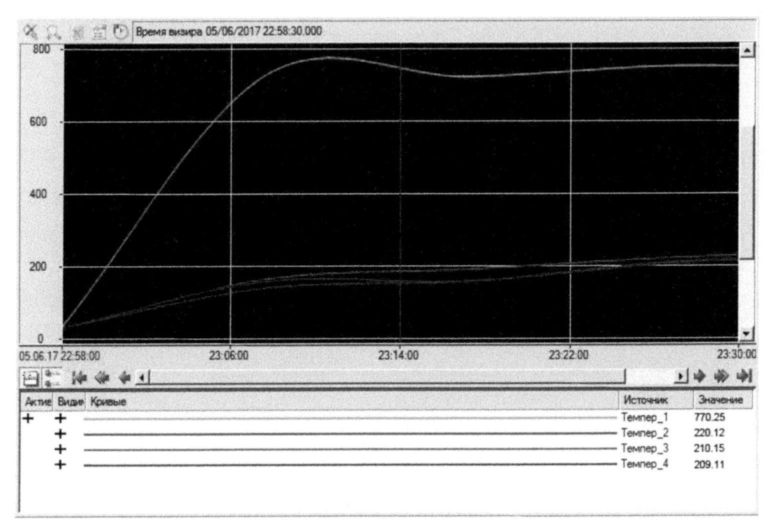

Figure 17.5 Graph screen of the temperature changes in the main control points of the MCP-5 reactor.

a set of temperature values in the key points of the SPC). This makes possible to increase the SPC productivity by 17% and to reduce the maximum value of its total power consumption by 12.5%. Moreover, the advantage of this IIoT system is the ability to operate with the personnel with low qualification, since all processes of control and data analysis are automated.

17.5 Conclusion

In this work, the authors developed the control system, based on the IoT methodology, for the monitoring and control processes' automation of the SPC. The proposed system allows monitoring, automatic control, and visualization of the SPCs' main process parameters with high accuracy in real-time mode at the local level and at the remote level via the Internet. In addition, the given IIoT system provides visual display of information about the state of the SPCs' main components with clear indication of current states and emergencies as well as the ability of integration into existing large-scaled systems of industrial plants, enterprises, or factories. The implemented reactor load level values calculations by the indirect method don't require complex software in these systems.

As the example of the proposed systems' application, the authors developed the IIoT system for the specialized pyrolysis complex MCP-5 that has modern and integrated programming environment, flexible network designs, and ability of connection of online expert system for more detailed data analysis.

The designed HMI of the IIoT system for the specialized pyrolysis complex MCP-5 gives the opportunity to show all the necessary information on the operator screen. In addition, it provides the ability to identify and expose the event of an emergency due to the developed control algorithms in real-time conditions.

As a result, the developed IIoT system for the MCP-5 complex provides high precision control of the SPC operating processes in the real-time mode, monitoring, and automatic control of its current technological parameters with high-quality indicators that lead to substantial energy and, respectively, increasing economic efficiency of the given SPC.

In future, the authors will consider the implementation of the cloud-based IIoT control system that will use the dedicated server for implementation of advanced control algorithms based on the real-time processing and analysis of collected data and low-power local control devices for data collection and control signals' generation.

In further research, it is planned to implement the proposed (for SPC automation [24, 25]) IIoT's approach for increasing efficiency of different real-time control and decision-making processes, objects, and systems, in particular, for automation of (a) green IT and fault-tolerant complex systems [21–23] in engineering, management, and economics, (b) remote level control of liquid tanks and reservoirs [26, 27], and (c) remote monitoring and control of thermoacoustic engines, refrigerators, and plants [28–30], non-stationary robotic objects and complexes [31], as well as implementation to marine practice for marine logistics [32], planning and optimization of bunkering operations at sea, increasing navigation safety of ships, and stabilization of marine mobile objects such as floating docks and ships [33, 34] .

References

[1] Ryzhkov, S. S., and Markina, L. M. (2007). "Experimental researches of organic waste recycling method of multiloop circulating pyrolysis," in *J. of Collected Works of NUS*, 5, 100–106.

[2] Kondratenko, Y. P., Korobko, O. V., and Kozlov, O. V. (2016). "PLC-based systems for data acquisition and supervisory control of environment-friendly energy-saving technologies," in *Green IT Engineering: Concepts, Models, Complex Systems Architectures, Studies in Systems, Decision and Control*, eds Kharchenko, V. Kondratenko, Y., and Kacprzyk, J. (Berlin: Springer International Publishing), 247–267.

[3] Kondratenko, Y. P., and Kozlov, O. V. (2012). "Mathematic modeling of reactor's temperature mode of multiloop pyrolysis plant," in *Lecture Notes in Business Information Processing: Modeling and Simulation in Engineering, Economics and Management*, (eds) Engemann, K. J., Gil-Lafuente, A. M., and Merigo, J. M. (Berlin: Springer-Verlag), 178–187.

[4] Aydogmus, Z., and Aydogmus, O. (2009). "A web-based remote access laboratory using SCADA," *IEEE Transactions on education*, 52, 126–132.

[5] Dulău, I. L., Abrudean, M., and Bică, D. (2015). "SCADA simulation of a distributed generation system with storage technologies," in *Procedia Technology*, 19, 665–672.

[6] Kim, H. J. (2012). "Security and vulnerability of SCADA systems over IP-based wireless sensor networks," in *Hindawi Publishing Corporation International Journal of Distributed Sensor Networks*.

[7] Erez, N., and Wool, A. (2015). "Control variable classification, modeling and anomaly detection in Modbus/TCP SCADA systems," in *Int. J. of Critical Infrastructure Protection*, 10, 59–70.

[8] Vermesan, O., and Friess, P. (eds). (2018). *Digitising the industry internet of things connecting the physical, digital and virtual worlds.* Denmark: River Publishers.

[9] Louis, E. F. Jr. (2017). *Electronics explained fundamentals for engineers, technicians, and makers*. New York, NY: Elsevier.

[10] Yang, G., Liang, H., and Wu, C. (2013). "Deflection and inclination measuring system for floating dock based on wireless networks," in *Ocean Engineering*, 69, 1–8.

[11] Topalov, A., Kozlov, O., and Kondratenko, Y. (2016). "Control processes of floating docks based on SCADA systems with wireless data transmission," in *Perspective Technologies and Methods in MEMS Design: Proceedings of the International Conference MEMSTECH 2016*, Lviv-Poljana, 57–61.

[12] Pasha, S. (2016). "Thingspeak based sensing and monitoring system for IoT with matlab analysis," in *International Journal of New Technology and Research (IJNTR)*, 2(6), 19–23.

[13] Kondratenko, Y. P., Kozlov, O. V., Klymenko, L. P., and Kondratenko, G. V. (2014). "Synthesis and research of neuro-fuzzy model of ecopyrogenesis multi-circuit circulatory system," in Advance Trends in Soft Computing, eds Jamshidi, M., Kreinovich, V., and Kazprzyk, J. (Berlin: Springer, Series: Studies in Fuzziness and Soft Computing), 312, 1–14.

[14] Zadeh, L. A. (1994). "The role of fuzzy logic in modeling, identification and control," in *Modeling Identification and Control*, 15(3), 191–203.

[15] Piegat, A. (2001). *Fuzzy modeling and control.* New York, NY: Physica-Verlag.

[16] Jang, J.-S. R., Sun, C.-T., and Mizutani, E. (1996). *Neuro-fuzzy and soft computing: a computational approach to learning and machine intelligence*. Upper Saddle River, NJ: Prentice Hall.

[17] Palagin, A. V., Opanasenko, V. M., and Kryvyi, S. L. (2017). "Resource and energy optimisation oriented development of FPGA-Based adaptive logical networks," in *Green IT Engineering: Components, Networks and Systems Implementation*, eds Kharchenko, V., Kondratenko, Y., and Kacprzyk, J. (Berlin: Springer), 195–218.

[18] Grafkin, A. V. (2010). *The principles of software control of ICP DAS I-7000 modules in the tasks of industrial automation*, Samara: SNTS RAN, 133.

[19] Kondratenko, Y. P., Kozlov, O. V., Korobko, O. V., and Topalov, A. M. (2017). "Internet of things approach for automation of the complex industrial systems," in *Proceedings of the 13th International Conference on Information and Communication Technologies in Education, Research, and Industrial Applications, Integration, Harmonization and Knowledge Transfer*, eds Ermolayev, V. et al. ICTERI'2017, CEUR-WS, 1844, Kyiv, 3–18.

[20] "TRACE MODE 6." *User Guide*. Adastra Research Group, Ltd., Moscow.

[21] Sklyar, V. (2017). "Vedic mathematics as fast algorithms in green computing for internet of things," in *Green IT Engineering: Components, Networks and Systems Implementation. Studies in Systems, Decision and Control*, eds Kharchenko, V., Kondratenko, Y., and Kacprzyk, J. (Cham: Springer), 105, 3–21.

[22] Atamanyuk, I., Kondratenko, Y., Shebanin, V., and Mirgorod, V. (2015). "Method of polynomial predictive control of fail-safe operation of technical systems," in *Proceedings of the XIIIth International Conference "The Experience of Designing and Application of CAD Systems in Microelectronics," CADSM 2015*, Polyana-Svalyava, 248–251.

[23] Atamanyuk, I., and Kondratenko, Y. (2015). "Computer's analysis method and reliability assessment of fault-tolerance operation of information systems ICT in education, research and industrial applications: integration, harmonization and knowledge transfer," in *Proceedings of the 11th International Conference ICTERI-2015*, eds S., Batsakis, H. C., Mayr, et al. (Lviv: CEUR-WS), 1356, 507–552.

[24] Kondratenko, Y. P., and Kozlov, O. V. (2016). "Mathematical model of ecopyrogenesis reactor with fuzzy parametrical identification," in *Recent Developments and New Direction in Soft-Computing Foundations and Applications. Studies in Fuzziness and Soft Computing*, eds L. A., Zadeh et al. (Berlin: Springer-Verlag), 342, 439–451.

[25] Kondratenko, Y. P., and Kozlov, O. V. (2012). "Mathematic modeling of reactor's temperature mode of multiloop pyrolysis plant," in *Modeling and Simulation in Engineering, Economics and Management*, in *International Conference MS 2012*, eds K. J. Engemann, A. M. Gil-Lafuente, J. L. Merigo, New Rochelle, NY, in *Proceedings* Lecture Notes in Business Information Processing (Berlin: Springer), 115, 178–187.

[26] Kondratenko, Y., Korobko, O., Kozlov, O., Gerasin, O., and Topalov, A. (2015). "PLC based system for remote liquids level control with

radar sensor," in *Proceedings of the 2015 IEEE 8th International Conference on Intelligent Data Acquisition and Advanced Computing Systems: Technology and Applications (IDAACS)*, Warsaw, 1, 47–52.

[27] Kondratenko, Y. P., Kozlov, O. V., Topalov, A. M., and Gerasin, O. S. "Computerized system for remote level control with discrete self-testing," in *ICTERI-2017, CEUR Workshop Proceedings Open Access*, 1844, 608–619. Available at: http://ceur-ws.org/Vol-1844/10000608.pdf

[28] Kondratenko, Y., Korobko, V., Korobko, O., Kondratenko, G., and Kozlov, O. (2017). "Green-IT approach to design and optimization of thermoacoustic waste heat utilization plant based on soft computing," *Green IT Engineering: Components, Networks and Systems Implementation. Studies in Systems, Decision and Control*, eds Kharchenko V., Kondratenko Y., Kacprzyk J. (Springer, Cham), 105, 287–311. DOI: 10.1007/978-3-319-55595-9_14.

[29] Kondratenko, Y. P., Korobko, O. V., and Kozlov, O. V. (2016). "Synthesis and optimization of fuzzy controller for thermoacoustic plant," in *Recent Developments and New Direction in Soft-Computing Foundations and Applications', Studies in Fuzziness and Soft Computing 342*, eds Lotfi A. Zadeh et al. (Berlin, Heidelberg: Springer-Verlag), 453–467. DOI: 10.1007/978-3-319-32229-2_31.

[30] Kozlov, O., Kondratenko, G., Gomolka, Z., and Kondratenko, Y. (2019). "Synthesis and optimization of green fuzzy controllers for the reactors of the specialized pyrolysis plants," *Green IT Engineering: Social, Business and Industrial Applications. Studies in Systems, Decision and Control*, eds Kharchenko V., Kondratenko Y., Kacprzyk J. (Springer Cham), 171, 373–396. DOI: 10.1007/978-3-319-55595-9_14.

[31] Tkachenko, A. N., Brovinskaya, N. M., and Kondratenko, Y. P. (1983). "Evolutionary adaptation of control processes in robots operating in non-stationary environments," *Mechanism and Machine Theory*, 18(4), 275–278. DOI: 10.1016/0094-114X(83)90118-0.

[32] Solesvik, M., Kondratenko, Y., Kondratenko, G., Sidenko, I., Kharchenko, V., and Boyarchuk, A. (2017). "Fuzzy decision support systems in marine practice," in *Fuzzy Systems (FUZZ-IEEE), 2017 IEEE International Conference on*, (IEEE), 9–12. DOI: 10.1109/FUZZ-IEEE.2017.8015471.

[33] Kondratenko, Y., Kozlov, O., Korobko, O., and Topalov, A. (2018). "Complex industrial systems automation based on the internet of things implementation," *Information and Communication Technologies in Education, Research, and Industrial Applications. ICTERI*

2017. Communications in Computer and Information Science, eds Bassiliades N. et al. (Springer Cham), 826, 164–187.

[34] Kondratenko, Yuriy P., Kozlov, Oleksiy V., Korobko, Oleksiy V., and Topalov, Andriy M. (2017). "Synthesis and optimization of fuzzy control systems for floating dock's docking operations," Book of Fuzzy Control Systems, Nova Science Publishers, Chapter 4, pp. 141–213.

18

Cloud-based IT Infrastructure for "Smart City" Projects

Oleksii Duda[1], Nataliia Kunanets[2], Oleksandr Matsiuk[1] and Volodymyr Pasichnyk[2]

[1]Ternopil Ivan Puluj National Technical University, Ternopil, Ukraine
[2]Lviv Polytechnic National University, Lviv, Ukraine
E-mail: oleksij.duda@gmail.com; nek.lviv@gmail.com; oleksandr.matsiuk@gmail.com; vpasichnyk@gmail.com;

The experience of cloud computing technology use in the "smart cities" projects is analyzed. In this case, the wide range of the interest group complexes for interested organizations, companies, firms, and a large population of the city residents concerning the delimitation of their duties and rights on formation and access to different information resources is defined. The main classes of problems solving of which requires the cloud information technology use while providing context-sensitive information services, which in turn needs the research of mechanisms for increasing their efficiency including functional Big Data profiling, security, segregation, processing, and storage, are investigated by the authors. The implementation of the "smart cities" projects based on information and communication technologies involves the integration and processing of different data from diverse sources for the search various relevant information. It is proposed that the use of systematic approach to information consolidation processes in the "smart cities" projects for "hard" – production-business and "soft" – socio-cultural, educational, scientific, and security domains is offered. It is emphasized that the main information flows in the "smart cities" are formed on the basis of data accompanying the various services supply. It is emphasized that the main information flows in the "smart cities" are formed on the basis of data accompanying the diverse services supply. During the investigations, the authors analyzed the multilayered architecture of the "smart city" information system for the accumulation and processing scale, multitype structured and poorly structured Big Data received from various smart sensors. The complex of basic characteristics and implementation of information technologies on the basis of cloud computing used in information-technological projects of "smart cities" is considered.

18.1 Introduction

The onrush of information technologies, the universal distribution of the data transmission, and the growth of their bandwidth capacity provide the improvement of the wide range of services offered to the residents and visitors of the large cities and megapolises.The "smart cities" concept as complex innovation system realizing the functions of receiving, storage, and analytical processing of data and open access to the data of various profiles concerned with the living standards' improvement and environmental protection in big cities is very important.

The "smart city" is the modern innovation environment where information and communication technologies are used for "intellectualization" of all spheres of the city and its residents' life. Under the "smart city" conditions, comfortable and technologically efficient services available anywhere at any time to the urban community due to any computer devices are provided. The city residents and visitors as consumers of intellectual services provided in the "smart city" require processing of big unstructured, poorly structured, and structured data formed as the informational result of the hyper-complex system which represents the modern city. In its turn, the urban authorities require the rapid and efficient processing of information flow using the methods and means of information and communication technologies in order to increase the urban environment management process efficiency. An active urban community and individual participation is the key component in the effective decision making process during the "smart city" planning and functioning [1]. The implementation of the "smart city" concept is intended to simplify and make easier the procedures of access to the information concerned with the characteristics and processes typical to the highly technological urban environment, especially to the information about the level and quality of local management, social-economic activities in megapolis, services provided, etc.

One of the basic information technologies providing implementation of such functions in the "smart cities" projects is particularly the cloud computing technology. The interest of municipalities and local administrations to their application for solving complicated problems of urban management is significantly growing [2]. Problems of reliability of the "smart city" model based on cloud computing are considered in [3]. Migration of smart city applications into cloud environment and peculiarities of their implementation are described in [4]. In [5], an image-retrieval method in a multicamera system of "smart city" based on cloud computing is given, while applications of mobile technologies and cloud computing for widespread municipal network of "smart city" are presented in [6]. The GIS spatial information sharing of "smart city" based on cloud computing is described in [7], and features of implementation and architecture of government affairs' service platform for "smart city" integrated on the basis of 3D GIS are investigated in [8]. In [9], approaches and challenges related to Big Data and machine learning in "smart cities" are considered, and importance of their integration on the basis of cloud platforms is underlined. The role of cloud computing in implementation of architecture of the "smart city" with the use of IoT devices is analyzed in [10], problems of security in integration of IoT devices and cloud computing are investigated, and a series of scientific articles are discussed [11]. The key requirements for cloud-based smart cities, such as security, scalability, reliability, productivity, and energy efficiency are considered in [12]. The priority of security issues of cloud datacenters is noted due to the possibility of DoS attacks, legislative and regional features of the delineation of rights and privileges of access to information collections, and the transfer of data outside of a separate data center. For these reasons, the conceptual and full-scale design of the architecture and structures of the cloud information and technology platform of the "smart city" is an actual direction of modern scientific research and requires detailed elaboration.

18.2 Socio-Communicative Component of the "Smart Cities" Projects

In the modern urban community, a stable, integrated complex of communication processes between citizens, their groups, and communities and representatives of the municipal government implementing data transmission due to information-analytical and communication means as well as making the effective managerial decisions at all levels of hierarchical infrastructure is formed. The city residents and visitors as active actors of the information system can function as information sources used to form the complexes of needs which occur in the urban environment; or collectors of the urban

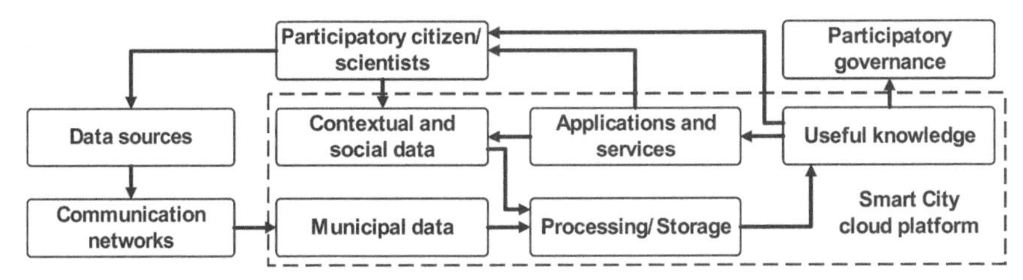

Figure 18.1 Functional information-communication structure of the "smart city."

data which can be structured, poorly structured or unstructured in general presentation Figure 18.1.

The context-sensitive and accessible information concerning the city functioning is intended to help the main groups of actors to make reasonable decisions. For example, data concerning air quality in certain locations can be useful for preventive measures by citizens with respiratory tract problems. The city residents and visitors can be involved in collecting environmental and other data (for example, data on environment quality in industrial areas and parks) that are technologically difficult or expensive to collect by other means.

The use of cloud information technology for providing context-sensitive urban services requires both research concerning the mechanisms effective implementation and analysis of the problems concerning functional profiling, security, segregation, processing, and storage of Big Data.

In this case, it is useful to develop the conceptual principles for cloud-based context-sensitive information technology use while providing the relevant services, and to formulate the relative roadmap that enables the development of the comprehensive cloud architecture of information services in "smart cities." In addition, the analytical processing of available implementations and projects such as "Bristol Open Data" [13] concerning formation of appropriate cloud architecture should be carried out.

18.3 Information-Technological Resources for the "Smart Cities" Projects Implementation

The implementation of "smart cities" projects based on information and communication technologies involves the integration and processing of data of different types from various sources for the search for diverse, relevant information. In this case, the main information flows are formed on the basis of data while supplying diverse services. The problems of informational-technological support of the "smart cities" life activity processes have different methods and means of solution. These are the processes of Big Date collection, their aggregation in various formats, actualization in solving practical tasks and scenarios for their proper use, analytical processing, selection of useful information and its visualization, as well as management of historical and operational data sets. The solution of these problems requires the use of a multidimensional system approach, standardization of data formats, mechanisms for their coordination, the development of appropriate computing algorithms, and the formation of extended storage infrastructure.

In the urban environment, the number of data sources is increasing, starting with the stored data concerning provided services – data of municipal organizations and received from citizens while implementing various social programs and public surveys. The lack of the common standardized platform for collecting, storing, using, analyzing, and distributing urban data significantly restricts the provision of information and services that can be implemented and creates the comfortable environment based on the "smart city" information-technological infrastructure. Therefore, one of the main tasks of the systematic implementation of "smart city" projects and the formation of innovative information and service complexes provided to a wide range of individuals (citizens, municipalities, and private enterprises) is the development of appropriate rules and conditions for standardization and coordination of procedures for the urban data use.

The important aspect information and communication technology use in the "smart cities" projects is particularly the city residents and guests' involvement into the processes of collecting and efficient information processing. Urban citizens will obviously be more effective in cases when the services have broad social and environmental context. At the same time, the processes of context Big Data integration are significantly complicated due to its heterogeneity and variety from diverse sources. The selection of entities from information resources in the process of contextual services supply depends on the effective implementation of procedures for the data integration, collection, processing, and distribution. Integration of information and communication technologies in the process of the formation of the "smart cities" integrated information and technology platforms is one of the priorities of the European Innovation Partnership Program [14].

Information technologies based on cloud computing, built on the concepts of converged infrastructure and public services, can be effectively used in the "smart cities" projects to solve a wide range of problems. Cloud computing is ideal to provide infrastructure support for such kinds of projects and to provide key characteristics of reliability, scalability, performance, and cost effectiveness. Cloud computing is ideally suited to provide infrastructure support to projects of this class and to provide key characteristics of reliability, scalability, effectiveness, and cost-effectiveness. Using external cloud computing infrastructure can reduce the load of urban computing infrastructure and management and administration means. At the same time, it reduces the negative impact of the heterogeneity factor resulting from the use of its own computing infrastructure by various municipal services.

18.4 The "Smart City" Project Tasks which can be Solved on the Basis of Cloud Computing

There is a wide range of opportunities for the implementation of the promising intelligent services and networks in the "smart cities" [15, 16] projects that can be introduced using cloud computing technologies [17].

Especially such tasks as the dynamic setting of energy prices can be solved; the transfer of potential peak demand for the provision of certain services at other time intervals; storage of heterogeneous data obtained from various sources such as sensors and AMI for SCADA system; stream processing of structured and poorly structured Big Data in real time, etc.

Using the platform based on cloud services contributes to the simplification of the networks infrastructure, provides qualitative solutions, and increases their energy efficiency. Some research results of a number of municipal tasks connected with cloud implementations of intellectual networks and services are given in [18–20]. At the same time, they do not have information about scalable cloud platforms to solve municipal data stream management tasks in real time. The problem of the effective use of smart networks for solving the problems in the "smart cities" projects, which can be solved due to cloud technologies, still remains unsolved.

The key problem resulted from the information-technological "smart cities" projects implementation is: real-time processing of Big Data obtained from diverse sources (such as measuring sensors, analytical models, maps, video cameras, vehicles, city residents and visitors, various management systems, etc.)

Urban smart networks require more scalable and consistent models' formation since the data from different sources interact with each other in the standard representation of the SCADA system. Devices transmitting data should receive the instructions of the same type regardless of the way they are connected to the SCADA municipal system and their interaction with the control system servers [21].

18.5 Generalized Architecture of Information-Technological Support of the "Smart City"

Having analyzed the results of theoretical and applied researches, one can conclude that the general presentation of the structure of the "smart city" in the information-technological context should be divided into three levels (see Figure 18.2):

- **The first level** is formed from infrastructure objects and networks of resources and services supply, waste management subsystems, and social components. At this level, smart counters, specialized devices, and smartphone sensors can be used as information sensors.
- **The second level**, implemented as a center for analytical data processing, includes software and modern information technology for collecting, storing, and processing information on key components of urban infrastructure and processes taking place in the city.

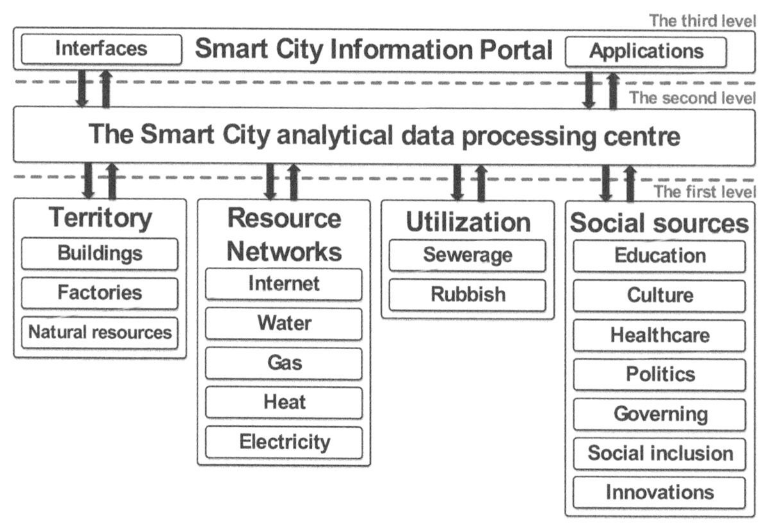

Figure 18.2 Three-level structure of the "smart city" in the information-technological context.

- **The third level** is based on the "smart city" information portal providing data representation, implements the main interfaces, and contains applications providing a wide range of service functions [22].

The interfaces and applications used to provide information are focused to the needs of the city residents, visitors, and managers of the "smart city." The portal functionally provides information-technological services and information using any device connected to the Internet, including personal computers, smartphones, tablets, and other modern mobile devices. The active use of cloud platforms and smart networks in the "smart cities" projects generates a number of important not completely solved at the moment, including: designing cloud architecture to improve resource efficiency and reliability; development of analytical models of demand forecasting and load with any degree of detail for the real-time energy efficiency optimization; and coordination of heterogeneous city data to support the scalability and availability of municipal services and applications.

It should be noted that failure-proof solutions based on cloud technologies have significant advantage over the system solutions in organizations focusing on the use of their own computing infrastructures. At the same time, it is evident that the increase in carbon emissions from the exponential growth of a number of data processing centers necessary for a full-scale implementation of the concept of cloud computing is the main disadvantage.

18.6 Infrastructure Platform for Cloud-Based "Smart City" Projects

Cloud architecture has general characters since different cloud services, components, standards, and technologies can be applied at different levels of implementation depending on the needs of various applications. Besides, it must be flexible enough for implementation of new services and procedures containing new technologies and standards.

Information technologies of cloud computing appear as a model that enables the transformation of applied applications, increasing their adaptability and efficiency [2]. The use of cloud services significantly simplifies the processes of the implementation of the smart cities' complex information systems, increasing their efficiency and providing new additional opportunities for implementing applications for the city services and information-technological services. System solutions for using the cloud computing technologies in the "smart cities" make it possible to implement different complexes of applications and services depending on the size of cities, their level of organization, and the financial completeness of city budgets.

The cloud computing model, as an information technology, is a fundamental transformation in the approach to create, develop, deploy, scale, update, maintain, and minimize hardware and software costs [23]. It serves as the foundation for using the paradigm of global computing in applied applications for engineering networks (such as water, gas, energy, heat supply, communication networks, etc.) [24], therefore changing the traditional model of such kind of system implementation where the data and applications are physically located in one place (in one LAN) into a new one, where the users can access the necessary data and applications distant on the Internet.

The American National Institute of Standards and Technology (NIST) defines the cloud computing as the *"universal, convenient, network access standard for a common pool of configurated computing resources (networks, servers, storage systems, applications, and services) that can be quickly provided with minimal administrative effort or interaction with the service provider"* [25]. Gartner's IT research and consulting firm defines the cloud computing as *"the style of computing, in which Elastic Computing is the computing IT support and is provided as the based on Internet-technologies"* [26]. For better understanding of the principle of the information technologies cloud computing, they can be associated with in comparison to the implementation of lease procedures and with the acquisition procedure, since the city municipality, instead of buying its own equipment and software, can lease the necessary resources (computing capacities, disk space, applications, etc.) from the cloud service provider and use them due to the Internet. At the same time, they pay only for the actually used resources.

The NIST of the United States introduced into the scientific circle the definition which can be considered as cloud hosting. According to the definition, the product that meets five basic principles [25] is called the cloud-hosting: "On-demand self-service" is the principle of the availability of arbitrary volume of services; "Ubiquitous network access" is the principle of network accessibility; "Elasticity "is the principle of computing elasticity; "Metered use" is the principle of payment for services provided; and "Resource pooling" is the principle of resources' combination.

This makes it possible to create highly efficient, scalable, and flexible computer environment particularly for the implementation of the "smart cities" business model, where municipalities purchase the services and computing capacities they need at a specific time. The implementation of the "payment for consumed resources only" allows the administration of the "smart city" to significantly reduce the initial and depreciation costs.

The factors increasing the efficiency of the implementation of the cloud computing information technologies in the "smart cities" are as follows [27, 28]: the relevance of the infrastructure to actual needs; effective use of resources; dynamic response to the situation changes; increase of reliability, productivity, and safety; use of more accurate information and aggregation of heterogeneous data; and evaluation and optimization of city-wide processes by means of integrated rapid analysis of applications and their infrastructure components. The use of the cloud computing technologies promotes entrepreneurship activity by reducing the risks of innovation implementation and providing a relatively cheap and universally accessible platform for testing new applications and services [29]. While planning the strategy of information technologies use in the "smart cities" projects, the possibilities for providing various categories of cloud services [25, 30] should be taken into account, particularly:

- Infrastructure-as-a-Service (IaaS): For the users to build a customized information environment, it is possible to initialize processes, storage systems, networks, and other resources. Providers of this class of cloud services are Amazon and Mosso.
- Platform as a Service (PaaS): The provider provides the users with the ability to use the platform and information environment to create, test, and deploy new applications. An example of such a cloud service class is Google App Engine providing users with a programming environment by abstracting developers from physical servers and other technical components of computer networks.
- Software as a Service (SaaS): The users are given the opportunity to use provider applications running on the cloud-based infrastructure. Examples of this class of cloud services are scalable distributed systems for Google's BigTable and Amazon's SimpleDB structured data storage.
- Network as a Service (NaaS): The users are given the opportunity to use a virtual network based on the network infrastructures of different owners integrated into the cloud service.An example ofcloud services of this class is IT & T MPLS VPN.
- Storage as a Service (STaaS): The users have the possibility to store data in logical pools formed from a certain number of physical servers under the hosting provider control. In this case, providers are responsible for the integrity and availability of the cloud, data storage, and the

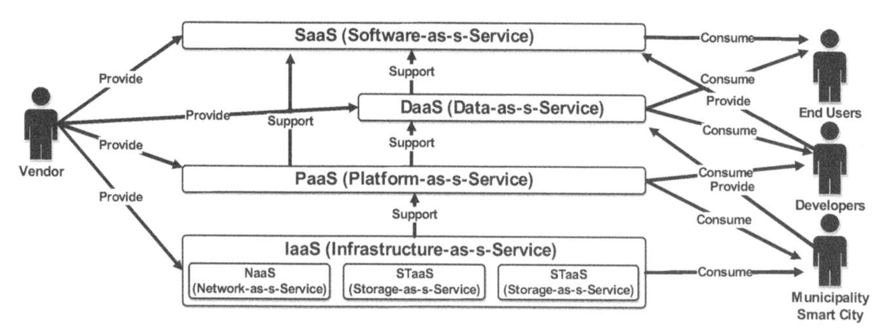

Figure 18.3 Categories of cloud services in the "smart cities."

information environment protection. Examples of cloud services of this class are CSC STaaS, PROACT STaaS, etc.

- Sensor as a Service (SSaaS): The users are given the opportunity to receive numeric rows and sensor metadata. Examples of cloud services of this class are the sensor observation service, built on the basis of the Open Geospatial Consortium standards.

A wide range of projects initiated by the European Commission are aimed at the development of all-European cloud services, and the creation of the new cloud technologies, including resource management, security, etc. [31]. The list of cloud services presented by separate layers [30] of the cloud service categories for the "smart cities" should be supplemented by data as a service (DaaS) (see Figure 18.3.) and the new actor – municipality smart city, because it is the provider of open-source urban data sets, restricting the procedure for their availability.

18.7 Architecture of the Center for the "Smart City" Analytical Data Processing

One of the possible alternatives of the architecture of the center for the "smart city" analytical data processing using the information technology based on cloud computing is shown in Figure 18.4.

The IaaS category is represented by two subcategories:

- System requirements contain the services necessary for the operation of the "smart city" analytical data processing center (WebHosting, data store, tools for the information archiving and retrieval, network performance monitoring tools, high-performance computing tools, etc.).

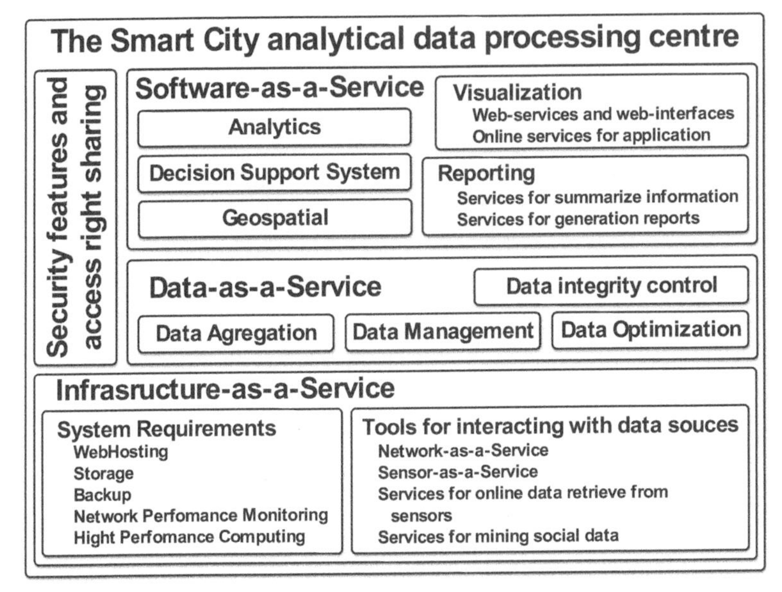

Figure 18.4 Architecture of the "smart city" analytical data processing center.

- Tools for interacting with data sources include leased services such as NaaS, SSaaS, services for online receiving of data from sensors (intelligent flowmeters and sensors), and services for extracting data from socially oriented sources.

The DaaS category provides services for data aggregation, data management, data optimization, and data integrity control.

The SaaS category is presented by the following services:

- Analytical processing of data (include BigData analytics and a comprehensive heterogeneous data analysis to find the hidden dependencies).
- Decision Support Systems are designed to solve the problems based on unstructured and poorly structured data.
- Geospatial is designed for geospatial positioning of the processed data and information entities.
- Visualization includes web services, web interfaces, and online services for mobile applications for the organization of smart city information portal.
- Reporting includes services concerning summarizing information and generation of reports.

A separate component of the center of the "smart town" analytical data processing is security features and access rights' sharing which should be used in the context of all constituent elements.

The abstract architecture of the system of analytical Big Data processing on the basis of the cloud technologies is analyzed in [32]. The cloud service analytics functioning implemented on the basis of cloud technologies is presented in detail. The architecture of the given system shown in Figure 18.5. is divided into three levels and functionally oriented to the formation of a unified knowledge base.

Each layer is focused on the application of the new ones and implementation of the common tasks. The lower architecture level consists of distributed and heterogeneous stores, connected to the system and sensors of different types. This layer is designed to collect, filter, and classify data using standard approaches such as the API, or OGC to provide and combine web services.

Tools such as The DataTank and CKAN implemented in the form of REST services are available for the data transformation and publication. To store poorly structured data, Cassandra and PostgreSQL for relational structured and virtual data – RDF structures are used.

The next layer is used to link data located on different resources and establish relationships using scenarios and provide support for processes and relationships that would not be possible in isolated data stores. Typically, the data are collected in different formats and semantic representations due to the heterogeneity of their sources and require additional arrangement. A vivid example of this is related or open data [33, 34] stored in databases supported by cross-referencing to related data. In addition, the semantic model of related data can be developed as a separate layer. Once the metadata from heterogeneous sources are stored in the corresponding sets, the links between the resources to be display are established, references are generated, and the data

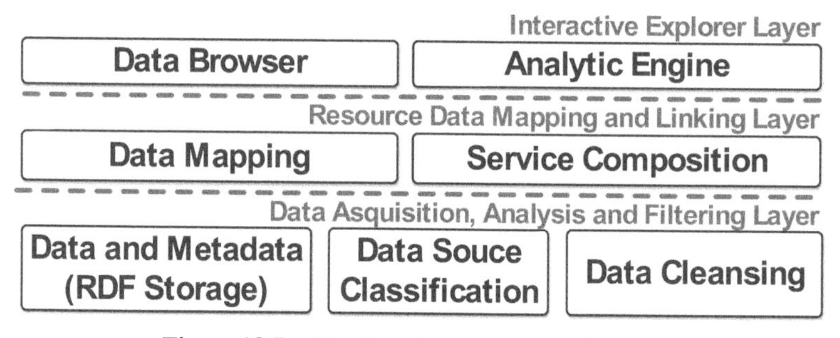

Figure 18.5 Cloud service analytics architecture.

become semantically relevant for viewing. The known metadata formats, such as EDM, Talis Aspire, Open library, DBLP, and Linked Data, can be used to describe and store information collections from different sources. The next step is to compare data using standardized semantic resource descriptions, for example, through an RDF reproducing the necessary connections established between entities and resources.

The top-level analytical engine processes the data to achieve specific goals. It uses the data available in the linked data layer and helps the users to form inquiries using specific algorithms and workflows processes to search Big Data stores. However, data mining technology is the new innovative trend used to process Big Data [35, 36]. Big Data mining is increasingly being used as an important and effective means of extracting Big Data in various data-driven applications, such as risk analysis and network traffic, business data analysis, etc. These methods are highly effective in identifying non-obvious relationships and associations of Big Data concerning provided services and services in the "smart cities" in future.

18.8 Computing Architecture for Providing Information and Technology Services in the "Smart Cities"

Today, in the cloud environment, significant methodological, technological, and instrumental capabilities [37] are developed for the integrated implementation of the "smart" municipal management systems, providing a solid basis for adequate responding to the demands of "smart cities" by developing context-sensitive components in the cloud environment. Extended due to the implementation of context-dependent opportunities to meet citizen demand for access to contextual information, is presented in Figure 18.6. The offered architecture [37] can be used to create a platform based on PaaS. In the case of building the service model, it becomes the PaaS solution allowing other services' creation on its basis. In case when the architecture is used to create a service interacting directly with users, for example, to visualize, then the application becomes the SaaS solution. The computing architecture of providing services in the "smart cities" consisting of five horizontal and two vertical layers is shown in Figure 18.6.

At the lower level, there is an integration platform for hybrid cloud infrastructure. At the thematic level, the collected data are analyzed and adapted to the specific tasks' requirements. One of the key principles of the architecture is the introduction of context-sensitive components at different

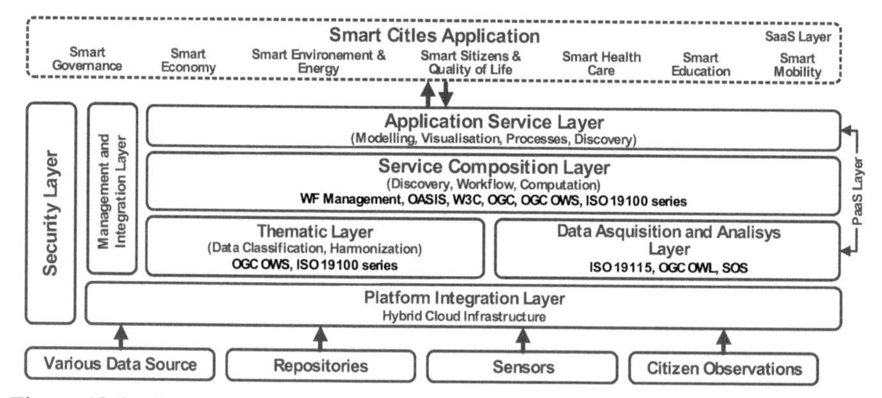

Figure 18.6 Computing architecture of providing context-sensitive services in the "smart cities."

levels in its structure in order to coordinate continuously the transmission of vertical data streams and procedures for storing associated information contexts. By transferring data to higher levels, each layer contributes to the processing of contextual information.

The integration platform level is a set of hardware and software components providing the necessary computing infrastructure, such as hybrid municipality and private cloud services supplying cross-platform data availability. In addition to physically implemented computing hardware and virtual cloud resources, this layer also provides integration of hardware and software sensors that form data sources. In this case, OGC observation sensors [38], providing standard APIs interfaces for controlled data management and receiving, can be used. These standards contain statements about the metadata formation both for existing and for new sensors in order to obtain context information about them.

The data acquisition and analysis layer allows collecting data about the urban environment from diverse sources such as databases, sensor networks, and citizen messages sent by smartphones. At the level of this layer, the authenticity of the data is checked and the need for filtering is determined. Context-sensitive components are used to filter unlinked data and select only the agreed context linked sets. The Open GIS Sensor data format [37] is determined as standard for simple, aggregated, and derived data entities with support for different codes (e.g., XML). Such features of the accumulation of context data allow aggregating information sets of complex physical, economical, and social processes.

At the level of the thematic layer, the classification of the received data according to certain thematic categories and the coordination and updating for further use in the processes of providing services are carried out. Thematic categorization of contextually annotated data helps to use them more efficiently in applications at the higher layers of the architecture.

The arrangement of services includes service and operating processes, identifies data sources, and links to the processing components providing computing platform for the implementation of the context-sensitive services. In this layer, an analytical analysis of the results of the processes is carried out and analytical processing of the data for the expert systems implemented at the highest level is provided. A service application layer uses the results and service of the lower layer in the form of specific tools such as modeling tools, implementation of "smart" mobile applications, and visual maps for contextual analysis for decision making. This layer allows using available tools and developing new applications for the implementation of specific components and services (at the SaaS level) to meet the contextual information needs of users. Information sets can be modeled in spatial, extended, and machine formats. The ContextML [39] format is used. This layer also supports analytical processing and knowledge extraction processes.

The management and integration level is used to automate the distribution of filtered data streams and information tuples between horizontal layers. This, in turn, guarantees that the results obtained from one layer will be adequately linked and syntactically correctly transmitted to another layer. This level also implements processes for changes' control occurring at different levels and reduces the cost of multilevel architecture control.

The security level implements the functions of authentication, authorization, and audit of data use and service provision processes. The personalization of user services on the basis of predefined rights and privileges for sampling and processing information in the cloud environment is control at this level.

18.9 Conclusions and Further Investigation

In this chapter, the socio-communicative component of the "smart city" projects is considered and information-technological resources for the "smart city" projects implementation are described. The "smart city" project tasks which can be solved on the basis of cloud computing technologies are highlighted. The generalized three-level architecture of information-technological support of the "smart city" and the conceptual principles of using cloud computing in the "smart city" projects are analyzed, various categories of cloud services are selected, and features of their interaction with actors of projects of this class are described. On the basis of the above-mentioned categories of cloud services, architecture of the center for the "smart city" analytical data processing, cloud service analytics architecture, and computing architecture for providing information and technology services in the "smart cities" are proposed.

It is planned to focus further investigations on the detailed presentation of the developed architecture, taking into account the peculiarities of the use and interaction of public and private cloud services in the future "smart cities." The processes for the systematic combination of cloud computing technologies with GRID, IoT, BigData, OLAP, and GIS technologies will be perspective in further researches since at present, there are no integrated solutions for given technologies' peculiarities' integration. For the effective implementation of the concept of the "smart city," it is necessary to implement complex principles of the system approach taking into account the peculiarities of the available IT infrastructure and relevant municipal information-analytical services.

In further research, it is planned to carry out the system modeling on the basis of the offered architecture, and to develop software-algorithmic prototypes of applications for personalized accounting of water and heat consumption and conducting online financial mobile calculations of individual consumers for the services they were provided with. In prototype models, the possibilities of function expansion and integration into a single set of subsystems of the "smart city" innovative information-technological project are predicted in the dumb prototypes.

References

[1] Khan, Z., Kiani, S.L., and Soomro, K. (2014). A framework for cloud-based context-aware information services for citizens in smart cities. *Journal of Cloud Computing: Advances, Systems and Applications*, 3:14.

[2] Kakderi, C., Komninos, N., and Tsarchopoulos, P. (2016). Smart cities and cloud computing: lessons from the STORM CLOUDS experiment. *Journal of Smart Cities*, 2(1), 4–13.

[3] Sarkar, M., Banerjee, S., Badr, Y., and Sangaiah, A. K. (2018). Configuring a trusted cloud service model for smart city exploration using

hybrid intelligence. *Cyber Security and Threats: Concepts, Methodologies, Tools, and Applications: Concepts, Methodologies, Tools, and Applications*, 337.

[4] Tsarchopoulos, P., Komninos, N., and Kakderi, C. (2017). Accelerating the uptake of smart city applications through cloud computing. *World Academy of Science, Engineering and Technology, International Journal of Social, Behavioral, Educational, Economic, Business and Industrial Engineering*, 11(1), 129–138.

[5] Yang, J., Jiang, B., and Song, H. (2018). A distributed image-retrieval method in multi-camera system of smart city based on cloud computing. *Future Generation Computer Systems*, 81, 244–251.

[6] Mazza, D., Tarchi, D., and Corazza, G. E. (2017). A unified urban mobile cloud computing offloading mechanism for smart cities. *IEEE Communications Magazine*, 55(3), 30–37.

[7] Cai, P., and Jiang, Q. (2018). GIS spatial information sharing of smart city based on cloud computing. *Cluster Computing*, 2018, 1–9.

[8] Lv, Z., Li, X., Wang, W., Zhang, B., Hu, J., and Feng, S. (2018). Government affairs service platform for smart city. *Future Generation Computer Systems*, 81, 443–451.

[9] Mohammadi, M., and Al-Fuqaha, A. (2018). Enabling cognitive smart cities using big data and machine learning: Approaches and challenges. *IEEE Communications Magazine*, 56(2), 94–101.

[10] Chen, N., and Chen, Y. (2018). "Smart city surveillance at the network edge in the era of IoT: opportunities and challenges," in *Smart Cities* (Cham: Springer), 153–176.

[11] Stergiou, C., Psannis, K. E., Kim, B. G., and Gupta, B. (2018). Secure integration of IoT and cloud computing. *Future Generation Computer Systems*, 78, 964–975.

[12] Booth, B. (2018). "The cloud: a critical smart city Asset," in *Smart Cities* (Cham: Springer), 97–105.

[13] Bristol City Council GIS Support (2018). *Bristol data profiles*. Available at: http://profiles.bristol.gov.uk [accessed Jan 7, 2018].

[14] Smart Cities and Communities. (2015). *The European Innovation Partnership on Smart Cities and Communities*. Available at: http://ec.europa.eu/eip/smartcities/ [accessed Jan 7, 2018].

[15] Fang, X., Misra, S., Xue, G., and Yang, D. (2012). "Managing smart grid information in the cloud: opportunities, model, and applications," in *IEEE Network*, 26(4), 32–38.

[16] Naphade, M., Banavar, G., Harrison, C., Paraszczak, J., and Morris, R. (2011). "Smarter cities and their innovation challenges," *IEEE Computer*, 44(6), 32–39.

[17] Cloud Computing for Smart Grids and Smart Cities. (2016). Available at: http://zeitgeistlab.ca/doc/cloud_computing_for_smart_grids_and_smart_cities.html [accessed Jan 7, 2018].

[18] Yamamoto, S., Matsumoto, S., and Nakamura, M. (2012). "Using cloud technologies for large-scale house data in smart city," in *Proceedings of the IEEE 4th International Conference on Cloud Computing Technology and Science*, 141–148.

[19] Rusitschka, S., Eger, K., and Gerdes, C. (2010). "Smart grid data cloud: a model for utilizing cloud computing in the smart grid DOMAIN," in *Proc. First IEEE International Conference on Smart Grid Communications*, 438–488.

[20] Khan, Z., and Kiani, S. L. (2012). "A cloud-based architecture for citizen services in smart cities," in *Proceedings of the IEEE 5th International Conference on Utility and Cloud Computing*, 315–320.

[21] Birman, R., Ganesh, L., and van Renesse, R. (2011). "Running smart grid control software on cloud computing architectures," in *Proceedings of the Computational Needs for the Next Generation Electric Grid*, 15–47, U. S. Department of Energy.

[22] Park, J. P., Yun, C. H., Jung, H. S., and Lee, Y. W. (2014). "Mobile cloud and grid web service in a smart city," in *Proceedings of the Fifth International Conference on Cloud Computing, GRIDs, and Virtualization*, 20–25.

[23] Marston, S., Li, Z., Bandyopadhyay, S., Zhang, J., and Ghalsasi, A. (2011). Cloud computing – the business perspective. *Decision Support Systems*, 51(1), 176–189.

[24] Buyya, R., Yeo, C. S., Venugopal, S., et al. (2009). Cloud computing and emerging IT platforms: vision, hype, and reality for delivering computing as the 5th utility. *Future Generation Computer Systems*, 25(6), 599–616.

[25] Mell, P., and Grance, N. (2011). "The NIST definition of cloud computing," *NIST Special Publication 800-145* Available at: https://csrc.nist.gov/publications/detail/sp/800-145/final [accessed Jan 7, 2018].

[26] Gartner IT Glossary. *Cloud Computing*. Available at: https://www.gartner.com/it-glossary/cloud-computing [accessed Jan 7, 2018].

[27] Jinesh, V. (2011). *Architecting for the cloud: best practices, amazon web services*. Available at: https://www.google.com.ua/url?

sa=t&rct=j&q=&esrc=s&source=web&cd=2&cad=rja&uact=8&ved=0ahUKEwj-ve-lrs7YAhXKAJoKHeZVAhsQFgg7MAE&url=https%3A%2F%2Fmedia.amazonwebservices.com%2FAWS_Cloud_Best_Practices.pdf&usg=AOvVaw1P3qe3dtM9_eHTcMTScMfA [accessed Jan 7, 2018].

[28] Professional Services Council. (2015). Best practices for federal agency adoption of commercial cloud solutions. Available at: https://www.google.com.ua/url?sa=t&rct=j&q=&esrc=s&source=web&cd=1&cad=rja&uact=8&ved=0ahUKEwiWo6fcr87YAhVIDZoKHb3yCBoQFggmMAA&url=http%3A%2F%2Fwww.pscouncil.org%2FDownloads%2Fdocuments%2FPSC-Cloud-WEB%2520-%252012-10-15.pdf&usg=AOvVaw3DECWMLVGzbk8cVMmy6cWR [accessed Jan 7, 2018].

[29] Kundra, V. (2011). "*U.S. Chief Information Officer, Federal cloud computing strategy*," White House. Available at: https://www.google.com.ua/url?sa=t&rct=j&q=&esrc=s&source=web&cd=1&cad=rja&uact=8&ved=0ahUKEwjO4IbhsM7YAhVLCpoKHYaFCeEQFggmMAA&url=https%3A%2F%2Fwww.dhs.gov%2Fsites%2Fdefault%2Ffiles%2Fpublications%2Fdigital-strategy%2Ffederal-cloud-computing-strategy.pdf&usg=AOvVaw1pqz_jUBh1rjMoR8javOL9 [accessed Jan 7, 2018].

[30] Rao, P. V. L. N., Abhilash, P. S., and Kumar, P. S. P. (2015). Wolkite smart city community cloud computing with cyber security. *International Journal of Innovative Science, Engineering & Technology (IJISET)*, ISSN 2348-7968, 2(3), 176–182.

[31] Vulpe, A., Todoran, G., Cropotova, J., Suciu, V., and Suciu, G. (2013). "Cloud computing and internet of things for smart city deployments," in *International Conference: CKS - Challenges of the Knowledge Soc*, 1409.

[32] Khan, Z., Anjum, A., Soomro, K., and Atif Tahir, M. (2015). Towards cloud based big data analytics for smart future cities. *Journal of Cloud Computing: Advances, Systems and Applications*, 4:2.

[33] Linked Data – Design Issues. Available at: http://www.w3.org/DesignIssues/LinkedData.html [accessed Jan 7, 2018].

[34] Linked Open Data: The Essentials – A Quick Start Guide for Decision Makers. (2012). Available at: https://semantic-web.com/2012/01/20/linked-open-data-the-essentials-a-quick-start-guide-for-decision-makers/ [accessed Jan 7, 2018].

[35] Wu, X., Zhu, X., Wu, G-Q., and Ding, W. (2014). "Data mining with big data," *Transactions on Knowledge and Data Engineering*, IEEE, 26(1), 97–107.

[36] Fan, W., and Bifet, A. (2012). "Mining big data: current status and forecast to the future," in *SIGKDD Explorations: ACM,* 14,1–5.

[37] Khan, Z., Ludlow, D., McClatchey, R., and Anjum, A. (2012). "An architecture for integrated intelligence in urban management using cloud computing," in *ITAAC 2011: International Workshop on Intelligent Techniques and Architectures for Autonomic Clouds in conjunction with the 4th IEEE/ACM International Conference on Utility and Cloud Computing (UCC 2011)*, Melbourne, 415–420. Available at: http://eprints.uwe.ac.uk/16041.

[38] Botts, M., Percivall, G., Reed, C., and Davidson, J. (2007). "OGC sensor web enablement: overview and high level architecture," in *OpenGIS® White Paper.* Available at: https://www.google.com.ua/url?sa=t&rct=j&q=&esrc=s&source=web&cd=1&cad=rja&uact=8&ved=0ahUKEwjox5ubus7YAhVBWywKHVjwDWMQFggmMAA&url=http%3A%2F%2Fportal.opengeospatial.org%2Ffiles%2F%3Fartifact_id%3D25562&usg=AOvVaw29g1C86EiBZlYY5DLAUy1p [accessed Jan 7, 2018].

[39] Knappmeyer, M., Kiani, S. L., Frà, C., Moltchanov, B., and Baker, N. (2010). "ContextML: a light-weight context representation and context management schema," in *2010 5th IEEE International Symposium o Wireless Pervasive Computing (ISWPC)*, IEEE, Modena, 367–372. Available at: http://ieeexplore.ieee. org/xpl/articleDetails.jsp?arnumber=5483753

19

A Framework for Real-Time Public Transport Information Acquisition and Arrival Time Prediction Based on GPS Data

Inna Skarga-Bandurova, Marina Derkach and Artem Velykzhanin

Volodymyr Dahl East Ukrainian National University, Severodonetsk, Ukraine
E-mail: skarga_bandurova@ukr.net; gln459@gmail.com; velykzhanin@snu.edu.ua

Real-time public transport information service infrastructure is a part of the core functionality of intelligent transport systems. This paper focuses on developing a framework for real-time data acquisition and choosing an efficient model for trolleybus arrival time prediction that can be easily implemented to improve public transport services by leveraging on the GPS data and data provided by the Internet of Things applications. An architecture model of information service infrastructure for public passenger transport was developed. As a use case of the proposed approach, eight methods combining historical average, Kalman filtering technique, and Google Maps API for trolleybus arrival time prediction were implemented and tested. An assessment of the model's performance and their effectiveness with real-time data are investigated. The results show that combinations of average travel speed (ATS) and distance from Google Maps API and ATS and distance from Google Maps API with Kalman filtering gave the best arrival time predictions for low-speed urban transport.

19.1 Introduction

Road traffic requires accurate and up-to-date information about the current situation and available services. The enhancement of the road conditions is essential to all infrastructures. However, only this improvement cannot meet the continually growing demands for safe, convenient, cost-effective, and comfortable road services.

In this context, public transport information service is one of the essential parts of intelligent transport systems (ITSs) aimed to make transportation system safer and more efficient. It provides the real-time travel information according to the people needs through appropriate tools. As mentioned in [1], applying ITS delivers several benefits by increasing traveler safety, improving the operational performance of the transportation network, mainly by reducing the traffic jam, enhancing personal convenience, providing better environmental conditions, and expanding economic and employment growth.

The emergence of the Internet of Things (IoT) adds new zest to ITS. IoT deals with different physical objects merging their data in a network in one form or the other. It mainly deals with RFID, infrared sensors, global positioning systems (GPSs), and laser scanners.

Electronic recorders in the vehicles can track and record the acceleration, speed, engine RPM (rotational speed of the crankshaft of the engine), and other parameters, and this information generated by traffic IoT and collected on all roads can be presented to fleet managers, travelers, and other users. Hence, ITSs, as well as IoT in ITSs, allow deploying the advanced transport systems that enable vehicles to share data about their positions so that the passengers aware the availability of their buses and trains in real time [2]. However, dependable and real-time communication in the scope of developing intelligent transport infrastructure is still a critical challenge and needs to be tackled for the success of its applications.

The current engineering practice for the development of such systems includes a set of different methodologies and therefore needs carefully designed approaches. Construction public transport information service infrastructure requires enforcement of existing standards in data collection, transmission, processing, and dissemination of information, and ensuring consistent service quality, which is fundamental to their convenient usage by people.

19.1.1 Real-Time Public Transport Information Service Infrastructure

Real-time public transport information service infrastructure as a part of the core functionality of ITS includes fleet management, dispatching and scheduling services, emergency alerts, security services, and passenger information services. Figure 19.1. shows how the different subsystems are organized and interact with each other. The model is based on international ITS architecture model standards ISO/TR 14813-2:1999(E) and ISO 14813-1:2015 and provides a framework for the constant enhancement of these systems, giving desirable properties such as dependability, flexibility, and integration [3–5].

This pilot project is designed to enhance the passenger information services and realize the people-centered paradigm by focusing on services for passengers. Passenger information services include message boards and kiosk boards at bus stops, web-based information services to deliver information about the public transport routes, scheduling, ticket information, estimated time of arrival/estimated time of departure, on-bus announcements, etc.

19.1.2 Objective and Challenges

The aim of this study is to develop a framework for real-time data acquisition and choosing an efficient model for trolleybus arrival time prediction as a part of passenger information service (blue boxes in Figure 19.1).

There are several challenges arising from our study and the current traveler information services that should be addressed. Among them, the strict requirements consider real-time data processing, availability of data, security, predictability, low power consumption, and many others. Despite the wide variety of issues, there are two tasks on which we will address in this paper.

Task 1: Calculation of the predicted arrival time of the vehicle to the specific bus stops

Ensure that an accurate arrival time is a fundamental task for efficient operation of public transport companies. The punctuality brings a vast improvement in public transport services. However, it is still a challenging task to find the best method to predict the accurate vehicle arrival time and as with anything better is a matter of the particular application. For the prediction problem, different approaches that have been proposed in [6–11] should be taken into account.

The task is to define the best arrival time prediction model for low-speed urban transport, as for the moment just trolleybus fleet has been equipped with GPS devices.

Task 2: Trolleybus routing

For a variety of reasons, there are no vehicles associated with a specific route for everyday use. By the start of working time, every trolleybus in

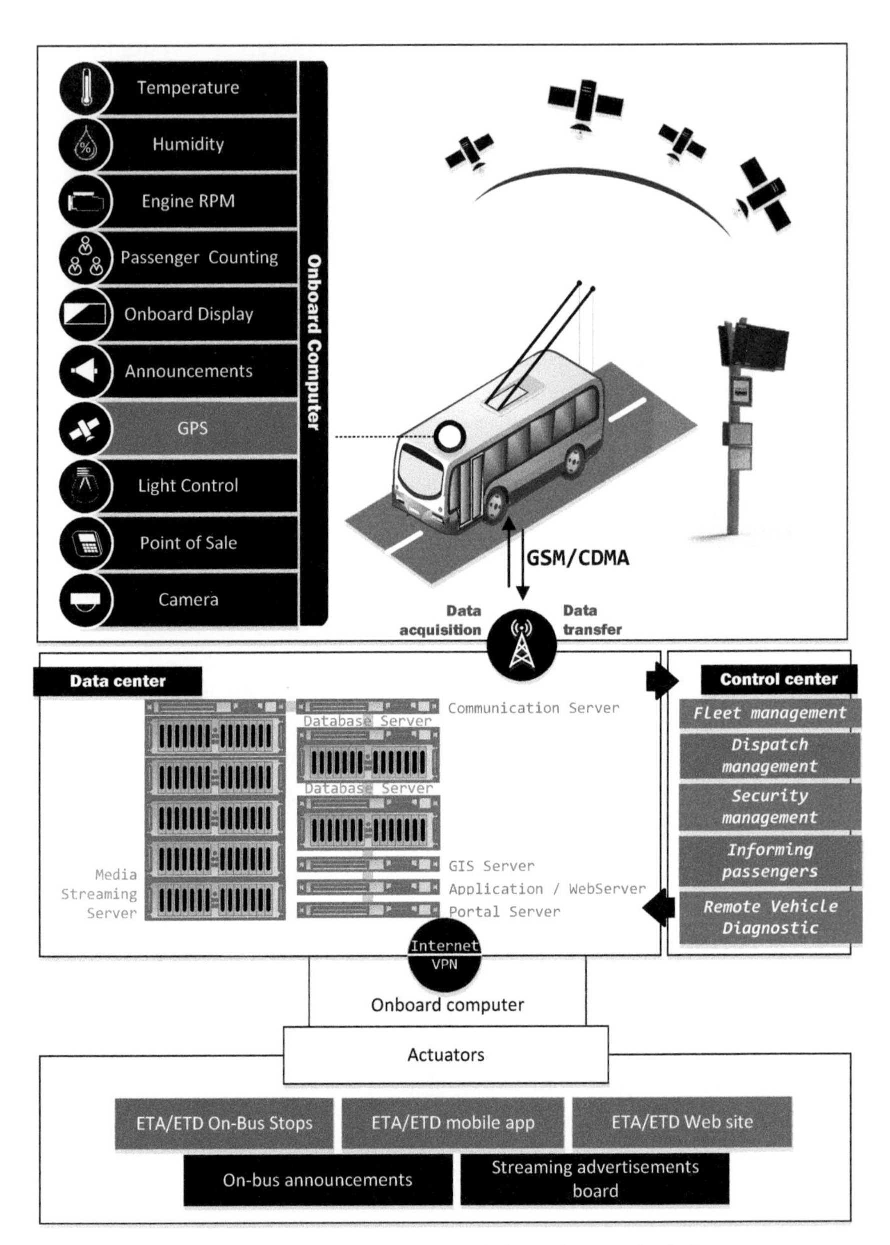

Figure 19.1 Real-time public transport information service infrastructure.

trolleybus depot is being given a new direction. Hence, to perform correct prediction, it is necessary to know which one trolleybus runs along that route on a particular day. There are at least two routing approaches for system input: daily manual route update and automatic route assignment. In the latter case, a unique technique to assign routes will be required.

The remainder of this paper is organized as follows. Section 19.2 describes arrival time prediction methodology including statistics with averages, Kalman filtering, and some accuracy metrics. In Section 19.3, a case study of utilizing this methodology is described, and the general strategy and main algorithmic components of public transport information service particularly for bus stop information boards are developed. Finally, the main conclusions and future lines of work are presented in Section 19.4.

19.2 Arrival Time Prediction Models

Several techniques were used in this study, including a simple statistical model (historical average) computing averaged travel speed and a model based on Kalman filter to calculate arrival time. The input variables are the travel speed of the vehicle, its location, and time of arrival to the target checkpoint (bus stop).

19.2.1 Prediction Methodology

To define the best arrival time prediction model, the following technique is proposed for predicting trolleybus arrival time.

1. Fix the number of time slots *n,* passed by the vehicle during the traveling time from the start point to the current location:

$$\mathrm{n} = \frac{T - T_0}{I} \tag{19.1}$$

where T is a real-time value, T_0 is a time of departure, and I is a unit of measured time-span.

2. Determine the average travel speed (ATS) between two consecutive stops.

$$\mathrm{s} = \frac{\sum_{i=1}^{n} \mathrm{s}_i}{n}, \tag{19.2}$$

where s_i is the actual speed of vehicle at a point in time *i*, *i* denotes the current time slot, and *n* is the number of time slots throughout the vehicle movement from the start point up to the current time slot.

3. Compute the distance from vehicle current location to the bus stop.

Information about the vehicle position is acquired from a GPS module, and then these values can be used to calculate the distance between the current and a fixed waypoint. The distance between current locations and a bus stop can be computed at least in two ways.

The first approach suggests that the distance *d* between two consecutive points can be computed from a haversine formula as follows:

$$d = \Delta\sigma \cdot \mathrm{R} \tag{19.3}$$

where *R* is the radius of the Earth (6378.1 km) and $\Delta\sigma$ is the angular difference.

When it comes to measuring relatively short linear distances between two geographic positions, the angular difference $\Delta\sigma$ is calculated by using the haversine formula as

$$\Delta\sigma = 2\arcsin\left\{\sqrt{\sin^2\left(\frac{\varphi_2 - \varphi_1}{2}\right) + \cos\varphi_1\cos\varphi_2\sin^2\left(\frac{\Delta\lambda}{2}\right)}\right\} \tag{19.4}$$

where $\varphi_1, \lambda_1; \varphi_2, \lambda_2$ indicate the latitude and longitude of point 1 and point 2, respectively; $\Delta\lambda$ is the longitude difference between two consecutive points. Haversine formula ensures a more straightforward computation, but it does not provide the high accuracy. The Vincenty's method can be utilized as an alternative to haversine, providing sufficient accuracy for any pair of points, but it is also more computationally intensive and, therefore, performs slower and increase battery usage. That why the haversine formula may be considered as a basis for calculating the distance between two points for IoT applications. In Android, the haversine formula is used in Google Map Utils [12].

The second way is to utilize the Google Maps API. It provides information about the real-time location and the distance calculation, and allows you to map GPS coordinates and get routes between numbers of points on a map. It is available for Android, iOS, web browsers, and via HTTP. In particular, Service Google Maps Directions API allows you to calculate routes between two known points using the HTTP request [13]. However, the Google API does not allow determining the external boundary based on time or distance from a location.

4. Calculate the expected arrival time of the vehicle to the particular bus stop according to the values of distance and speed obtained at the previous stages [14].

$$t = \frac{d}{s} \tag{19.5}$$

where *t* denotes the expected arrival time of the vehicle, *d* is a distance between two points, and *s* is the average speed of the vehicle.

The remaining distance is divided by the speed previously measured to roughly estimate the arrival time. To improve the results of the predicted arrival time of the vehicle to the city's stops, the Kalman filtering technique can be applied [15, 16].

19.2.2 Kalman Filtering

The Kalman filter is a linear recursive prediction algorithm used to evaluate process model parameters. Starting from the initial estimates, the Kalman filter allows us to predict the model parameters and adjust it with each new dimension. Its ability to combine both noise and process noise with measurements has made it very popular in many research and application areas, including autonomous and auxiliary navigation. For solving our task, the algorithm [17] consisting of five stages is applied.

Stage 1: Initialization

Set $k = 0, x_0^-, P_0^-$.

Stage 2: Extrapolation

a) State estimate extrapolation

$$x_{k+1}^- = F \cdot x_k^- + B \cdot u_k \tag{19.6}$$

where x_{k+1}^- denotes the predicted travel time at $k+1$; variable x_k^- is the travel time to be predicted at time *k*; *F* is the state transition matrix (dynamical model of system) which describes the time dependent relationship between x_k and x_{k+1} obtained from historical data; *B* is controllability matrix; and u_k is the controlling action in antecedent timepoint.

b) Error covariance extrapolation

$$P_{k+1}^- = F \cdot P_k \cdot F^T + Q_k \tag{19.7}$$

where P_{k+1}^- is the error covariance extrapolation; P_k^- is the error covariance at time *k*; and Q_k is the noise covariance.

Stage 3: Kalman gain calculation

$$K_k = \frac{P_k^- \cdot H^T}{H \cdot P_k^- \cdot H^T + R'} \tag{19.8}$$

where K_k denotes the Kalman gain; *H* is the measurement matrix, measurement-to-state ratio; and R' is the estimation error covariance.

Stage 4: Updating

a) State estimate updating with regard to z_k:

$$x_k = x_k^- + K_k \cdot (z_k - H \cdot x_k^-) \tag{19.9}$$

where z_k is the current observation, i.e., the average of the travel times of the vehicles at time *k*.

b) Error covariance updating

$$P_k = (I - K_k \cdot H) \cdot P_k^- \tag{19.10}$$

where *I* is an identity matrix.

If the system state is described just one variable, then $I = 1$ and matrices degenerate into the simple equation.

Stage 5: Until a vehicle reached the finish of route

Assign $k = k + 1$ and return to Stage 2.

For the case study, presented in the next section, the following function notation for Kalman filtering is used.

```
function filter($z, $u =0.1) { if (is_nan($this->x)) {
            $this->x = (1 / $this->H) * $z;
            $this->P = (1 / $this->H) * $this->R * (1 / $this->H);
        }
        else {
            // Compute prediction
            $predX = ($this->F * $this->x) + ($this->B * $u);
            $predP = (($this->F * $this->P) * $this->F) + $this->Q;
            // Kalman gain
            $K = $predP * $this->H * (1 / (($this->H * $predP *
$this->H) + $this->R));
            // Correction
            $this->x = $predX + $K * ($z - ($this->H * $predX));
            $this->P = $predP - ($K * $this->H * $predP);
        }
        return $this->x;
    }
endmodule
```

19.2.3 Accuracy Metrics

There are different performance metrics to evaluate the prediction accuracy; they are MAE function that calculates the mean absolute error, mean absolute percentage error (MAPE), root-mean-square error (RMSE), etc. [18, 19].

Absolute error (absolute deviation) of sample i

$$AE_{k,i} = t_{m,k} - \hat{t}_{m,k} \tag{19.11}$$

where $t_{m,k}$ is the actual arrival time collected by trolleybus equipment for trolleybus m at stop k and $\hat{t}_{m,k}$ is the predicted arrival time of trolleybus m at stop k.

Mean absolute error

$$MAE = \frac{1}{N}\sum_{i=1}^{N}\left|\hat{t}_{m,k} - t_{m,k}\right| \cdot 100 \tag{19.12}$$

where N is the sample size for prediction.

Absolute percentage error (APE) of sample i

$$APE_i = \frac{AE_{k,i}}{t_{m,k}} \cdot 100 \tag{19.13}$$

MAPE

$$MAPE = \frac{1}{N}\sum_{i=1}^{N}\left|\frac{t_{m,k} - \hat{t}_{m,k}}{t_{m,k}}\right| \cdot 100 = \frac{1}{N}\sum_{i=1}^{N} APE_i \tag{19.14}$$

RMSE

$$RMSE = \sqrt{\frac{\sum_{i=1}^{N}\left(t_{m,k} - \hat{t}_{m,k}\right)^2}{N-1}}. \tag{19.15}$$

To compare the accuracy between different prediction methods and quantify the prediction error in this study, we use MAPE as a performance criterion which measures the magnitude of relative error over an observed time range [20, 21]. In the experimental section, different combinations of arrival time prediction models were tested. A lower MAPE value indicates a higher precision of the model.

19.3 Case Study

This section describes how the proposed methodology and provisions discussed above has been applied for the development of advanced traveler information service particularly trolleybus arrival notification system (TANS).

We use general GPS and the method of uploading cellular signal information to the Google Maps server to acquire the trolleybus position and monitor their movement along a route with checkpoints. As a GPS tracking system, Wialon Hosting is used.

Wialon is a fleet management system [22], also served for tracking moving and stationary objects, observing dynamic changes their parameters, e.g., travel speed, voltage, temperature, etc.

19.3.1 General Strategy of Public Transport Information Service Delivering

A TANS is designed as part of public transport information service in Severodonetsk, Ukraine. To adjust and ensure the availability of TANS, the following steps are required:

1. Ensure the remote access to the GPS for obtaining spatial and temporal coordinates of the vehicle.
2. Collect information from the sensors on the location and travel speed of each vehicle included to the system. Add sensor data to the database. The sensors are surveyed during the working day.
3. Assign the routes to trolleybus (once at the start of working time).
4. Calculate the predicted arrival time of the specific vehicles for the specified bus stops (check-points). This task is performed for each vehicle, and the result is recorded in the database by their route.
5. Display the information about nearest vehicles at the bus stop information board. The trolleybus number and minimum arrival time for the bus stop are taken from the database concerning their predicted values.

19.3.2 The Remote Access Configuration

For working with Wialon RemoteApi, the library cURL is deployed. It provides functions for generating requests for the vehicle identifier on spatial and temporal coordinates and entry them into the database. A new entry is done as follows:

```
$result = $wialon_api->login($token);
$json = json_decode($result, true);
if(!isset($json['error']))
{
$result=$wialon_api-core_search_item('{"id":14157051,"flags":1024}');
```

```
$jn_result = json_decode($result, true);
$jn_x = $jn_result['item']['pos']['x'];
$jn_y = $jn_result['item']['pos']['y'];
$jn_s = $jn_result['item']['pos']['s];
$time = date('H:i:s');
$result = mysqli_query ($db, "INSERT INTO tr (id,x,y,s,time)
VALUES('14157051','$jn_x','$jn_y','$jn_s','$time')");
$wialon_api->logout();
}
else echo WialonError::error($json['error']);
```

19.3.3 GPS Data Acquisition

The software developed for this system takes the information from the Wialon. Information about geographic coordinates of a vehicle, travel speed, and sampling time in coordinated universal time is displayed in real time and saved in the database. The row of the database table corresponds to the point on the map.

An example row is: ('38 .4522149', '48 .9351065', 5, '09: 00: 53','6').

The fields in this row are latitude, longitude, speed (m/s), time, and route number.

These data are used for further calculations. It is assumed that a system provides measurements with a maximum error of 10 m under fair weather conditions and when the GPS receiver can acquire the signal from a minimum of four satellites [23].

19.3.4 Assigning Route Number to Each Trolleybus

Since trolleybuses do not have permanent routes and follow along the assigned route only within one working day, at the beginning of each work shift, it is necessary to perform the routing setup.

For all routes, the set of control points is selected, and their geozones are determined. A geozone is thought of as a predetermined map area in which we can access data. They are depicted as colored polygonal outlines or circle areas of radius R on the fleet tracking map (see Figure 19.2). The number of geozones equals the number of existing routes in the city.

Each control point is selected so that it belongs to only one route. This placement of control points avoids the problem of crossing routes and allows assigning the route to the trolleybus when it gains the specific zone. The assignment of the route is very simple and is as follows. In the first route, at the beginning of the working day, arrival time calculations are not performed.

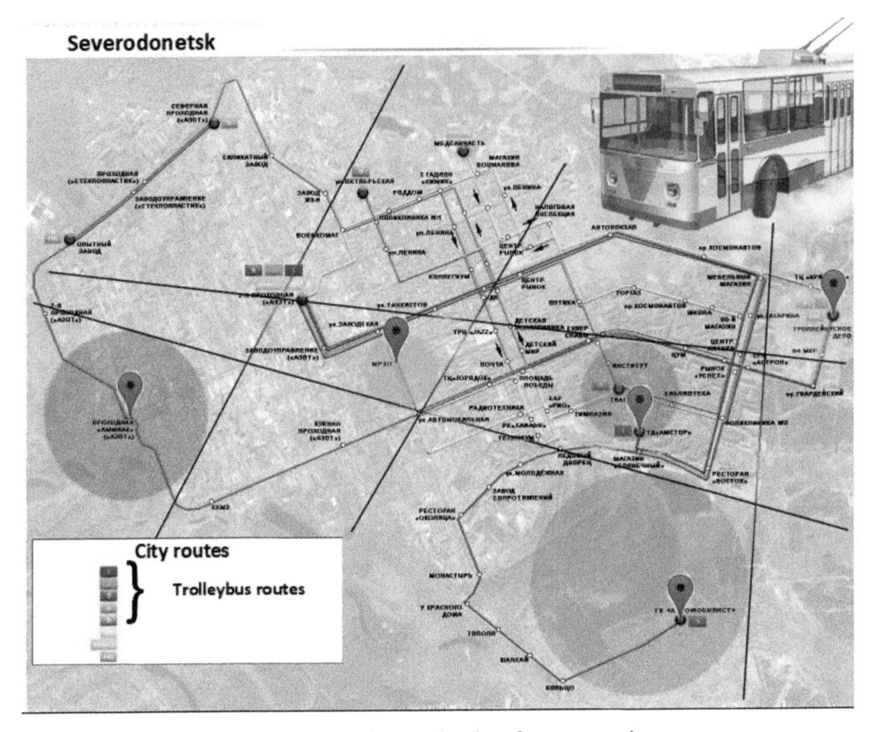

Figure 19.2 A set of control points for route assignment.

The objective of this phase is to fill an array of data with trolleybus IDs and numbers of geofence rooms where they have entered.

When a vehicle crosses a specific geozone, it is assigned a route. The assignment algorithm is shown in Figure 19.3.

19.3.5 Calculate the Predicted Arrival Time

For our purposes, we introduce the concept of a segment which is represented by the distance between two adjacent checkpoints (stops) with one or several bus stops. Thus, the segment notion provides useful flexibility of real-time information provided to the passengers. There is no need to install the information board at every bus stop.

The prediction of trolleybus arrival time at a certain checkpoint, in this case, is equal to the forecast of travel time as follows:

$$T_{a,j} = T_{d,i} + t_{ij} \tag{19.16}$$

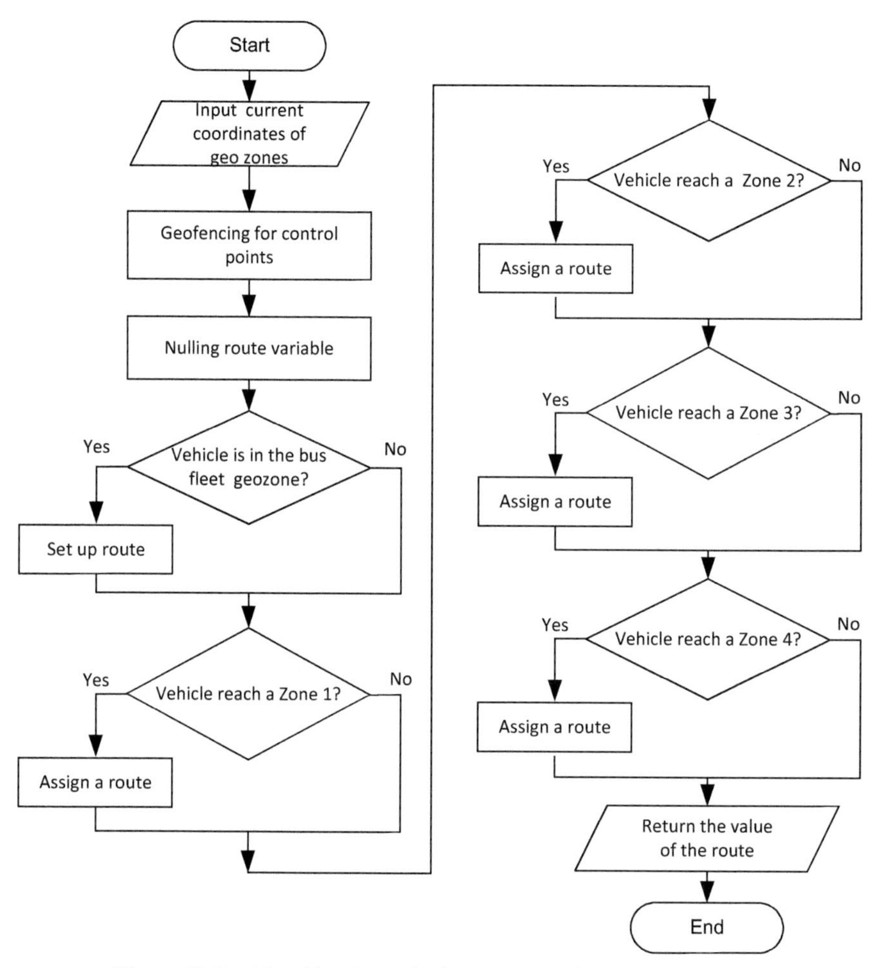

Figure 19.3 Algorithm for assigning route number to the trolleybus.

where $T_{a,j}$ denotes the trolleybus arrival time at a certain checkpoint (stop) *j*; $T_{d,i}$ is the time of departure from the checkpoint *j*; and t_{ij} is the travel time between checkpoint *i* and checkpoint *j*. It is assumed that there may be one or several bus stops between *i* and *j*. The remaining calculations are performed by formulas (19.1)–(19.10). When the calculations are over, a series of parameters for the predicted time of the selected trolleybuses are available.

19.3.6 Data Acquisition and Information Processing Algorithm

The data acquisition, information processing, and displaying strategies are described in the following flowchart (Figure 19.4). The trolleybus number and minimum arrival time for the bus stop are taken from the database concerning their predicted values. The bus stop information board is limited to four rows; thus, it seems reasonable to display four nearest incoming trolleybuses.

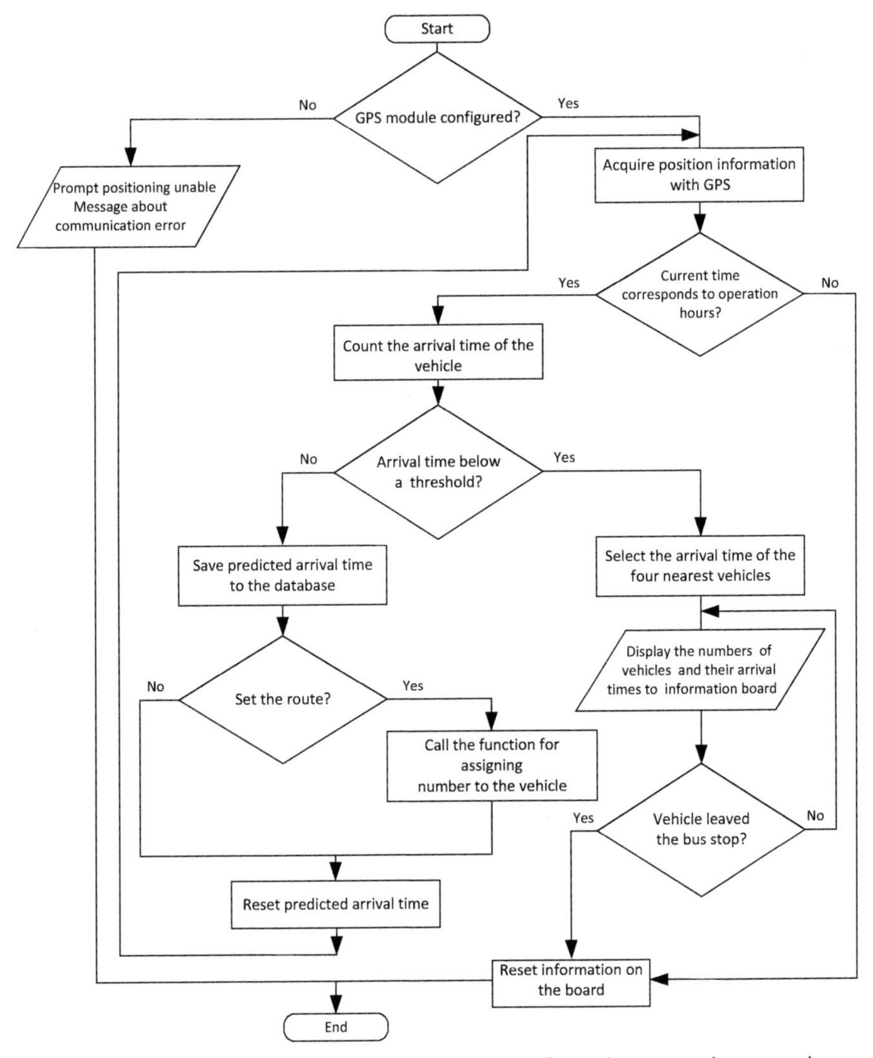

Figure 19.4 The flowchart of data acquisition and information processing strategies.

For better prediction accuracy, it is necessary to keep calculations updated every 15–30 s as the new information is received.

19.3.7 Experimental Results and Model Predictions for Trolleybus Arrival Time

The experiment was carried out with real data from GPS sensors installed on trolleybuses. The maximum length of the route is 12 km, the average speed of the trolleybus is 16 km/h, and the maximum speed is 25 km/h. The approximate travel time along the longest route is 45 min. Following recommendations [24], calculations were performed with for distances not exceeding 4 km, which corresponds to an interval of 15 min. From these parameters, the actual speed and the geographic location of the trolleybus were measured. The scanning of sensors was performed every 15 s while the trolley was moving from the beginning to the end of the route.

The trolleybus arrival time calculations are performed involving all techniques discussed in Section 19.2. The goal is to compare the efficiency different approaches and produce a method with

- Maximized accuracy
- Minimized prediction errors
- Minimized computational cost.

With all notions discussed above, there are at least eight combinations in which the predictor may be constructed.

- Method 1 aims to solve the prediction problem using ATS and computing the distance between the current locations of the trolleybus and the bus stop by the haversine formula.
- Method 2 comprises an ATS with Kalman filtering and the distance between the current locations of the trolleybus and the bus stop is computed using the haversine formula.
- Method 3 is grounded on ATS and the distance between the current locations of the trolleybus and the bus stop is obtained from Google Maps API.
- Method 4 uses the ATS with Kalman filtering and the distance between the current locations of the trolleybus and the bus stop is obtained from Google Maps API;
- Method 5 includes an ATS and the distance between the current locations of the trolleybus and the bus stop is computed using the haversine formula with Kalman filtering;

- Method 6 activates an ATS with Kalman filtering and distance between current locations of trolleybus and bus stop is calculated using the haversine formula with Kalman filtering;
- Method 7 uses ATS and the distance between the current locations of the trolleybus and the bus stop is obtained from Google Maps API with Kalman filtering.
- Method 8 involves the Kalman filtering both for the computing the average traffic speed of the trolleybus and the distance traveled by the trolleybus obtained from the Google Maps API.

Data were obtained from one route (Figure 19.5) during 10 days in December 2017. The trolleybus arrival time was computed from the start of the route, through two control points and to the terminal stop by formulas (19.1)–(19.10).

Every 15 s the ATS was re-computed. Table 19.1 shows the main characteristics of each route segment.

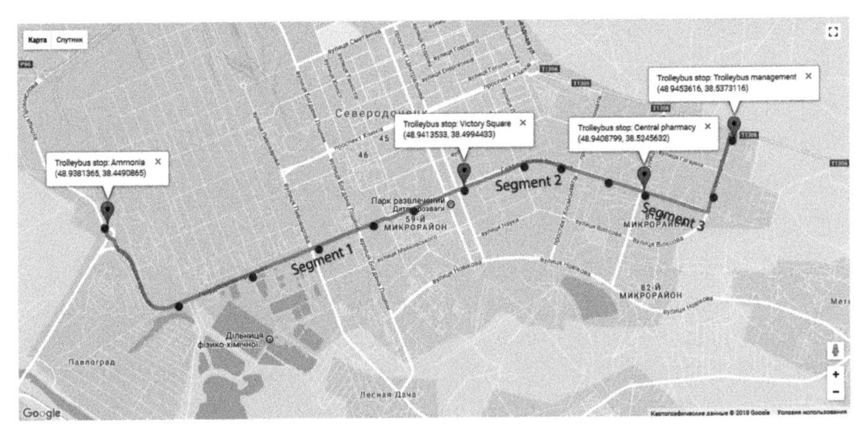

Figure 19.5 Test road segmentation.

Table 19.1 Characteristics of three different segments

	Length, m	Haversine Length, m	Number of Bus Stops Between Checkpoints	Number of Crossroads
Segment 1	4 500	3 696	5	5
Segment 2	1 900	1 836	3	3
Segment 3	1 600	1 056	1	0

Table 19.2 Experimental results of trolleybus arrival time for three road segments

	Parameters	Segment 1	Segment 2	Segment 3	MAPE
Actual arrival time, s		765	482	449	
Method 1	Arrival time, s	647	491	274	18,76
	AD	+118	–9	+175	
	APE	15,4	1,87	38,96	
Method 2	Arrival time, s	647	489	273	18,69
	AD	+118	–7	+176	
	APE	15,42	1,45	39,19	
Method 3	Arrival time, s	788	508	415	5,32
	AD	–23	–26	+34	
	APE	3,00	5,39	7,57	
Method 4	Arrival time, s	788	506	413	5,33
	AD	–23	–24	+36	
	APE	3,01	4,98	8,01	
Method 5	Arrival time, s	646	493	275	18,86
	AD	+119	–11	+174	
	APE	15,56	2,28	38,75	
Method 6	Arrival time, s	+645	491	273	18,92
	AD	120	–9	+176	
	APE	15,69	1,87	39,19	
Method 7	Arrival time, s	788	508	415	5,32
	AD	–23	–26	+34	
	APE	3,01	5,39	7,57	
Method 8	Arrival time, s	788	506	413	5,33
	AD	–23	–24	+36	
	APE	3,01	4,98	8,02	

The calculation of the average speed was carried out for each segment separately, without taking into account the average speed on the previous segments. Table 19.2 summarizes several route discovery test cases from three road segments to realize comparative analysis for different implementations.

The main parameters that are taken into account to carry out the experimental comparison are the arrival time to the checkpoint of a route, absolute deviation (AD), APE, and MAPE. For the largest segment with five bus stops, the AD varies from 23 s to 2 min, arrival time predictions for the second segment give us AD up to 26 s lag, and the shortest segment shows the worst results where AD varies from 34 s to almost 3 min ahead of real time.

The MAPE between the predicted travel time and the measured travel time of the test trolleybuses for consecutive bus stops shows variability for these 10 days from 5.33% to 18.92%.

From the field experiments, it was found that the response of the Kalman filter to the jump in the parameter values (the travel speed and the location of the trolleybus) was insignificant. Minor improvements (a maximum of 2 s) have been achieved when the Kalman filter was applied on the second and third segments of the route, that is, at the end of the route (see Table 19.2). Moreover, the use of the Kalman filter to calculate the distance on the shortest segment using the Google Maps API did not yield any improvements, since this service does not take into account the difference of 10 m, namely, this difference was shown by the haversine method.

The diagram of actual and predicted arrival tile (Figure 19.6) shows that the highest error was observed at the beginning and the end of the route (segments 1 and 3).

The most effective are methods 3 and 7, the MAPE of which is 5.32% much better than any of haversine. This is because the Google Maps API returns the true length of the path segment, taking into account all the turns of the route. In contrast, the haversine methods give us the length of the straight segment. However, this fact does not exclude this technique, since it is possible to linearize the route.

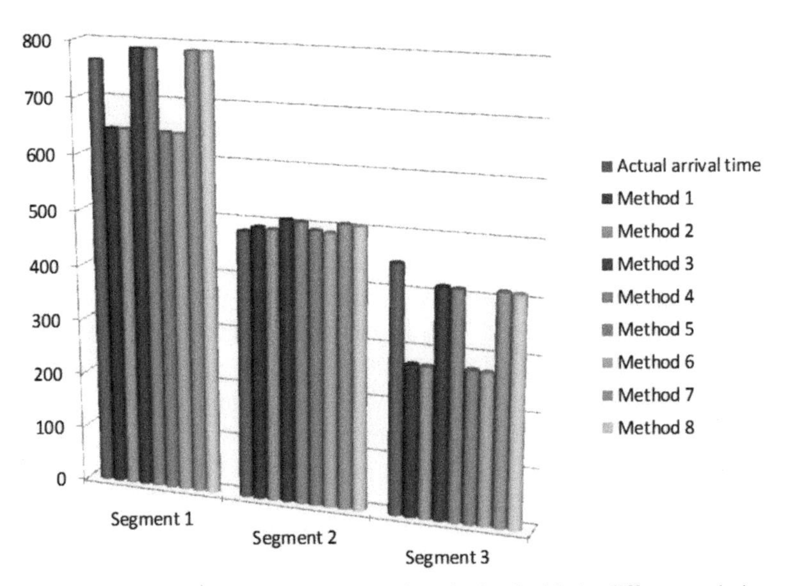

Figure 19.6 Trolleybus arrival time prediction obtained with the different techniques.

19.4 Conclusion and Future Work

Advanced public transport information services have significant benefits ensuring suitable information about services and logistics available on the road and making urban passenger transport more efficient and reliable. The major contribution of this paper is an integrated, formal, and automated methodology for public transport information services. In this study, an architecture model for the deploying information service infrastructure for public passenger transport was developed. A concept for real-time data acquisition and choosing an efficient model for trolleybus arrival time prediction was evolved.

As a use case of the proposed technique, we have implemented and tested different models for the trolleybus arrival time prediction on the existing routes. The efficiency evaluation of the models has been performed with respect to the actual arrival time and prognosis time.

The best arrival time predictions for low-speed urban transport were obtained using ATS and distance from Google Maps API and ATS and distance from Google Maps API with Kalman filtering with MAPE 5.32%. The calculation has also demonstrated that for low-speed transport, considerable reduction of prediction errors using Kalman filter has not been achieved.

All the above results make it sure that these techniques can be easily implemented in real time to improve public transport services by leveraging on the GPS data and data provided by the IoT applications. Information about vehicle arrival time can be disseminated via the various services, e.g., information boards and kiosks at bus stops, displays in a public transport, and smartphones through the Internet and by SMS.

These results are meant to be considered for the next series of experiments on two pilot information boards. Future work will be aimed at implementing this technology in the public transport infrastructure and validating whether it improves the passenger experience. It is also expected that follow-up study will expand on full employment of IoT technologies to improve traffic and passenger services and involve both vehicle-to-infrastructure and vehicle-to-vehicle communication along with their enhancement.

References

[1] Ezell, S. (2010). 'Explaining International IT Application Leadership: Intelligent Transportation Systems,' *The Information Technology and Innovation Foundation*, 58.

[2] Keramidas, G., Voros, N., and Hübner, M. (eds). (2017). *Components and Services for IoT Platforms: Paving the Way for IoT Standards.* Basel: Springer International Publishing, 383. doi: 10.1007/978-3-319-42304-3

[3] Festag, A. (2014). Cooperative intelligent transport systems standards in Europe. *IEEE Commun. Mag.*, 52(12), 166–172. doi: 10.1109/MCOM.2014.6979970

[4] Williams. B. (2008). *Intelligent Transport System Standards.* London: Artech House, Inc., 827.

[5] ETSI ITS-G5 Standard. (2011). Final draft ETSI ES 202 663 V1.1.0, Intelligent Transport Systems (ITS); European profile standard for the physical and medium access control layer of Intelligent Transport Systems operating in the 5 GHz frequency band. *Technical report ETSI.*

[6] Chung, E.-H., and Shalaby, A. (2007). Expected time of arrival model for school bus transit using real-time global positioning system-based automatic vehicle location data. *Journal of Intelligent Transportation Systems*, 11(4), 157–167.

[7] Sun, D., Luo, H., Fu, L., Liu, W., Liao, X., and Zhao, M. (2007). Predicting Bus Arrival Time on the Basis of Global Positioning System Data. *Transportation Research Record: Journal of the Transportation Research Board*, No. 2034, 62–72, 2007.

[8] Bai, C., Peng, Z.-R., Lu, Q.-C., and Sun, J. (2015). Dynamic Bus Travel Time Prediction Models on Road with Multiple Bus Routes. *Computational Intelligence and Neuroscience*, 2015. doi: 10.1155/2015/432389.

[9] Yu, B., Zhongzhen, Y., and Baozhen, Y. (2006). Bus arrival time prediction using support vector machines. *Journal of Intelligent Transportation Systems*, 10(4), 151–158.

[10] Chien, S.I.J., and Kuchipudi, C. M. (2003). Dynamic Travel Time Prediction with Real-Time and Historic Data. *Journal of Transportation Engineering*, 129(6), 608–616.

[11] Cathey, F. W., and Dailey, D. J. (2003). A prescription for transit arrival/departure prediction using automatic vehicle location data. *Transportation Research. Part C*, 11, 241–264.

[12] Broadfoot, C. *Googlemaps* Android-maps-utils/*MathUtil.java*. (Sept. 2013) Available at: https://github.com/googlemaps/android-maps-utils/blob/master/library/src/com/google/maps/ android/MathUtil. java.

[13] *Using the Google Maps API to book the perfect ride.* Available at: https://static.googleusercontent.com/media/enterprise.google.com/ru//

maps/resources/UsingtheGoogleMapsAPIstobooktheperfectride.pdf [accessed Jan 7, 2018].

[14] Asfandyar, M. (2015). "Real-Time Public Transport Arrival Information System," in *Student Research Paper Conference*, 2(13), 67–71.

[15] Huang., Y., Xu, L., Luo, Q., and Kuang, X. (2013). Urban Expressway Travel Time Prediction Method Based on Fuzzy Adaptive Kalman Filter. *An International Journal Applied Mathematics and Information Sciences*, 7(2L), 625–630.

[16] Zaki, M., Ashour, I., Zorkany, M., and Hesham, B. (2013). Online Bus Arrival Time Prediction Using Hybrid Neural Network and Kalman filter Techniques. *International Journal of Modern Engineering Research (IJMER)*, 3(4), 2035–2041.

[17] Yang, J.-S. (2005). "Travel Time Prediction Using the GPS Test Vehicle and Kalman Filtering Techniques," in *American Control Conference*, Portland, OR, 2128–2133.

[18] Li, B., Liu, M. (2014). "Performance Evaluation of Arrival Time Prediction Models," in *CICTP 2014: Safe, Smart, and Sustainable Multimodal Transportation Systems*, 2867–2873.

[19] Velásquez, J. D., Ríos, S. A., Howlett, R. J., and Jain, L.C. (eds). (2009). *Knowledge-Based and Intelligent Information and Engineering Systems.* Berlin: Springer International Publishing, 383. doi: 10.1007/978-3-319-42304-3

[20] Chai, T., and Draxler, R. R. (2014). Root mean square error (RMSE) or mean absolute error (MAE)? – Arguments against avoiding RMSE in the literature. *Geoscientific Model Development*, 7, 1247–1250.

[21] Willmott, C. J., and Matsuura, K. (2005). Advantages of the mean absolute error (MAE) over the root mean square error (RMSE) in assessing average model performance. *Climat Research*, 30, 79–82.

[22] Wialon. Available at: https://gurtam.com/en/wialon

[23] Covaciu, D., Preda, I., Ciolan, G., and Câmpian, O.-V. (2010). "Data acquisition system based on gps technology, for vehicle dynamics analysis," in *Conference: CONAT 2010 - International Automotive Congress*, At Brasov, 31–36.

[24] Biagioni, J., Gerlich, T., Merrifield, T., and Eriksson, J. (2011). "EasyTracker: Automatic Transit Tracking, Mapping, and Arrival Time Prediction Using Smartphones," in *Proceedings of the 9th ACM Conference on Embedded Networked Sensor Systems*, 68–81.

20

Scalable Smart Transducer Networks Using Power-over-Ethernet and Neural Networks

Ivan Lobachev[1,2]

[1]Intel Corporation, USA
[2]Odessa National Polytechnic University, Odesa, Ukraine
E-mail: lobachev@ieee.org

In the age of rapid technological advancement, many telecommunication applications are being integrated into our lives, including smart phones and Internet of Things (IoT). Smart buildings (and houses) use these technologies to reduce energy consumption and increase safety and add supporting features, where applicable. The need for these buildings is growing as urbanization continues and resources dwindle. According to a report by the United Nations, by 2050, 66% of the world's population will live in cities. This will cause the size and number of megacities to expand drastically in the near future. Complex communication networks, controls, and other services will allow us to build smart cities to manage and improve the public's quality of life. The necessity of smart buildings and, eventually, cities, has stimulated growth of sensor networks for these purposes. This paper discusses the research on creating a network architecture concept that uses Power over Ethernet (PoE) as a method for transferring data and power over a single medium, conjoined with the principals of neural networks as well as decentralized and remote computing for data processing. The concept used a Cisco Catalyst 4507R+E switch and utilized cloud and on-board computing to provide an easily scalable and adaptable architecture that can be modularly integrated into existing solutions. The setup was tested on RaspberryPi microcontroller boards as sensor hubs, and used DigitalOcean as the cloud computing service of choice. The Movidius Neural Compute chip was used to deploy the neural networks for the relevant data processing and deep learning. The server in this implementation acts as the user interface, the front end, and the console unit back end. The proposed architecture shows great promise on the feasibility of creating modular systems that unite the concepts of IoT, PoE, and neural network approaches.

20.1 Introduction

Sensors have become indispensable elements in today's society, being utilized in nearly every electronic device that is being used. Due to this situation, there has been a great advancement in the field, and today, there is an innumerable variety of sensors for every application imaginable. As the technology moves forward, so do the trends and user demands. One of such trends is smart buildings IoT as well as the new "mega buildings" [1] we are seeing more and more of, along with the need to gather data on the state of the building and its affairs. Another trend is "ecological buildings" or tall wooden buildings which have been an object of interest particularly in areas with a high supply of lumber such as British Columbia. Furthermore as noted in a report by the United Nations, by 2050, over 65% of the world's population will live in urban areas according to their projections [2]. This means that the cities will continue to evolve into smart cities, expressing a need for smart and green-oriented buildings, and infrastructure. This in turn will require a system with a data-oriented approach, along with a sufficient degree of autonomy, as well as adaptability to different scenarios. With currently present approaches, such large-scale systems risk being infrastructure dependent, and potentially very costly both in installation and maintenance. By devising an architecture that is able to modularly integrate elements of technology available today, a system that maximizes the strength of each individual component can be constructed. This approach allows us to raise the degree of scalability and deployability, as well as minimize redundant closed-loop data transmission to improve resource efficiency by means of sharing the load via a distributed computing approach as well as by minimizing the need to have custom connections and interface.

20.2 Research Objectives and Related Work

The main objectives of this work are to develop a flexible and scalable architecture concept of a system that would be able to unify the benefits provided by IoT, along with power and installation efficiency of PoE as well as to investigate the feasibility of integration of neural network approaches and relevant hardware. PoE has demonstrated the possibility to reduce installation costs for certain types of applications, such as LAN, IP, and VoIP [3]. Furthermore, the necessary infrastructure for a PoE-enabled system is already in place for any building that has internet access thorough Ethernet ports, as the same Ethernet connections can be used for both power and data transmission. The system architecture needs to be flexible enough to adapt to the vast variety of sensor types and configurations, in order to allow various peripherals to be connected or disconnected to an existing, operating network. One of objectives includes investigating the applicability and usability of some off-the-shelf smart sensors, applicable to structural monitoring applications. In addition, the research aims to base the new concept on the lessons learned from previous iterations, in order to improve the aspects such as scalability, adjustability, and ease of deployment.

Once the prototype of the architecture concept has been developed, it was tested to verify the feasibility of use of such a system, and its possible deployment schemes. In particular, the testing attempted to answer whether the new design concept is able to adhere to the paradigm of high degrees of scalability and adjustability. The prototype system passed the testing stage, the details of which can be found in Section 20.5. As a result, the developed concept can be used to later create a commercial system, a hybrid sensor network (as enabled by the designed system architecture), combining the abilities and properties of both wireless sensor networks as well as hard-wired sensor networks and their respective sensor modules.

This area has been previously investigated from a number of different directions, primarily from the point of view of the type of network in questions, for example, wired vs. wireless. Furthermore, a separate criterion is the deployment scale, a single smart room or house versus an entire network, for example, smart energy grid. A good example of a smart sensor that has been previously developed is the Imote2 smart sensor platform developed by Jennifer Rice et al. [4]. Their work focused on a wireless sensor network setup, differentiated from other solutions by employing a decentralized computing scheme to reduce the amount of data communication within the network; as a result, the architecture has reduced the amount of power consumed by the system through limiting the power usage of individual elements. However, while the system does offer benefits over traditional wireless setups (where the design would attempt to mimic a wired configuration using wireless transmission), it still has its flaws. While the group managed to reduce the overall power consumption, the main problem becomes the maintenance of the energy sources, which are in these cases batteries, regardless of whether they are primary or secondary cells. A number of other recently developed solutions target more specific applications, such as wireless sensor networks for home lighting systems [5], industrial wireless sensor network data transmission scheme in [6], or the case study on Greenorbs in [7]. Which looked into the issues of scalability and system dependence on individual nodes, an issue addressed in this work. Yet another solution for a semi-wireless approach is for structural health monitoring, which has been implemented by Xu et al. with their system named Wisden [8]. The system was then later improved and tested again by Chintalapudi et al. who published their results in [9]. However, while the system uses wireless sensors, which send the data to a local hub, or node as titled by the authors, it is then connected to a local PC via serial port, a hard-wired connection. The computer would then handle the bulk of the data processing and can be connected as a network to other computers employing a similar setup. This design shows a number of drawbacks, such as the need for the wiring, which defeats the purpose of using a wireless setup as well as the delay due to the serial connection employed when compared to faster protocols such as TCP/IP which can be employed in the architecture presented in this work. A number of attempts to optimize the wireless configuration using neural networks have also been approached by Kulkarni et al. with their approach to utilize neural networks for a different clustering algorithm in [10]; however, this approach would present its own challenges when attempting to increase the deployment scale.

20.3 Advantages and Improvements

One major disadvantage of wireless sensor networks is that in big concrete buildings, the signal faces a lot of interference, from both other electronic equipment and due to signal bouncing off of the walls, ceilings, and floors of the rooms. These effects were also demonstrated by Sato et al. [11] as well as by Wallace et al. [9] where measurements were taken to gage the behavior of wireless systems in such environments. This increases the difficulty of having a high concentration of sensors in a single given area. Furthermore, current systems such as Wisden, while wireless in sensor-to-gateway transmission, still require a wired connection for gateway to computer as well as separate wiring for power. By utilizing PoE and cloud computing as well as taking advantage of the computational power available on modern microcontrollers such as Raspberry Pi, BeagleBone, or the vast selection of TI MCUs, it is possible to create an easy to set up, efficient system, with minimal wiring and local processing of time-sensitive data for prompt response to events. Lastly, the use of Ethernet-based data transfer protocols directly, for the sake of the prototype UDP, was employed, and the transfer rates are more optimal than a serial connection such as RS 232, for example [12]. The introduction of neural networks in turn, when properly configured, can help offload the data processing from the main processor, and methods similar to [13], as discussed by Rao et al. can be applied for the collected non-linear sensor data.

20.4 System Architecture

The concept required the architecture to take the matters of saleability, configurability, deployability, and power efficacy into account. These were achieved by employing a hierarchical structure to ease expansion and saleability, while modular sensor hubs and PoE were used for deployability and configurability aspects, with the addition of a software control layer. Further power efficiency was added by implementing varying sensor hub classes, with the corresponding varying operation schedules, and semi-distributed data processing with a heavy emphasis on cloud computing. The general overview of the architecture can be seen in Figure 20.1. The inter-system communication and linkage, in addition to providing the possibility of hierarchical and remote control, allows for system to be optimized with regard to power consumption and processing cycles, by only powering on the necessary hubs on a per-case basis; the role of a hub is to act as a local sensor module interface, with some additional configuration and local signal processing tasks. Further details on

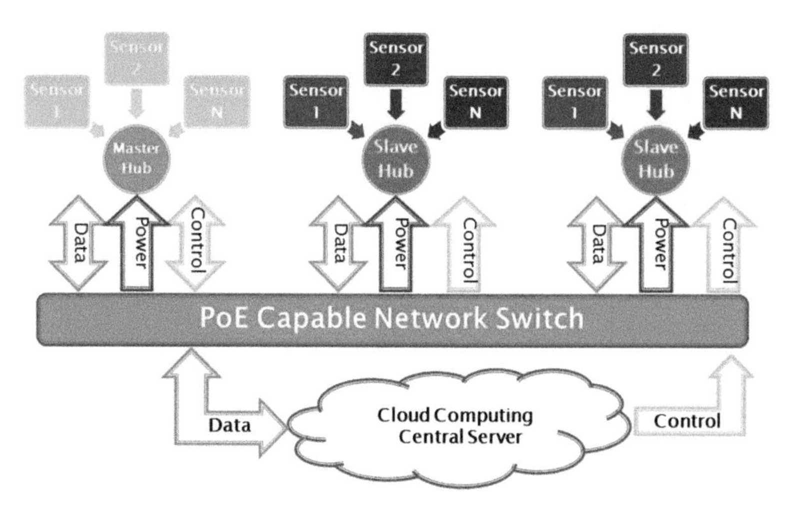

Figure 20.1 General architecture view.

the operation models of operation and classes of the hubs will be discussed in the sections below, and the principle behind the case-dependent activation can be observed in Figure 20.3. The use of PoE allowed for both the data, power and control commands to be sent via a single cable, which is already present in almost any building that has an internet connection infrastructure.

The setup of the prototype used to test the validity and feasibility of the concept proposed in this work has used a Cisco Catalyst 4507R+E switch [14] to handle the routing and supply the power to the sensor hub network [15].

The catalyst switches of these series are capable of delivering up to 60 W per port using Universal Power Over Ethernet (UPOE), which essentially doubles the per-port power output specified by the PoE standard [16]. In this setup, a hub is a microcontroller board (MCB) capable of receiving PoE, either by means of a shield[1] or an integrated adapter, and is capable of housing the Linux kernel. While the current prototype used an Ubuntu distribution, a lighter operating system, such as Puppy Linux or ArchLinux, can be used on MCBs with less resources, as well as scenarios where the size of the hub has higher priority over the amount of local processing that is necessary. A good example of such a scenario would be the slave class hubs, which will be discussed later in this chapter. Due to their task being non-resource-demanding, the proposed scheme of optimization can be used to reduce costs. The booting sequence of the operating system on the MCBs used in the experiment was modified to reduce booting time, by preventing programs and services unnecessary for the functionality of the system from loading. A high-level overview diagram of the system concept can be found in Figure 20.2. The figure outlines the overall setup that was assembled, while later a lower level description will be provided to accompany the discussed details.

20.4.1 System Requirements

There is a short list of requirements for the components that can be used to set up the system. In particular, the MCB should have the following:

1. **Sufficient processing capabilities** – The board, which will serve as a sensor hub module, needs to be able to run either Linux, or some operation system that is capable of connecting to the Internet and executing python scripts (in the current iteration). It also needs to able to process and package the data into HDF5 format for sending.

[1]https://www.pi-supply.com/product/pi-poe-switch-hat-power-over-ethernet-for-raspberry-pi/

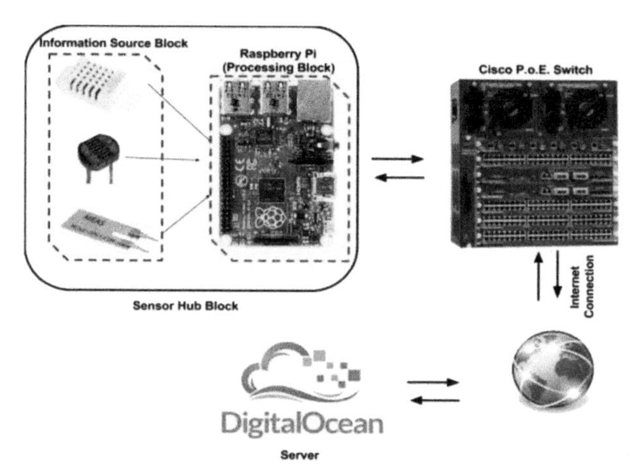

Figure 20.2 A high-level diagram of the implemented system set-up.

2. **An interface to connect the sensors** – The board needs the capability to connect the various types of sensors that are necessary for a given task. The connection and corresponding peripherals can be either wired or wireless by nature, depending on the boards' capabilities and other requirements and restraints of a given application.
3. **Sufficient storage** – This aspect needs to be considered for MCBs that will operate in semi-autonomous and autonomous modes as described in Section 20.4.4. The storage requirement in this case will be dictated by the amount of information to be processed and stored, which in turn is determined by the application. In our setup, the dedicated storage for the data was 4 Gb.
4. **PoE capable** – The last major requirement is the ability to operate with PoE. This can be accomplished either by a built-in circuit that the board has or by a means of a shield, such as the Pi POE Switch hat for Raspberry pi, or a Cisco Catalyst Power Splitter for many of the other boards.

20.4.2 Sensor Hub Classes

There are two general classes of sensor hubs in this system setup: a master class sensor hub and a slave class sensor hub. The naming convention was used due to the behavior of the system as described below.

1. **Master** is a sensor hub class which can operate in either fully autonomous or semi-autonomous mode of operation. In order to assign this class to a hub, it needs sufficient processing power, and the amount

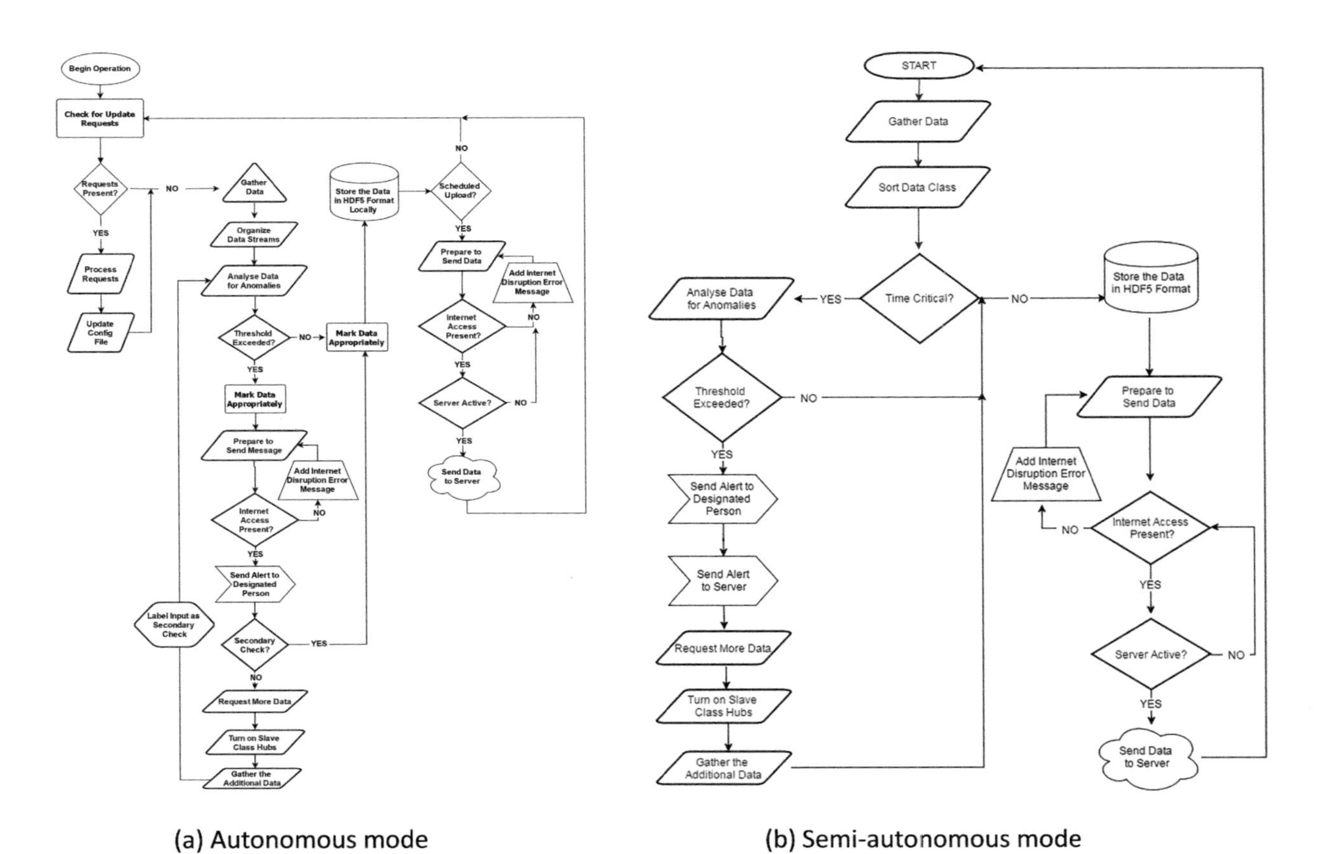

(a) Autonomous mode (b) Semi-autonomous mode

Figure 20.3 A diagram displaying the high-level algorithm of semi-autonomous and autonomous modes of operation.

is dictated by the exact application and performance constraints. While in possession of this class, the hub may issue commands to the slave class hubs assigned to it, such as to gather more data, and is also capable of bi-directional communication with the server.

2. **Slave** is a sensor hub class which is capable only of one mode of operation. However, while it is unable to perform complex or resource demanding tasks locally, it provides the benefit of a lighter build. This in turn means that the hub may have much smaller physical dimensions and may be more economical, which provides a more financially efficient solution on larger scales.

20.4.3 Configuration

As seen in Figure 20.4, the edit menu provides the user with the ability to configure all of the sensor hub modules via the web interface, remotely. The changes made there will be carried over into the master configuration file, which is stored on the server. The configuration file uses JavaScript Object Notation, JSON, format to store all the settings and relay them to

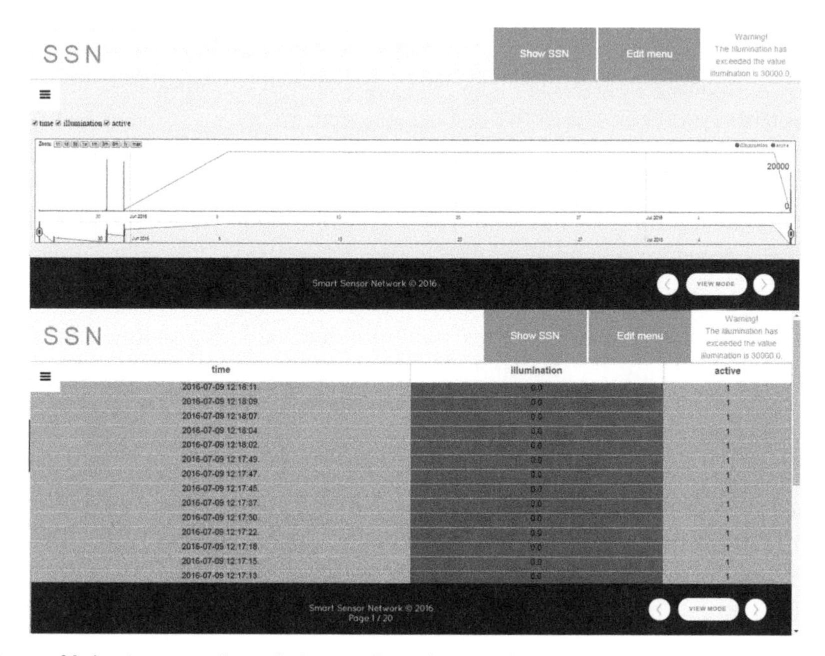

Figure 20.4 A screenshot of the varying views and representation of numerical data collected.

the respective modules. Once any changes have been made, the server will issue an update notification to the MCBs that have been affected; once the MCBs receive the notification and have prepared to receive information, the new configuration file for them will be created, sent, and applied, upon which the changes will have taken effect. The individual configuration file is created by taking a relevant snippet of the JSON master list file that contains the information of all the MCBs, including the ones that were affected. That snippet is then used to create the file that will be sent. If desired, all of the settings can be altered by directly accessing the back end as well, assuming that sufficient access privileges are present.

20.4.4 Requirements and Modes of Operation

While an MCB acts a sensor hub, it can house a number of logical clusters of sensors, for example, a sensor that provides multiple types of data, such as an accelerometer or a humidity and temperature sensor such as DHT22 or even a visual data stream. The hierarchy of the system takes into account what is considered a "parameter," as seen in Figure 20.5. The MCB, which, as long as it meets the requirements outlined in Section 20.4.1, is non-discriminant of the type and model used, communicates directly with the server.

The server is located on a cloud computing service host; in the case of this implementation, DigitalOcean was used; however, depending on the application, the server can be both outsourced to a third party or hosted locally depending on the available resources and performance needs. In addition to the previously implemented functionality of a master/slave status of the sensor hubs, which focused more on power consumption control, the system now also has three modes of operation, designed to allow it to adapt to any scenario. These modes are:

1. **Autonomous** – The sensor hub functions individually, conducting all of data processing locally, and performing actuation and/or executing scenario scripts as is dictated by its local code. The hub will periodically send its stored information to the server. When operating in this mode, the hub will also periodically probe the server to determine whether it is operational and whether Internet is present. If either one of them does not respond, the error will be logged and brought to the attention of the person in charge once the connection is restored.
2. **Semi-autonomous** – The most common mode of operation, typically used by the master class sensor hubs. When operating in this mode, the hub will process only time-critical information locally, and act based on the results. Data that are categorized as non-time-critical will simply be

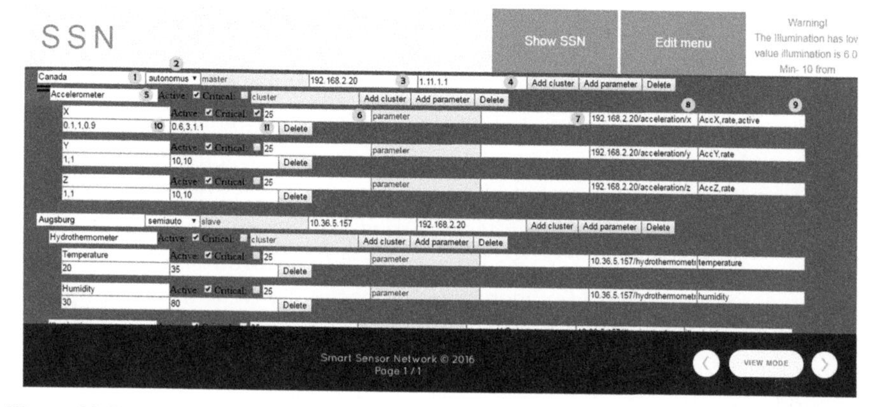

Figure 20.5 A screenshot of the web view of the user interface and configuration window. *The numbers highlighted in the figure are used to reference the fields they are located at, where: **1** is the name of the master or slave module, **2** is the mode of operation, **3** is the IP address of the module, **4** is the IP address of the corresponding switch or master module, **5** is the name of the cluster, **6** is the pin or port that the cluster is connected to on the board module, **7** is the e-mail address of the person to contact in case of anomalies, **8** is the path within the HDF5 file, **9** are the headings in the HDF5 file that will be monitored, **10** are the minimum thresholds for the specified parameters, and **11** are the maximum thresholds for the specified parameter.*

periodically forwarded to the server by the hub. A diagram outlining the algorithm of this mode of operation can be seen in Figure 20.3(b).

3. **Follower** – The most basic mode in which the hub will simply execute the commands it was given, while forwarding all of the collected data to the server for processing.

20.4.5 Parameters, Organization, and Data Processing

Each parameter has maximum and minimum threshold values; if the current value of the reading exceeds either of the defined boundaries, an alarm is triggered. The numerical data are stored and can be inspected via two modes, as seen in Figure 20.4. The analysis algorithm marks the concerning values on the web page for manual inspection if needed and sends a notification to the corresponding e-mail; in addition, the error is kept in a log of alarms which is present on the main header tab, as can be seen in Figures 20.4 and 20.5. The alarm will also trigger an additional data request. This request will wake up all of the slave hubs that are under the master hub which recorded the anomaly. Once awake, the slave hubs will collect an additional set of data to either confirm or deny the presence of the anomaly that was picked up from the readings provided by the master class hub.

By unifying the main control under one server, and allowing it to communicate with any of the switches or hubs from a single location, the scalability becomes a trivial matter. To facilitate direct access, the switch must first be configured to allow Secure Shell (SSH) [14, 15]. Once that is done, the switch can be accessed remotely by any machine that has access to the Internet and the terminal, through the server virtual machine. To further automate the process, a macro can be saved for each switch if necessary by writing a simple script which will pull the necessary credentials from a configuration file containing the information on the switches, hubs, and sensors [17, 18]. Due to the need for security, as the switches may potentially have a lot of control within a building, measures need to be taken to secure the information. However, due to the focus of this work being the improvement of scalability, adaptability, and responsiveness, this aspect will be further discussed in Section 20.5 as future work. One of the primary uses for automatic SSH scripts is to issue commands to the switch remotely and autonomously, such as in the case where additional slave hubs need to be activated. For this purpose, Tool Command Language (TCL) is used on the switch side. It is invoked by the server once an SSH connection is established and can be configured to perform a variety of tasks in addition to controlling the power supply. A CRON job is used to perform a series of regular, periodic checks on the non-time-critical data; the server also constantly listens to any alerts from the hubs. The script can be activated in two cases, either an alert about an anomaly is received directly from the master hub, which processed the critical data locally, or derived from processing the information on the server. Each hub, if requesting additional data, will transmit its ID along with the request, which will be used to look up the required credentials. Once found, a script will be run to issue the command to the corresponding switch and slave hubs. When the data are processed on the server, the IP address of the hub that detected the anomaly will be used to find the required credentials.

20.4.6 Data Processing and Presentation

As mentioned previously, data are collected and stored in HDF5 format. The use of the format allows assigning meta data to the gathered measurements, which simplifies the process of parsing the data for further processing, as well as cross platform import, in cases where different parties want to use different software to analyze the data locally. In addition, once the data have been

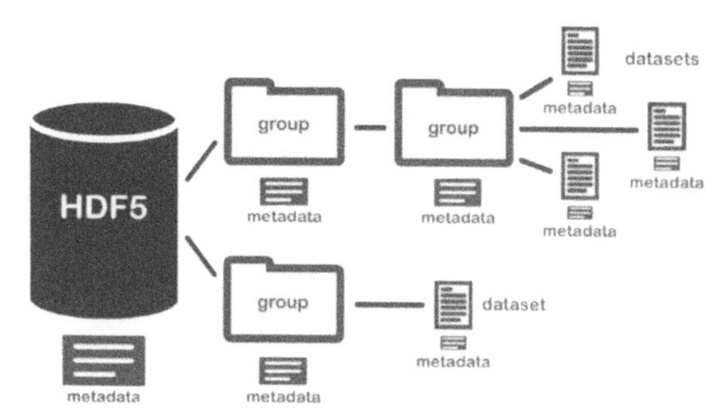

Figure 20.6 A diagram displaying the structure of the HDF5 data format [19].

arranged correctly, the new transition times of file transfer become shorter, as the files are a lot more compact and light weight. In order to achieve this, the raw data are first gathered from the individual sensors. A preassigned name is used to label the source of information; the name is obtained from the JSON configuration file. In addition, the type of data that are being gathered and its safety constraints are also used as the meta data for the measurement. Once formatted correctly, the newly arranged data are stored in its own subdirectory of the local database, as seen in Figure 20.6 along with the other measurements. On the server side, the data are extracted, parsed into arrays, and displayed to the user.

20.4.7 The Hierarchy

Hierarchy is a very important aspect in a project that emphasizes scalability such as the one discussed in this work. As such, a few approaches are used to introduce the concept of hierarchy into the system. The database format that is used to send, store, and process the data is Hierarchical Data Format – version 5 (HDF5), which is great for fast transfer of data due to its ability to keep the file sizes low and well organized. The latter point also makes it great for processing, as accessing any piece of information from the database or parsing it becomes simple when using HDF5.

Hierarchy is also employed in other aspects of the design, as can be seen in Figure 20.7; the hierarchy determines the amount of control a unit has. For example, a server has full authority, while a single salve hub has none. This "chain of command" makes propagation of commands rather simple and clean.

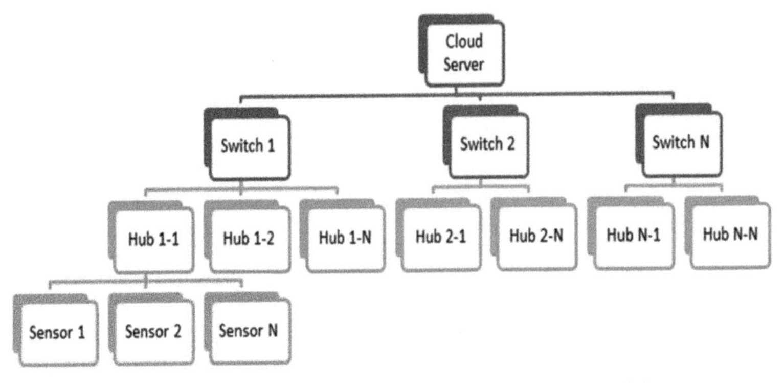

Figure 20.7 A high-level diagram outlining the hierarchy of the system.

20.4.8 Incorporation of Neural Networks

To demonstrate the modularity of the system, and to further enhance its applicability and local data processing performance, the possibility of introducing the concept of neural networks was investigated. The Movidius Neural Compute Chip by Intel [20] was used as the mobile host for the neural network due to its low power consumption, USB interface, and mobility. As the main goal is to have a mobile, distributed *ad hoc* network, the hubs need to be modular, and consume the least amount of power possible. That was the reason for choosing a low-power processor MCB; however, typical hosts for neural networks in addition to having rather strict hardware and interface protocol requirements also have high levels of power consumption, such as the NVDIA GPUs cards which can consume in the hundreds of watts per unit. Furthermore, the available interface methods would allow at most one unit per hub, with an elaborate adaptation hardware, which does not fit the needs of the system. Instead, the Movidius NCS consumes approximately 1 W of power, and we are able to deploy multiple units per hub, which in turns allows us to have a flexible system where multiple neural networks, for different data types, can be deployed nearly simultaneously. Furthermore, by being able to offload the actual processing of the data to the neural network hosts, the processor's primary jobs becomes managing and parsing the data flows, as well as interaction with the adjacent hubs in the network.

The on-going development of this aspect has so far allowed to local data processing such as image processing and pattern recognition which can be employed for resource optimization within a building, in the case of green building applications. While the efficiency of employment of neural networks and deep learning is still being investigated, currently the benefits of sharing the processing load with the neural chip when processing numerical or graphical data have already been seen, due to the repetitive nature of the received data. For the setup, the binary result data employed the $14 \times 40 \times 2$ architecture (14 inputs, 40 hidden neurons, and 2 outputs), where the learning process is carried out using the back propagation algorithm with the employment of the ReLU function within the hidden neurons. In order to facilitate the non-linearity output layer, contrary to the hidden layer, it uses Sigmoid, which allows obtaining the probability of class assignment. The training of the neural network used the Adam optimization method, as it showed the best results based on trial runs. In addition, the binary signal analysis used the mean squared error approach for the weighting function, as seen below:

```
#Define loss and optimizer
cost = tf.losses.mean_squared_error(y, pred)
optimizer = tf.train.AdamOptimizer(learning_rate).
minimize(cost)
```

The two currently investigated frameworks that have been applied are Tensorflow and CAFFE, where in the former, the DNNClassifier was employed. The CAFFE framework is being tailored more toward processing of the graphical input data, since the framework also includes a number of established algorithms which can be used as basis, such as AlexNet, for example. The diversified use model is supported by the mechanical properties of Movidius sticks, which can have more than one unit attached to a single hub, while having different trained matrix employed.

20.5 Testing

For the network testing, the experimental setup that can be seen in Figure 20.8 was used. The units of the Y-axis in Figure 20.9 are in milliseconds, and display the latency statistics as will be discussed below. A number of components in this system were tested. The connection speeds between the machine and the server. Due to the fact that the main limitations of the connection speed are the Internet service provider imposed limitations and the number of hops the transmission has to go through, which usually correlates with the geographic location of the communicating parties. The average time for the European setup was 42 ms, while the average Canadian time was 190 ms. Due to the server being located in Germany, these results are to be expected. For more optimal times, either a service that is geographically close or personal servers can be used if necessary.

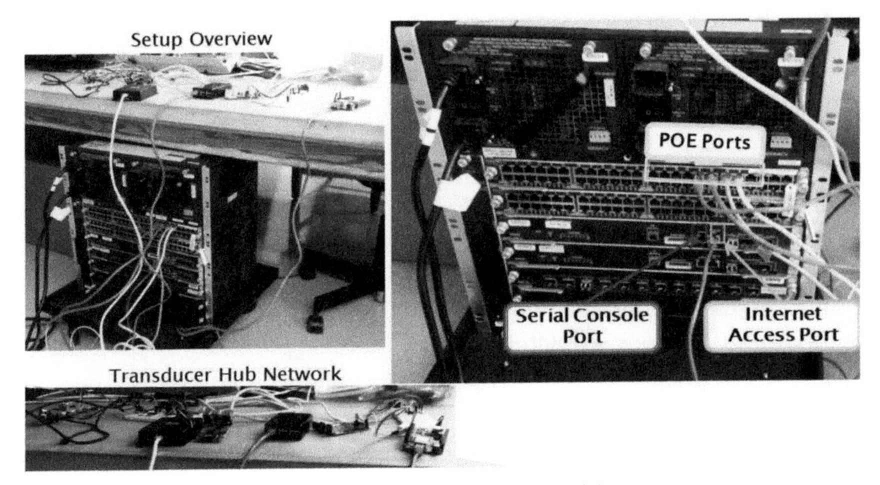

Figure 20.8 Experimental setup for local testing of the prototype system.

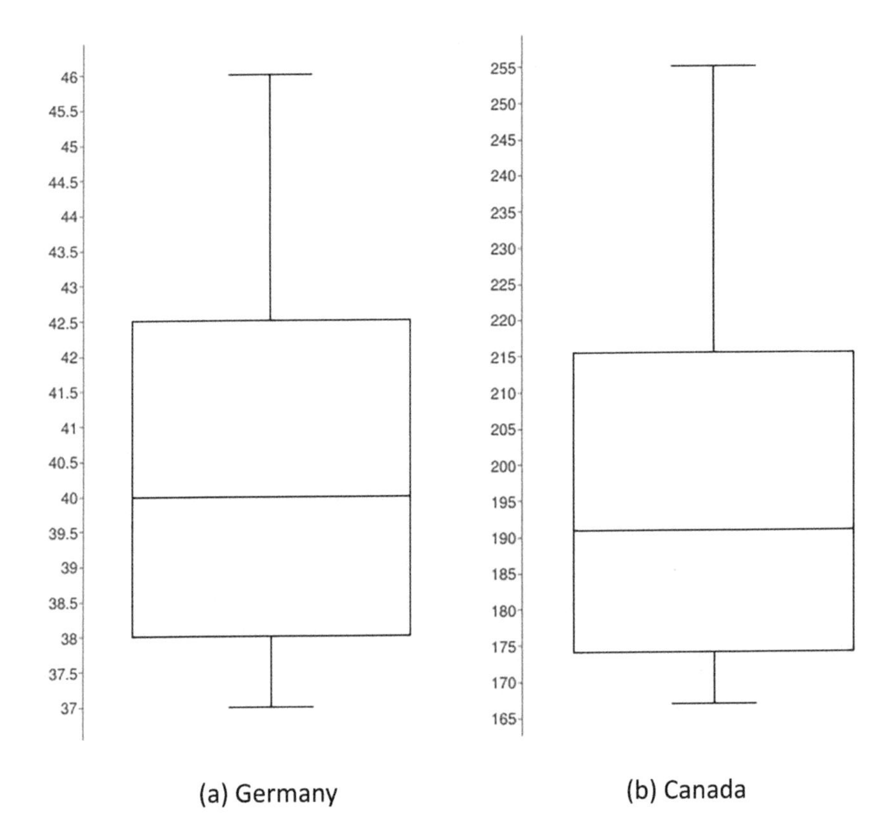

Figure 20.9 Latency test results for the two cities where the tests were conducted.

The system was also tested for scalability by connecting two setups from two continents. The first setup was done in Vancouver, Canada, while the second one was done in via access in Germany. Each system consisted of at least one master and at least one slave module. Each of the modules had a number of sensors connected to them. While some of the sensors varied, an illumination sensor, a vibration sensor, a temperature and humidity sensor, and a power sensor were used in different configurations on all of the MCBs. After setup, to test operational stability and performance, a latency test was performed. The test was carried out over the course of 15 days, at approximately the same time according to the local time, the peak demand period of afternoon was used as studies have shown that around 14:00 is when usage is the heaviest according to [21]. During the measurement three samples were taken and recorded. The resulting latency statistics can be found in Figure 20.9 where (a) displays the latency measurement results obtained in Germany and (b) displays the ones obtained in Canada. Over the course of testing, there has been minimal package loss, even on long transitions in case of Vancouver measurement (11%). During the testing, there were no errors detected on the server side with neither the data nor the request handling. The system had no issues handling the requests from both setups and operating with both sets of MCBs, even if their requests arrived nearly simultaneously.

As a test of adaptability, a different MCB was used BeagleBone. After a simple configuration of the output pins, no other changes were necessary. The sending, receiving, and processing schemes as well as the control modes all performed as expected without any errors, and did not require any excessive remodeling or re-coding.

A full-scale test of the system is planned to be conducted as part of the Green Campus initiative at UBC, Canada. There the system is planned to be used to monitor the energy consumption and generation around a section of the new building that has been built, as well as its structural integrity of one of the sections and the corresponding interior climate.

For this purpose, vibration sensor, LDT0-028K, was chosen as the sensor of choice as it has shown to perform well for the purposes of earthquake detection and building vibration [22] as well as parasitic power generation as discussed by P. Glynne-Jones in their work outline in [23].

Large area of effect strain sensors, such as a long-gage fiber optic sensor, which have been shown to perform well as civil structure monitoring device as explained in their overview paper by H. N. Li et al. [24] as well as in the paper by S. Li et al. who discussed the feasibility of a distributed system of such sensors in [25]. As well as temperature and humidity sensors and power measuring sensors will be used. The alpha prototype has tested the correctness of operation of the vibration sensor, the temperature and humidity sensor, and the power measurement sensor by assembling a circuit for each and calibrating their outputs. Once the obtained readings matched the expected values, which were obtained by performing a control measurement by industry manufactured devices, the operation level was deemed feasible.

20.6 Conclusion

As a result, a system that matches the set goals has been developed. The system is capable of operating by utilizing PoE, which drastically simplifies installation procedures as well as reduces power consumption costs. The removal of power bank configuration, used by similar alternatives, allows for much longer maintenance periods of one to two years on average for electrical inspections compared to approximately six months to replace power sources in the aforementioned solutions. The dependability of the verdicts issued by the hubs is reinforced by validating any spotted anomaly either via the subordinate slave hubs, which will collect the same data, and a statistical result can be obtained to avoid false positives, or via other master hubs which can perform the same action in their area of effect. Also as discussed previously, the system can perform "self-checks" for connectivity to both the higher members of the hierarchy and its subordinates. This allows it to spot any units or modules that may have gone "offline" and therefore have they serviced in a much more timely manner rather than by a more manual inspection method. The introduction of neural network modules allows performing a lot more of the data processing locally, which will greatly benefit any data flows that are time critical, for example, the ones that would issue an evacuation warning or signal. This in turns allows reducing the amount of required bandwidth, and as a result, may also reduce the dependence on an external internet connection which may be an issue in certain geographical locations. However, the general inter-communication algorithm presented in this work allows integrating a number of application-specific failsafe algorithms to improve the reliability of the system's operation in a given environment. Furthermore, by placing the central client of the system in the cloud, an online processing service, a high degree of scalability has been achieved, as well as a high degree of controllability, where the control addressee can be targeted from a single hub to a whole branch enveloping a building or a floor. This in turn is a very important aspect for the IoT field. Furthermore, the data transfer speeds as well as throughput have been increased by integrating an

Ethernet-based transfer protocol, which by nature provides higher bandwidth and can be driven at higher data rates when compared to the earlier models utilizing RS-232. Nevertheless, nowadays, many other models also utilize TCP/IP-based communication from their systems, where the distinguishing factor becomes the organization and positioning of the nodes. This matter feeds back into and is resolved by the hierarchical structure discussed in this work, while the system could be further improved by increasing the security of the transmitted data, as well reducing the size of the MCBs used. It can nevertheless benefit smart buildings and large structures that wish to reduce the amount of power consumed by the heating ventilation and air conditioning systems, as well as buildings that apply new building technology and materials and need to analyze their performance. As can be expected, the system functions well with the big data approach and has been tested to be operational in a tans-continental setup, where the gathered data were accessible from any machine that had access to the Internet, and the processing and transfer of the data were completed in a timely and efficient manner, due to high processing capabilities of cloud computing machines and small size of the files that used HDF5 database format.

20.7 Future Work

The system has improved its aspects of usability, adaptability, and scalability; however, for commercial use, there are still a number of improvements that could be made that would benefit the performance.

The major aspect to be considered is the enhancement of security and, in case of in-house servers, more elaborate data routing and handling. While a certain level of security it is obtained by securing the server, the system can benefit from employing software-defined networking (SDN) to improve internal routing, which would in turn improve and optimize data handling as well as packet control to ensure no tampering or unauthorized access occurs. Kapil [26] outlines how the employment of SDNs can benefit a network, and quantify the performance benefits of such a transition from the traditional hardware-defined networks. Another improvement that can be made to the system is further segmentation and automation of access to different levels of the system. This includes adding the capability to select an "area of effect" which can be defined based on the application, but in case of buildings, it would define what portion of the building is affected by a certain action.These functions,well as more diverse data processing and automation can be further expanded by adding more layers to the learning algorithms and developing the classes of the classes and matrixes of the neural networks.

References

[1] Goldstein, R., and Neuman, D. (2010). "Mega-buildings: Benefits and opportunities of renewal and reuse the essential role of existing buildings in a carbon neutral world," in *Proceedings of the American Institute of Architects National Convention and Design Exposition*, Miami, Florida.

[2] United Nations, Department of Economic and Social Affairs, Population Division. (2014). *World Urbanization Prospects: The 2014 Revision, Highlights (st/esa/ser.a/352).* Available at: https://esa.un.org/unpd/wup/Publications/Files/WUP2014-Highlights.pdf

[3] Commscope Inc. (2015). *Laying the groundwork for a new level of Power over Ethernet.* Available at: http://www.commscope.com/Docs/POE_Groundwork_WP-107291.pdf

[4] Rice, J. A., Mechitov, K., Sim, S.-H., Nagayama, T., Jang, S., Kim, R., et al. (2010). Flexible smart sensor framework for autonomous structural health monitoring. *Smart Structures and Systems*, 6(5–6), 423–438.

[5] Magno, M., Polonelli, T., Benini, L., and Popovici, E. (2015). A low cost, highly scalable wireless sensor network solution to achieve smart led light control for green buildings. *IEEE Sensors Journal*, 15(5), 2963–2973.

[6] Kreibich, O., Neuzil, J., and Smid, R. (2014). Quality-based multiple-sensor fusion in an industrial wireless sensor network for mcm. *IEEE Transactions on Industrial Electronics*, 61(9), 4903–4911.

[7] Liu, Y., He, Y., Li, M., Wang, J., Liu, K., and Li, X. (2013). Does wireless sensor network scale? a measurement study on greenorbs. *IEEE Transactions on Parallel and Distributed Systems*, 24(10), 1983–1993.

[8] Xu, N., Rangwala, S., Chintalapudi, K. K., Ganesan, D., Broad, A., Govindan, R., et al. (2004). "A wireless sensor network for structural monitoring," in *Proceedings of the 2nd international conference on Embedded networked sensor systems.* ACM, 2004, 13–24.

[9] Chintalapudi, K., Fu, T., Paek, J., Kothari, N., Rangwala, S., Caffrey, J., et al. (2006). Monitoring civil structures with a wireless sensor network. *Internet Computing, IEEE*, 10(2), 26–34.

[10] Kulkarni, U. M., Kulkarni, D. V., and Kenchannavar, H. H. (2017). "Neural network based energy conservation for wireless sensor network," in 2017 *International Conference On Smart Technologies For Smart Nation (SmartTechCon)*, IEEE.

[11] Sato, H., Domae, H., Takahashi, M., and Abe, M. (2000). "Reflection and transmission control of electromagnetic wave for concrete walls," *Electronics and Communications in Japan (Part II: Electronics)*, 83(11), 12–21.

[12] Townsend, W. (2000). The barretthand grasper-programmably flexible part handling and assembly. *Industrial Robot: An International Journal*, 27(3), 181–188.

[13] Rao, H.-X., Xu, Y., Zhang, B., and Yao, D. (2018). Robust estimator design for periodic neural networks with polytopic uncertain weight matrices and randomly occurred sensor nonlinearities. *IET Control Theory and Applications*, 12(9), 1299–1305.

[14] Barnes, D., and Sakandar, B. (2004). *Cisco LAN Switching Fundamentals*. Indianapolis, IN: Cisco Press.

[15] Cisco Inc. (2014). *Cisco Catalyst 4500E Supervisor Engine 8-E Configuration Guide (Wireless), Cisco IOS XE Release 3.7E*, 2nd ed. Available at: http://www.cisco.com/c/en/us/td/docs/switches/lan/catalyst4500/XE3-7-0E/wireless/configuration-guide/b_37e_4500sup8e_cg.html

[16] Cisco Systems Inc., (2015). *Cisco Catalyst UPOE Power Splitter*. Available at: https://developer.cisco.com/fileMedia/download/99c67d92-8089-44b9-980a-9bc304abf739

[17] Barrett, D. J., Silverman, R. E., and Byrnes, R. G. (2005). *SSH, The Secure Shell.* O'Reilly, 2005.

[18] Available at: http://www.openssh.com/manual.html

[19] Osorio, F. G., Xinran, M., Liu, Y., Lusina, P., and Cretu, E. (2015). "Sensor network using power-over-ethernet," in 2015 *International Conference and Workshop on Computing and Communication (IEMCON)*, IEEE, 1–7.

[20] Intel Corporation. (2017). *Movidius Neural Compute Stick User Guide.* Available at: https://movidius.github.io/ncsdk/

[21] Brownlee, N., and Claffy, K. C. (2002). Understanding internet traffic streams: dragonflies and tortoises. *IEEE Communications Magazine*, 40(10), 110–117.

[22] Alves, F. R. d. C. (2015). "*Low-cost vibration sensors: tendencies and applications in condition monitoring of machines and structures*," Ph.D. Dissertation, Instituto Superior de Engenharia de Lisboa.

[23] Glynne-Jones, P., Tudor, M., Beeby, S., and White, N. (2004). An electromagnetic, vibration-powered generator for intelligent sensor systems. *Sensors and Actuators A:* Physical, 110, 344–349.

[24] Li, H.-N., Li, D.-S., and Song, G.-B. (2004). Recent applications of fiber optic sensors to health monitoring in civil engineering. *Engineering Structures*, 26(11), 1647–1657.

[25] Li, S., and Wu, Z. (2007). Development of distributed long-gage fiber optic sensing system for structural health monitoring. *Structural Health Monitoring*, 6(2), 133–143.

[26] Bakshi, K. (2013). "Considerations for software defined networking (sdn): approaches and use cases," in *2013 IEEE Aerospace Conference*, IEEE, 1–9.

21

IoT Systems of the AAL Sector: Application, Business Model, and Data Privacy

Jelena Bleja[1], Uwe Grossmann[1], Bettina Horster[2], Andree Roß[2], Enrico Löhrke[3], Christof Röhrig[1], Jan Oelker[1], Aylin Celik[1] and Reiner Hormann[1]

[1]University of Applied Sciences and Arts Dortmund, Dortmund, Germany
[2]VIVAI Software AG, Dortmund, Germany
[3]inHaus GmbH, Duisburg, Germany
E-mail: jelena.bleja@fh-dortmund.de; uwe.grossmann@fh-dortmund.de; bettina.horster@vivai.de; Andree.Ross@vivai.de; Loehrke@inhaus.de; cristof.roehrig@fh-dortmund.de; Jan.Oelker@fh-dortmund.de; Aylin.Celik@fh-dortmund.de; reiner.hormann@fh-dortmund.de

This Chapter presents results of three projects aiming to provide the requirements for enabling elderly people to stay longer in their flats or houses before moving to senior citizens' residences or nursing homes. Using market available devices, ambient assisted living systems are developed for recording vital data. A minimal invasive radio transmission system and smart home sensors are installed. Data from various sensors and devices will be recorded, collected, and analyzed automatically in one single IoT platform. On top of the recorded data, a collaborative system business model is developed enabling fair allocation of revenues and cost savings. Data-driven services will be developed and deployed. The implementation of data protection against the background of the European General Data Protection Regulation is taken into account.

21.1 Introduction

Internet of Things (IoT) refers to a technology designed to increase comfort and quality of life, enabling new ways of interacting with people and things.

In IoT systems, devices can communicate with each other over the Internet and learn from each other, devices that are equipped with sensors can perceive the environment, and the sensor data can be read out from anywhere. The main goal of IoT is to connect things and physical objects like machines but also people so that people can provide information regardless of their location and time independent, thus, among other things, providing support in everyday lives [1].

The private sector represents a big area for IoT systems. Within smart home systems, houses and flats were equipped with sensors and actors enabling the owner to monitor and control window blinds, heating, or even locks of doors. The intelligent devices that are used in the context of AAL systems are partially equipped with sensors and connected to each other and, for example, with a nursing service or health center. Consequently, with the use of IoT, the interaction between various healthcare devices and the nursing service or health center can be significantly simplified. In addition, safety and efficiency in the AAL area can be increased [1].

In recent years, several research projects have been carried out at the European level (for example, HOMEBUTLER and I-stay@home) as well as nationally (for example, SmartSenior and SmartAssist) in the area of ambient assisted living (AAL, age-appropriate assistance systems for a self-determined life). Above all, solutions were developed that focused on technology development. Various AAL projects have not reached the practice and the market beyond the funded project duration. This was mainly due to missing business models.

In addition to the projects, there are various integration platforms in the field of Smart Home and AAL, such as QIVICON (Telekom), EmotionAAL, universAAL, WieDAS-Projekt1 (D-Link and Partners), TOPIC (The Online Platform for Informal Caregivers), or meinPAUL (CIBEK), which usually bring together individual components and sensors on the platform, or focus on social support. However, these systems are only about process data – data that are needed for evaluation purposes and for documentation are not collected and archived. However, these are important for intelligent systems that perform an evaluation, for example, when it comes to comparing the activity patterns of the residents. Often only very few basic data are processed, the control of the actuators is the focus.

IoT systems are at the verge of becoming increasingly important in the health context. Because the demographic change is a major challenge for our society, an increasing share of the population needs long time care and care achievements. This leads on the one hand to financial restrictions and on the other hand to supply-side obstacles: The nursing and social consequential costs are increasing and there is a lack of full-fledged nursing staff. The demand for rationalizing and supporting technical assistance systems, especially for the home environment, is increasing. Within the care market, a trend for individual value-added services is currently recognizable. Many care providers translate their offer from a structured care offer into an individualized counselling and service model. They are looking for ways to provide home care with more quality care and additional personalized services. Technical assistance systems can make an important contribution to this market development. They enable a nursing provider to offer a new personalized service model from the assistance system for more security at home and individual support and nursing care. Therefore, technical, age-appropriate solutions are becoming more and more important in this context. They have the potential to ensure the quality of care in Germany and to reduce costs. Besides, they offer the possibility to relieve nursing staff, so that they can increasingly focus on the interpersonal, social component of care [2].

In recent years, a lot of AAL systems have been developed in a prototypical manner. A major problem is the sustainable establishment of these projects in the market. Despite the high potential assessment of AAL systems and the increased development of AAL solutions in recent years, getting the developed solutions market-ready and establishing them permanently in the market are still the challenge. The reasons are, for example, the healthcare system with its specific requirements especially in the area of data protection and data security and the fears, difficulties, and the lack of acceptance of elderly people to use technical innovations [3]. However, the lack of adequate and profitable cooperative business models for the introduction of age-appropriate systems was the core problem in most cases.

This Chapter will present first results of three projects in the area of AAL for elderly people. In Sections 21.2 and 21.5, we present the project Smart Service Power (SSP), in Section 21.3 SOLION, and in Section 21.4 covibo. While the SOLION and Open iCare Assistant projects have already been completed, the SSP project will run until September 2019. The projects aim predominantly to enable elderly people to live longer in their flats or houses before moving to a senior citizens' residence or a nursing home. All of these AAL projects are service-oriented and have the challenge of getting the developed solutions market-ready and establishing them permanently in the market. Therefore, Section 21.5 discusses the development of business models for AAL applications, which are to be developed in particular within the framework of the SSP project. Moreover, the collected data will contribute the profitability of the business model. Therefore, privacy and data security aspects will be addressed as they pose a major challenge for IoT systems.

21.2 Smart Service Power (SSP) – Ambient Assisted Living for Elderly People

21.2.1 The Ambient Assisted Living (AAL) of SSP

Within SSP, the living environment is provided with market available recording devices for vital signs, a minimal invasive radio transmission system, and smart home sensors [4].

A major innovation of the project is recording, collecting, and analyzing data from various sensors and devices automatically in one single IoT platform. SSP mostly uses commercially available sensors and devices in addition to a standard IoT-software platform. Due to all the components and the data that can be derived, it is able to integrate the horizontal services of AAL like nursing services, emergency call systems, concierge, and smart home services. The whole system is interoperable, flexible, and expandable, and aggregates the different data sources.

Activity and behavioral profiles will be created and an alarm will be forwarded in case of emergencies. The potential needs and context-based requirements of (elderly) people will be derived on the basis of big data analysis and machine learning – necessary support can be offered automatically and demand-driven.

The profiles could also be very valuable, for instance, for diabetics, if the drug treatment changes and the resident moves less and less in his or her home and changes the drinking habits. The residents get security and assistance through reminders and notifications regarding, for example, the intake of medicines, food, and drinks.

Along with smoke alarms and motion detectors, sensors at doors, windows, light switches, refrigerators, ovens, water taps up to special pressure sensors for determining the presence in bed can be installed. A fall detection system can find out, if the resident is collapsed and lies on the floor.

The software system using big data methods recognizes deviations from the normal daily routine based on vital signs and activity data. It then communicates this matter with the resident through the natural language assistant. If the plaster which measures the skin tension detects that the person did not drink enough, it triggers a reminder to the resident.

Possibly dangerous situations like a detection of a fall may be derived and an action series is started, e.g., dropping an emergency call or informing a service provider about the situation in the home. Detailed information about the present situation is provided to the organization in charge via a control center.

21.2.2 Smart Service Power Top-Level System Architecture

In general, a technical solution in the scope of an AAL scenario needs to meet a wide range of different requirements. A lot of technical prerequisites (related to interoperability, high availability, security, etc.), non-technical demands (reliability, privacy protection, ease of use, etc.), as well as quality and comfort aspects (accuracy of speech understanding, quality of machine learning, etc.) need to be harmonized in one user-centric system that is still capable of adjusting itself to a user's living and environmental conditions (and not vice versa) [5, 6]. The amount of processed data in such a system can be very high (e.g., due to the continuous processing and fusion of sensor data input, frequent language-based user interactions, etc.) but appropriate response times (dependent on the AAL application scenario) of the system must still be ensured. Furthermore, an AAL system is in touch with a lot of confidential user data, so it is also mandatory that the system facilitates a person's right of informational self-determination through a data usage control system. These examples illustrate the wide range of different requirements that a technical solution for AAL has to fulfill.

A technical solution for a system in the context of AAL is shown in Figure 21.1. It shows the top-level system and software architecture as developed in the scope of the SSP project [4]. The SSP system aims to enable elderly or disabled people (residents) to live in their home (flat, house, etc.) as long as possible. For that purpose, the SSP system can be installed at home.

The selection of sensors (vital data, movement data, etc.) is depending on the resident's needs and the desired scope of the SSP application. An installed SSP Home Collector records and preprocesses SSP sensor data (e.g., sensor data fusion) via edge computing [7]. Furthermore, it autonomously manages the communication (via Internet) to the SSP Service Provider, which is the main access point to get in touch with the wide range of cloud-based [7] services the SSP system brings along.

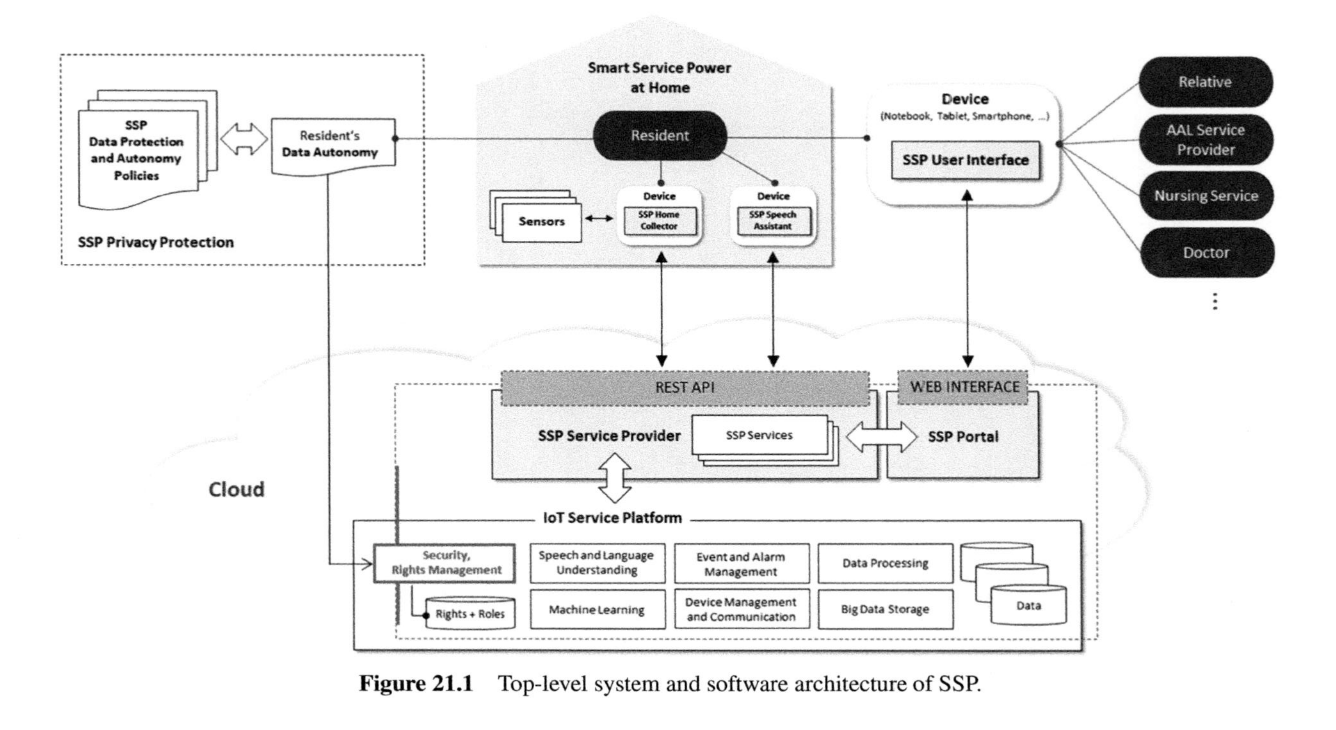

Figure 21.1 Top-level system and software architecture of SSP.

Since language is a very comfortable and natural way for system control as well as for user interaction, an autonomous SSP speech assistant consisting of a speaker, a microphone, and a processing unit is managing the language-based communication between the resident and the SSP system and vice versa. For this purpose, the SSP Language Assistant is used, which is based on a standard chatbot for speech analysis and understanding as part of the IoT service platform [8]. In order to ensure data autonomy, each SSP resident has individual and full control over the way how the system handles its data. An SSP data usage control system which includes the rights and roles concept compliant to the SSP data protection and autonomy policy is used. It is the basis for security and data protection along all levels across the SSP system.

The most important means of communication for residents is the natural language interface, but of course, there is also the possibility to access the system with other devices like notebooks, smartphones, tablets, etc. For other involved parties like relatives, doctors, nursing services, and concierge services, the means of communication is a graphic HTML-based web interface [9]. Through the data usage control system, the SSP rights and roles concept is imposed on each connected party. Depending on the role and data access rights, a user can securely access his or her view of the SSP cloud-based services and SSP data. The SSP user interface is oriented toward the partly significantly varying needs of SSP target user groups. For example, many residents only want their relatives to know that they are safe and they can manage their daily routines properly. Therefore, there is only a traffic light system available for them. A lot of the users do not want the relatives to know a lot of the details. The nursing service needs information about a resident's daily activities while an emergency service is only interested in the fact that the resident has collapsed and is not able to get up. So the view and functionality of the user interface differs depending on the user's role and also the rights. The basis of the web interface is the SSP Portal, which manages the protected communication to the SSP service provider.

For all locally installed SSP home collectors, the SSP service provider is the main access point to SSP cloud-based services (such as, e.g., device and communication management, data processing, event and alert management, machine learning, data storage, etc.). From the technical point of view, the SSP service provider represents micro service-based software architecture [10] for machine-to-machine communication that provides a representational state transfer (REST) interface [11] which is protected by state-of-the-art security mechanisms (e.g., industry-standard protocols for authorization, single sign-on identity and access management, etc.) strictly obeying the SSP data protection and autonomy policies. All SSP cloud services and operations are based on an IoT service platform, so that the SSP service provider in a broader sense works as a kind of middleware [12] between user-centric SSP applications (SSP home collector, SSP speech assistant, and SSP user interfaces) and IoT service platform.

As the result of an extensive selection process which included the examination of 28 features, Microsoft's IoT platform "Azure" [13] has been chosen as the most appropriate IoT service platform for the SSP project. This IoT platform supports state-of-the-art cloud services, e.g., device communication and configuration management, data operations for (big) data processing and database storage, and strong security and data protection technologies [14]. Furthermore, Microsoft Azure provides a wide range of additional services and features, e.g., machine learning, language understanding, speech recognition, and much more. It also involves the tools for building a bespoke SSP natural language assistant. It is possible to run the assistant in a private cloud – strictly separated from Microsoft's Cortana System. From the perspective of IoT dependability, Microsoft invests a lot of effort in Azure to provide reliability, security, and maintenance on a high level [15]. Especially the use of IoT edge technology [16] in SSP enables SSP to provide a wide scope of maintenance functions (dependent on the requirements that need to be met for all SSP user roles). With the Microsoft Deutschland cloud, the data stored exclusively in Germany [17]. Deutsche Telekom's subsidiary T-Systems operates as the data trustee under German law and monitors access to the very sensitive user data.

21.3 Solion – A Radio-based Assistance System

Mobile radio-based assistance systems are simply integrated into the existing living environment, automatically recognize situations requiring support, and signal these, for example, to a nursing service, so that it can act as part of the individually tailored service.

Solion Assistance is a radio-based, retrofittable assistance system (no emergency call system) developed jointly with a nursing service [18].

It consists of various radio-based smart home sensors and actuators and a central home station, which records the radio telegrams and evaluates them by means of software and determines deviations from the normal daily routine (Figure 21.2).

Figure 21.2 SOLION Assistenz – radio-based assistance system (inHaus GmbH) [18].

Figure 21.3 Assistance system components in a living environment (left: motion-based night light; middle: motion sensor bath (ceiling); and right: door opening contact) [19].

The assistance system can be easily and quickly integrated within 15 min into any living environment without changing the existing electrical installation. In addition, it is not necessary to have the assistance system integrated by a technician, as this can easily be done by a caregiver because it comes preconfigured [18].

It provides information about the everyday life at home, like movement, attendances, use of domestic appliances, opening of cabinets or doors, etc. (Figure 21.3).

For example, the resident wakes up during the night because he has to go to the bathroom. Getting up from the bed will be detected and the night orientation light will be automatically switched on. However, if the resident does not return from the bathroom, the caregiver will be informed within a short time, because it indicates a possible downfall [19].

On the other hand, movements in the kitchen or the opening of the refrigerator are signs of regular and normal daily routines and activities.

The equipment supports clients' day-to-day operations (such as cooking and nightly walk to the toilet) while providing 24/7 security for clients and reassurance for relatives. Danger situations are recognized by deviations from the normal daily routine (without emergency callers and without cameras) and forwarded to caregivers (via telephone, e-mail, message, or mobile visualization.), so they can decide what to do based on the information they provide.

No personal data from the apartment will be transferred to the nursing service. These remain in the apartment on the home station and will be deleted after a coordinated time (for example, four weeks). Based on the requirements of data protection, deviations from normal daily routines are transmitted to caregiver via a traffic light (red, yellow, and green) signalization.

With the help of intuitive visualization, each system can be adapted and adjusted individually by the nursing staff to the circumstances of the respective client. The adaptation is also intuitive, so that no prior technical knowledge is necessary.

21.4 Covibo – Vital Data Acquisition

Many people prefer to stay in their private environment even in the need of care. Automatic vital data acquisition offers more clarity about the current physical state of the resident. The daily record and check of the vital data increases the safety and makes living at home for elderly people easier. Processes in nursing care are automated. The following introduced system is developed based on the research of the Open iCare Assistant Project [20].

The application field of the system is in private living areas. Measurement data, like weight, blood pressure, and glucose, are automatically acquisitioned and documented after each measurement. No further actions are required. In addition to the vital data acquisition, a therapy plan is offered. Every intake of medication and measurement can be entered in the therapy plan. This way, reminders show up for every task and ensure that nothing is forgotten and no double medication and interactions between medications are taking place. The control over the data remains always with the user. Only he decides who has access to the data. It can be shared once or permanently with relatives, caregiver, and doctors. In the first step, all sensitive data are stored locally at home on the base station. If the resident wants to share the data durably, it has to be stored on a web server. Multiple users can use the system parallel. Therefore, they can use the different profiles on one measuring device or every resident uses his own device [21].

21.4.1 System for Vital Data Acquisition

The system contains a base station, an easy to use app, a web service, and the manufacturer-specific measuring devices. The base station only needs power supply and wireless local area network (WLAN). It communicates via its interfaces with all devices and sensors. The system operates with manufacturer-specific devices, whereas the system itself stays vendor-independent. It is always listening to receive data and is like a digital file folder for local data storage.

The actual state and history of the data is viewable in an app. The vital data are displayed as a graph and the therapy as a plan. The design of the app offers a high usability, which is important for the regular use of the system. In case the resident rejects or is not able to use the app, he still benefits from the whole functionality of the vital data acquisition. In such a way, the app is optional for the resident and can be used by the relatives independently. Via the app, the system can be used easily and on the way.

A web service is provided for sharing the data durable and enables access from everywhere. The web service takes into account all data privacy acts, like the EU General Data Protection Regulation.

The system has to be set up once at the beginning. The WLAN is set up via a step-by-step guide in the app. Afterward, the first user account is registered. Sharing of the data for other users can now take place. With automatic device recognition, the devices are added to the system easily and without any prior technical knowledge.

21.4.2 Communication Structure

The near-field technology Bluetooth Low Energy (BLE) has successfully established itself in the market in the recent years. The number of measuring devices and sensors for this technology is increasing steadily. Since the choice of devices to be included is vendor-independent, the system remains an open, easily extendable system. The BLE protocol offers different profiles for blood pressure or weight measurements.

After a measurement, the device starts to advertise in BLE as a peripheral slave. The base station is acting as a central master and listens for these broadcasts. If an advertisement of a registered device is detected, the base station establishes the BLE connection as shown in Figure 21.4. As soon as the connection is established, the base station is using the generic attribute profile to read the specific characteristics. After all data are transmitted, the device closes the connection and is going into standby. Now the data are

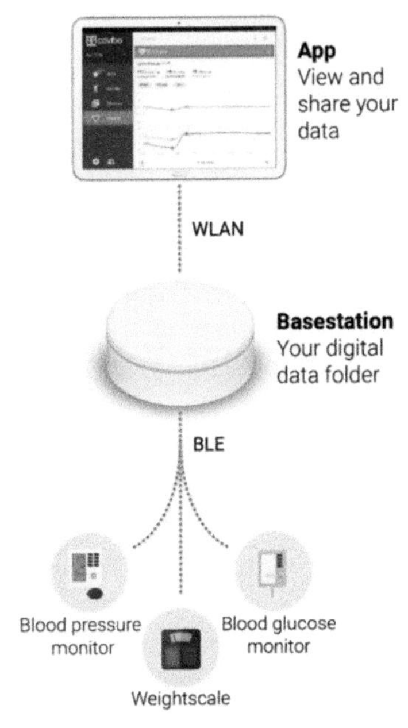

Figure 21.4 Covibo – Communication structure.

stored in the database of the base station. Depending on the current settings, the data can now be shared via WLAN to the app or the web service [22].

21.5 Business Models for AAL Applications

One well-known method beside others for describing a business model for a single company is the business model canvas (BMC) of Osterwalder and Pigneur [23]. Figure 21.5 shows the essential compartments of the BMC.

The Business Model Canvas				
Key Partners	*Key Activities*	*Value Proposition*	*Customer Relationships*	*Customer Segments*
	Key Resources		*Channels*	
Cost Structure		*Revenue Streams*		

Figure 21.5 The **BMC** (according to [23]).

The aim of the BMC is to clearly visualize the most important components of a business model. An intensive and creative examination of the model in the team should contribute to new approaches and possibly to the emergence of business model innovations. The left side of the BMC represents the resource-based view, while the right side depicts the market-driven, customer-oriented view [24].

The model has the nine key components: key partners, key activities, key resources, customer segments, value promises, channels, customer relationships, cost structure, and revenue stream. Key Partners work together to fulfill the value proposition, i.e., the services performed for the customers, who are described within the customer segment. Key Activities covers all activities necessary to perform the value proposition. Key Resources contains all necessary physical, intellectual, human, and financial resources to execute the necessary activities. Channels comprise all communication channels the company uses to contact the customer. Customer Relationships describes the relations a company develops to its customers, e.g., personal support. Last but not least there is the section Cost Structure with all costs arising while fulfilling the value proposition. Revenue Streams contains all revenues received, e.g., from customers for delivering the value proposition.

AAL scenario as used for the time being within the project SSP (see Section 21.2) is shown in Figure 21.6.

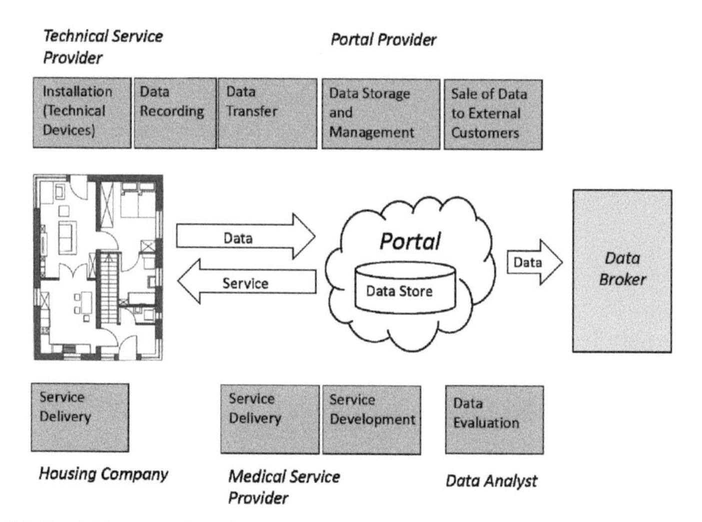

Figure 21.6 AAL-scenario of partners, activities, and services used within project SSP.

Presented are key partners (technical service provider, portal provider, data analyst, medical service provider, and housing company), key activities (technical installation, data recording, data transfer, data storage and management, data analysis, service development, service delivery, and sale of data to external customers). Potential customers are the resident or client and/or his/her relatives and the external data broker, who buys data released by the resident or client, but also the medical service provider or the technical service provider. Key resources are the technical equipment of the flat or house (sensors, etc.), the portal (data storage, etc.), algorithms for evaluating the data and detecting conditions, or events for delivering services. The value proposition to the resident/client or the medical service provider contains information delivered after detection of specific situations and events, e.g., emergency situations. Other services to clients/relatives/medical service providers tentatively comprise continuous reports regarding the vital status, to external customers the sale of released data.

Revenue streams may include the fees the customers pays for specific services, the financial contributions of health insurance companies for medical services, or the payment of external companies for received data.

In the first place, the BMC represents a model to describe all aspects of the business context of a single company in order to reach profitability. The AAL business scenario shows that several companies are involved and additional external stakeholders with financial contributions as health insurances, etc. In order to reach a sustainable business model for the whole scenario, it is necessary to introduce a broader view on the whole scenario including several companies and external stakeholders. For these purposes, the BMC has to be extended. A higher level view is necessary and, therefore, a business model is required for a collaboration of several enterprises. Figure 21.7 shows a first approach to extend the BMC to a higher level collaborative system business model.

First, each company has its own business model. Collaboration between multiple companies, however, creates a shared value proposition. Therefore, this requires an additional common business model. The resulting dynamic system business models with partially changing partners must take a superordinate view of the different participants. At the first level, the individual companies involved from the various industries are considered with their individual corporate business models. At the second level, the individual business model components of the participating companies form a superordinate collaborative system business model. Not all components of the

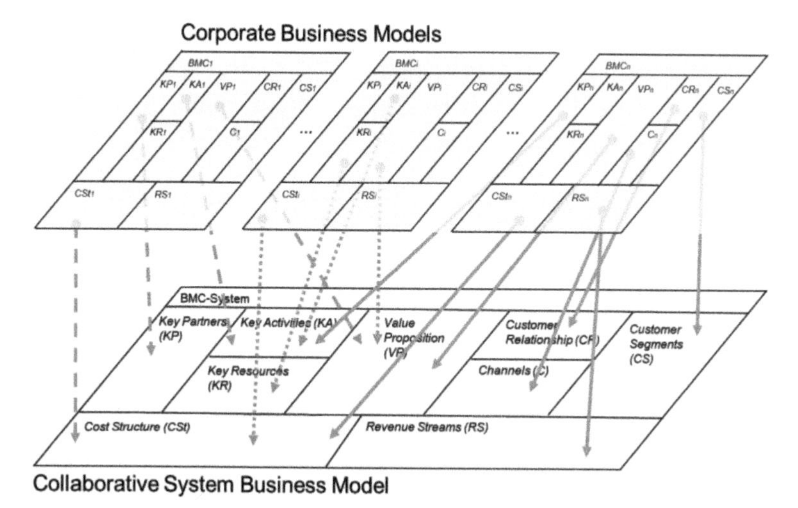

Figure 21.7 Corporate business models and collaborative system business model.

participating companies pass into the collaborative system business model, but only those that are relevant to the project.

In the following step, how incoming revenues or other external financial contributions will be distributed to the collaborating business partners must be clarified. If there is one dominant partner leading the value chain and choosing his partners, the question of allocating revenues does not appear as a problem. But if several partners having equal rights share the value chain and contribute to deliver a service to the customer/client, this question becomes more important. Are there methodologies for allocating revenues or cost savings between independent business partners fairly? And moreover, is there a suitable definition of the term 'fairly'?

There are several methods from accounting dealing with the problem of allocating costs, like indirect costing, transfer pricing, etc., but they all show weaknesses, if it comes to fair allocation between independent business partners being on equal footing. A different, promising approach depicts the marginal principle of allocation, which may be used together with methods of cooperative game theory, e.g., Shapley-value or τ-value. Such methods have been applied in the past on scenarios in different industrial sectors, e.g., the transport/logistics sector [25] or the agricultural sector [26].

In addition to the discussed extensions of the BMC, further aspects have to be included in the model, for example, data protection aspects and the location of the service.

Fachinger et al. [3] are working on similar problems concerning sustainable business models for neighborhood networks. They are using meta business models including business models of several companies and external stakeholders. The external stakeholders may be communal offices or institutions.

21.6 Data Privacy and Data Usage Control

The entire collaborative business model outlined above is based on the evaluation of data. Mostly these data are of highly sensitive and private nature. The sensors deliver profiles, which allow inferences onto the state of health and the personal lifestyle of the resident/client, i.e., the data are highly interesting for a lot of companies of the industrial, the insurance/financial, or the service sector. One challenge of AAL application is the consideration of privacy requirements of the resident/client. Exclusively, data will be recorded, which are necessary to deliver specific services to the resident/customer. It is up to the resident/client to decide, which of his data may be released for sale to external companies.

An important aspect of SSP is the compliance with the European General Data Protection Regulation (EU GDPR), which came into force on 25 May 2016. After a transitional period of two years, the EU GDPR will be applied directly in the EU Member States in May 2018. It applies to both European companies and non-European providers of products or services to EU citizens. Thus, the provisions of the EU GDPR also affect companies such as Facebook and Google [27].

The objective of the EU GDPR is to establish a high level of uniform rules for the processing of personal data across the EU, which will meet the new challenges of increasing digitization better [27]. The EU GDPR strengthens the fundamental right to informational self-determination by allowing people in the EU greater data transparency and participation in the use of their data by third parties [28].

In the case of violations of the EU GDPR, significantly higher fines of up to 20 million euros or four percent of global annual turnover are possible (Article 83 (5) GDPR) [29].

Therefore, in order to meet these new requirements, a clear and transparent procedure for assessing the processing of personal data is needed [30].

Privacy metrics can evaluate the IoT platform's compliance with data protection requirements based on a large number of measurements. For this purpose, the requirements, some of which cannot be measured directly, must first be made measurable. Automated checks, to what extent the requirements within the IoT platform are fulfilled, are useful in order to fulfill necessary control obligations more easily and continuously. In order to meet the requirements of the EU GDPR, it is therefore necessary to analyze and select suitable privacy metrics for the described IoT services [31].

SSP follows EU GDPR regulation and is designed with its core principles in mind. The standard data protection model released by the German Independent Data Protection Authorities [32] defines several data protection goals. The fundamental protection goal of "data minimization" is extended by the classic protection goals of data security (availability, integrity, and confidentiality) and further data protection-specific protection goals (unlinkability, transparency, and intervenability).

Users of AAL applications demand the highest level of safety and security regarding their personal information stored and transported. Communication between all nodes needs to be secured against eavesdropping and manipulation with end-to-end encryption. This can be achieved by different means, depending on the communication scenarios. Since AAL applications uses nodes in the homes of all kinds of users, network environments will vastly vary.

One way to enforce a given environment is by using virtual private networks (VPNs). While these networks are rather complicated to set up, they offer a closed network environment for communication between participating nodes. Another way to secure communication between nodes is by using an HTTPS-secured REST-API. HTTPS also (very likely to VPN) offers mutual authentication [32] by assigning additional client certificates.

Both methods rely on the principle of asymmetric encryption, where every node participating in the communication comes with its own key pair of private and public keys used for encryption. The intention is to implement these security features with open and established standards.

Furthermore, there are some approaches, which allow the resident/customer to keep certain data private and release other data under well-defined circumstances.

Besides encryption as a typical way to protect user data, in the scope of SSP an implemented single sign-on identity and token-based access delegation management realizes a rights and roles concept compliant to the SSP data protection and autonomy policy. Each SSP role (e.g., resident, doctor, nursing service, system administration, etc.) is associated with rights that define the set of authorizations for (role) specific operations, (REST)

APIs, or resources within the SSP system. Additionally, each SSP resident individually may grant or even withdraw rights (e.g., by using the SSP portal as user interface) for the usage of its data along all levels across the SSP system. So, a resident has individual and full control over the way how the SSP system handles its data. Dependent on the skills of a resident, the configuration of user data control mandatorily needs to be kept on an understandable and manageable level (e.g., from a very simple (guided) beginner level (without details) until expert level (with details)).

Another approach is a data usage control system as described by Steinebach et al. [33]. The security framework Integrated Distributed Data Usage Control Enforcement allows the exploitation of data usage control for practical purposes. The data owner is enabled to define and control precisely and finely granulated the data usage by security policies, e.g., which data may be used or under which circumstances and how often data may be read, copied, or transferred.

Schuette and Brost [34] propose a mechanism to control the ways in which data may be processed, thereby limiting the information which can be gained from data sets to the specific needs of a service. They model data analytics as a data flow problem and apply dynamic taint analysis for monitoring the processing of individual records. A policy language is used to define how data are processed and enforce measures to ensure that critical data are not revealed.

Azaria et al. [35] present MedRec, a novel, decentralized record management system to handle electronic medical records, using blockchain technology. The system gives patients a comprehensive, immutable log, and easy access to their medical information across providers and treatment sites. Leveraging unique blockchain properties, MedRec manages authentication, confidentiality, accountability, and data sharing. Using blockchain technology the development of smart contracts is possible.

21.7 Conclusion

Several AAL applications like SSP, SOLION, and covibo are aiming to set up the basic requirements for extending the period of time elderly people can stay in their flats or houses before moving to senior citizen's residences or nursing homes.

For example, a daily record and check of the vital data increases the safety and makes living at home easier for elderly people. In addition, assistance systems enable nursing providers to offer a new personalized service model of the assistance system for more security at home and individual support and nursing care.

Technically market available recording devices for vital signs, a minimal invasive radio transmission system and smart home sensors will be used in the project. Data will be stored and managed in a private cloud-portal. Services will be developed and deployed after detecting special events or conditions when evaluating the data. With the consent of the data owner (resident/client), data can be sold to external companies.

In addition to an intelligent data aggregation and the development of innovative functions in the age-appropriate technology-based living, suitable, profitable, and fair collaborative business models are necessary in order to achieve a sustainable establishment. Therefore, on top of the recorded data, a collaborative business model is developed providing fair allocation of revenues, cost savings, and a data fair share concept. Standard business models like the BMC will be extended to a higher level collaborative system business model integrating several service companies and external stakeholders.

In addition, data protection and data security will become increasingly important in the future. In order to meet the requirements of the EU GDPR applicable in May 2018, appropriate technical and organizational measures must be taken. With the described security measures, system operation can be done ensuring confidentiality, integrity, as well as availability. The compliance of data security policies will be achieved either by a data usage control system or by blockchain-based smart contracts.

Acknowledgments

The presented work was done in the research project "Smart Service Power" funded by the state government of North-Rhine-Westphalia and European Union Fund for regional development (EUROPÄISCHE UNION – Europäischer Fonds für regionale Entwicklung).

References

[1] Hail, M. A., and Fischer, S. (2015). "IoT for AAL: An Architecture via Information-Centric Networking," in *Proceedings of IEEE Globecom Workshops (GG wkshps)*, San Diego, CA: Piscataway, NJ: IEEE, 1–6.

[2] Barth, A., and Doblhammer, G. (2017). "Physische Mobilität und Gesundheit im Alter – Ansätze zur Reduktion von Pflegebed—"urftigkeit und Demenz in einer alternden Gesellschaft," in *Die transformative Macht der Demografie*, ed. T. Mayer, Springer Fachmedien Wiesbaden, 207–244.

[3] Fachinger, U., Helten, S., Nobis, S., and Schöpke, B. (2016). "Meta-Geschäftsmodelle - eine Möglichkeit zur erfolgreichen Einbindung von assistierenden Techniken in Quartiersnetze," *Discussion Paper* 23/2016, Institut für Gerontologie – Ökonomie und Demographischer Wandel der Universität Vechta, 2016.

[4] Smart Service Power Homepage. Available at: http://www.smartservicepower.de [acccssed Jan 23, 2018].

[5] Homepage. Available at: www.werpflegtwie.de, retrieved from https://www.werpflegtwie.de/leitfaden-pflege/ambient-assisted-living-aal/ [accessed Jan 23, 2018].

[6] Maske, D. (2018). *Anforderungen für mobile Endgeräte im AAL-Bereich - Neue Norm zu Ambient Assisted Living*. Available at: http://www.elektro.net/50658/anforderungen-fuer-mobile-endgeraete-im-aal-bereich/ [accessed Jan 23, 2018].

[7] Wanner, W. (2018). *Datenverarbeitung im IoT: Cloud- versus Edge-Computing*. Available at: http://www.funkschau.de/telekommunikation/artikel/130928/ [accessed Jan 23, 2018].

[8] Microsoft Azure. *About Language Understanding*. Available at: https://docs.microsoft.com/en-us/azure/cognitive-services/luis/home [accessed Jan 23, 2018].

[9] HTML5. (2014). *A vocabulary and associated APIs for HTML and XHTML*. W3C Recommendation. Available at: https://www.w3.org/TR/2014/REC-html5-20141028/ [accessed Jan 23, 2018].

[10] Microservices. Available at: from http://microservices.io/ [accessed Jan 23, 2018].

[11] Fielding, R. T. (2000). *Architectural Styles and the Design of Network-based Software Architectures*. Ph.D. Dissertation, University of California, Irvine.

[12] Microsoft Azure. *Was ist Middleware?* Available at: https://azure.microsoft.com/de-de/overview/what-is-middleware/ [accessed Jan 23, 2018].

[13] Microsoft Azure. Available at: https://azure.microsoft.com/de-de/, last accessed Jan 23, 2018.

[14] Microsoft Azure Trust Center. Available at: https://www.microsoft.com/de-de/trustcenter/cloudservices/azure [accessed Jan 23, 2018].

[15] Microsoft. "*The Cloud and the state of modern security and reliability.* Available at: https://partner.microsoft.com/en-pk/case-studies/security-reliability?wt.mc_id=CORP_HP_security-reliability [accessed May 9, 2018].

[16] Microsoft. '*Azure IoT Edge.*' Available at: https://azure.microsoft.com/de-de/services/iot-edge/ [accessed May 9, 2018].

[17] Microsoft Deutschland Cloud. Available at: https://www.microsoft.com/de-de/cloud/deutsche-cloud [accessed Jan 23, 2018].

[18] inHaus GmbH. *Pflegeimmobilien/Pflegedienste – Ambulante Betreuung (im Quartier)/Wohngemeinschaft.* Available at: http://www.inhaus-gmbh.de/pflegeimmobilienpflegedienste/ambulante-betreuung-im-quartier-wohngemeinschaften.html [accessed Jan 24, 2018].

[19] inHaus GmbH. *Ambient Assisted Living (AAL) in einer Demenzwohngemeinschaft.* Available at: http://inhaus-gmbh.tops.net/index.php?id=111, last accessed Jan 24, 2018.

[20] Celik, A., Oelker, J., Künemund, F., and Röhrig, C. (2015). "Automatic vital data acquisition via a Bluetooth Low Energy – Network," *Paper for the 8*, AAL-Kongress.

[21] covibo GmbH. Available at: http://covibo.com/ [accessed Jan 18, 2018].

[22] Townsend, K., Akiba, C., and Davidson, R. (2014). *Getting Started with Bluetooth Low Energy'*. O'Reilly Media.

[23] Osterwalder, A., and Pigneur, Y. (2010). *Business model generation: a handbook for visionaries, game changers, and challengers.* Amsterdam: Modderman Druckwerk.

[24] Kreutzer, R. T., Neugebauer, T., and Pattloch, A. (2017). *Digital Business Leadership.* Wiesbaden: Springer Gabler.

[25] Strangmeier, R., and Fiedler, M. (2011). "Lösungskonzepte zur Allokation von Kooperationsvorteilen in der kooperativen Transportdisposition," *Diskussionsbeiträge der Fakultät für Wirtschaftswissenschaft der FernUniversität Hagen*, 1–100.

[26] Grossmann, U., Gansemer, S., and Pauls, A. (2015). "Fair Allocation of Benefits for M2M Communication based Collaborative Business Models," in *Proceedings of the 8th IEEE Inter-national Conference on Intelligent Data Acquisition and Advanced Computing Systems (IDAACS): Technology and Applications.*

[27] Für soziales Leben e.V. *Europäische Datenschutz-Grundverordnung - DSGVO-EU.* Available at: http://www.europaeische-datenschutz-grundverordnung.de/ [accessed Aug 18, 2017].

[28] BMWi. (2016). *Europäische Datenschutzgrundverordnung.* Available at: https://www.bmwi.de/Redaktion/DE/Artikel/Digitale-Welt/europaeische-datenschutzgrundverordnung.html [accessed Aug 18, 2017].

[29] iT-CUBE SYSTEMS AG. (2017). *GDPR - Die neue EU-Datenschutz-Grundverordnung (DSGVO).* Available at: https://www.it-cube.net/cubespotter/gdpr-die-neue-eu-datenschutz-grundverordnung-dsgvo/ [accessed Aug 18, 2017].

[30] AK Technik der Konferenz der unabhängigen Datenschutzbehörden des Bundes und der Länder. *Das Standard-Datenschutzmodell: Eine Methode zur Datenschutzberatung und -prüfung auf der Basis einheitlicher Gewährleistungsziele [V.1.1].*

[31] Luhn, S., and Hils, M. (2015). "Ein Rahmenwerk für Datenschutz-Metriken in der Cloud," in *INFORMATIK*, eds D. W. Cunningham, P. Hofstedt, K. Meer, I. Schmitt (Bonn: Gesellschaft für Informatik e.V.), 511–523.

[32] Dierks, T., and Rescorla, E. (2008). *RFC 5246 - The Transport Layer Security (TLS) Protocol Version 1.2.* Available at: https://tools.ietf.org/html/rfc5246#section-7.4.6 [accessed Jan 4, 2018].

[33] Steinebach, M., Krempel, E., Jung, C., Hoffmann, M. (2016). "Datenschutz und Datenanalyse," in *Datenschutz und Datensicherheit – DuD*, 40/7, 440–445.

[34] Schütte, J., and Brost, G. (2016). "A Data Usage Control System using Dynamic Taint Tracking," in *Proceedings of the IEEE 30th International Conference on Advanced Information Networking and Applications*, 909–916.

[35] Azaria, A., Ekblaw, A., Vieira, T., and Lippman, A. (2016). "MedRec, Using Blockchain for Medical Data Access and Permission Management," in *Proceedings of the IEEE 2nd International Conference on Open and Big Data*, 25–30.

PART V

Education and Training

22

Internet/Web of Things: A Survey of Technologies and Educational Programs

Volodymyr Tkachenko[1] and Eugene Brezhniev[2]

[1]National Technical University "KhPI," Kharkiv, Ukraine
[2]National Aerospace University "KhAI," Kharkiv, Ukraine
E-mail: proftva@ukr.net; e.brezhnev@csn.khai.edu

This chapter is devoted to review of Internet of Things/Web of Things (IoT/WoT) technologies, curriculum (case studies of WoT) for future specialists on IoT/WoT technologies' development, and experience of teaching of discipline: "Technologies and tools of WoT application development" in different variations. The review of existing WoT/IoT technologies demonstrates that server-side programming language of Web Thing API built with consideration of architecture of REST is JS environment in Node.JS. Considering Web Thing API programming language, the curriculum suggests the stack of technologies for development of WoT applications using JavaScript/Node.JS. The discipline "Technology and tools for developing Web applications," supplemented by the sections of "Cloud Computing," "IoT/WoT," and related reviewed curriculum is taught in National Technical University "Kharkiv Polytechnic Institute" at the Department of Information Systems and National Aerospace University "Kharkiv Aviation Institute" at the Department of Computer Systems, Networks, and Cyber Security under the courses "Web of Things" and "Industrial Internet of Things."

22.1 Introduction

Development of technology machine-to-machine (M2M)/IoT, information processing tools (Big Data), and decision-making (cognitive analytics) leads to changes in the technological, economic, and social development models society. Areas of use of IoT are expan- ding in energy, transport, medicine, agriculture, housing, Smart City, Smart Home, etc. IoT focuses only on connecting physical objects to the network and their interaction with each other. Cisco introduced a new concept – Internet of Everything (IoE), which is based on the integration of people, things, data, and processes. Thus, the next stage in the development of IoT/WoT is the Internet of all (IoE). In the future, the orbit of IoT will include technology deep machine learning, artificial intelligence, technology blockchain, robotics, etc. IoT is characterized by large changes in the infrastructure of the Internet and new communication models Smart Things or connections: "Thing–Thing," "Thing–User," and "Thing–Web Object." The IoT Infrastructure consists of various networks of physical objects based on heterogeneous hardware and software platforms and protocol stacks, which are generally incompatible with each other. So, the IoT is a collection of isolated physical networks that cannot communicate with each other via the Internet.

The concept of a WoT based on Web and its new technologies [1] enables the integration of all kinds of Smart Things and applications with which they interact. The concept WoT introduced such a notion as "Web Thing," which is a digital representation of a physical or virtual object that is accessible through Web API RESTful. One of the major development issues for this new concept is creating efficient hypermedia-enriched application programming interfaces (APIs) [2]. Web API RESTful or Web API built with consideration of the REST architecture for a virtual representation of the physical objects are identified by URL and use application layer protocols such as HTTP, WebSocket, CoAP, MQTT in JSON format, and TLS/DTLS cryptographic streaming protocols. Thus, the virtual equivalent of physical objects (Web Thing), which were assigned a URL via Web API, can communicate with each other or with applications by using application-level protocols and share data in text-based JSON. In addition to the software interface, Web Thing can be equipped with custom interfaces to ensure interoperability model "Thing–User." The WoT reuses existing and well-known Web standards used in the programmable Web (e.g., REST, HTTP, and JSON), semantic Web (e.g., JSON-LD, Microdata, etc.), the real-time Web (e.g., web sockets), and the social Web (e.g., OAuth or social networks) [3].

Thus, WoT provides the integration of Internet-connected physical devices of different producers on application level regardless of how they are connected on a network level and ensures the creation of a single global ecosystem of the IoT, which is open and compatible.

Currently, WoT standardization has engaged WoT community and such international organizations for Standardization, as W3C (https://www.w3.org/WoT/), IETF (https://www.ietf.org/), ETSI (http://www.etsi.org/), OCF (https://openconnectivity.org/), and OGC (http://www.opengeospatial.org/), and supported by European research projects on the IoT, such as Sensei-IoT (http://www.sensei-iot.org/), IoT-A (https://iota.org/), and SmartSantander (http://www.smartsantander.eu/). WoT Interest Group published draft standards [4–8]. In addition to set out draft standards, Mozilla IoT community published its draft standards [9].

22.1.1 Motivation

Together with the development of IoT technology and its network service WoT, there is an increasing need for specialists for development (software and hardware) and integration of technical solutions in the field of IoT, maintenance, and operation of IoT networks. The problem of training (training and retraining) of IoT/WoT specialists is becoming urgent. The issue of training current and future engineers and researchers, technology application development, and integration of modern IoT/WoT solutions can be solved jointly by the companies that design and manufacture tools for IoT and institutions of higher education. For example, the aim of the project ALIOT (Internet of Things: Emerging Curriculum for Industry and Human Applications) [10] is to integrate all available and prepared training programs, manuals, and tools for the provision of training and advisory services in the field of systems based on the IoT for applications in different areas.

Companies involved into developing and manufacturing tools for IoT are interested in higher-education-obtained professional skills creation and exploitation of IoT/WoT in a timely manner, in order to remain competitive in the field of development and production of IoT/WoT. In their turn, higher education institutions are interested in teaching students the basics of the IoT/WoT design and operation to be competitive on the labor market in the field.

22.1.2 State of Art

It should be noted that companies develop and produce tools for IoT and universities prepare specialists in the field of development and integration of modern IoT solutions. For example, the company has created a University for Telit IoT [11]. The program Telit IoT University currently includes six courses, one of which is the IoT for developers. In IoT University course [12], students look at user interface and user experience design strategies common to the industry and apply those strategies to building applications in ThingWorx using the Mash Up Builder. This course is focused on the IoT project, which does consider the WoT technologies. The PTC IoT Academic Program [13] consists of the ThingWorx^{TM} application enablement platform in a PTC hosted environment where students and educators can build their own IoT applications. PTC works with corporate customers as well as market partners to ensure that students from all disciplines are better prepared to meet the needs of today's IoT world.

In the article [14], the information is presented on many bachelor's and master's programs on IoT. WoT technology is not considered. The IoT M.Sc. program [15] is available at the Queen Mary University of London. M.Sc. Internet of Things (Data) is currently available for one-year full-time study, two years' part-time study (Introduction to IoT, Enabling Communication Technologies for IoT). WoT technology is not considered. The article [16] presents the best universities that offer courses in the field of "IoT," and studies in detail what they offer their students. WoT technology is not considered. The project "IoT Academy Samsung" [17] is organized on the base of Moscow Institute of Physics and Technology. In accordance with the experts of the research center of Samsung's teaching materials, students will undergo a year-long training course on examining real case studies on IoT technologies in various industries and will be able to create their own IoT devices prototypes. WoT technology is not considered. The Cisco IoT [18] certifications and training are job-role-based programs designed to help meet the growing need for specialized talent. This education portfolio provides Internet Protocol (IP) networking expertise, with a focus on automation, manufacturing, and energy and future expansion to include equally transformative industries. WoT technologies are not considered.

National Aerospace University n. a. N.E. Zhukovsky "KhAI" prepares specialists on programmable mobile systems and IoT [19]. IoT-based systems are developed and investigated as well. WoT technologies are not considered in the discipline of "Industrial IoT." Lviv IT Cluster and National University "Lvivska polytechnica" have launched a "Bachelor program Internet of Things" [20]. Goal: to prepare specialists in the field of designing elements and applications of IoT, WoT is not considered. The work [21] deals with possible reflection of the theme of the IoT and M2M in higher education curriculum (programs), but it does not reflect the technology of WoT. The purpose of such a curriculum is to consider issues related to information and communication technologies used in IoT and M2M. The proposed course aimed at listeners acquainted with modern information technologies, which stand behind such directions as inter-machine interaction and IoT.

Currently, the authors have not seen in open papers mentions of any existing training programs on teaching students about technology for Web Things application development. Practical guide “Building the Web of Things” [1] is a basic training manual, which presents key technologies and concepts necessary to build application-level IoT and architecture WoT, as well as defines the methodology of application development for the Web Things on JS/Node.JS. This manual is intended for trained professionals in the field of Web application development technologies at JS/Node.JS. Step-by-step tutorial can help professionals use WoT and semantic network to develop applications of semantic WoT [22], but does not solve the problem of student learning of all the necessary technologies for the development of WoT. The book “Web application development Framework” [23] is designed for inexperienced Web developers, it outlines the creation of user interfaces, and it discusses ways to develop the server-side application on Node.JS and methods of cloud services usage for deploying Web applications. The book “Web development with Node and Express” [24] is intended for programmers who want to build Web applications (regular sites that embody REST APIs) using JavaScript, Node.JS, and Express. The curriculum “Technology and development tools WoT applications” was prepared with consideration of all teaching materials that are prepared for programmers’ learning on development technologies of advanced Web applications based on the JS/Node.JS, that will give the listeners the starting point for the WoT concept development and developing real applications’ Web Things.

22.1.3 Goals and Structure

The goal of this chapter is to review the IoT/WoT technologies and educational programs, the presentation of the syllabus structure (case studies of WoT) of the academic discipline “Technologies and tools for developing WoT applications,” which is drawn up in the direction of the development of WoT API technologies proposed in the WoT concept, and tested in NTU “KhPI”. This program offers technology development of WoT applications, that is, the stack of MEAN technologies (Mongo, Express, Angular, and Node) for development of WoT applications using JavaScript/Node.JS. Duration of training – one semester, the course is designed to train professionals (for master students) in “Computer Science,” which is intended to form the students' theoretical knowledge and practical framework in design and exploitation of WoT applications.

The task of the study course "Technology and tools for developing Web applications" is a theoretical and practical training of future specialists on such matters as:

- technologies for the application of markup languages and languages for description and programming in client Web applications;
- technology and tools for creating interactive Web interfaces;
- technology application in Node.JS server applications;
- technology exchange messages between Web Apps in a mode of Real Time (Ajax, WebSockets) in XML messaging formats, JSON;
- technology for building applications with SOA architecture (architecture, REST);
- cloud computing technology and application deployment model for cloud platforms;
- architecture and technology of IoT;
- cloud platforms and services for the WoT;
- technology for application development based on Web-based Things Raspberry Pi using the Node.js;
- security, privacy, and access control to the physical devices in the IoT/WoT.

The structure of this chapter is the following: Introduction; Survey of IoT/WoT Technologies; Structure of the training program "Technologies and tools for developing WoT applications"; and Conclusions.

22.2 Survey of IoT/WoT Technologies

Existing M2M-technologies allow machines to exchange information with each other. M2M is a subset of IoT. IoT is the Internet of People (IoP), extended by computing networks of physical items (Smart Things), which can independently organize various connection models. IoT is a concept of the network infrastructure development (physical basis) online, in which "smart" things without human intervention are able to connect to the network for remote interaction with other devices (Thing–Thing) or interaction with autonomous or cloud data processing centers, or DATA centers (Thing–Web Objects) for data transmission, storage, processing, analysis, and management decisions aimed at changing the environment Wednesday, or to interact with user terminals (Thing–User) for the control and management of these devices [25]. The article [26] is the first to present the correlations among M2M, wireless sensor networks, cyber-physical systems (CPS), and IoT. The authors suggest that CPS is an evolution of M2M by the introduction of more intelligent and interactive operations, under the architecture of IoT. Cisco believes [27] that the IoE is the next step in the evolution of smart objects-interconnected things in which the line between the physical object and digital information about that object is blurred. The WoT is a refinement of the IoT by integrating smart things not only into the Internet (network), but into the Web architecture (application) [28].

22.2.1 IoT Global Network Architecture

This paper provides the overview of the IoT: concepts, architectures, development technologies, physical devices, programming languages, protocols, and application [29–33]. The IoT consists of the networks of physical objects, the traditional network of the Internet, and various devices (gateway, border router, etc.) that connect these networks. Figure 22.1 presents the components of the IoT architecture, which consists of several computer networks of physical objects connected to the Internet.

As seen in Figure 22.1, the network of IoT consists of: the computer networks of physical objects (Smart Objects), traditional IP Internet, and various devices (gateway, a border router, and router), integrating these networks. It should be noted that Smart Objects are the sensors or actuators (sensors or actuators), equipped with a microcontroller with real-time operating system with a stack of protocols, memory, and communication devices embedded into various objects, such as in electricity or gas meters, pressure sensors, vibration or temperature switches, etc. Smart Objects can be organized in computer network physical objects that can be connected via gateways (hubs or specialized IoT platform) to the traditional Internet. In IoT, there is not a single universal protocol for the integration of physical objects. Therefore, to create a network of physical devices, one shall acquire all the components of one manufacturer. As a result, the network of physical objects is fragmented and the provision of integration of physical devices connected to the Internet with incompatible protocol stacks is expensive.

Gateways are used to integrate the networks of IoT (for example, Z-Wave, ZigBee, etc.), protocol stacks, which are incompatible with the TCP/IP stack of the Internet. Edge routers are used to integrate the Internet with networks of IoT, based on network protocol 6 LoWPAN (IPv6 over low-power wireless personal area networks), where IPv6 is a version for wireless personal area

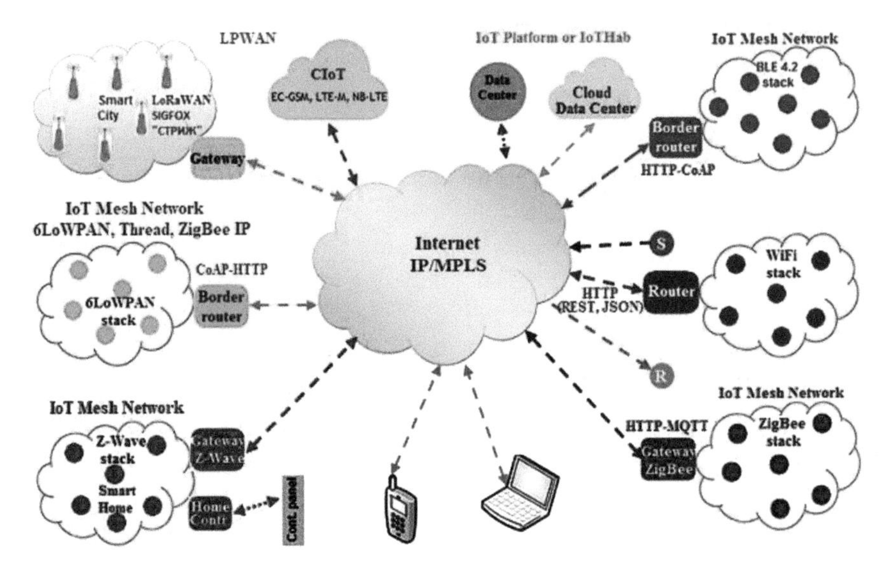

Figure 22.1 Components of IoT architecture [25].

sensor networks with low power consumption IEEE 802.15.4. A proxy is used to harmonize protocols HTTP-CoAP. Technology-based network Thread, ZigBee, and 6LoWPAN self-organizing nature are IP networks, and may not have an exit to external IP network using 6loWPAN protocol stack for the organization of the work of autonomous networks and data transmission between the autonomous network nodes.

Wireless networks, used in LPWAN (low-power wide-area network): IoT, WLAN, and WPAN. LPWAN Technology. Key long-range LPWAN network technologies include: LoRaWAN; SIGFOX; Swift; and CIoT (EC-GSM, LTE-M, and NB-IoT). According to experts' estimates, more than 50% of IoT solutions would use LPWAN network. WLAN Technology. Medium-range technology, WLAN refers Wi-Fi (www.wi-fi.org) – a set of wireless standards IEEE 802.11, which can be used to build a wireless local area WLAN-based network objects on the TCP/IP stack. To build local wireless computer network items, Wi-Fi Alliance has created a new IEEE 802.11 specification, which provides technology to build cellular networks. In addition, a new standard Wi-Fi HaLow (IEEE 802.11 specification ah for the IoT) was created with low power consumption. Wireless personal area networks (WPAN). Key WPAN short-range wireless network technologies:

6LoWPAN, Thread, ZigBee, Wireless IP, Z-Wave, EnOcean, RFID/NFC, and BLE 4.2. Controllers and mini computers in the IoT. Today to manage the physical devices, the IoT uses controllers and mini computers: Arduino, Espruino, Tessel, Intel Edison and Galileo, and Raspberry Pi, whose applications are created in c/C++, Java, JavaScript, Python, etc.

IoT application layer protocols. In the networks of physical objects, the interaction between components is done using the application layer protocols: DDS [34], CoAP, MQTT, XMPP, AMQP, JMS, REST/HTTP [35], etc. DDS is the core technology for Industrial IoT. CoAP Protocol (Constrained Application Protocol) – limited data transfer protocol similar to HTTP, but adapted to work with "smart" devices. MQTT protocols, XMPP, AMQP, JMS – these messaging protocols are based on broker scheme: publish/subscribe.

Security considerations for IoT. Security of IoT must be addressed at all stages of the development cycle and operation of hardware and software, communication channels, protocols stack, cloud components, etc., which is currently given a lot of attention in the field of IoT security. In [36], several security and privacy concerns related to IoT are mentioned. The protection of data and privacy of users has been identified as one of the key challenges in the IoT. The survey presents IoT with architecture and design goals. In addition, a review and analysis of security and confidentiality issues at different levels in the IoT was performed. It should be noted that for the security of the IoT, standards [37] and guidance [38] have been created that provide manufacturers of tools with a set of guidelines for improving IoT security. The document "State-of-the-Art and Challenges for the Internet of Things Security" [37] can be used by implementers and authors of IoT specifications as a reference for details about security considerations while documenting their specific security challenges, threat models, and mitigations. The goal of guidance [38] is to help manufacturers build more secure products in the IoT area. Royal Academy of Engineering (London) [39] is a leader in Cybersecurity of the IoT. The PETRAS Cybersecurity of the IoT Research Hub brings together nine leading UK universities.

The development of IoT depends on many factors: technology, low-power wireless networks; Smart Objects technology; the pace of 5G networks adoption; operating systems for microcontrollers sensors and actuators; widespread use of 6LoWPAN/IPv6 protocol stack; M2M technology; effective use of Cloud computing for IoT platforms; Misty technology computing (fog computing); and software-defined networks, ensuring hardware and software cyber resilience.

22.2.2 Web of Things

IoT focuses on the lower layers of the network stack and the WoT service on the upper layers, application tiers. By using web technologies, protocols, programming languages, and formats [40] such as REST, XML, JSON, MQTT, XMPP, Atom, WADL, Open ID, and OAuth, the WoT has contributed to reducing the barriers for common understanding and smooth interplay between heterogeneous real-world devices, services, and data. The WoT concept, based on the Web and its new technologies [1], provides integration of all types of Smart Things and applications with which they interact. It is known that WoT uses standards applied in such technologies as programmable Web (HTTP, REST, and JSON), semantic Web (JSON-LD, Microdata, etc.), real-time Web (WebSockets), and social Web (OAuth or social networking APIs). The problems of the WoT architecture, development technologies, programming languages, APIs, and application-level protocols based on RESTful principles are described in many articles [41–44].

Thus, WoT provides integration of devices in the Internet. The WoT is a service similar to the IoT infrastructure service, World Wide Web (WWW), of the Internet infrastructure. WWW is a distributed information system based on the use of hypertext documents in HTML format, access and transfer of which are achieved using the HTTP application. WoT is an extended service Web. By analogy with the Web architecture, the architecture of the WoT is the World Wide Web or distributed system of Web Things virtual resources (virtual representations of Things) that provide access to the physical objects, i.e., applications that are hosted on Smart Objects or intermediate IoT network devices through Web Thing API. The essence of WoT is that Web Thing physical objects or intermediate gateway devices, given that they have their own URL (Web address) and software interface with the RESTful Web API can communicate in text-based JSON both with each other and with applications based on SOA. To ensure interaction model "Thing–User" Web Things applications must have user interfaces.

Due to limited resources, not all Web Things can offer their own Web API is RESTful, based on the concept of WoT: for integration of Smart Objects in the Internet, three different integration templates are provided: direct connectivity, gateway-based connectivity, and cloud-based connectivity [1]. Implementation of Web Thing API on its own platform can be performed on the basis of a Web server that is hosted on the controller embedded in Things. Web application for the Web Things can consist of frontend and backend, i.e., can be implemented as user interfaces for users to interact with Things via Web browsers (for example, site sensors: http://devices.webofthings.io/pi/)

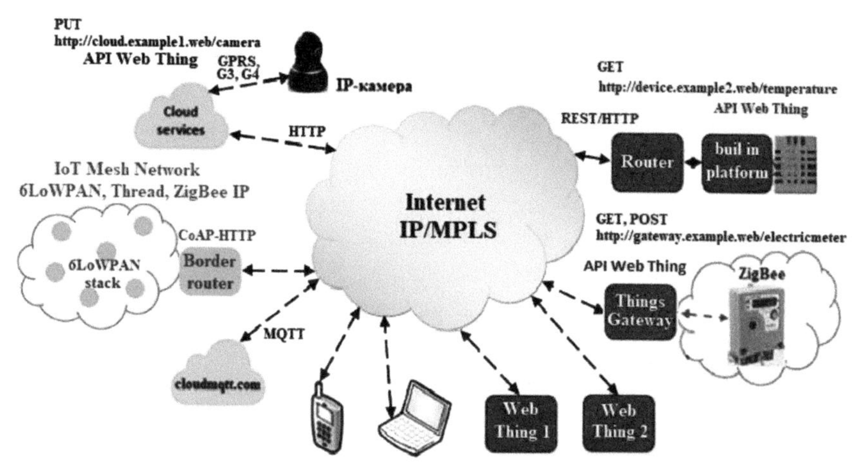

Figure 22.2 Components of WoT architecture [45].

and mobile applications and interfaces (API) applications using the RESTful architecture for data exchange between devices. As a controller, you can apply, for example, single-board computer Raspberry Pi based on Linux. The latest version of the computer (Raspberry Pi 3 Model B) has a built-in support for Wi-Fi and Bluetooth 4.1. In addition, Raspberry Pi GPIO ports available for direct connection to devices (e.g., temperature sensors, displacement, etc.). To implement Web Server, Node.js can be applied (for example, Node v 7.10.1). To develop the server-side application, it is advisable to choose the programming language JavaScript in Node.js. Software code of client side of the application is developed in HTML, CSS, and JavaScript. For the data exchange between devices in application, the interfaces of Client APIs and API Server, built with consideration of the REST architecture, are implemented.

Figure 22.2 presents Web API Thing, which can be placed either directly on the device itself, and intermediate Web Things gateway network or in cloud service.

In the case of implementation of a Web Thing API on an intermediate device such as a gateway, you can use the prototype gateway Things Gateway [46].

Things Gateway is created by the developers of the Mozilla community in JavaScript using Node.js server platform, and is available as a ready to install on the Raspberry Pi board assemblies. If you implement a Web API Thing on the cloud server, Web Thing Clients (Web devices or users) communicate to the cloud-based server (cloud-based server analogous Things Gateway) by the domain address of devices, which runs the application hosted on that server, and the application accesses devices, such as a camcorder, and manages them. The EVRYTHNG Platform [47] is a cloud platform-as-a-service (PaaS) for storing, sharing, and analyzing data generated by physical objects. The platform gives a unique and permanent digital identity (also known as ADIs) to each individual object and allows authorized applications and users to access it via REST and Pub/Sub (MQTT) APIs.

Security in WoT is provided by certificates, encryption, and authentication. Cryptographic streaming protocols TLS/DTLS [48, 49] are the basis of secure HTTP protocols (HTTPS), WebSocket (WSS), MQTT (MQTTS), and CoAP, which are used in WoT. To do this, you can install the OpenSSL library on the server (sudo apt-get install openssl). In addition, you can apply recommendations and methods for authorizing and controlling access to the server. Authentication is one of the means of protecting WoT applications [50–53]. You can set the API token with Node.js (install Node.js: node-oauth2-server). Authentication OAuth2 is designed to protect the Web API using a token-based authentication process. The token will be used to authenticate Smart Objects for each request to the server. You can use the OAuth 2.0 social media tools for WoT authentication.

22.3 Structure of the Training Program "Technologies and Tools for Developing WoT Applications"

All of the technologies outlined in the proposed themes are used in the development of Web Thing applications in line with the concept of WoT. Thus, the training program proposed in Table 22.1 is designed to prepare future specialists develop real Web Thing applications.

NTU "KhPI Information Systems Department, trains students in the specialty "Computer Science" according to the curriculum presented in Table 22.1.

Figures 22.3 and 22.4 shows the implementation materials of the web interface API Web Thing for the integration template direct connectivity. From the one shown in Figure 22.3, it follows that the HTTP and WebSocket servers are functioning. The Web Thing application installed in the Node.js environment on the Raspberry Pi 3 Model B provides data on the functioning of the motion sensors (PIR sensors), temperature and humidity (DH22), and the actuator (LED 1). Data on temperature, humidity, and movement are constantly updated.

Table 22.1 Curriculum structure

The Title of the Topic	Content
Development tools for Web applications: IDE, browser, and version control system (VCS)	At present, the only IDE for creating client and server applications on JS/Node.js is WebStorm [54]. But the own IDE (for JS development) could be built based on a text editor Sublime [55] with the plugins. In this program, Git [56] is used as a VCS of the Web application files; Git is used in many famous projects as VCS. In addition, familiarity with Git gives students the opportunity to explore GitHub, the largest Web service for hosting IT projects and their joint development.
HTML and XML technologies in client-side Web applications.	HTML document structure, logical languages HTML5 markup [57] and XML [58], technologies used in creating layouts or templates on HTML5 for Web sites, technology HTML-layout technology in editor Sublime Text.
CSS Technology And CSS3 in client Web applications and the use of the Bootstrap framework	CSS [59] – language for describing the appearance of documents (style declaration, selector types, block and line elements, style preprocessors, CSS frameworks, and Emmet LiveStyle). The layout technology of a web application with adaptive design based on Bootstrap [60] in the Sublime Text editor.
Technology and tools for creating interactive Web interfaces	To develop interactive Web interfaces, one must apply the basic triad HTML technologies, CSS and JavaScript [60–62], which form the structure, style, and behavior of Web applications. One of the components of the triad technologies is: the JavaScript programming language (syntax, set of technologies for creating interactive Web applications with JavaScript).
Technology of using jQuery to create interactive Web interfaces	jQuery [63] is a JavaScript-based library that contains ready-made JavaScript functions. jQuery manipulates the html elements of the document and uses the DOM to change its structure. There are two methods for connecting the jQuery library to the client application: local and remote connections. JQuery has a large number of third-party plug-ins with which make it possible to significantly improve the interface of the client side Web-applications or WoT.
The exchange of messages between Web Apps in a real-time mode (Ajax)	Four network technologies for interacting Web Apps based on client-side JavaScript scenarios (AJAX, COMET, SSE, and WebSocket). This topic is dedicated to AJAX [64] (Ajax technology and data transfer formats, ajax requests for "pure" JavaScript; ajax and jQuery).

(*Continued*)

Table 22.1 Continued

The Title of the Topic	Content
Exchange of messages between web apps in real-time mode (WebSocket)	WebSocket [65] is a technology of asynchronous interaction between the Web client and the Web server. WebSocket is a protocol of full-duplex communication over a TCP connection, intended for the exchange of messages between a web client and a web server in real time. WebSockets have an API that can be used in web applications and is called the WebSockets API [66, 67].
Web servers and application servers	It is proposed to consider the HTTP protocol, the client/server model, the architecture of the Web server, the application server, as well as to form an idea about the technology of these tools. In addition to traditional Web servers, Node.js technology is considered [68], which enables to create event-driven servers using JavaScript.
DBMS. Technology and software for creating databases	There are six main data models: lists (flat) relational databases, hierarchical, network structures, object-oriented databases, and document-oriented data model. Currently, they are the most widely used when designing a relational database model (MySQL, PostgreSQL, and MSSQL Server). It should be noted that the most popular database management system for Node.js is currently the MongoDB [69], which is a NoSQL. MongoDB is a document-oriented management system (DBMS) open source software. MongoDB is a new approach to build databases without SQL queries, tables, foreign keys, etc. In MongoDB, JavaScript is used as the query language, and the data are stored in the BSON format, i.e., binary JSON.
WebRTC is a technology for creating Web communications applications.	WebRTC [70] is an open source technology for building peer-to-peer networks, which allows us to send text and multimedia data directly between browsers. Signaling server is used only for setting up p2p connection between the two browsers. The WebRTC technology is implemented by three JavaScript APIs: RTCPeerConnection, Media Stream (getUserMedia), and RTCDataChannel.
Cloud technologies' development tools for Web applications and messaging service in real time mode	Cloud computing [71] is the delivery of computing services – servers, storage, databases, networking, software, analytics, and more – over the Internet ("the cloud"). Companies offering these computing services are called cloud providers and typically charge for cloud computing services based on usage, similar to how you are billed for water or electricity at home.

Table 22.1 Continued

Technologies for creating applications with SOA architecture and use of SaaS with APIs	A service-oriented architecture [72] is essentially a collection of services. These services communicate with each other. The communication can involve either simple data passing or two or more services coordinating some activity. Some means of connecting services to each other is needed.
IoT technologies	The WoT [29–35] is a refinement of the IoT by integrating smart things not only into the Internet (network), but into the Web architecture (application). In this section, it is necessary to consider: the IoT architecture, controllers and mini computers, the IoT platform, application programming languages, wireless network technologies, protocols, and IoT security issues.
Protocols and technologies of creating applications for Web things based on Raspberry Pi 3 Model B using the Node.js platform	The Semantic Web [1, 40]. The WoT [41–44] is a high-level application protocol designed to maximize interoperability in the IoT. The WoT architecture stack is not composed of layers in the strict sense, but rather of levels that add extra functionality. Each layer helps to integrate Things to the Web even more intimately and hence making those devices more accessible for applications and humans. The following shall be considered in this section: Linux-based mini computers; versions of Raspbian for Raspberry Pi; and implementation of the Web Thing API for the direct connectivity integration template and for the gateway-based connectivity integration template.
Cloud platforms and services for the WoT	Data Analytics. The cloud-based connectivity integration template allows the Web platform to act as a gateway to implement API Web Thing on the staging device. EVRYTHNG platform [73] is a cloud-based PaaS. The platform provides a unique and permanent identifier for each individual object and enables authorized applications and users to access it via the REST API and Pub/Sub (MQTT). It is proposed to consider the technology for implementing of the Web Thing API for the cloud-based connectivity integration template.
Security, privacy, and access control to the physical devices on the Internet	Security in IoT [36–38] should be provided at different levels of the network. Security in WoT is provided by certificates, encryption, and authentication [48–53]. The security issues: protecting Web Thing (encryption, enable HTTPS, WSS, and TLS on the server); authentication and access control; and the use of social networking tools OAuth 2.0 for WoT authentication.

```
pi@raspberrypi: ~/wot-pi
Файл Правка Вкладки Справка
pi@raspberrypi:~ $ cd wot-pi
pi@raspberrypi:~/wot-pi $ node wot-server.js
Hardware Passive Infrared sensor started!
Hardware LED 1 actuator started!
Hardware Temperature & Humidity sensor started!
Temperature: 23.9 C, humidity 30.2 %
HTTP server started...
WebSocket server started...
Your WoT Pi is up and running on port 8484
Temperature: 23.9 C, humidity 30.6 %
Temperature: 23.9 C, humidity 30.1 %
Temperature: 23.9 C, humidity 30.1 %
not anymore!
Temperature: 23.9 C, humidity 30.1 %
there is someone!
Temperature: 23.9 C, humidity 30.1 %
```

Figure 22.3 A screenshot of the application's operation Web Thing for the direct connectivity integration template.

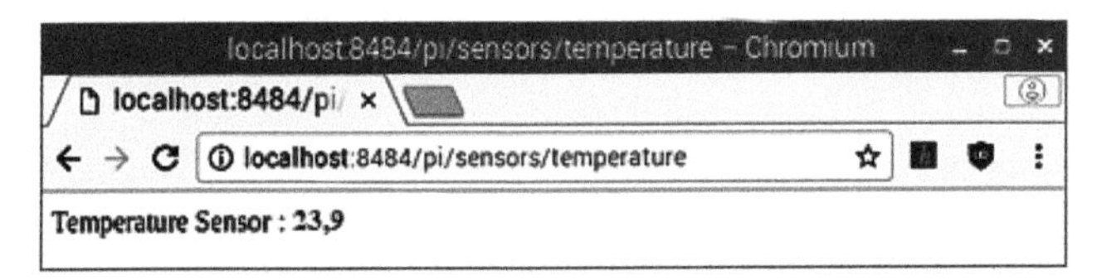

Figure 22.4 A screenshot displaying the values of the temperature sensor in the Web browser.

According to the Figure 22.4, on request you can view physical devices parameter values in a Web browser, for example, the values of the temperature sensor.

22.4 Conclusion

This work reviews the existing technologies in the IoT/WoT, for which draft standards are being developed. At the 8th International Workshop on Web Objects (WoT 2017), it was noted that the REST architecture was *de facto* basis for building the software interface of intelligent physical objects connected to the Internet. WoT, based on the Web and its new technologies, provides integration of all kinds of Smart Things and applications with which they interact, and transforms the world of physical objects into a distributed information system. Nowadays, a creation of RESTful Web API for Web Thing on JS in Node.JS is preferred over other programming languages. In this regard, the proposed curriculum technologies are relevant and aimed to prepare future professionals to develop real Web Thing applications. The technologies mentioned are tested in the teaching process of students.

The industrial IoT is part of IoT. The IoT is a network of computers, devices, and objects that collect and share the industrial data. It allows operating of industrial systems in a more efficient and safe manner. Besides, industrial IoT is used for smart grid to improve efficiency of power generation and distribution. The many challenges and risks caused by industrial IoT shall be addressed by the development of methods and tools. This will be done under the course that is to be developed by National Aerospace University.

In this chapter, we proposed and discussed the structure of the curriculum "Technology and development tools WoT applications" designed for training the development of modern web applications based on JS/Node.JS, which will give the listeners a starting point for mastering the WoT concept and developing Web Things applications. Because in the future, the orbit IoT/WoT will include the technologies of deep machine learning, artificial intelligence, technology blockchain, and robotics, the curricula for training of future specialists in the field of IoT/WoT will be updated and filled with new content. In the future, it is planned to deploy a specialty "Architecture and technology IoT/WoT" and propose the disciplines that will be taught within the framework of this specialty.

References

[1] Guinard, D. D., and Trifa, V. M. (2016). *Building the Web of Things.* United States: Manning Publications, 344.

[2] Martins, J. A., Mazayev, A., and Correia, N. (2017). Hypermedia APIS for the Web of Things. *IEEE Journals and Magazines*, 5, 20058–20067.

[3] Trifa, M. V. (2011). *Building blocks for a participatory web of things: devices, infrastructures, and programming frameworks* (Doctoral dissertation, ETH Zurich), 190.

[4] Kajimoto, K., Kovatsch, M., and Davuluru, U. (2017). "Web of Things (WoT) Architecture," *W3C First Public Working Draft.* Available at: https://w3c.github.io/wot-architecture/ [accessed May 3, 2018].

[5] Kaebisch, S., and Kamiya, T. (2018). "Web of Things (WoT) Thing Description," *W3C Editor's Draft.* Available at: https://w3c.github.io/wot-thing-description/ [accessed May 3, 2018].

[6] Kis, Z., Nimura, K., Peintner, D., and Hund, J. (2018). "Web of Things (WoT) Scripting API," *W3C Editor's Draft.* Available at: https://w3c.github.io/wot-scripting-api/ [accessed May 3, 2018].

[7] Koster, M. (2018). "Web of Things (WoT) Protocol Binding," *W3C Editor's Draft*. Available at: https://w3c.github.io/wot-binding-templates/ [accessed May 3, 2018].

[8] Trifa, V., Guinard, D., and Carrera, D. (2015). "Web Thing Model," *W3C Member Submission*. Available at: https://www.w3.org/Submission/wot-model/ [accessed May 3, 2018].

[9] Francis, B. (2018). "Web Thing API," *Unofficial Draft*. Available at: https://iot.mozilla.org/wot/ [accessed May 3, 2018].

[10] The name of project is ALIOT, which is acronym from official name. (n.d.). *Internet of Things: Emerging Curriculum for Industry and Human Applications*. Available at: http://aliot.eu.org/ [accessed May 3, 2018].

[11] Telit IoT University. (n.d.). *Things-to-Apps Made Easy*. Available at: https://www.telit.com/iot-university/ [accessed May 3, 2018].

[12] IoT University. (n.d.). *IoT UI Development with ThingWorx: Course Summary, Course Milestones*. Available at: https://www.iotu.com/enrollment/student/iot-ui-development-with-thingworx [accessed May 3, 2018].

[13] IoT Academic Program. (n.d.). *The PTC IoT Academic Program is a "passport" to the future for students, makers, and researchers*. Available at: https://www.ptc.com/-/media/Files/PDFs/Academic/iot_academic-program_EN.pdf?la=en&hash=6AAA287FED90220A9B4104D100CBA9BF94B1CF3B [accessed May 3, 2018].

[14] Kortuem, G., Bandara, A. K., Smith, N., Richards, M., and Petre, M. (2013). Educating the internet-of-things generation. *Computer*, 46(2), 53–61.

[15] Queen Mary University of London. (n.d.). *MSc Internet of Things – Queen Mary University of London*. Available at: https://www.qmul.ac.uk/postgraduate/taught/coursefinder/courses/173148.html [accessed May 3, 2018].

[16] IoT India Magazine. (2016).*10 Leading University Courses on IoT (Worldwide)*. Available at: https://www.iotindiamag.com/2016/08/10-leading-university-courses-iot-worldwide/ [accessed May 3, 2018].

[17] MIPT. (n.d.). *The IOT Academy of Samsung opens in MFTI*. Available at: https://mipt.ru/news/v_mfti_otkryvaetsya_iot_akademiya_samsung. [accessed May 3, 2018].

[18] Cisco. (n.d.). *Internet of Things Specialists*. Available at: https://www.cisco.com/c/en/us/training-events/training-certifications/certifications/specialist/internet-of-things.html [accessed May 3, 2018].

[19] Department of Computer Systems, Networks and Cybersecurity (n.d.). *Programmable mobile systems and Internet of Things*. Available at: https://csn.khai.edu/speciality/programmiruemye-mobilnye-sistemy-i-internet-veschej [accessed May 3, 2018].

[20] Internet of Things (n.d.). *Bachelor program in Internet of Things*. Available at: http://iot.lviv.ua/ [accessed May 3, 2018].

[21] Namiot, D. E. (2015). About the Internet of Things training programs. *International Journal of Open Information Technologies,* 3, 2307–8162.

[22] Gyrard, A., Patel, P., Datta, S. K., and Ali, M. I. (2017). "Semantic Web Meets Internet of Things and Web of Things: [2nd Edition]," in *WWW 2017 - 26th International World Wide Web Conference*, Perth, 917–920.

[23] Purewal, S. (2014). *Learning Web App Development*. O'Reilly Media, 306.

[24] Brown, E. (2014). *Web Development with Node and Express*. O'Reilly Media, 336.

[25] Tkachenko, V. (2016). *IoT - modern telecommunication technologies*. Available at: http://www.lessons-tva.info/articles/net/013.html [accessed May 3, 2018].

[26] Wan, J., Chen, M., Xia, F., Li, D., and Zhou, K. (2013). From Machine-to-Machine Communications towards Cyber-Physical Systems. *Computer Science and Information Systems*, 10(3), 1105–1128.

[27] Selinger, M., Sepulveda, A., and Buchan, J. (2013). "Education and the internet of everything: How ubiquitous connectedness can help transform pedagogy," *Cisco Consulting Service and Cisco EMEAR Education Team*. Available at: https://www.cisco.com/c/dam/en_us/solutions/industries/docs/education/education_internet.pdf [accessed May 3, 2018].

[28] Guinard, D. (2017). *What is the Web of Things*? Available at: https://web ofthings.org/2017/04/08/what-is-the-web-of-things/ [accessed May 3, 2018].

[29] Fleisch, E. (2010). What is the Internet of Things? *An Economic Perspective. Economics, Management, and Financial Markets*, 5(2), 125–157.

[30] Atzori, L., Iera, A., and Morabito, G. (2010). The Internet of Things: A survey. *Computer Networks*, 54, 2787–2805.

[31] Parashar, R., Khan, A., and Neha. (2016). A survey: The internet of things. *International Journal of Technical Research and Applications*, 4(3), 251–257.

[32] Sethi, P., and Sarangi, S. R. (2017). Internet of Things: Architectures, Protocols, and Applications. *Journal of Electrical and Computer Engineering*, 2017:9324035, 25.

[33] Lin, J., Yu, W., Zhang, N., Yang, X., Zhang, H., Zhao, W. (2017). A Survey on Internet of Things: Architecture, Enabling Technologies, Security and Privacy, and Applications. *IEEE Internet of Things Journal*, 4(5), 1125–1142.

[34] DDS the proven data connectivity standard for the IoT. (n.d.). *What is DDS*? Available at: http://portals.omg.org/dds/what-is-dds-3 [accessed May 3, 2018].

[35] Al-Fuqaha, A., Guizani, M., Mohammadi. M., Aledhari, M., and Ayyash, M. (2015). Internet of Things: A Survey on Enabling Technologies, Protocols and Applications. *IEEE Communications Surveys Tutorials*, 17(4), 2347–2376.

[36] Kumar, S., and Patel, D. (2014). A Survey on Internet of Things: Security and Privacy Issues. *International Journal of Computer Applications*, 90(11), 20–26.

[37] Garcia-Morchon, O., Kumar, S., and Sethi, M. (2018). State-of-the-Art and Challenges for the Internet of Things Security draft-irtf-t2trg-iot-seccons-14. Available at: https://datatracker.ietf.org/doc/draft-irtf-t2trg-iot-seccons/ [accessed May 3, 2018].

[38] OWASP Internet of Things Project. (n.d.). *IoT Security Guidance*. Available at: https://www.owasp.org/index.php/IoT_Security_Guidance [accessed May, 3, 2018].

[39] Oxford e-Research Centre. (n.d.). *The Internet of Things: realising the potential of a trusted smart world*. Available at: http://www.oerc.ox.ac.uk/news/Centre-contribution-IoT-reports [accessed May 3, 2018].

[40] Kamilaris, A., Yumusak, S., and Ali, M. I. (2016). "WOTS2E: A search engine for a Semantic Web of Things," in *2016 IEEE 3rd World Forum on Internet of Things (WF-IoT)*, Reston, VA, 436–441.

[41] Guinard, D., Trifa, V., and Wilde, E. (2010). "A Resource Oriented Architecture for the Web of Things," in *Proc. of the 2nd International Conference on the Internet of Things (IoT 2010)*, LNCS, Tokyo. Berlin: Springer.

[42] Guinard, D., Trifa, V., and Wilde, E. (2010). Architecting a Mashable Open World Wide Web of Things. *Technical Report No. 663*, Department of Computer Science, ETH Zürich.

[43] Zhou, Y., De, S., Wang, W., and Moessner, K. (2016). Search Techniques for the Web of Things: A Taxonomy and Survey. *Sensors*, 16(5), 600.

[44] Kamilaris, A., Pitsillides, A., Prenafeta-Bold, F. X., Ali, M. I. (2017). "A Web of Things based eco-system for urban computing - towards smarter cities," in 24th International Conference on *Telecommunications (ICT) 2017,* 1–7.

[45] Tkachenko, V. (2017). *Web of Things - IoT Network Service.* Available at: http://www.lessons-tva.info/articles/net/014.html [accessed May 3, 2018].

[46] Francis, B. (2017). *Building the Web of Things.* Available at: https://hacks.mozilla.org/2017/06/building-the-web-of-things [accessed May 3, 2018].

[47] EVRYTHNG Developer Hub. (n.d.). *Welcome.* Available at: https://developers.evrythng.com/docs [accessed May 3, 2018].

[48] Friedl, S., Popov, A., Langley, A., Stephan, E. (2014). Transport Layer Security (TLS) Application-Layer Protocol Negotiation Extension. Available at: https://tools.ietf.org/html/rfc7301 [accessed May 3, 2018].

[49] Peck, M., and Igoe, K. (2013). *Suite B Profile for Datagram Transport Layer Security/Secure Real-time Transport Protocol (DTLS-SRTP) draft-peck-suiteb-dtls-srtp-02.* Available at: https://tools.ietf.org/html/draft-peck-suiteb-dtls-srtp-02 [accessed May 3, 2018].

[50] Ferry, E., O'Raw, J., and Curran, K. (2014). Security Evaluation of the OAuth 2.0 Framework. *Information Management and Computer Security*, 23(1), 73.

[51] OAuth Community Site. (n.d.). *OAuth 2.0.* Available at: https://oauth.net/2/ [accessed May 3, 2018].

[52] Borgohain, T., Borgohain, A., Kumar, U., and Sanyal, S. (2015). Authentication systems in Internet of Things. *arXiv preprint arXiv*:1502.00870.

[53] Reshetova, E., and McCool, M. (2017). *Web of Things (WoT) Security and Privacy Considerations.* Available at: https://www.w3.org/TR/wot-security/ [accessed May 3, 2018].

[54] WebStorm. (n.d.). *The smartest JavaScript IDE. Powerful IDE for modern JavaScript development.* Available at: http://www.jetbrains.com/web storm/ [accessed May 3, 2018].

[55] Sublime text. (n.d.). *A sophisticated text editor for code, markup and prose.* Available at: https://www.sublimetext.com/ [accessed May 3, 2018].

[56] Git. (n.d.). *Git-local-branching-on-the-cheap.* Available at: https://git-scm.com/ [accessed May 3, 2018].

[57] Resources for developers. (n.d.). *HTML5 - Web developer guides.* Available at: https://developer.mozilla.org/en-US/docs/Web/Guide/HTML/HTML5 [accessed May 3, 2018].
[58] World Wide Web Consortium (W3C). (2016). *Extensible Markup Language (XML).* Available at: https://www.w3.org/XML/ [accessed May 3, 2018].
[59] Resources for developers. (n.d.). *CSS3 - CSS: Cascading Style Sheets.* Available at: https://developer.mozilla.org/en-US/docs/Web/CSS/CSS3 [accessed May 3, 2018].
[60] Bootstrap. (n.d.). *The most popular HTML, CSS, and JS library in the world.* Available at: https://getbootstrap.com/ [accessed May 3, 2018].
[61] Resources for developers. (n.d.). *JavaScript.* Available at: https://developer.mozilla.org/en-US/docs/Web/JavaScript [accessed May 3, 2018].
[62] JavaScript. (n.d.). *Ready to try JavaScript?* Available at: https://www.javascript.com/ [accessed May 3, 2018].
[63] jQuery. (n.d.). *What is jQuery?* Available at: https://jquery.com/ [accessed May 3, 2018].
[64] Resources for developers. (n.d.). *Ajax.* Available at: https://developer.mozilla.org/en-US/docs/Web/Guide/AJAX [accessed May 3, 2018].
[65] Websocket.org - Powered by Kaazing. (n.d.). *About HTML5 WebSocket.* Available at: http://websocket.org/aboutwebsocket.html [accessed May 3, 2018].
[66] Socket.IO. (n.d.). *Socket.io 2.0 is here. Featuring the fastest and most reliable real-time engine.* Available at: https://socket.io/ [accessed May 3, 2018].
[67] World Wide Web Consortium (W3C). (n.d.). *The WebSocket API.* Available at: https://www.w3.org/TR/websockets/ [accessed May 3, 2018].
[68] Node.js®. (n.d.). *Node.js v8.11.1 Documentation.* Available at: https://nodejs.org/en/ [accessed May 3, 2018].
[69] MongoDB for GIANT Ideas. (n.d.). *MongoDB Atlas. Database as a Service.* Available at: https://www.mongodb.com/ [accessed May 3, 2018].
[70] Johnston, A.B., Burnett, D. C. (2013). *WebRTC: APIs and RTCWEB Protocols of the HTML5 Real - Time Web.* St. Louis, Digital Codex LLC, Smashwords Edition, 247.
[71] Microsoft Azure. (n.d.). *What is cloud computing?* Available at: https://azure.microsoft.com/en-in/overview/what-is-cloud-computing/ [accessed May 3, 2018].

[72] Service Architecture. (n.d.). *Service-Oriented Architecture (SOA) Definition.* Available at: https://www.service-architecture.com/articles/web-services/service-oriented_architecture_soa_definition.html [accessed May 3, 2018].

[73] EVRYTHNG Developer Hub. (n.d.). *Standard API Introduction.* Available at: https://developers.evrythng.com/v3.0/reference [accessed May 3, 2018]

23

Prospects for Constructing Remote Laboratories to Study Cognitive IoT Systems

Mykhailo Poliakov[1], Karsten Henke[2] and Heinz-Dietrich Wuttke[2]

[1]Zaporizhzhia National Technical University, Zhukovskogo, Zaporizhzhia, Ukraine
[2]Ilmenau University of Technology, TU Ilmenau, Ilmenau, Germany
E-mail: polyakov@zntu.edu.ua; karsten.henke@tu-ilmenau.de; dieter.wuttke@tu-ilmenau.de

Cognitive systems are a perspective object for distance learning. Futurologists talk about the massive use of cognitive systems including in the sphere of security and reliability of critical infrastructure facilities in one or two decades. Remote laboratories are an important element of distance learning systems. Currently, these laboratories are used to study information control systems and require development and improvement to study cognitive systems. A refined structure of the cognitive system is proposed, which includes a knowledge base setup on the following three items: the pyramid of knowledge forms, a knowledge conversion/extraction subsystem, and a subsystem of activity. In this structure, elements that are missing in information control systems are identified and, accordingly, are not studied using existing remote laboratories. First, this is the finite automata of targets and scenarios, as well as knowledge in understanding and understanding in wisdom. The functionality of a number of existing elements will need to be expanded. This concerns, for example, automatons of behavior, which must be hybrid automata, combining the means of describing both discrete and continuous behavior. The knowledge obtained by the system being studied is suggested to be described as the values of fuzzy, linguistic, binary, and ternary variables. To obtain models of understanding and wisdom based on this knowledge, the use of logical schemes that are processed with the help of inference systems, for example, the PROLOG, is suggested.

23.1 Introduction

Computerization of the world population, development of the Internet, achievements in the field of teaching methods, increasing students' mobility, and development of the system of continuous education are the reasons for the development and growing application of distance learning, including engineering education. It is quite logical that distance learning extends to new products of technological progress.

Futurologists note that the cognitive revolution will be an important result of technological progress in the next 10–20 years [1]. Its essence lies in expanding the range of things and processes, the effectiveness of which will be enhanced by computer-based knowledge management. Previously, computers controlled the spacecraft, then the washing machine, the turn of the light bulb, the wires, and everything that surrounds us (All of Things) with a built-in computer. The desire to expand the information base to justify the solution in management tasks led to the use of trends, post-history and forecasts, and parameter change scenarios that took the form of knowledge. Cognitive systems are developed by integrating IoT technologies and advances in the direction of artificial intelligence. These systems tend to solve the tasks of artificial intelligence in real time. In the structure and functions of security systems and reliability of critical infrastructure facilities, there is also a tendency to increase cognitive level [2]. Such solutions promise to significantly improve the effectiveness of the application of human systems, things, and processes and change the everyday way of man and the technological way of humankind.

The coming cognitive revolution puts before the system of engineering education tasks that need to be addressed now. In particular, this also applies to the improvement of the laboratory-training base. Remote laboratories are an important element of distance learning systems. The number of these laboratories in engineering education systems is continuously growing, their characteristics are continuously improving, and the fields of application are expanding. In view of the foregoing, it is legitimate to raise the question of the applicability of existing laboratories to the study of cognitive systems.

23.2 State of the Art

Well-known surveys characterize remote laboratories in terms of their structure, purpose, subject area, and types of applied models of objects of study [3, 4]. The remote student interface plays an important role in the structure of the remote laboratory. The initial functionality of such an interface is the ability to input the student's project input data and output the results in the form of an animation of the study object model. The extended functionality of the remote student interface includes the demonstration of WEB images of the position of the physical object, which is realized, for example, in the remote laboratory of the Grid of Online Laboratory Devices Ilmenau (GOLDi) [5]. The limited functionality of these physical objects is compensated by the use of added functionality in the laboratory [6].

With the help of remote laboratories, physical phenomena, the characteristics of electronic nodes, and the behavior of systems of continuous [7] and event-based [8] control are studied. Moreover, when forming the response of the control system to changes in the control object, information is used about the current values of its parameters and the states of the finite automaton of the behavior of the control object. Such systems are usually called information control systems. In the well-known papers, there are no estimates of the feasibility of using remote laboratories to study cognitive control systems.

The purpose of this paper is to determine the prospects for using remote laboratories to study cognitive control systems. The goal is to find answers to questions about the nature, structure of the cognitive control system, knowledge and activities that the student should use when designing it, the services, and user interfaces required for this in the remote laboratory.

The next sections of this paper deal with the cognitive control system model and the prospects for building remote laboratories for studying such systems.

23.3 Cognitive Control System Model

The model is based on the cognitive control system [9], whose structure is shown in Figure 23.1. It is based on two main functions of such a system, which are

- First, the study of a control object by processing;
- Second, storing in the database information, coming from sensors and knowledge-based activities expressed in various forms [10, 11].

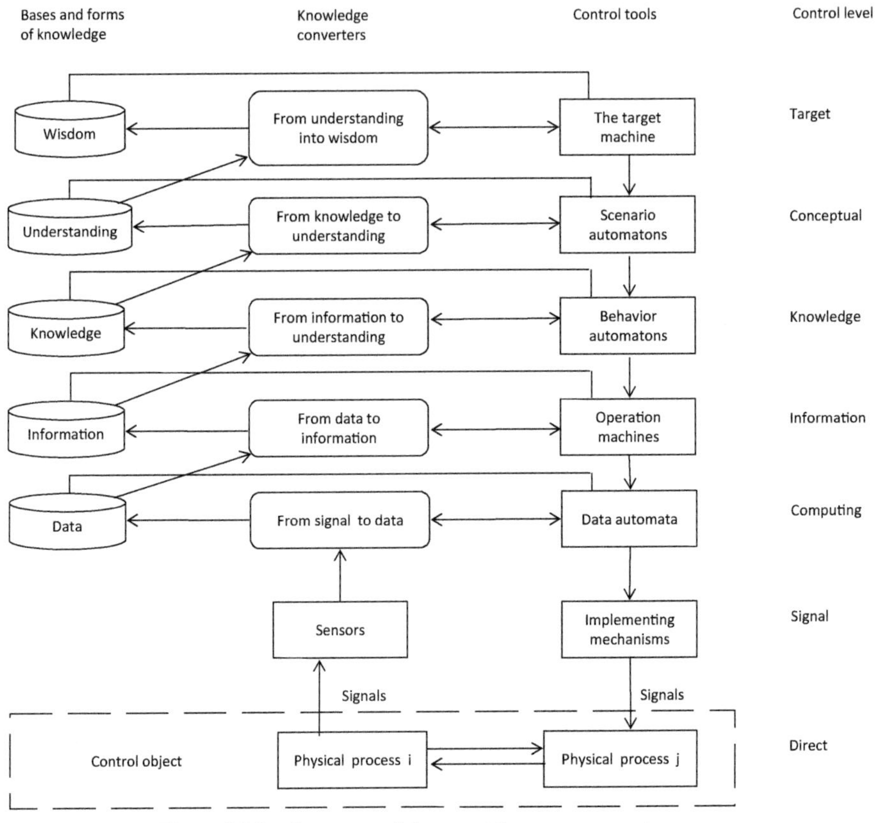

Figure 23.1 Structure of the cognitive management system.

The activities of the control system ultimately form influences on the control object and thereby change its parameters in the desired direction. In the cognitive system, we will distinguish a number of levels.

Each level integrates a certain part of the knowledge base with a part of the management tools and forms subsystems of a certain type. These subsystems are combined into a hierarchical structure, which is presented in Figure 23.2.

The function of control tools is to control the conversion of knowledge into a higher level form and execute the functions of an automaton at this level. The functions of the automaton include the determination of the next

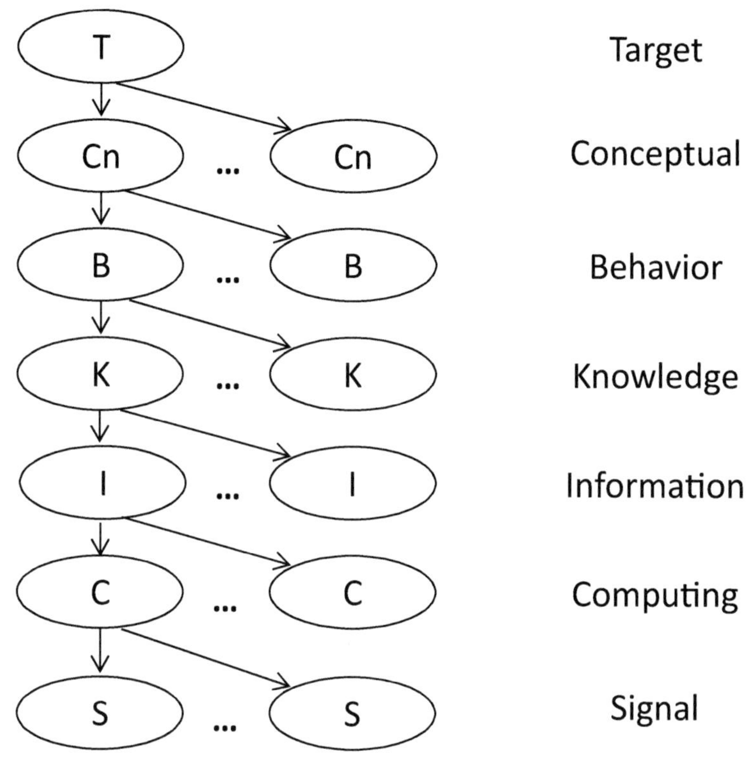

Figure 23.2 Hierarchy of subsystems of the cognitive control system.

state of the automaton (the transition function) and the formation of its outputs (output function). The new state of the machine is determined based on its current state and the actual values of the input variables from a database. The outputs of the signal level automata control the actuators of the control object. At other levels, outputs are used for initialization, start (stop), and parametric and structural adaptation of the submachine. Examples of values of elements of the control system are given in Table 23.1.

The proposed model of the cognitive control system allows determining the requirements for a remote laboratory for its study, decomposing it into simpler systems, identifying typical elements, detailing their interfaces, and ultimately, simplifying the design and study of the system.

Table 23.1 Values of elements of the cognitive control system

Element of the System	Value
Analog input # 2 «Temperature sensor»	Voltage: 745 mV
The N record in the database associated with input # 2	System Time: 12145964; Value: 2279
Record N in the database associated with input # 2	The time now is 5:18 AM. Date: May 31, 2017; Working temperature: 78 degrees Celsius.
The N record in the knowledge base associated with the temperature of the working area	Temperature parameters are normal and correspond to the forecast
Recording N in the understanding base associated with the trend of temperature variation of the working area	In the near future, a sharp increase of the temperature of the working area is expected, which will lead to undesirable overheating of the control object.
The entry of N in the base of wisdom associated with the understanding of trends in the temperature of the working area	It is necessary to reduce the consequences of the expected overheating of the control object (CO)
State of the target automaton	Assign a priority objective to prevent overheating of the CO
State of the scripting machine	Scenario advanced control of cooling control object to assign priority scenario.
The state of the automaton of behavior	Start the thermodynamic model of the control object. Connect the temperature forecast channel of the c to the cooling system controller
State of operation machine	Select the switching point and the operating mode of the control system cooling system
State of data machine	On the discrete output number 7 to form a "logical 1" at the time of the system time 12146000
Discrete output # 7 "Pump drive"	Voltage 24 V

23.4 Prospects for Constructing Remote Laboratories

Prospects for constructing remote laboratories for the study of cognitive systems are examined using the example of the GOLDi laboratory. This laboratory allows learning remotely the principles of controlling electromechanical devices based on behavior. For this purpose, GOLDi lab has the appropriate physical models shown in Figure 23.3 and behavior design models (Figure 23.4).

Figure 23.3 Physical models of the GOLDi lab.

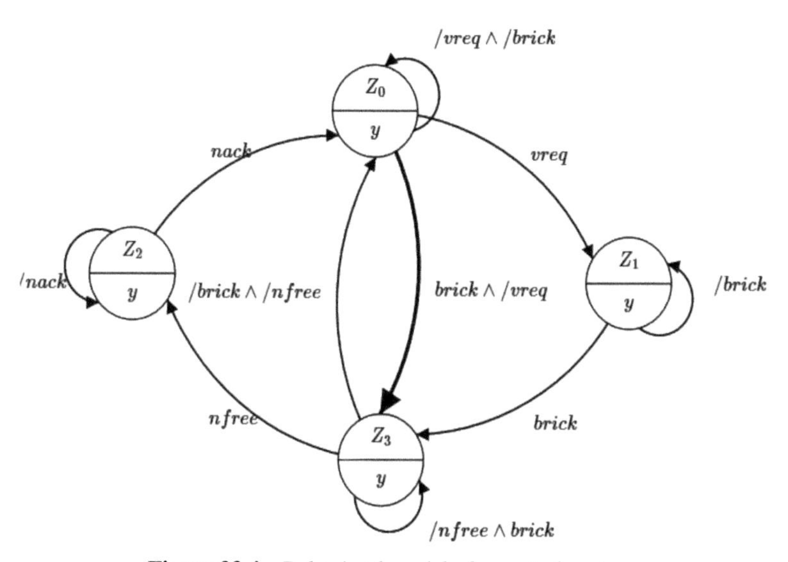

Figure 23.4 Behavioral model of a control system.

Comparison of the instruments of the laboratory with the structure of the cognitive control system allows us to conclude that GOLDi lab does not have the tools for designing converters from knowledge in understanding and from understanding in wisdom. Moreover, there are no environments for designing goal automata and scenarios as well.

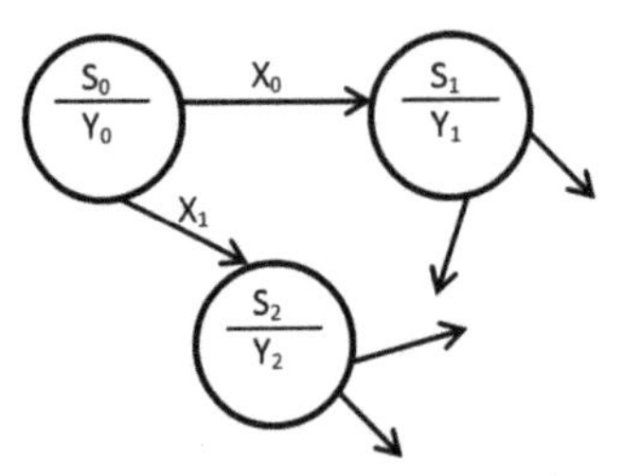

S_0 – target "Efficiency"; S_1 – target "Keep working";
S_2 - target "Prevent overheating";
Y_0, Y_1, Y_2 - – output "Activate scenario 0, 1 or 2";
X_0 – event of wisdom "A parametric failure is detected";
X_1 - event of wisdom "It is more important to prevent overheating"

Figure 23.5 Fragment of the target AG.

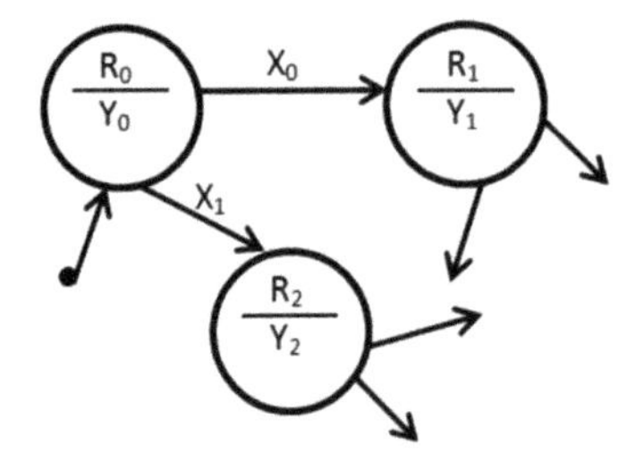

R0, R1, R2- episode "Optimize the load", "Optimization of cooling taking into account the forecast" and "Changing the cooling parameters"; Y0, Y1, Y2 - - output "Activate behavior automate 0, 1 or 2"; X0, X1-event of outstanding "Use forecast is preferable" and "Advanced cooling is not effective".

Figure 23.6 Fragment of the scenario AG.

Figures 23.5 and 23.6 show fragments of target and scenario automata graphs (AGs) of the cognitive control system.

As can be seen from the figures, the visual form of these automata is similar to the image of the automaton of behavior in the environment of the Graphical Interactive FSM Tool. The refinement of this tool will include interpreting the inputs of the automaton as variables from the databases of wisdom and understanding, as well as interpreting the outputs of automata as start-up signals and configuring automata of the corresponding lower levels. To design hybrid behavior automata, it is advisable to use the Matlab – Simulink – Stateflow system [12] or more specialized software packages.

In the cognitive system, the conversion functions of the knowledge acquired by the system from one form to another are performed by converters, the number of which is equal to the number of levels in the DIKW-pyramid by [10] or additional of the DIKUW pyramid (data – information – knowledge – understanding – wisdom) knowledge forms. The cognitive system contains two converters that are not available in information control systems. These are the converters of "knowledge – understanding" and "understanding – wisdom." When designing these converters, it is supposed to formalize knowledge using fuzzy, linguistic, binary, and ternary variables. Formalized knowledge is structured using knowledge bases, cognitive situation maps, and decision logic. It is assumed that the analysis of these knowledge structures will be carried out by the methods of fuzzy production systems and inference systems.

Based on the presented forecast of the structure, it is advisable to develop and include in the software of a remote laboratory for studying cognitive systems tools/environments for the design of target automata, scenario automata, hybrid behavior automata, cognitive maps, and knowledge logic schemes. These tools should include translators from the visual image of the automaton model, cognitive map, and logic scheme to the program code in standard text programming languages, for example, in C# or PROLOG.

The study of cognitive systems with the help of remote laboratories will require an increase in the added functionality of physical models of the object of study. Inclusion in the software remote laboratory simulators of this functionality may allow the use of existing non-cognitive physical models of objects of study in the study of cognitive systems.

It is assumed that the complexity of projects of students on the subject of designing elements of cognitive systems will increase. In addition, the complexity and duration of experiments on the approbation of these projects using remote laboratories will increase. Perhaps, students will repeatedly experiment with their project for a long period, such as a study semester. At the same time, in the pauses between experiments in a remote laboratory, the context of the project and, related to the project, the added functionality of the physical models of the objects of study should be preserved.

23.5 Conclusion

The requirements for a remote laboratory for the study of cognitive systems are determined by the structure and characteristics of these systems. In describing the structures and elements of cognitive systems, the proposed model, which contains the knowledge pyramid and activity pyramid, is useful. Elements of these pyramids located on one level form one or several control subsystems of a certain type: direct, signal, computational, informational, behavioral, conceptual, and target. The last two levels are absent in information management systems that are studied by an existing remote laboratory. In addition, the level of behavioral management in the cognitive system may need to have the properties of hybrid control. Designers of these levels will use the formalisms of finite state machines of targets, scenarios, and hybrid controls. At the same time, for the construction of knowledge converters, it is required to conduct an additional analysis of the formalisms used in artificial intelligence systems. In general, the analysis showed that the remote GOLDi laboratory could be upgraded to study cognitive systems. In the future, it is supposed to develop tools for supporting the educational design of cognitive control systems using a remote laboratory.

Acknowledgments

This work was supported in part by the European Commission within the program "Tempus," "ICo-op," Grant No. 530278-TEMPUS-1-2012-1-DE-TEMPUS-JPHES as well as "DesIRE," Grant No. 544091-TEMPUS-1-2013-1-BE-TEMPUS-JPCR.

References

[1] Miller, G. A. (2003). The cognitive revolution. *Trends in Cognitive Sciences*, 7, 141–144.

[2] Kharchenko, V., Sklyar, V., and Brezhnev, E. (2013). "Security of information-control systems and infrastructures: models, methods and technologies", [Bezopasnost' informatsionno-upravlyayushchikh sistem i infrastruktur: Modeli, metody i tekhnologii (Russian)], *Paperback*, Palmarium Academic Publishing.

[3] Azad, K. M., Auer, M. E., and Harward, V. J. (2012). *Internet Accessible Remote Laboratories*, Hershey, Pa: IGI Global.

[4] Gravier, J. Fayolle, B. Bayard, M. Ates und J. Lardon, State of the art about remote laboratories paradigms – foundations of ongoing mutations. *International Journal of Online Engineering*, 4:1.

[5] Henke, K., Vietzke, T., Hutschenreuter, R., and Wuttke, H.-D. (2016). *The remote lab cloud GOLDi-labs.net*. IEEE: Piscataway, NJ, 37–42.

[6] Poliakov, M., Henke, K., and Wuttke, H.-D. (2018). "The Augmented Functionality of the Physical Models of Objects of Study for Remote Laboratories," in *Online Engineering and Internet of Things*, Cham, Springer International Publishing, 151–159.

[7] Gustavsson, I., Zackrisson, J., Nilsson, K., Garcia-Zubia, J., Håkansson, L., Claesson, I., and Lagö, T. L. (2008). A Flexible Instructional Electronics Laboratory with Local and Remote Lab Workbenches in a Grid. *International Journal of Online Engineering*, 4(2), 12–16.

[8] Henke, K., Fäth, T., Hutschenreuter, R. and Wuttke, H.-D. (2018). "GIFT – An Integrated Development and Training System for Finite State Machine Based Approaches," in *Online Engineering and Internet of Things*, 22, Cham, Springer International Publishing, 2018, pp. 743–757.

[9] Poliakov, M. (2017). *Cognitive Control Systems: Structures and Models*. Electrotechnic and Computer Systems, Odessa National Polytechnic University, 101, 387–393.

[10] Ackoff, R. (1989). From data to wisdom. *Journal of Applied Systems Analysis*, 16, 3–9.

[11] Rowley, J. (2007). The wisdom hierarchy. *Journal of Information Science*, 33, 163–180.

[12] Stateflow. Available at: https://www.mathworks.com/products/stateflow.html. [accessed: May 14, 2018].

24

Project-Oriented Teaching Approach for IoT Education

Peter Arras[1], Dirk Van Merode[2] and Galyna Tabunshchyk[3]

[1]Faculty of Engineering Technology, KU Leuven, Belgium
[2]Department of Electronics–ICT, Artesis-Plantijn University College, Belgium
[3]Software Tools Department, Zaporizhzhia National Technical University, Ukraine
E-mail: peter.arras@kuleuven.be; iiw.kuleuven.be;
dirk.vanmerode@ap.be; galina.tabunshchik@gmail.com

Considering the importance of new disruptive evolutions in the digital world and their impact on society, the authors in this chapter comment on examples of project-oriented teaching/learning approaches for engineering studies in Internet of Things (IoT). These educational methods are intended to stimulate a more effective and diversified self-study of engineering students in the area of IoT from hardware development at the embedded laboratory to the front-end software development. The authors show five different case studies which used a project-oriented approach. The 21st century engineer needs more skills than just technical knowledge but should also be competent in the area of teamwork, project management, and communication. In the conceive–design–implement–operate (CDIO) approach, these skills are taught to students in project work. The different projects, remote labs, and software tools described in the chapter were all realized with and by students in a CDIO-like approach. The (student) projects were conceived and serve to learn how to make use of projects in the field of IoT and related study fields so that they are effective and efficient in achieving the necessary learning outcomes to make professionals in IoT.

24.1 Introduction

A set of new disruptions in our 21st century society – social, technological, and business – are being provoked and facilitated by the evolution of digital technologies. Terms as "Internet of Things" (IoT) or "Cyber-Physical Systems" describe a new generation of highly distributed, intelligent digital systems. These terms refer to a global approach for the functioning and controlling of "Things." A "Thing" in this respect is anything which is used, controlled, measured, and is connectable to the Internet: all kinds of devices and microcontroller systems that can read sensors do some (preliminary) digital signal processing and send output over the Internet. The availability of powerful multicore microcontrollers, large size memories, and a wide variety of commercial of the shelf sensors enables this new and challenging market. The average amount of microcontrollers per person is rapidly growing and will continue to grow in the next few years [1, 2]. In their new "IoT report," Businessinsider.com projects, there will be 34 billion devices connected to the Internet by 2020 [2].

It is clear that there are great job opportunities for technicians and engineers in this specific high-skilled field of expertise. These specialists should have a profound knowledge of both hardware and software aspects of the system, in interfacing with sensors, in using embedded operating systems or real-time operating systems, but also on networking. Furthermore they need knowledge in systems' dependability which is described by attributes as availability (readiness for correct service), reliability (continuity of correct service), safety (absence of catastrophic consequences for the users and the environment), security (availability of the system only for authorized users), confidentiality (absence of unauthorized disclosure of information), integrity (absence of improper system state alterations), and maintainability (ability to undergo repairs and modifications) [3].

The task for higher educational institutes (HEIs) is to deliver to the labor market highly skilled engineers and developers who have the knowledge to design, build, operate, maintain, and problem shoot these devices [4]. Especially when one deals with IoT, where these systems are deployed in an industrial environment to run process-critical applications, quality issues of the combined hardware and software become extremely important.

To introduce the knowledge of IoT in the study curricula of engineering students, the authors looked at project-based learning as proposed in (CDIO – conceive–design–implement–operate) [5]. Introducing projects and learning through projects to students in which they are challenged to explore and extend their engineering skills and knowledge use transversal skills to communicate to technicians and non-technicians, and entrepreneurial skills. Creativity is enabled both within the classroom and oftentimes outside it, sometimes even without a formal curriculum. In this paper, we go through a number of projects related to this approach.

To introduce the knowledge of IoT in the study curricula of engineering students, a number of projects about remotely controlled experiments [6–8] were run to offer to students possibilities for experimenting and for shaping up their knowledge on the theme. All remote and virtual labs were constructed with the input and collaboration of students. This approach makes the remote labs a student project in the CDIO-philosophy itself.

In Section 24.2, we consider some existing examples of a practical-oriented remote experiment, which substitutes a real hands-on lab and a project oriented remote lab, for training students in a specific field of study and checking the achieved learning outcomes. In Section 24.3, we also show some case studies of what is achieved with students in the field of IoT through project-oriented teaching/learning. We also look at a project of re-engineering software for sustainability of software. In Section 24.4, we discuss a university factory for the production of (embedded systems) hardware as a supporting tool for student project in IoT and as a cornerstone for the CDIO-implement tasks.

The goal is ultimately to learn how to make use of projects in the field of IoT and related study fields so that they are effective and efficient in achieving the necessary learning outcomes to make professionals in IoT.

24.2 Remotely Controlled Experiments

The remote laboratory (RL) is an automated laboratory (experiment) in which students can experiment to find out about laws of nature or other phenomena. Remotely means that the user is not physically present near the experiment, but he is controlling it from a distance (over the Internet/intranet, with no need for manual interference).

As any lab – RLs – should offer enough possibilities for experimenting and offer measurable learning outcomes, associated with these experiments, care should be taken to ensure that the RL is more than a demonstration lab, making sure that it is a real experiment – although controlled from a distance. Over the last years, a great variety of remote and virtual labs in different fields of study were developed [8–11]. When developing an RL as a teaching/learning aid, we are faced with the same questions as when developing any other didactical method: namely, to think carefully on the learning outcomes, teaching, and learning approach. The learning outcomes will point out what and how students will need to learn and also point on how to evaluate [12].

The advantages of an RL for students are clear: 24/7 availability to experiment and repeatable experiments can motivate to achieve a deeper learning on the topics. The challenges for the construction of the RL are to make it user friendly and efficient in achieving the learning outcomes and motivating and attractive to students. Another major challenge is potential distant evaluation and feedback for the students on mistakes or good and bad practices they used. Finally, maintenance of the remote experiment is also challenging.

It is clear that the development of an RL is the work of a team of specialists. Creating a remote experiment, which will meet the requirements for the safe and reliable operation, as well as requirements arising from the didactics of education, is a challenging task. Complexity and specificity of this task require participation and close cooperation of a team of experts – specialists (Figure 24.1) [13].

For the development of the remote labs from the cases, we used a project-oriented (CDIO) approach. The implementation team for the construction consisted of teachers and students, working on it for their thesis project. The teacher mainly generates the ideas and contents, and students were responsible for the design and implementation of the hardware and software.

We examine here two case studies:

The ISRT-lab (Informational Systems on Reliability Tasks-lab) at Zaporizhzhia National Technical University for project-oriented eLearning. The ISRT software development consists of three-year bachelor student project work and theoretical part of Ph.D. student research.

The CALM (Computer-Aided Learning Module) eLearning system for material sciences at KU Leuven as an example of practical-oriented eLearning. The software for RL for CALM was a Master Thesis works and the Web interface was a bachelor thesis work. The hardware was built in the framework of master thesis works.

24.2.1 Informational Systems on Reliability Tasks-lab (ISRT)

ISRT-lab [14] consists of a number of smaller dedicated experiments which allow students to study and experiment on different aspects of embedded systems and communication tools over the Internet. The aim is to prepare

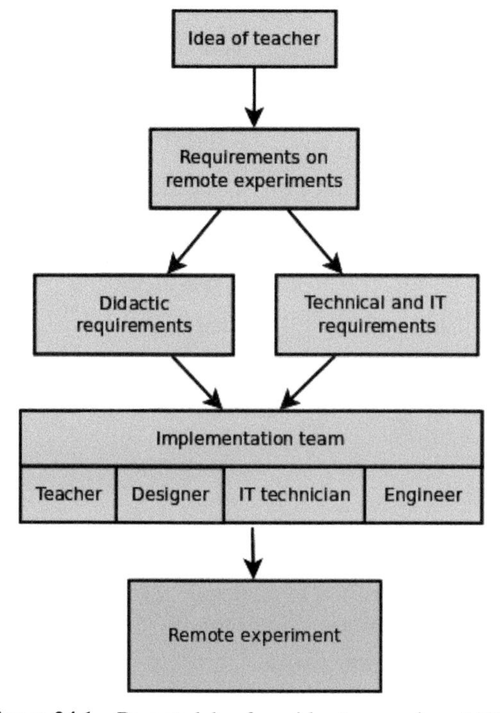

Figure 24.1 Remote labs: from idea to experiment [13].

students for IoT. The series of experiments include experiments on the manipulation of components (LED lights and stepper motors), on communication (mobile phone manipulation), on security (face detection through image detection), and programming (in C++ and Python) (Figure 24.2).

The ISRT was built to let students of bachelor and master studies in software development experiment and self-study the different aspect of programming and controlling. After the self-study, students get a project assignment in which they use the different skills they acquired using the ISRT. Evaluation of the learning outcomes was done on the project results.

Tasks include transformation of data, connecting and using different sensors for physical parameters (temperature, light intensity, luminosity, and distance), image recognition, detecting time-delays in the execution of programs, and access to remote working systems with different protocols like Wi-Fi, Bluetooth Low Energy and GSM. The goal of the predefined tasks is that students later on will work on an own-defined project in which they combine and use the knowledge to make a physical remote sensing device

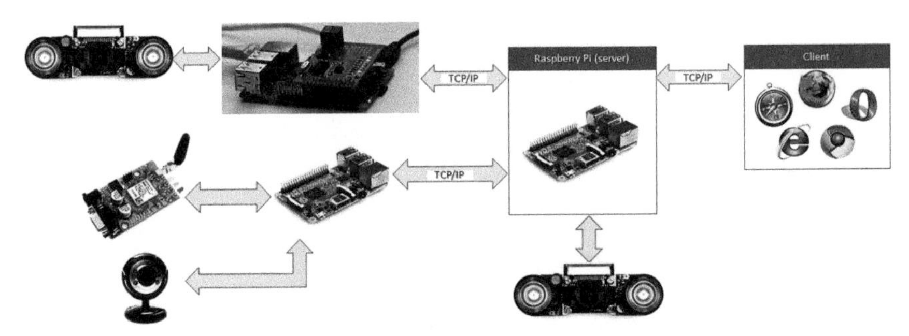

Figure 24.2 ISRT architecture.

for some physical status (e.g., ecological measurements and climate control measurements) [15].

A short description of some of the remote experiments:

- RL manipulation with LEDs on the expansion board: A Raspberry Pi is fitted with an expansion board. The task is to convert a number from one system into another (binary to hex, hex to binary, and oct to binary). A camera is used to display the expansion board display containing the number in hex or oct and the number in the binary system is displayed by the LEDs.
- RL manipulation with a light sensor: The expansion board contains light and temperature sensors. This RL allows us to change the distance between light and sensor and to measure luminosity and to build a chart representing the relation between distance and luminosity.
- RL face detection demo: The face detection demo lets students check the time which is needed for the face detection algorithm on the OpenCV Python libraries. There are two possibilities: to work either with the Raspberry Pi Professional Infrared Camera OV5647 (internal) or with a standard Web camera (external).
- RL GSM module manipulation: One of the common tasks in ES is to provide 24/7 access to remote working systems. To provide robustness of such a system, it should provide access to it by all possible protocols like Wi-Fi, BLE, and GSM.

Examples of the tasks are:

- To develop a program in Python for adding binary numbers and displaying the results of the addition on the display of the TMMA expansion board;
- To measure the delay in the response of the SIM900;

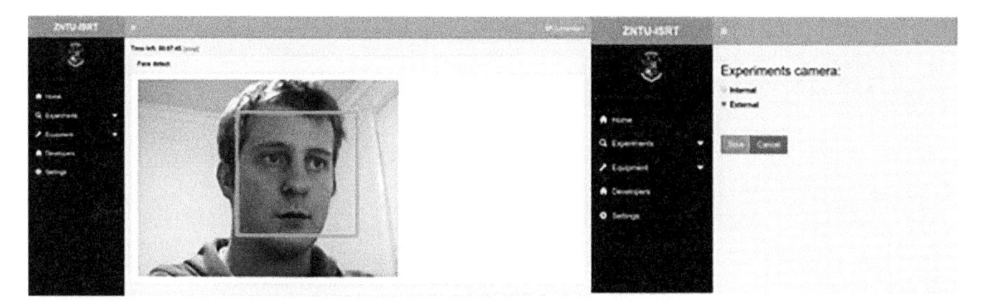

Figure 24.3 Face detection example.

- To create a system loop and measure the mean time to failure for the SIM900;
- To create a program for face detection and compare the response time to the same algorithms from OpenCV library (Figure 24.3).

In the case of the ISRT, no formal testing of knowledge of the students was done. The ISRT is used for training purposes and as a step-by-step build-up of competences for the own project. With each preparatory task, online feedback and a step-by-step lead through the task are foreseen. If students fail to answer the questions/perform the tasks, they cannot continue through the task list. Testing of the outcomes is not formally done, but evaluated in the students' own project afterward.

Remote experiments are used by bachelor and master students. Bachelor students provide experiments with the C++ programming language and master students in Python. Questionaries' of satisfaction showed that 86% of students were interested in providing remote experiments and 32% of students purchased Raspberry Pi for further individual projects. Master students implement achieved knowledge in the Master Thesis.

24.2.2 Computer-Aided Learning Module (CALM)

24.2.2.1 Aims and usage of the CALM

The CALM is an e-learning module for the study of material sciences for bachelor students in mechanical and construction engineering. CALM contains sections on theoretical knowledge, a virtual lab and remote lab, and the lab instructions for the physical hands-on labs which are used for the study of material sciences (Figure 24.4). The aim is to use the CALM in classroom teaching, in self-study for the students, and in the labs [16].

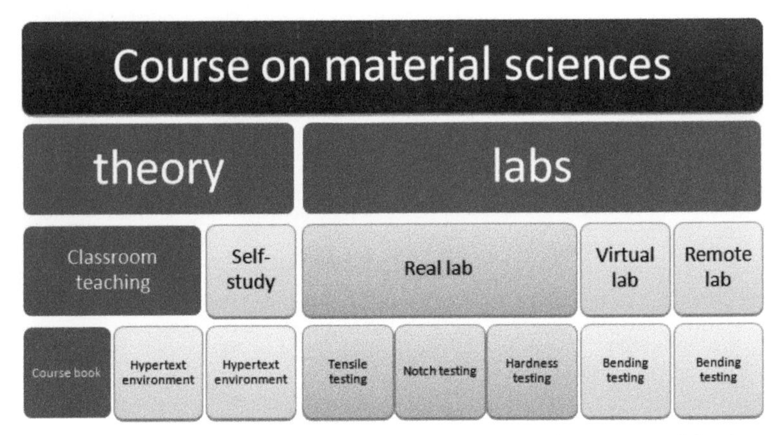

Figure 24.4 Structure of CALM, integrated courses, and labs on material science.

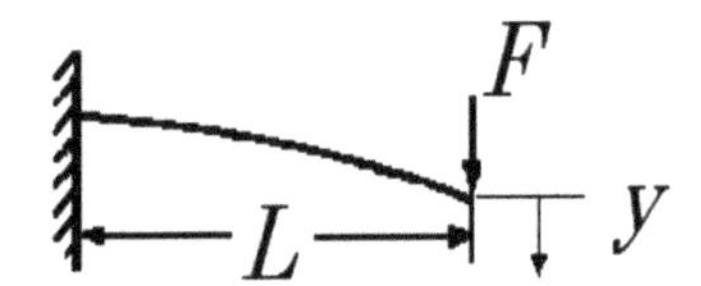

Figure 24.5 Two-point bending setup.

The learning outcomes envisaged for the CALM are that students use the theoretical part of the CALM as a support for their self-study of the course of material sciences. It contains the same contents as the course book and is also used during theoretical classrooms sessions as an illustration by the lecturer. Furthermore, it contains all lab-session manuals and assignments for the hands-on lab.

Incorporated in the CALM is a remote [17] and a virtual lab (Figure 24.6). They are used to study of the difference between material and shape stiffness of a structure. The remote lab (and virtual lab) uses a two-point bending test (Figure 24.5). They are available 24/7 to allow students to experiment on and check the theoretical formulas.

The goal is to determine the material of the bended beam by calculation through formulas with known and measured parameters. Second, students should estimate the errors they accumulated in the experiment (measuring errors and the effect on the result).

The remote lab is used as one of the four compulsory lab sessions in the material sciences courses for bachelor 2 students in engineering technology.

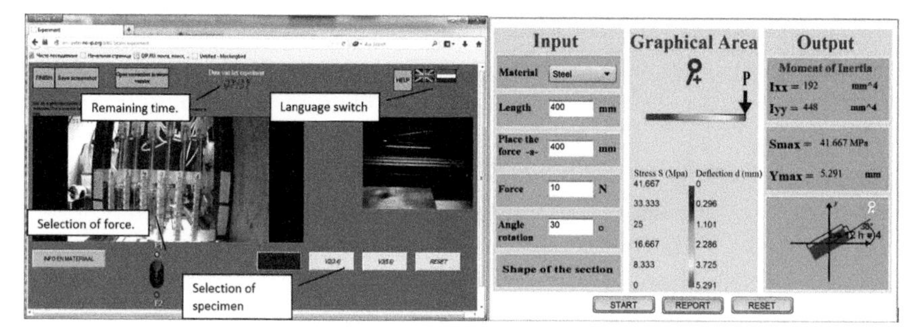

Figure 24.6 Remote lab, screenshot with view of two cameras, virtual lab.

Data gathered in the remote lab are combined with the test-data from the other labs to report on material properties of the tested specimen.

24.2.2.2 Project-oriented approach

As the remote lab and CALM are used by students, it was decided as the better solution to include students in the preparation and construction of both. However careful and understanding the development team of teachers can be, students have a different view on usability and user friendliness of experiments and as such contributed a lot to make the user interface better.

The eLearning platform and remote labs were built in collaboration between universities. The hardware of the remote lab was built during two student projects and software development involved a mixed team of staff of KU Leuven and ZNTU, and of two master students in ZNTU.

Several student projects were defined (as projects or thesis works):

- construction of the hardware of the remote lab (KU Leuven);
- development of the virtual lab (KU Leuven + ZNTU);
- development of the eLearning module of the CALM (KU Leuven + ZNTU);
- integration of the remote and virtual labs into the CALM (ZNTU).

Every student project was coached by a teacher and consisted of all major development steps and project steps: problem definition, solution finding, development, testing, and roll-out. Each project was result driven as it should be usable with students immediately after delivery, so the projects included also pilot usage with fellow students.

As it was a collaboration, an intercultural component was also present (communication in another language, time difference, different teaching, and learning approaches).

24.2.2.2.1 *Evaluation of the students on the use of the remote lab*

For an efficient pedagogical usage of remote labs and e-learning, there is the problem measuring the results on learning outcomes and comprehension measurement.

Doing online testing – which would be obvious in case of remote labs – presented too many problems. One of the more complex ones is the identification of the student who is doing the online test. Second was the timing of testing. Each student should have equal opportunities and possibilities for a test. This is also not guaranteed in an online test use of the labs.

Therefore, in the CALM, formal testing is used. The knowledge gathered in the remote and virtual lab is tested in a real physical exam, which is taken from all students at the same time.

24.3 IoT Projects for Education

A project-oriented approach is very efficient for modifying the existing solutions as for adding additional functionality as for re-engineering existing software.

These solutions allow the educator to reach more sustainability in engineering education. Both in the case of creating new software for existing IoT platform as for the creation of new software based on existing engineering approaches but with the usage of new platforms. Reusing previous knowledge and expertise makes it a sustainable solution.

24.3.1 Smart-campus Project

Within the framework of DESIRE Project [18], a start-up, called Smart-Beacons, was created. It consists of the realization of two applications on one platform: "Smart Campus" and "Smart City."

The general idea of a "Smart Campus" for universities is that the campus talks to you. Individual information for students, teachers, and visitors is delivered, depending on their profile and time of day through the use of beacons. For "Smart City," the same concept is used in an urban context, where the city interacts with the visitors in the city.

The student project was to make an actual product which is open to external users and is published both as an Android and as an iOS version. As the student developers are facing a wide variety of users, ranging from app users from all ages and backgrounds, to administrators of the beacon content and IT specialists to host and maintain the server application, they needed to keep user friendliness in mind. The students also learned to create a content management system (CMS), to attribute different roles to different users. The mobile application also needed to be attractive as the goal was to actually penetrate the consumers market, to get actual content providers on board, and to get rewarded and punished on the market. The students learned all steps which needed to be taken to get an app published with Google and Apple. Finally, the project showed that privacy issues and having a privacy policy as an HEI are absolutely necessary as the rules on privacy are getting more and more stringent.

All tasks were distributed between ZNTU-TMMA (Thomas More University College) teams. It involved three students from ZNTU, one Chinese exchange student at TMMA, and supporting staff for server administration at TMMA.

The Smart Campus Application consists of three main parts: a mobile application for different operational systems iOS and Android; a custom-made CMS for updating advertisement information, and an administration system, which consists of different components aimed to adjust hardware characteristics.

The Smart Campus Mobile Application provides users a variety of functions, allowing working in both online and offline modes, enrolling in favorite groups and blocking others and detecting buzz from the beacons [15, 19].

The work consists of three bachelor work for Android and iOS Mobile Applications and CMS development.

And further down the road, it transformed into three master thesis works – investigation and development of web-oriented mapping tools, mobile in-door route definition, and smart interfaces [19, 20].

The same data coming from the central CMS server were used by a Chinese exchange student from the Jiangnan University, but displayed in another fashion. The goal is to have a very user-friendly, easy-to-use, and attractive system. As the owners and content providers of the beacons are real external customers, the quality of the delivered work should answer to the highest demands of modern-day cloud-based systems sharpening the relevant skills of the students involved.

The server part of the developed platform was also used for development of in-door navigation system for Smart Campus [21].

24.3.2 Re-engineering of Existing Engineering Software for a New Platform (COPTURN Project)

Implementation of the Industrial IoT involves great changes in system architecture, technologies, software, and management. Nevertheless, it is critical for new developed systems not to lose achievements from earlier explored fields of study. Re-engineering of existing software can allow to reach sustainable goals in industry, research, and education but it is often prone to difficulties due to changed OS, changed programming languages, archived media-files on carriers which is no longer in use, or difficult to access.

The next project was devoted for the re-engineering of existing educational software for the new hardware and software platforms.

The existing program named COPTURN was a software package for the determination and optimization of cutting conditions and the calculation of cutting times and cutting costs. This system – originally dating back to the 1970s with a major upgrade near 1998 – is reengineered, rewritten in another programming language, and optimized with a new user interface. The system – in comparison to some commercial software – is open for students to research and work upon, and the IP in it is also owned by the university itself. So after years of losing functionality due to outdated (software) technology, a complete makeover was decided upon.

The new software system for the determination of economic optimum machining conditions was designed and built with the use of the technology and strategy of COPTURN. It is based on the minimum cost criteria as a main concept and relies on fundamental cutting laws, terms, and constraining factors (restrictions imposed by the machine, workpiece material, workpiece geometry and fixture, etc.). The new program uses the scenarios and static data of the COPTURN software that have been updated and re-engineered.

Also, the software provides an interface that allows managing information stored in the DB, i.e., adding new objects, removing, or editing existing. The appearance of the new program was improved to make it more modern and user friendly (Figure 24.7). New images that describe all operations were implemented and a guidance for users with respect to the physical meaning of the imposed values provided. A "Help" section was added as well.

The new implementation is a Python desktop application, based on PyQt5 – a Python binding of the cross-platform application framework Qt. Its GUI was built with the use of QtDesigner application. The resulting Qt forms were used to generate the UI code with the help of PyQt5's pyuic5 utility. The standard appearance of Qt forms was changed and stylized by setting

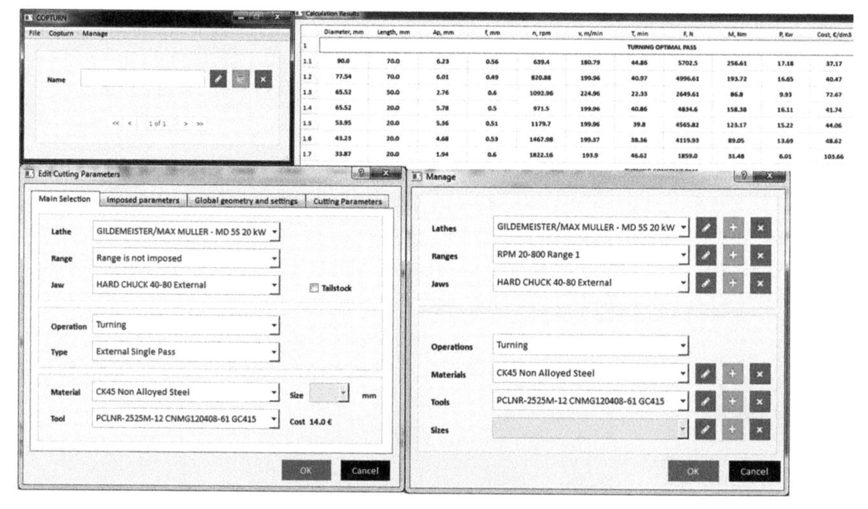

Figure 24.7 New COPTURN interface.

a designed style sheet and QtAwesome library. The process of development was performed in PyCharm IDE. CXFreeze is a library that allows to "freeze" Python scripts into executable files under any platform Python works on itself.

Within the software sustainability concept, the functionality-centric approach was implemented. In order to reproduce the operation of the old software system, the migration process has been made. This approach means re-coding the software to run it on new hardware or perform with new and reliable software. It also gives an excellent opportunity to enhance the old program's performance by solving issues and adding new features. The code was completely rewritten to another programming language with the use of modern technologies that keeps a code up to date with the latest changes that the program relies on. This migration is in fact a development of new software, but constrained by the old architecture. Migrating the software that completely produces a new code package and at the same time recreates in a reliable manner the work of the old software is required to make the program stay in use over a long period of time [22].

24.4 The Embedded Factory as a Tool for Implementation

In the CDIO project approach, projects lead to prototypes and products. Specifically for coping with the production of industry-standard products in the field of electronics, the "embedded factory" was set up to give possibilities to students to build real embedded systems.

With the increasing demand for autonomous IoT applications, which get sensor data from remote location and store these data in the cloud on sensor systems, a whole new boost is given to the already fast growing market of embedded systems. It is clear that a fast evolving set of skills need to be taught to students, who need to have a decent knowledge of the complete design process from idea to realization. Most of the time, the work which can be done at an HEI, limits itself to making a prototype or proof of concept, due to the decreasing size of electronic components and the increased complexity for assembly, as a consequence. This is why most HEIs cannot cope with the newest technology in hardware design, while the labor market is very much in need of these specialists. Although the general public has the wrong opinion that all electronics are only designed and produced in the Far-East, there is abundance of SMEs that develop and produce locally in Western Europe. A survey conducted at the labor market at the premises of Campus De Nayer in June 2017 questioning local developing industry – both of hardware and software – showed a dramatic shortage of capable technologists in both fields. Moreover, there is a lot of specific technologic knowledge between producing and testing a prototype compared to actual automated production and going to the market with an electronic end-product.

A lot of applied research and projects in the field of IoT need small numbers of embedded devices, especially when you need several sensor nodes, in remote locations or on a wide area to collect sensor data. It is easy to imagine static sensors in a vast area of a modern city or dynamic sensors in the postal services' car park. This calls for the need for automated assembly of these sensor nodes. Producing electronic hardware, with surface mount technology in numbers, calls for the need of specific technology, which in turn requires a specific set of competences for the students. The use of solder past, stencils, a pick and place machine, and a reflow oven, and some hand soldering and testing all need careful attention in the design process. The development and ordering of printed circuit boards (PCBs) also come with a certain set of rules which need to be taken into account during design. The students learn the different competences needed in targeted courses on advanced PCB design and their projects during Practice Enterprise, project-oriented courses in which technologic aspects come together with the economic reality of producing electronics. The students learn how to design for test, automated assembly, maintenance, cost effectivity, legal demands on EMI/EMC, lead-free difficulties, and so on. All these learning outcomes are assessed in the output of student projects.

To cope with the initial high investment cost and to enable the students to do project-oriented learning, public–private cooperation is set up at TMMA, during the DESIRE project, in the form of the Embedded Factory (Figure 24.8). This is a small factory to automatically produce a small series of embedded systems, capable of placing up to the smallest electronic components and automatically soldering the systems in a small reflow oven. The quality of the assembly can be tested by visual and X-ray systems, and CE certification can be done at a spin-of company on site. Due to the intense cooperation with local enterprises, the factory can be made available both for students and private businesses.

The first part of the factory is a manual solder paste application station, which uses stencils with apertures to place the solder paste on the pads of the electronic components of the design. Technologic difficulties here are the features of the paste; it should stick to the pads, but not to the stencil. Also the thickness of the stencil, the surface finishing of the apertures, and the movement of the stencil determine whether the applied solder paste can successfully be used later on in the process. All these challenges are fed back to the designers, being students and researchers.

The pick and place robot is there for putting components on the PCB board. This robot should be programmed, which takes some time, and has limitations on the sort of packages which can be placed on the PCB. These limitations are fed back to the designers in the projects.

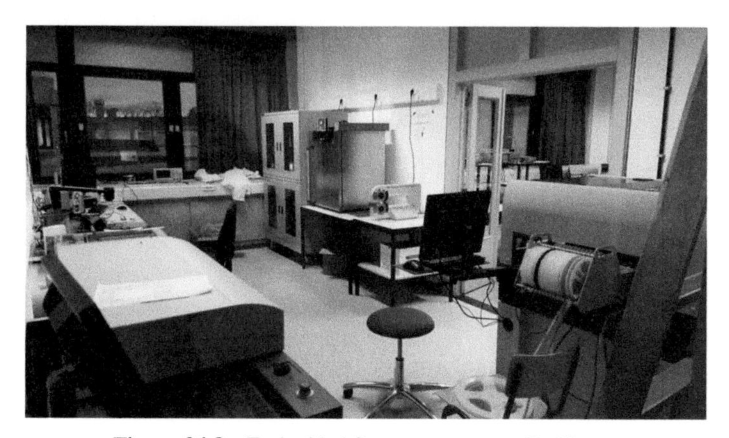

Figure 24.8 Embedded factory at campus De Nayer.

The reflow oven melts the solder balls in the solder paste. The melted solder places the components on their place and produces a conductive joint between the leads of the component and the PCB. The temperature curve, in relation to the throughput time, which is applied to the PCB and the components sticking in the applied solder paste, has a great influence on the process. The necessary parameters depend on the components used, which again is a design choice.

Next, the soldering station is used for manually placing components. This induces the need to design in some room around the components in order to reach the leads with a soldering iron.

Testing stations are used to make visual and electrical tests. Here the designer should place test pads and test circuits on the PCB for efficient and effective testing of the electric features and functionality of the project.

An important learning outcome for the designers is the proper choice of components, which are suitable for the complete production process and the design of a decent PCB to take the production process into account. The technology used is also a design driver, whether it is flex-rigid, high-speed, low-power, or small footprint design. Keeping all the parameters and limitations into account within the design phase is a difficult task and this is the main learning outcome which is envisaged when doing hardware projects with students.

24.5 Conclusions

Implementing IoT and industrial IoT requires special knowledge, multidisciplinary knowledge, team work, and soft skills. The boundaries between different fields of studies become more blurred. A project-based learning track (based on CDIO) gives the opportunity to integrate all of these competences.

Implementation of remotely controlled experiments into study allows not only teachers to give students the possibility to have access to the unique equipment 24/7 but also students to self-study and (re)experiment different aspects without much time limits. On top, the equipment is used more effective (longer time) without much personnel, as such cutting costs. Involvement of students for the development implements project-based teaching approach and increases multidisciplinary knowledge.

The collaboration of students and staff in the different institutes showed the power of international cooperation. Collaboration from different countries gave new insights and approaches, as solutions offered by the different participants also reflect their (educational and national) background. The same applies to the (formal) evaluation of student work in these projects: systems and tradition of evaluations are different in different countries. This gives insight and room for discussion between teaching staff on the other systems.

Acknowledgments

Projects were done within the framework of Tempus projects "Modernization of two cycles (MA, BA) of competence-based curricula in Material Engineering according to the best experience of Bologna Process" (543994-TEMPUS-1-2013-1-BE-TEMPUS JPCR) [23] and "Development of Embedded System Courses with implementation of Innovative Virtual approaches for Integration of Research, Education, and Production in UA, GE, AM" (544091-TEMPUS-1-2013-1-BE-TEMPUS-JPCR) [23], and the Erasmus+ project 573818-EPP-1-2016-1-UK-EPPKA2-CBHE-JP "Internet of Things: Emerging Curriculum [24].

References

[1] Mukhopadhyay, S. C. (ed.) (2014). *Internet of Things. Challenges and Opportunities*. Switzelends: Springer.

[2] Stastny, S., Farshchian, B. A., and Vilarinho, T. (2015). Designing an Application Store for the Internet of Things: Requirements and Challenges, *AmI*, 2015, 313–327.

[3] Zamojski, W., Kacprzyk, J., Mazurkiewicz, J., and Sugier, J. (eds) (2011). *Dependable Computer Systems*. Berlin: Springer-Verlag, 314.

[4] Bibri, S. El. (2015). *The Shaping of Ambient Intelligence and the Internet of Things Historico-epistemic, Socio-cultural, Politicoinstitutional and Eco-environmental Dimensions*. ©Atlantis Press, 320.

[5] Poliakov, M., Henke, K., and Wuttke, H.-D. (2017). "The augmented functionality of the physical models of objects of study for remote laboratories," in *Proc. REV2017 – 14th International Conference on Remote Engineering and Virtual Instrumentation*, Columbia University, New York, 148–157.

[6] Arras, P., Van Merode, D., and Tabunshchyk, G. (2017). "Project Oriented Teaching Approaches for E-learning Environment," in *Proc.*

IEEE 9th International Conference on Intelligent Data Acquisition and Advanced Computing Systems (IDAACS), 317–320.

[7] Maiti, A., Zutin, D. G., Wuttke, H.-D., Henke, K., Maxwell, A. D., and Kist, A. A. (2017). "A Framework for Analyzing and Evaluating Architectures and Control Strategies in Distributed Remote Laboratories," in *IEEE Transactions on Learning Technologies*, IEEE.

[8] Parkhomenko, A., Gladkova, O., Sokolyanskii, A., Shepelenko, V., and Zalyubovskiy, Y. (2016). Implementation of Reusable Solutions for Remote Laboratory Developmen. *iJOE*, 12, 24–29.

[9] Henke, K., Vietzke, T., Wuttke, H.-D., Ostendorff, S. (2015). Safety in Interactive Hybrid Online Labs. *International Journal of Online Engineering (iJOE)*, 11, 1861–2121.

[10] Sancristobal, E., Martín, S., Gil, R., Orduña, P., Tawfik, M., Pesquera, A., et al. (2012). "State of art, Initiatives and New challenges for Virtual and Remote Labs," in *Proceedings of 12th IEEE International Conference on Advanced Learning Technologies*, 714–715.

[11] Kozik, T., Simon, M., Arras, P., Olvecky, M., and Kuna, P. (2016). *Remotely controlled experiments,* eds Noga, H., Cernansky, P., Hrmo, R., Nitra, Slovacia: Univerzity Konstantina Filozofa v Nitre.

[12] Kozík, T., and Šimon, M. (2012). "Preparing and managing the remote experiment in education," in *ICL 2012: 15th International Conference on Interactive Collaborative Learning* (pp. 1–4), Villach, CORDIS.

[13] Arras, P. (2011). "Construction of a remote laboratory aimed at augmenting knowledge on properties of materials," in *14th international conference on interactive collaborative learning* (pp. 248–252). Piestany, Slovak Republic: Slovak universit of technology in Bratislava.

[14] Arras, P., Tabunshchyk, G., Kolot, Y., and Tanghe, B. (2014). Architectural Characteristics and Educational Possibilities of the Remote Laboratory in Materials Properties. *2014 11th International Conference on Remote Engineering and Virtual Instrumentation (REV)* (pp. 94–97). Porto, Portugal: Polytechnic of Porto (ISEP) in Porto, Portugal.

[15] Tabunshchyk, G., Van Merode, D., Arras, P., and Henke, K. (2016). "Remote Experiments For Reliability Studies of Embedded Systems," in *Proceedings of XIII Int. Conf. REV2016*, Madrid, UNED, 68–71.

[16] *Desire Project.* Available at: http://www.tempus-desire.eu/ [accessed january 2018].

[17] Tabunshchyk, G., Van Merode, D., Arras P., Henke K., and Okhmak V. (2018) "*Interactive platform for Embedded Software Development Study*" in Online Engineering and Internet of Things ed. M. Auer,

D. Zhutin, Book series Lecture Notes in Network and Systems, Vol. 22. Chapter No: 31. Springer International Publishing, pp. 315–321. DOI: 10.1009/978-3-319-64352-6_30.

[18] Tabunshchyk, G., Van Merode, D., Patrakhalko, K., and Goncharov, Y. (2016). "Flexible Technologies for Smart Campus," in *Proceedings of XIII International Conference on Remote Engineering and Virtual Instrumentation (REV2016)*, 58–62.

[19] Tabunshchyk, G., and Van Merode, D. (2017). "*Intellectual Flexible Platform for Smart Beacons*" Online Engineering and Internet of Things ed. M. Auer, D. Zhutin, Book series Lecture Notes in Network and Systems, Vol. 22. Chapter No: 31. Springer International Publishing, 895–900.

[20] Petrova, O., and Tabunshchyk, G. (2017). "Modelling of location detection for indoor navigation systems," in *IEEE 9th International Conference on Intelligent Data Acquisition and Advanced Computing Systems (IDAACS)*, 961–964.

[21] Arras, P., Shynkarenko, P., Tabunshchyk, G. (2017). "Sustainability in the Educational Process through Sustainable Software," in *Proceedings of the International Research Conference*, Dortmund, 113–117.

[22] *MMATENG Project*. Available at: http://www.mmateng.eu [accessed january 2018].

[23] *Aliot Project*. Available at: Available: http://aliot.eu.org [accessed january 2018].

[24] *CDIO*. Available at: website: http://www.cdio.org/

25

Internet of Things for Industry and Human Applications: ALIOT-Based Vertically Integrated Education

Artem Boyarchuk[1], Oleg Illiashenko[1], Vyacheslav Kharchenko[1], Dmytro Maevsky[2], Chris Phillips[3], Anatoliy Plakhteev[1] and Lolita Vystorobska[2]

[1]Department of Computer Systems, Networks and Cybersecurity, National Aerospace University "KhAI", Kharkiv, Ukraine
[2]Odessa National Polytechnic University, Ukraine
[3]University of Newcastle upon Tyne, United Kingdom
E-mail: a.boyarchuk@csn.khai.edu; o.illiashenko@csn.khai.edu; v.kharchenko@csn.khai.edu; dmitry.a.maevsky@opu.ua; chris.phillips@newcastle.ac.uk; a.plahteev@csn.khai.edu; lolitav1998@gmail.com

The need of new curricula in the Internet of things (IoT) for M.Sc., Ph.D., and engineering levels of education is described. The joint project on curricula development ALIOT, financed in the frame of Erasmus+ program, is discussed. The project ensures adaptation of academic programs in Ukraine and other countries to the needs of the European labor market, thus enhancing the opportunities of academic and labor abundant. The ALIOT covers hot domains of IoT applications such as health systems, intellectual transport systems, ecology and industry 4.0 systems, smart grid, and smart buildings and cities. The description of interdisciplinary multidomain and transnational programs of M.Sc. and Ph.D. levels is introduced with the mechanisms of intensive capacity building measures as well as the establishment of multidomain IoT cluster network in Ukraine is given.

25.1 Introduction

25.1.1 Motivation

The International collaboration between organizations for research and education purposes plays a key role in establishing and obtaining of practical and useful results in both areas. Modern economics presumes fast time-to-market and constant changes and development of new technologies. There are numbers of international programs which support cooperation between universities, and between university and IT industry companies. The most known and experienced are former Tempus, Erasmus+ program. The aim of Erasmus+ is to contribute to the Europe 2020 strategy for growth, jobs, social equity, and inclusion, as well as the aims of ET2020, the EU's strategic framework for education and training in modern technologies [1].

The Internet of Things (IoT) is an emergency topic of technical, social, and economic significance. Consumer products, durable goods, cars and trucks, industrial and utility components, sensors, and other everyday objects are being combined with Internet connectivity and powerful data analytic capabilities that promise to transform the way people work and live.

Nowadays, we are told that the machine-to-machine (M2M) connections will represent 46% of connected devices by 2020 as well as 73% of companies use IoT project data to improve their business and 95% of execs surveyed plan to launch an IoT business within three years [2, 3]. A number of worldwide companies and research organizations have offered a range of projections about the potential impact of IoT on the Internet and the economy during the next ten years: Cisco predicts more than 24 billion Internet-connected objects by 2019 [4] and Huawei forecasts 100 billion IoT connections by 2025 [5]. McKinsey Global Institute suggests that the financial impact of IoT on the global economy may be as much as $3.9–$11.1 trillion by 2025 [6].

Main challenges and questions in this technology are related to security and privacy issues, device interoperability, regulatory and rights' domain, emerging economy, and development issues.

25.1.2 State of the Art and Publication Statistics

In this section, we present a brief history of the origin and development of the IoT. Wikipedia says that the first publication, in which the idea of the IoT is presented, was published in 2000 [7]. This is an article by a large group of authors (13 authors): "People, places, things:

Web presence for the real world" [8]. It is based on their report at the 2000 Third IEEE Workshop on Mobile Computing Systems and Applications. Authors wrote that "... the convergence of Web technology, wireless networks, and portable client devices provides new design opportunities for computer/communications systems. ... we have been exploring these opportunities through an infrastructure to support Web presence for people, places, and things. Using URLs for addressing, physical URL beaconing and sensing of URLs for discovery, and localized Web servers for directories, we can create a location-aware but ubiquitous system to support nomadic users. On top of this infrastructure, we can leverage Internet connectivity to support communication services. Web presence bridges the World Wide Web and the physical world we inhabit, providing a model for supporting nomadic users without a central control point." Despite the fact that the words "IoT" do not appear in this annotation, one can see that the basic concept of the future IoT here is already clearly indicated.

It should be noted that the article "That 'Internet of Things' Thing" [9], which was published in January 2009 by Kevin Ashton, affirms another. The author shifts the date of the appearance of the term "Internet of things" one year ago. In particular, he writes: "I could be wrong, but I'm fairly sure the phrase "Internet of Things" started life as the title of a presentation I made at Procter & Gamble (P&G) in 1999. Linking the new idea of RFID in P&G's supply chain to the then-red-hot topic of the Internet was more than just a good way to get executive attention. It summed up an important insight – one that 10 years later, after the IoT has become the title of everything from an article in Scientific American to the name of a European Union conference, is still often misunderstood."

The presentation of which Kevin Ashton says is not published anywhere. The first publication, in the name of which the abbreviation "IoT" occurs, appeared in March 2008 [10]. It was presented in Zurich, at one of the workshops of the conference "Internet of Things 2008 International Conference for Industry and Academia." The workshop was named "Workshop on Designing the Internet of Things for Workplace Realities." It is noteworthy that initially at the same conference, a workshop "Sketchtools – Creative Tools for Prototyping Smart Devices" was announced, which did not take place due to a lack of interest of the participants in this problem [11].

Current state of the art and dynamics of IoT development can be analyzed on the basis of processing of statistical data about publications. To process such data, the software tool "accounting publication manager" (APM) has been developed using 1C: Enterprise 8.2 system. The capabilities of the

1C: Enterprise system allow us to accumulate and organize this information according to the chosen criteria. APM collects and presents information about the name, publication year, authors, and type of publication such as "Article", "Book," "Proceeding," "Techreport," etc. APM allows you to upload a list of publications in the BibTeX format [12] similar to the IEEE Explore Digital Library, The ACM Digital Library, ELSEVIER, and others. The ability to import data from a BibTeX file allows you to reduce time to enter information in APM and reduce the number of errors made when entering.

The IEEE Explore libraries and the ACM Digital Library were used to compile a list of IoT references. According to the key words "IoT," 12,406 publications were found in the library of IEEE Explore and 1714 publications were found in the ACM Digital Library. Thus, only 14,120 publications are downloaded in the APM. Figure 25.1 shows part of the list of publications received.

A large number of publications allows us to judge the status and main trends of IoT development.

First of all, the dynamics of the growth in the number of publications by years is of interest. The data obtained from the APM program for changing the number of publications are presented in Table 25.1.

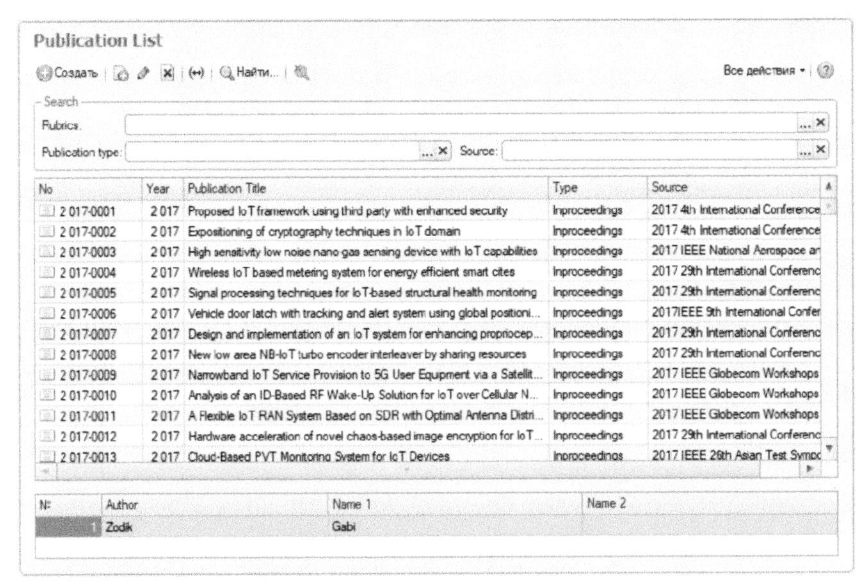

Figure 25.1 Publication list in APM.

Table 25.1 Number of IoT-devoted publications by years

No.	Year	Number of Publications
1	2008	5
2	2009	7
3	2010	127
4	2011	249
5	2012	396
6	2013	508
7	2014	1135
8	2015	2152
9	2016	3885
10	2017	5520

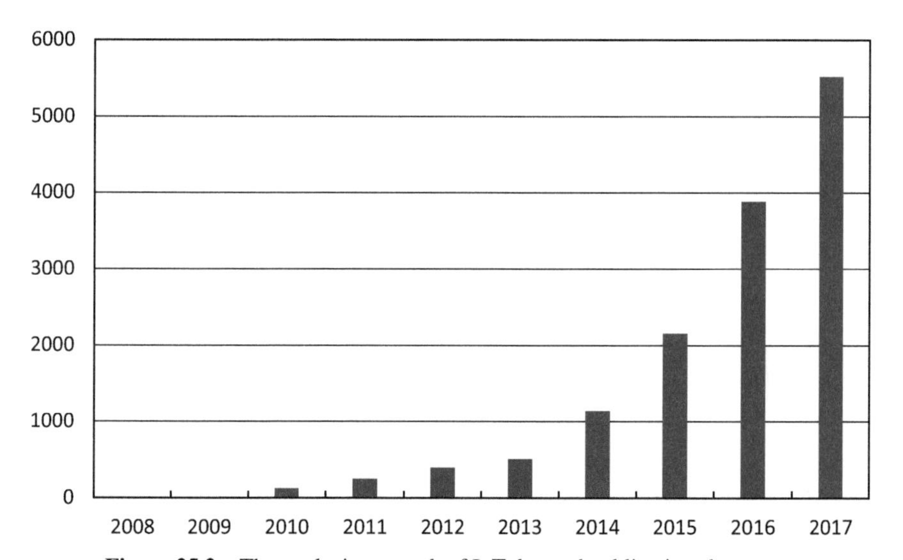

Figure 25.2 The explosive growth of IoT-devoted publications by years.

As can be seen from this table, the explosive growth in the number of publications began in 2014 and since then, the number of publications has been increasing from year to year. This explosive growth is clearly visible in Figure 25.2.

The average growth rate is 1250 publications per year. If this trend continues, then about 6800 publications should be expected in 2018, and in 2020, the number of publications should exceed 10,000. Whether this is true or not, time will tell. However, we can conclude that IOT is becoming more and more popular every year and can completely change our life. But can we be sure that this will be a change for the better?

Table 25.2 Number of types of IoT-devoted publications by years

Year	Article	Book	In Book	In Proceedings	Tech Report
2008	0	0	0	5	0
2009	0	0	0	7	0
2010	7	0	0	120	0
2011	11	0	0	238	0
2012	8	0	0	388	0
2013	51	0	6	451	0
2014	120	0	3	1012	0
2015	236	1	23	1892	0
2016	447	1	19	3417	1
2017	1167	3	77	4273	0

Table 25.3 Distribution of conferences with IoT presentations by year

Year	Conferences
2008	5
2009	7
2010	64
2011	124
2012	193
2013	229
2014	446
2015	697
2016	1048
2017	982

A more complete picture of IoT development trends can be obtained by analyzing how the number of types of publications has changed over the years (Table 25.2).

Most of the work was published in the conferences' proceedings. It is caused by fast feedback on presented results during conferences and workshops, possibilities of discussion in real time, and face to face in comparing with journal paper.

With the help of the APM software, the distribution by years of the number of conferences in which IoT reports were presented was obtained. This distribution is shown in Table 25.3 and Figure 25.3.

Figure 25.3 shows that the number of IoT conferences held in the world shows the same explosive growth as the number of publications. It is caused by the fact that, first, an increase in the number of conferences automatically leads to an increase in the number of publications, and second, an increase

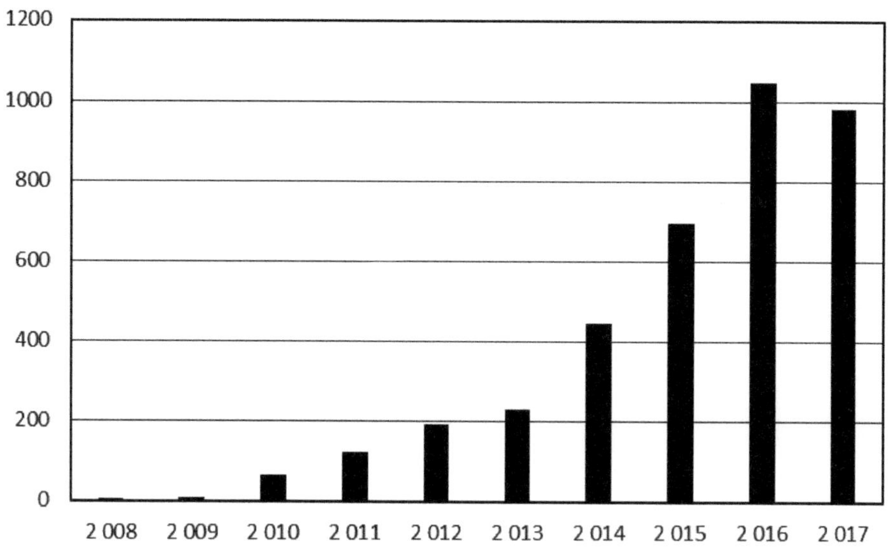

Figure 25.3 The diagram of the distribution of conferences by year.

in the number of publications reflects the growing interest of researchers in the IoT problem. And this, in turn, leads to an increase in the number of conferences. Hence, there is a kind of positive feedback. This can explain the nature of the diagrams in Figures 25.2 and 25.3.

However, we should pay attention to a small reduction (by 66) in the number of conferences in 2017. Indirectly, this may indicate a kind of "saturation" of the conference market. And, accordingly, it can lead to a decrease in the growth rate of the number of publications. However, the increasing number of publications from year to year can be predicted with certainty.

It is interesting to analyze the distribution by year of the number of authors of publications. This distribution, obtained from the APM software, is shown in Figure 25.4.

In constructing this diagram, only "authors" were taken into account. That is, if the author published several works within a year, he was counted only once. This diagram clearly demonstrates the growing popularity of IoT among the scientific community. So, in 2008, 23 authors published works on IoT subjects. In nine years, in 2017, the number of authors increased to 20,440 (more than 880 times).

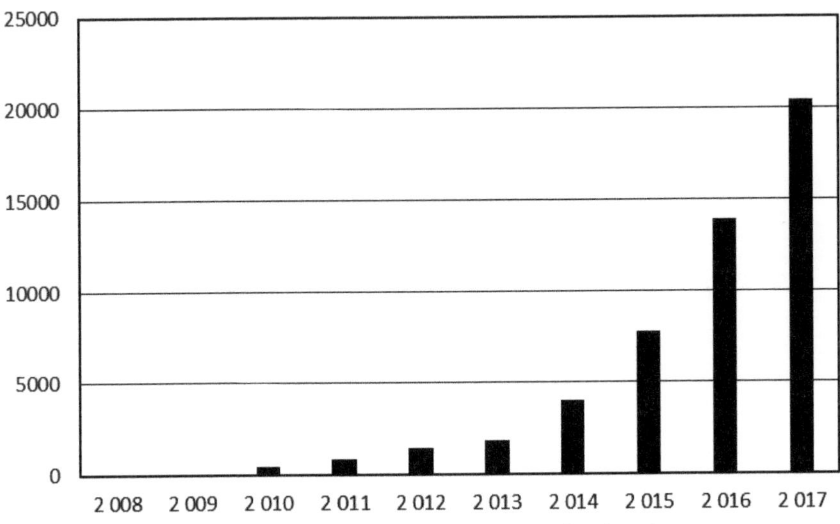

Figure 25.4 The diagram of the distribution of authors by year.

Thus, summarizing the short historical analysis, we can draw the following conclusions:

- The history of the IoT does not last more than ten years, except for the first work of this year related to this topic. This is probably the youngest area of research.
- IoT demonstrates explosive growth in all indicators: the number of people involved in development and research, the number of scientific publications, and the number of innovative developments.
- The results of work in the field of IoT will be global in nature. In this regard, the consequences of implementing IoT in various areas can surpass the consequences of the appearance of a personal computer.
- In accordance with the previous paragraph, it is clear that these consequences will not only be positive. Already now, a positive answer to the question "But whether your phone is spying on you" does not surprise anyone. Who knows, maybe in a few years we will be interested, is our electric light spying on us?

25.1.3 Objectives and Approach

The objectives of this chapter are the following:

- to analyze challenges in IoT-related education;

- to describe the Erasmus+ ALIOT project (project reference number 573818-EPP-1-2016-1-UK-EPPKA2-CBHE-JP) which ensures adaptation of academic programs in Ukraine and other countries to the needs of the European labor market in context of IoT;
- to overview an experience of EU and USA universities and content of modules and courses for some IoT-related domains such as smart building and smart cities, simulation of the IoT systems and their dependability, and security assessment and assurance;
- to present some preliminary results of the ALIOT project and compare with international experience.

We suggest vertically integrated program of education on IoT technologies and application of IoT-based systems. This program covers basing on B.Sc. Computer Engineering (or other IT Engineering curriculum) three levels of education and training:

- M.Sc. program;
- Ph.D. program;
- Engineering training for several key domains.

25.2 The Aliot Project for Vertically Integrated Education

25.2.1 Challenges and Rationale

The wise and intensive application of IoT on regional and country levels for key domains (health and aging, energy grid and smart cities, transport, and industry) will help to intensify the effort taken to overcome the pointed problems.

For their successful decision, it is extremely important to combine related efforts in IoT education, research, and engineering. Currently, the subject area of IoT is almost absent from B.Sc. and M.Sc. programs of Ukrainian universities, while the national and international IT markets require specialists in this area. There is no single specialty on computing science or engineering that covers all aspects of IoT research, development, and production implementation. The available learning laboratories, computer, network equipment, and software are not completely suitable for the educational processes in the described specialties due to the absence of specialized software applications/instrumental tools/subscriptions on specialized databases for testing and modeling of smart IoT-based infrastructures, such as advanced traffic management systems, smart lighting, forest fire detection, and smart-grid systems.

To sum up the existing challenges, it is important to provide the system analysis of the problems on the basis of universities' internal surveys:

- An innovative approach on IoT studies is not present within the M.Sc. curricula for the target and related specialties;
- There is a complete absence of doctoral courses in the area of IoT research and development;
- Urgent demand from most of the industrial actors for certified specialists in IoT area;
- Total absence of public awareness in the field of modern concepts and current approaches in IoT engineering.

International cooperation provides creation of a modernized EU innovative learning system for training and professional development in the emerging field of IoT, robotics, computer networks, and microcontrollers including the development of smart devices for traffic system with adapted academic programs to the requirements of UA and EU employers. The training resources (M.Sc., Ph.D., and industrial training modules) will be developed by leading experts in the EU and Ukraine. The EdX-based platform will have an internationalization dimension and will be made accessible to both ICT community in Ukraine and the EU. The multidomain IoT cluster network will help strengthen relations between the academia and industries (e.g., Samsung, Microsoft, Cisco, etc.) through academic and research collaborations, knowledge exchange, and transfer.

Unfortunately, through the national projects and programs, the following results cannot be obtained:

- It is not possible to provide fast response to changing market conditions without the development of a cutting edge IoT curriculum which has extensive applications for the industries and societal benefits.
- Without mobility, it will be difficult for the transfer and exchange of knowledge, skills, competences, and best practices among experts.
- The EU has extensive experience in social adaptation of persons with disabilities, but Ukraine has no such experience.
- The EU has research groups and experience in development and implementation of safe traffic systems which is almost absent in Ukraine.
- The European Union has implemented credit-transfer system of accumulation of academic credits ECTS (European Credit Transfer and Accumulation System) [13] long ago and it is tied to the EQF (European Qualifications Framework) [14]. This experience is very important for Ukraine.

25.2.2 Innovative Character

The Joint Project on Curriculum Development implements a new approach to the delivery of educational services through ongoing feedback from employers and correction of the educational process, and methodological and logistical support of the educational process. It also provides creation of professional community in IoT, robotics, computer networks, and microcontrollers. The project ensures adaptation of academic programs to the needs of the European labor market, thus enhancing the opportunities of academic and labor abundant. The other novelty is usage of the concept of ECTS and the concept of learning throughout life. In Ukraine, these concepts do not work in fact.

Among the main innovating elements, the following positions are taking place:

- Training courses on development and implementation of techniques of IoT, robotics, networks, and microcontrollers for the social adaptation of persons with disabilities (such programs in Ukraine are not available);
- Training courses on development and implementation of IoT, robotics, networks, and microcontrollers for the smart and safe traffic systems (such programs are not available in Ukraine);
- New interdisciplinary, and transnational MSc program on IoT (which is an emerging field) adapted to the modern Ukrainian and European labor market ensure the labor mobility;
- Providing the social adaptation of unemployed people and people with inadequate qualifications, by obtaining a new qualification on the principles of accumulation and ECTS, which is especially important for people with disabilities.

In order to achieve the indicated objectives the consortia members from five countries (UK, Ukraine, Sweden, Portugal, and Italy) agreed to apply for a joint project in curricula development named “Internet of Things: Emerging Curriculum for Industry and Human Applications” (ALIOT). The detailed description of the project consortia could be found on the official website of the ALIOT in section “Project Consortium” [15]. The consortia members had a mutual collaboration last years through other educational projects funded under the Tempus program – GREENCO [16], CABRIOLET [17], and SEREIN [18].

The wider objective of Curriculum Development project ALIOT financed under Erasmus+ program is to provide studies in the emerging field of IoT according to the needs of the modern society; to bring the universities closer to changes in global ICT labor market and world education sphere; and to give students an idea of various job profiles in different IoT domains. ALIOT will strengthen the internationalization dimension of the postgraduate program for higher education systems through the incorporation of Bologna objectives which ensure the transparency of the quality assurance systems, governance and management systems.

The specific objectives of Curriculum Development project ALIOT are established as:

- Introduction of a multidomain and integrated IoT program for master students in UA universities by September 2019;
- Introduction of a multidomain and integrated IoT program for doctoral students in UA universities by September 2019;
- Providing of the mechanism for intensive capacity building measures for UA CT tutors by September 2019;
- Establishment of the multidomain IoT cluster network in Ukraine by September 2019. This network will provide an environment for knowledge sharing and transfer as well as cross-fertilization of innovative IoT-related research ideas and practices between the academic and industrial sectors.

The aim of the multidomain IoT cluster network is to integrate all available and produced curriculum, methods, and tools for providing training and consultancy services in the area of IoT-based systems for different application domains: human, business-critical, and safety-critical. The network will be a means for knowledge sharing, exchange, and transfer. It will also promote public awareness of the cutting edge IoT-related concepts, technologies, and applications.

25.2.3 Project Activities and Methodology

In order to reach the described goal, the new M.Sc. program, Ph.D. program, and in-service training program on IoT will be developed and introduced, IoT cluster network offices will be established in seven Ukrainian universities, and intensive capacity building scheme for course developers, lectures, and IoT Cluster offices will be launched. This will be achieved through the implementation of the following activities (named key workpackages, or WPs):

- WP1: Development of master curriculum on IOT – strategy for the UA needs analysis, development of master curriculum, lecture books

and teaching plan, purchase and installation of needed hardware and software, scheme for the implementation of curriculum, and delivery of guest lectures.
- WP2: Development of doctoral curriculum on IOT – strategy for the development and introduction of doctoral modules which is similar to master one, described in WP1, with some specific measures applied to doctoral level curriculum.
- WP3: Capacity building measures – system for comprehensive training in the relevant theoretical and analytical skills needed to design and introduce the above approach for involved E&C engineering departments.
- WP4: Establishment of multidomain IoT cluster network on the base of involved ICT departments of seven Ukrainian universities. Each office will be specialized for the specific application domain and thus be responsible for networking and cooperating of R&D, academic, and industrial partners acting in the respective domain.

25.2.4 Expected Impact of the Project

Who will benefit from the project:

- Enrollees and students who wish to get an education in the field of IoT, computer networks, microcontrollers, and robotics (training with using new adapted to the modern UA and EU labor market training programs) – approximately 120 customers per year at one university.
- Recent graduates and young professionals on manufacturing processes' automation, computer networks, microcontrollers, and robotics who wish to adapt to the requirements of the employers (the possibility of studying individual disciplines) – approximately 80 customers per year.
- Regional and national industrial, profit and non-profit organizations and companies that are interested in improving the skills of their employees (training in the IoT centers, platform for online teaching, and web conferencing) – approximately 40 customers per year at one region where universities are located.
- Individual customers (development of individual training modules on request) – approximately 60 customers per year.

How these results will be useful for these target groups:

- *Local level* – training of specialists who meet the modern requirements in ICT industry, development of students start-up projects related to smart university infrastructure, and E-identification;

- *Regional level* – improvement of national industrial and agricultural standards, quality and production, implementation to E-government program, and medical and social services;
- *European level* – employment in the EU, development of joint educational and R&D projects with EU enterprises, and establishment subsidiary of EU enterprises in Ukraine.
- Before the recruitment of the target groups, prospective students will be reached through: publicizing ALIOT program to all partners' undergraduate and master's degree students; via ALIOT website [9] and social media accounts (e.g., Facebook, Twitter, etc.); and via Erasmus Mundus Plus interest groups.
- The target groups will also be reached with the help of electronic means: IoT web platform and a project website presenting full information about the project will be developed. Tempus networks (e.g., IITN [19], IoT, GREENCO [16], etc.) and another Erasmus Mundus programs (e.g., PERCCOM [20]).

After the end of the project, target groups will be reached with internal and external dissemination events: academic and promotional seminars, online meetings and workshops of the IoT cluster network, and regular face-to-face meetings at departmental and faculty levels to guarantee support for project activities.

Talented youth: Thanks to the nature of IoT and IoE (Internet of Everything), such kinds of teaching directions are considered as highly attractive for young generation of university graduates and in some years – generation of secondary school graduates. This will be resulted in more deep penetration of IoT ideas and methodology to different groups of active youth of Ukraine and Europe.

ICT business sector: Feedback from businesses and engineering companies who are interested in improving the skills of their employees, with the help of KhAI-maintained CIDECS by Tempus ECOTESY web platform [21] to adjust curricula of training specialists according to the new requirements; cooperation agreements with universities and centers of career and technology transfer.

Wide ICT community: Tempus networks, established in Ukraine and supported in last five years (International Technology Transfer Network by Tempus UNI4INNO [19] and CIDECS [21]) will be also used to reach Ukrainian ICT community with the help of graphical dissemination materials.

25.2.5 ALIOT Curriculum

Target *master courses* to be developed (MC1–MC4): MC1 Fundamentals of IoT and IoE (initial technologies, synergy of technologies, architectures, communications, and standards); MC2 Data science for IoT and IoE; MC3 Mobile and hybrid IoT-based computing; and MC4 IoT technologies for cyber-physical systems.

Target doctoral courses to be developed (PC1–PC4): PC1 Simulation of IoT and IoE-based system; PC2 software-defined networks and IoT; PC3 dependability and security of IoT; and PC4 development and implementation of IoT-based systems (sensors, actuators, and networking).

List of target *modules for industrial trainings* (ITM-1–ITM-6): ITM1: IoT for Smart energy grid; ITM2: IoT for Smart building and city; ITM3: IoT for UAV fleet; ITM4: IoT for automotive and intelligent transport systems; ITM5: IoT for ecomonitoring; and ITM6: IoT for industrial systems.

The development of tailor-made curricula will be based on knowledge transfer from ALIOT European partners, thus enabling development of European up-to-date curricula, in accordance with all current standards, including the Bologna process. Master courses will be based on the input made by partners (mainly from Royal Institute of Technology KTH, Sweden); doctoral curriculum will reflect theoretical knowledge and cases given by Leeds Beckett University and University of Newcastle upon Tyne, UK; training modules for business sector will be based on contribution from the Institute of Information Science and Technology, Italian National Research Council.

25.3 Overview of the IoT Courses in Europe and the United States

Before the discussion about opportunities in field of IoT for students and the future experts in the western countries, we could not ignore Ukrainian experience, and thus we attempted to search for any courses in the universities of our country. However, the result was predictable and, unfortunately, there is an apparent defect what is related to courses in IoT.

25.3.1 Overview of IoT Courses in ALIOT Project Partners

In this situation, we decided to consider an alternative European study experience in this subject. For sure, first of all, the members of consortium

ALIOT were examined, among which there were Newcastle University, Leeds Beckett University, Royal Institute of Technology, and University of Coimbra.

As we talk about study programs, it is relatively complicated to intend some kinds of compulsory methods of examination in a wide range of characteristics. So the table format was selected for the systematization and generalization of the obtained information.

It is also necessary to explain the outline of our table, and to give a comment to a few rows in it. In the beginning, we name the course with the university and the level of competence for this course (Master, Ph.D., etc.). After that, it is important to give some numbers about an amount of hours or credits (ECTS). The more hours require higher aims, which signals what and where we aspire for students to be by the end, what we want the students know.

As the followed part, we cite as an example two courses, among which are Internet of Things with Sensor Networks and Sensor-Based Systems, where it worth to emphasize the importance of the selected courses. Since a various number of connected devices are already added to the Internet, a multitude of sensors and mobile users' terminals are designed to interact in order to offer novel services in smart cities and territories in general. These devices, in the so-called IoT, have very specific characteristics both in terms of hardware (a very little amount of memory and computational power) and software and management (few system updates). Being able to understand and to simulate the IoT became essential. The only source of data, we collected, was performed by information on the web pages of the universities. However, the total amount of revealed course information fluctuates from announcements about opening to full bunch of disciplines and required references. One of the most informative site outline revealed in Newcastle University, where visitors could easily reach the learning outcomes, graduate skills framework, teaching activities, reading lists, assessment methods, and timetable.

Below we provide the results for both courses in Tables 25.4 and 25.5.

25.3.2 Metrics-Based Approach of IoT Courses Analysis

Unfortunately, the table-organized information is not pictured a broad look at the research we occupied with. Thus, for the main objective as ubiquitous analysis of courses, it was convenient to use numerical indicators and metrics. Therefore, we developed our own metric system based on works and articles on this subject. Our set of metrics represented the following:

Table 25.4 Course characteristic for Master Study (wireless embedded systems)

Name	Hours/Credits	Course Objective	Competence of the Graduates
EEE8092: IoT and sensor networks (NC)	100/5.0	Practical experience of wireless networking for computers, embedded devices or sensors, building upon the complementary taught module "Wireless Networks."	Wireless network protocols. Technologies for the implementation of wireless networks and sensor networks. Sensor systems and circuit design.

Table 25.5 Course characteristic for Master Study (embedded systems)

Name	Hours/Credits	Course Objective	Competence of the Graduates
II2302 sensor-based systems (KTH)	150/7.5	An introduction to sensor enabled systems, with an emphasis on embedded platforms, broad sensor technologies, the physical properties of measurement, and the usage in embedded designs.	Design a network topology for communicating sensor nodes that satisfies stated requirements of robustness, security, performance, and cost; the usage of sensor based architectures to design advanced applications that use context awareness, personalization, augmented, and virtual spaces.

1. *Course duration*: Collecting the actual amount of credits from each course, we took the highest number, and divided all other credits on it, so that there was a range 0–1; after that, we converted these credits according to the schema below: 5 points – 0.8–1; 4 points – 0.6–0.8; 3 points – 0.4–0.6; 2 points – 0.2–0.4; and 1 point – 0–0.2.
2. *Form of education*: 5 points – full-time courses; 4 points – remote; and 3 points – extramural studies.
   If it is mixed, then the corresponding points summarized.
3. *Basics of a course (coverage width)*: It is estimated at quantity of other courses and disciplines which are based on this course. When we found the highest number of connections, we assessed the basics the same way as in the first clause (division and range conversion).

4. *Completeness of a course on the website*: 5 points – on the website are completely available the program, the abstract of lectures, control questions and instructions for independent work; 4 points – are the program, the abstract of lectures, and control questions; 3 points – the program, the abstract of lectures; 2 points – only the program; 1 point – only the announcement of a course.
5. *Availability of a course*: 5 points – all materials in free access, record on a course are not required; 4 points – all materials in free access (free of charge) after record on a course; 3 points – materials paid; and 2 points – materials given by the professor personally.

Tables 25.6 and 25.7 provide the summary of metrics for two most widely known courses – simulation of IoT and IoE-based systems and IoT for Smart building and city.

Table 25.6 Metric for simulation of IoT and IoE-based systems' course

University Code/ Course Name	Credits	Form	Basics	Completeness	Availability	Sum
IoT and sensor networks (NC)	2	5	5	3	3	18
Sensor-based systems (KTH)	2	5	3	2	3	15
M2M technology IoT (NC)	2	5	1	3	3	14
Human–computer interaction (UC)	1	5	1	2	3	12
Intelligent sensors (UC)	2	5	3	2	3	15
Smart grid communications (UC)	2	5	1	2	3	13
Simulation and modeling (LBU)	5	5	1	1	2	14
IoT (UPU)	2	5	2	3	3	15
Sensor data fusion (PU)	3	5	3	2	3	16
Wireless, sensor, and actuator networks (RLU)	5	5	2	2	2	16
Smart cards, RFIDs, and embedded systems' security (RLU)	5	5	1	2	2	15
Interconnected devices (RLU)	3	5	1	2	2	13
Wireless, sensor, and actuator networks (BU)	5	5	2	2	2	16

Table 25.7 Metric for IoT for Smart building and city course

University Code/ Course Name	Credits	Form	Basics	Completeness	Availability	Sum
Power distribution engineering (NC)	3	5	1	3	3	15
Planning and forecasting in energy sector; energy for smart cities (KTH)	1	5	1	2	2	11
Energy simulation of buildings (UC)	2	5	1	2	3	13
Energy planning and sustainable development (UC)	2	5	1	2	3	13
Building energy management systems and intelligent buildings (LBU)	5	5	1	1	2	14

Thus, summing up the results of the metric analysis, it is possible to draw the following conclusions.

For the IoT programs, it is possible to find around 100 opportunities related to this subject area, moreover in the various faculties among which are science and technology, computer science, wireless embedded systems, advanced computing, engineering, and built environment.

However, at the European universities, there is no abundance and a variety of direct courses on IoT for today. The education system, as the most inertial part of a sheaf the science-university-industry, is late here.

The direct program with full-course performed only in 15 universities in Europe. The lead country with the biggest number of opportunities for those, who interests in being graduate in IoT, is the UK.

The most common modules for IoT courses are wireless, sensor, and actuator networks, embedded-systems security, engineering, intelligent systems, robotics, introduction to smart grids, human–computer interaction, systems engineering, data processing, signals and systems, and C programming.

25.4 ALIOT Project Case Studies

25.4.1 Control Unit for Mini Plotter

The technology of building IoT elements is based on the use of typical circuit-based solutions (platforms) based on microcontrollers and software-implemented functions – sensor interrogation, control of executive devices, information display, wired and wireless information exchange in sensor networks and with mobile devices, Internet access, cloud services, and others.

The control of actuators is usually realized by forming a sequence of combinations of control signals and is very often used in various projects. For project-based learning, a mini plotter is used with two-coordinate control of drawing simple closed contours [Figures 25.5(a) and (b)].

A universal way of describing the movement of the plotter's writing node is to build a G-file (Figure 25.6). The construction of the G-file interpreter based on the microcontroller is a rather difficult task for the subsequent stages of the project development.

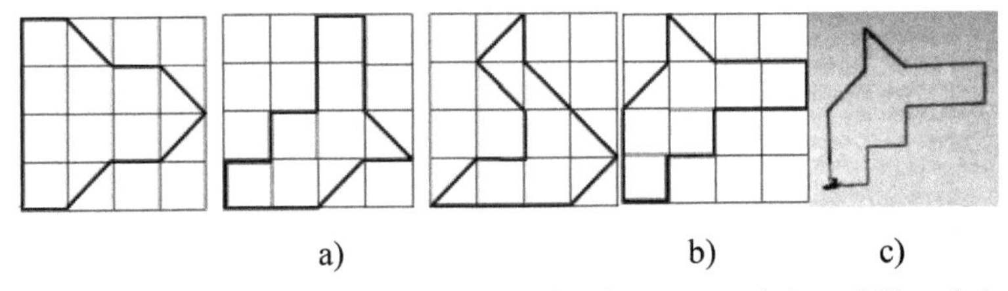

Figure 25.5 Examples of options for setting the plotter control (a and b) and the final result (c).

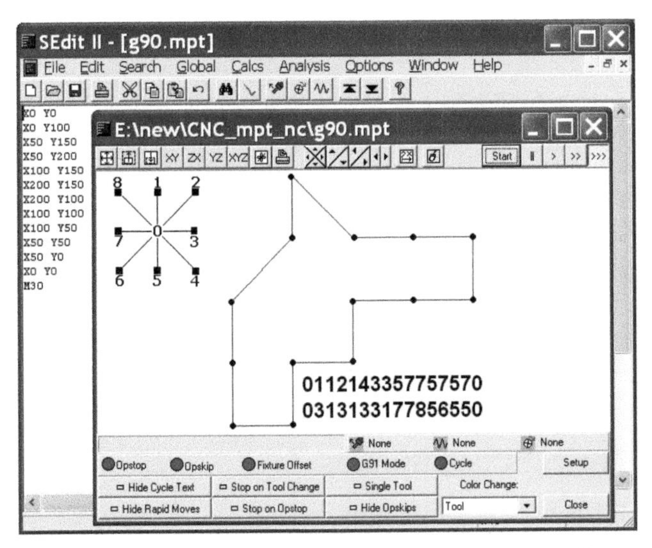

Figure 25.6 G-file for the variant of the task [Figure 25.5(b)].

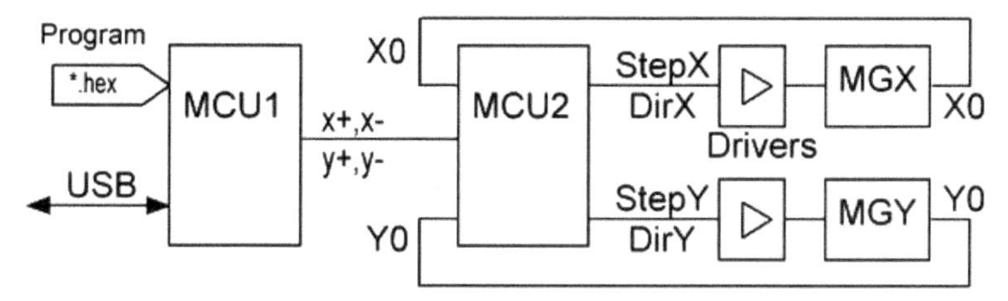

Figure 25.7 Block diagram of a mini plotter.

At the initial stage, a number of simplifications are introduced:

- Movements are performed between the nodes of the 4 × 4 grid [Figures 25.5(a) and (b)].
- Moving at each step of fixed duration corresponds to vectors 1–8 (Figure 25.6), and stop – vector 0.
- Each vector corresponds to a combination of motion control signals along the X-axis ($x+$, $x-$) and the Y-axis ($y+$, $y-$).
- For any contour, there is a representation of a row of vectors.

For the contour in Figure 25.6 when moving clockwise, we get the line "0112143357757570," and when moving counter-clockwise – the line "0313133177856550." These lines can be used to control the progress of the project.

The structure of the mini plotter (Figure 25.7) includes the target microcontroller MCU1, for which the program for generating signals ($x+$, $x-$, $y+$, $y-$) is being developed.

Direct control of stepper motors MGX and MGY with sensors of initial position X0 and Y0 is carried out by microcontroller MCU2 by means of drivers. The MCU2 is assigned the function of returning the writing node to the initial state ($X0$, $Y0$).

Figure 25.8 shows the mini-plotter view, for which various MCU1 target microcontrollers and software development tools can be used. The project implementation includes the following stages:

a. Analysis of an individual variant of the task and its formalization is the representation in the form of a row of vectors, tables, and time diagrams of output states;

a. Selecting the target microcontroller and control lines ($x+$, $x-$, $y+$, $y-$);
b. Construction of an algorithm for the functioning of MCU1;
c. Selection of development tools for MCU1;
d. Debugging the program using simulators (Proteus, etc.);

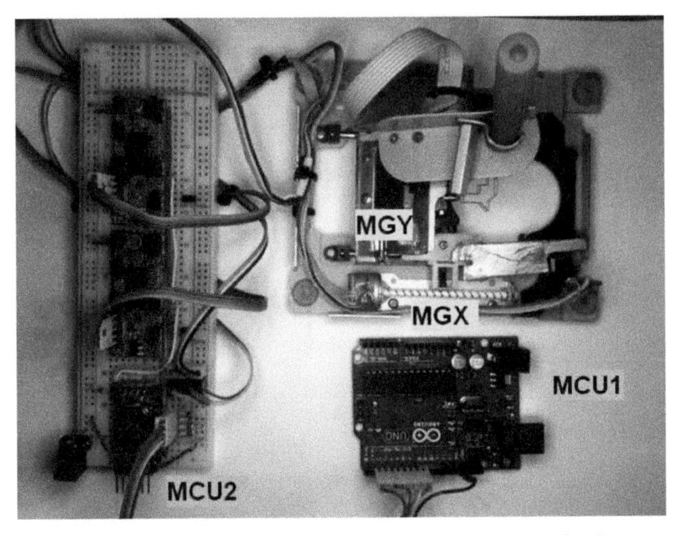

Figure 25.8 Type of mini plotter (Homemade DIY CNC).

e. Loading the program into MCU1;
f. Set the initial state of the mini plotter, connect MCU1 and start the program;
g. The representation of the final result is the image of the specified contour.

An intermediate monitoring of project implementation is carried out (stages A and E).

Further development of the project can include the use of more complex two- or three-axis drives, the implementation of a prototype drawing service (cutting, engraving, and burning) with the transmission of vector lines or G-files via the Internet.

25.4.2 Control Unit for the LED Ribbon with Pixel Addressing

LED strip with pixel addressing based on WS2812b is used in "smart" lighting, lighting, advertising, etc. Single-wire control of series-connected pixels is used. Each pixel receives 3 bytes – RGB. By setting the levels of 0–255 color components of the RGB of each pixel, you can implement various static and dynamic color effects. Matrix color panels can be constructed from the strip segments.

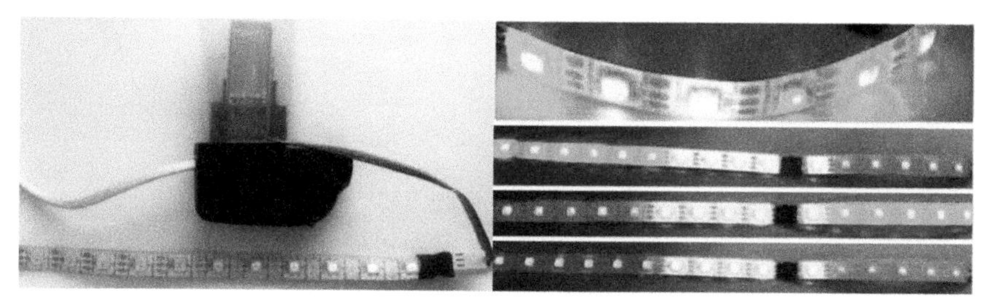

Figure 25.9 LED ribbon control module (a) and coloring options (b).

The microcontroller module [Figure 25.9(a)], which can work independently, or accept and execute commands via standard wired (USB, RS485, and Ethernet) or wireless (Bluetooth and WiFi) interfaces, manages the LED tape. Various proprietary interfaces can be used.

For the proposed projects, the asynchronous serial interface of the target microcontroller is used, which, with the help of converters, is converted to USB, Bluetooth. The microcontroller is able to receive commands:

- Select the length of the tape (*N* <1..30> *n*);
- Brightness setpoints of red (R <0..255> *r*), green (*G* <0..255> *r*), or blue (*B* <0..255> *r*) color components (RGB).

The length *N* can refer to the total number of pixels in the tape, or to determine the length of the initial portion of the tape. To this range of pixels will have the action of commands to change the brightness of the RGB.

Three-color coloring (green–white–red) of a 16-pixel LED strip [Figure 25.9(b)] is implemented using a sequence of command lines:

N16nR0rG255gB0b
N10nR255rG255gB255b
N5nR255rG0gB0b

The prepared command standards for job variants allow for a formal check of the progress of the task. The commands are transmitted to the module using a terminal program via USB, or an application – a Bluetooth terminal for a mobile device. To develop variants of the task of varying complexity, one can change the length of the tape and the color scheme (splitting into groups of pixels and group colors).

Further directions of tasks may include addition of a set of instructions executed by the microcontroller and the implementation of dynamic color schemes.

25.5 Conclusions

Market research and analysis has shown that there are bottlenecks in the education system and industry in Ukraine which brings this country to the lack of specialists in IoT area, and it could be covered via introduction of the new curricula for students and developed staff in the field of constantly accelerating IoT technology. The description of an Erasmus+ funding program, which supports the ALIOT project, is given.

The teaching courses for M.Sc. and Ph.D. students as well as capacity building in the field of training modules that will be developed in frame of Erasmus+ project ALIOT form an integrated vertical structure of education, training, and research space in different areas of IoT-based systems' industry and human applications.

To date, the teaching courses and training modules are under development and the process of their discussion within the consortium universities is under way. Successful project implementation will ensure sustainable and comprehensive staff provision in IoT education and engineering for Ukrainian enterprises and institutions.

The project team expresses confidence that the obtained results will be useful both for Ukrainian universities and other countries' higher educational establishments acting in the field of training specialists in the area IoT technologies.

Acknowledgments

This research was supported by the project STARC (Methodology of SusTAinable Development and InfoRmation Technologies of Green Computing and Communication) funded by the Department of Education and Science of Ukraine. Besides, this paper results from the Erasmus+ program educational project ALIOT «Internet of Things: Emerging Curriculum for Industry and Human Applications» (reference number 573818-EPP-1-2016-1-UK-EPPKA2-CBHE-JP, web-site http://aliot.eu.org) in which the appropriate course is developed (ITM4 – IoT for health systems) within its framework, we have developed modules related to IoT systems modeling. The authors would like to thank colleagues on this project, within the framework of which the results of this work were discussed. The authors also would like to show their deep gratitude to colleagues from the Department of Computer Systems, Networks, and Cyber Security, National Aerospace University «KhAI» and colleagues from Odessa National Polytechnic University for their patient guidance, enthusiastic encouragement, and useful critiques of this paper.

References

[1] *Erasmus+ Programme*. Available at: http://ec.europa.eu/programmes/erasmus-plus/node_en

[2] Cisco. (2014). *Annual Security Report*. Available at: http://www.cisco.com/web/offer/gist_ty2_asset/Cisco_2014_ASR.pdf

[3] Internet of Things (IoT). (2017). *The IoT links objects to the Internet, enabling data and insights never available before*. Available at: http://www.cisco.com/c/en/us/solutions/

[4] Cisco. (2015). *Cloud and Mobile Network Traffic Forecast - Visual Networking Index (VNI)*. Available at: http://cisco.com/c/en/us/solutions/serviceprovider/visual-networking-index-vni/index.html

[5] Huawei. (2017). *Harnessing the Power of Connectivity*. Available at: http://www.huawei.com/minisite/gci/files/gci_2017_whitepaper_en.pdf?v=20170

[6] Manyika, J., Chui, M., Bisson, P., Woetzel, J., Dobbs, R., Bughin, J., et al. (2015). Unlocking the potential of the Internet of Things. Available at: http://www.mckinsey.com/business-functions/digital-mckinsey/our-insights/

[7] *Web of Things*. Available at: https://en.wikipedia.org/wiki/Web_of_Things

[8] Kindberg, T., Barton, J., Morgan, J., Becker, G., Caswell, D., Debaty, P., et al. (2000). People, places, things: Web presence for the real world. doi:10.1109/MCSA.2000.895378

[9] Ashton, K. (2002). *That 'Internet of Things' Thing*. Available at: http://www.rfidjournal.com/articles/view?4986 [accessed January 2009].

[10] Ramirez, G., Munoz, M., and Delgado, C. (2008). "IoT early possibilities in learning scenarios," in *Proceedings of the Workshop on Designing the Internet of Things for Workplace Realities: Social and Cultural Aspects in Design and Organization (Social-IoT)*, Zurich.

[11] Internet of Things. (2008). *Workshops*. Available at: http://www.iot-conference.org/iot2008/cfp/workshops.html [accessed March 2008].

[12] Patashnik, O. (1988). *Designing BIBTEX Styles*. Available at: https://pctex.com/files/downloads/manuals/btxhak.pdf [accessed February 2018].

[13] *European Credit Transfer and Accumulation System (ECTS).* Available at: http://ec.europa.eu/education/resources/european-credit-transfer-accumulation

[14] *Descriptors defining levels in the European Qualifications Framework (EQF).* Available at: http://ec.europa.eu/ploteus/content/descriptors-page

[15] Erasmus+ ALIOT Project. (2016). *Internet of Things: Emerging Curriculum for Industry and Human Applications.* Available at: website http://aliot.eu.org/

[16] Tempus GREENCO project. (2018). *Green computing and communications.* Available at: http://my-greenco.eu/

[17] Tempus CABRIOLET project. (2013). *Model-oriented approach and intelligent knowledge-based system for evolvable academia-industry cooperation in electronic and computer engineering.* Available at: http://my-cabriolet.eu/

[18] Tempus SEREIN project. (2013). *Modernization of postgraduate studies on security and resilience for human and industry related domains.* Available at: http://serein.eu.org/

[19] Tempus project. (2012). *Innovation Offices in Ukrainian Higher Education Institutions.* Available at: http://www.uni4inno.eu/

[20] *Erasmus Mundus project "Pervasive computing & communications for sustainable development" PERCCOMM project.* Available at: http://perccom.univ-lorraine.fr/

[21] *Centers of Innovations in Ecosystems Technosphere.* Available at: http://cidecs.net/about/

Index

5G 1, 37, 173, 487
6LoWPAN 12, 173, 237, 486

A

AAL services 456, 472
academic programs 482, 535, 543, 545
actuators 18, 173, 353, 552
ambient assisted living 13, 455, 456, 458
architecture of software-algorithmic complex 405
Arduino Uno 5, 36, 133, 487
arrival time 411, 415, 427, 429
artificial intelligence 25, 266, 274, 512
assistance system 296, 456, 462, 472
assurance case 308, 318, 320, 325
attacks 12, 113, 174, 312
augmented functionality of physical systems 5, 25, 37, 170
autonomous control 274, 276, 283
autonomous organization 88, 351, 352
availability 6, 107, 301, 552

B

big data 9, 25, 246, 370
blockchain 13, 351, 358, 480

C

case studies of WoT 480, 495
cloud 21, 55, 111, 255
cloud services 170, 394, 462, 554
cloud computing 9, 151, 254, 492
cognitive control systems 505, 507, 509, 512
collaborative business model 470, 473
collaboratory 7, 51, 58
competitive vs. cooperative behavior 83, 86
conceptual architects 50, 52, 55, 58
control processes automation 367, 372, 383
control station 276
criticality 111, 129, 139, 198
critical domains 3, 28, 113, 450
critical embedded systems 20, 28
curricula development 545, 546
curriculum and education programs 479, 483
cyber space 2, 111, 148, 406
cyber-physical systems (CPS) 332, 485, 516, 558

D
data analytics 12, 32, 243, 472
data center 242, 371, 391, 484
data fair share 473
data usage control 459, 470, 473
decentralized 88, 265, 351, 472
deep city 250
dependability 12, 129, 263, 324
design 2, 28, 201, 311
design for globalization 31, 33
Digital Avatar 56
digital ecosystem 9, 53, 58
distributed sensor networks 29, 151

E
eBusiness 51, 56
edge computing 12, 170, 241, 459
eGovernment 51, 56
eGuild 50, 51, 56
eHealth 7, 51, 53, 251
eLearning 51, 57, 518
emb::6 226, 229, 237
embedded systems 17, 63, 221, 551
energy and mobility systems 335, 346, 348
energy efficiency 31, 67, 267, 396
Energy-Efficient Routing 65, 67
engineering education 11, 29, 504, 524
Ethereum 352, 358, 364
Ethics 11, 47, 58

F
fair allocation 455, 469, 473
fault tree 105, 110, 113, 123
FMECA 134, 139, 140, 142
fog computing 170, 241, 267, 487

G
Google Maps API 411, 416, 425, 429
Grand Challenges 1, 53, 58
green technology 96, 143, 338, 557

H
hard real-time 336
hardware and software 5, 132, 369, 528
haversine formula 416, 425, 426
healthcare 105, 108, 395, 457
Hierarchical Protocols 66, 67, 70
higher educational establishments 558
human–machine interface 367

I
ICT 241, 294, 364, 515
industrial internet 25, 222, 367, 479
Industrial Internet of Things 222, 367, 479
information-technological platform 392, 395, 404
innovation 1, 17, 221, 369
insulin pump 106, 109, 115, 123
intelligent sensor 552
intelligent transport system 12, 295, 411, 412
Internet of Drones 12, 200, 204
Internet of Things (IoT) 1, 83, 367, 535
Internet of Vital Things 17, 25, 35

Intrusion 107, 134, 152, 176
IoT 1, 106, 221, 302
IoT system 47, 105, 247, 376
IoT-based physical security system 12, 127, 131
IoT-based system 333, 482, 543, 546

K

Kalman filtration 411, 417, 425, 429

L

light fidelity 202

M

Markov Model 12, 117, 120, 190
Markov process 12, 107, 183, 195
measurement 112, 214, 232, 551
mobile object 273, 279, 289, 384
multi-rate 19

N

neural networks 13, 151, 260, 446

O

object recognition 285
obstacles avoiding 275, 286, 457
operator-controller-module (OCM) 337
organic rankine cycle (ORC) 331, 333

P

path planning 286
peer-to-peer interactions 170, 353, 492
performance 6, 223, 451, 551
Physical Security Modes and Effect Analysis (PSMECA) 127, 139, 141, 143
PLC 340, 378, 379, 381
PoE 433, 435, 439, 450
power management 272
prediction 411, 415, 422, 428
project-oriented approach 515, 523, 524
PSOC 4, 35, 40
pyrolysis complex 367, 376, 382

Q

quality measurement tool 282, 291
queueing theory 105, 114, 115

R

radio channel 274, 277
Raspberry Pi 5, 132, 173, 489
real-time 10, 230, 491, 516
redundancy 201, 204, 212
Regional Infrastructure (RI) 127, 129, 131, 142
reliability 6, 197, 204, 518
reliability block diagram 197, 206
reliability models 197, 204, 206
remote control 181, 274, 380, 437
remote experiments 520, 521
remote laboratories 503, 508, 511
router 169, 185, 190, 486

S

safety and security 107, 247, 319, 471
SAGE 26, 27
SCADA 368, 371, 381, 395
security 101, 110, 175, 414

security event correlation 12, 147, 165
security events 147, 154, 162, 166
security monitoring 129, 149, 164
semi-automatic ground environment 26, 27
sensor- and drone-based communication networks 12, 197, 213, 215
sensor networks 28, 63, 436, 554
sensors 3, 132, 261, 372
SIEM 148, 150, 152
signal processing 25, 437
smart city 7, 243, 295, 389
smart buildings 10, 249, 433, 535
Smart Business Center 12, 169, 172
smart environment 10, 241, 249
smartification process 331, 344, 347
social objects 83, 88, 100
software metrics 1, 136, 169, 320
structural analysis 12, 147, 154
survivability 12, 201, 311, 324
sustainability 86, 242, 356, 527
system of systems 27, 50, 55
systems engineering 12, 331, 347, 553

T

teaching programs 479, 483, 515, 558
telemedicine 49, 52, 250
telepresence 49
trolleybus 13, 411, 425, 429
trusted systems 30, 57, 179
T-shaped learning 7, 51, 52

U

university 169, 293, 332, 558

V

vehicle 19, 21, 303, 416
verification 28, 221, 318, 345
virtual testbed 222, 225, 228, 230
vital data 199, 455, 464, 472
Vital Electronics 3, 35, 39
VITAL-iISOLVE 17, 35, 38, 40
VitalNet 25
vulnerability 116, 119, 174, 191

W

waste management 13, 53, 351, 395
Web of Things 13, 100, 479, 488
web thing API 479, 483, 493
Wi-Fi 173, 213, 486, 520
wired/wireless network subsystems 197, 204, 211, 215
wireless sensor networks (WSNs) 63, 170, 356, 485
wireless sensor nodes 64, 267, 528, 551
World Wide Web 47, 488, 537
WOT application development technologies 13, 479, 483, 495

About the Editors

Vyacheslav Kharchenko, Head of Computer Systems, Networks and Cybersecurity, National Aerospace University KhAI, Head of Centre for Safety Infrastructure Research and Analysis, RPC Radiy. *Education and degrees*: Kharkiv High Military Engineering Rocket College (1969–1974), PhD (1981), Central Patent Institute (1983–1986), Honor Inventor of Ukraine (1990), Professor on Control Systems Department (1991), Doctor of Science on Engineering (1995). *Experience*: supervisor and developer of 12 national standards on NPP and aerospace I&C safety; supervisor of 45 PhD and DrS dissertations; invited researcher and professor (UK, 2004; Germany, 2006; Slovakia, 2010; USA, 2011; UK, 2016, 2017); national coordinator of 8 EU projects; consultant of R&D companies on critical software and systems development and certification. Conferences: invited speaker of 21 conferences (Estonia, Poland, Slovakia, United Kingdom, etc); 19 awards for the best papers; General Chair of Dependable Systems, Services and Technologies (DESSERT) Conferences (2006–2018), PC Chair of the WS CrISS, GreenSCom, TheRMIT (2011–2018); supervisor of standing CriCTechS Seminar (2001–2018). *Publications:* 32 books and chapters (Springer, IGI-Global, etc); 225 papers in Journals and Proceedings indexed in Web of Science and Scopus; more 800 patents. *R&D interests*: software and FPGA systems dependability; diversity for safety and security; IT-Infrastructure (Cloud, IoT) intrusion-tolerance and resilience, green IT engineering.

Dr Ah-Lian Kor received the degree of MSc in 1995 (Malaya University, Malaysia) and PhD in 2002 (Leeds Beckett University, UK). She is part of Leeds Beckett MSc Sustainable Computing Curriculum Development Team. She has been involved in several EU projects for Green Computing, Innovative Training Model for Social Enterprises Professional Qualifications, Integrated System for Learning and Education Services, and Internet of Things. She has published more 100 papers, chapters and other works on ontology, Semantics Web, Web Services, Portal, semantics for GIS etc.

She is active in AI research and has developed an intelligent map understanding system and reasoning system. She is an Editorial Board Member of the International Journal of Web Portals; a member of IQN (international quality network for spatial cognition); sits on many international conference program/technical committees (e.g. IEEE Cloud Computing, Cyberlaws Conference, ICT-EurAsia, etc.); an active paper reviewer for journals and conferences (e.g. AMCIS, International Journal of Emergency Services, Interscience Journal, etc.); Editorial Advisory Board Member for International Journal on Advances in Intelligent Systems and International Journal on Advances in Security; an associate member of the EPSRC funded e-GISE (e-Government and System Evaluation) Network) and has helped organise an international workshop (eGOV05) for the network.

Andrzej Rucinski, Professor Emeritus, Department of Electrical and Computer Engineering, Institute for the Study of Earth, Ocean, and Space, Space Science Center, University of New Hampshire, USA. Ambassador of International Society of Service Innovation Professionals, USA. Ambassador of Polish Congress Ambassadors, Poland. Member of eCollaborative Ventures (eCV) Collaboratory, USA. He was educated both in Poland and the former Soviet Union and has conducted his academic career in both the United States, Europe (France, Germany, Hungary, Poland, Russia, Ukraine, United Kingdom), Africa (Ethiopia), Asia (India, Kazakhstan). His service has been with high tech industry, governments, ranging from the state level (National Infrastructure Institute, serving as the Chief Scientist) to a global level (NATO, United Nations Organization). He has been a member of the IEEE Computer Society's Design Automation Technical Committee. He chaired the leading conference in microelectronics education, the 2009 Conference on Microelectronics Systems Education (MSE'09) in San Francisco. At the University of New Hampshire, he is the founding Director of the Internet of Things Research and Development Laboratory, a former Critical Infrastructure Dependability Laboratory. Member of ASEE, ACM, IEEE (Senior Member), IEEE-SA, IGIP, SEFI, and ACM Distinguished Speaker. He was the Member of the US State Department/Fulbright National Screening Committee.